Forensic Biotechnology

Forensic Biotechnology: The Basics and Beyond provides a comprehensive exploration of biotechnology's transformative role in modern forensic science. This essential resource covers the field through four carefully structured modules, beginning with foundational concepts including sample collection, preservation, and data interpretation, then progressing through practical applications such as microbial forensics, wildlife forensics, and DNA barcoding. The book advances to sophisticated topics including DNA mixture interpretation, forensic DNA phenotyping, epigenetic signatures, and proteomics. It concludes with a critical examination of DNA evidence admissibility in courtroom proceedings.

Beyond technical expertise, the book tackles the ethical, legal, and social implications of biotechnology in forensic investigations, including privacy concerns and scientific evidence admissibility. It provides in-depth analysis of cutting-edge techniques including polymerase chain reaction (PCR), short tandem repeat (STR) analysis, mitochondrial DNA analysis, next-generation sequencing (NGS), and advanced molecular biology methods. Coverage examines both the strengths and limitations of various forensic biotechnology methodologies, ensuring readers develop a balanced understanding of current capabilities and constraints in that field.

Enhanced with real-world case studies and practical examples, the book serves as an educational resource and professional reference guide of vital importance to forensic scientists, attorneys, researchers, and students exploring the impact of biotechnology on justice and criminal investigations.

Forensic Biotechnology

The Basics and Beyond

Edited by
Ritesh K. Shukla, Kamayani Vajpayee,
and Hirak Ranjan Dash

CRC Press
Taylor & Francis Group
Boca Raton London New York

CRC Press is an imprint of the
Taylor & Francis Group, an **informa** business

Designed cover image: Getty Images

First edition published 2027
by CRC Press
2385 NW Executive Center Drive, Suite 320, Boca Raton FL 33431

and by CRC Press
4 Park Square, Milton Park, Abingdon, Oxon, OX14 4RN

CRC Press is an imprint of Taylor & Francis Group, LLC

ISBN: 9781032755007 (hbk)
ISBN: 9781032754994 (pbk)
ISBN: 9781003474296 (ebk)

DOI: 10.4324/9781003474296

Typeset in Garamond
by KnowledgeWorks Global Ltd.

Contents

Preface . . . . **ix**
About the Editors . . . . **xi**
Contributors . . . . **xiii**

1 Introduction to Forensic Biotechnology **1**
Kamayani Vajpayee, Rajendra Prasad Pawar, and Ritesh K. Shukla

2 Collection and Preservation of Samples for Forensic DNA Analysis . . **20**
Anna Barbaro

3 Insights into Body Fluid Identification for Criminal Investigations **34**
Kamayani Vajpayee and Ritesh K. Shukla

4 Introduction to DNA Profiling Techniques **61**
Poonam Moon, Ashish Badiye, and Neeti Kapoor

5 Next-Generation Sequencing in Forensic Science: Advancements, Limitations, and Applications **80**
Vidhi Paida, Kamayani Vajpayee, and Ritesh K. Shukla

6 From Data to Interpretation: Statistical Software in Forensic DNA Typing. . . 104
Ranjeet Kumar Singh and Afreen Tarannum

7 Challenges and Limitations of Forensic DNA Profiling Technique 123
Shrijoy Banerjee, Tanurup Das, and Ankit Srivastava

8 Ethical and Normative Implications of Forensic DNA Technology 136
M.C.A. De Ungria and E.B. Jimenez

9 DNA Barcoding and Its Potential Applications in Forensic Investigations 161
Amandeep Kaur Bhambara and Mukesh Kumar Thakar

10 Microbial Forensics 175
Neeti Kapoor, Girraj Sharma, Chhote Raja Patle, and Ashish Badiye

11 The Rising Path of Forensic Archaeology: An Indian Perspective 195
Gyaneshwer Chaubey, Rahul Mishra, Pankaj Shrivastava, and Chandana Basu Mallick

12 Wildlife Forensics: From the Science of Conservation to the Courtroom . . 204
Ajit Kumar, Dimple Bhatia, and Rajinder Singh

13 Forensic DNA Phenotyping: Techniques, Successes, and Future Directions . . . 226
Charles E. "Chuck" MacLean

14 Streamlining Forensic DNA Analysis with RAPID DNA Typing: Current Developments and Future Directions 244
D.S. Yadav, Roshan Kumar, and Tanya Gupta

15 Exploring the Role of Epigenetic Factors in Forensic Science 256
Chavi Bhardwaj, Jayapriya Panicker, and Ajay Kumar Rana

16 Mass Spectrometry in Forensic DNA Analysis: A New Frontier 287
Bharti Jain, Rajeev Jain, and Ritesh K. Shukla

17 Role of Proteomics in Criminal Investigations 296
Satish Kumar, Neelkant Verma, and Aaliya Sachdev

18 Unraveling Injustice: The Impact of DNA Exonerations 315
Avantika Johari and Purvi Pokhariyal

Index 335

Preface

Forensic science stands at the crossroads of biology, technology, and justice. Over the past few decades, biotechnology has transformed this field from a discipline focused primarily on observable traits and serological methods into one driven by molecular precision, high-throughput platforms, and data-intensive interpretation. *Forensic Biotechnology: Basics and Beyond* was conceived with the vision of providing a comprehensive, accessible, and forward-looking resource that reflects this remarkable evolution.

This book brings together fundamental principles, methodological advancements, and emerging frontiers that are shaping modern forensic investigations. Each chapter has been carefully curated to guide readers—from students stepping into the world of forensic biology to researchers, practitioners, and academicians seeking a deeper understanding of contemporary techniques.

The journey begins with foundational concepts in forensic biotechnology and progresses through real-world processes such as sample collection, preservation, body fluid identification, and DNA profiling. The text then expands into sophisticated areas, including next-generation sequencing, statistical interpretation, proteomics, microbial forensics, DNA barcoding, and mass spectrometry, offering readers a holistic appreciation of modern forensic workflows.

As the field advances, so do its challenges. This book addresses the technical limitations, interpretational hurdles, ethical considerations, and global normative concerns that accompany the use of DNA and proteomic technologies in the justice system. Chapters on rapid DNA typing, forensic ancestry, epigenetics, wildlife forensics, and DNA phenotyping highlight the growing intersections between biotechnology, society, and emerging investigative needs. The inclusion of a dedicated section on DNA exonerations and The Innocence Project underscores the profound impact that forensic biotechnology can have—not only in solving crimes but also in rectifying miscarriages of justice.

This book is the result of collaborative scholarship, rigorous research, and the collective experience of experts from diverse subfields of forensic science. Their contributions reflect both established knowledge and the innovations that will continue to redefine forensic practice in the years to come.

It is my hope that *Forensic Biotechnology: The Basics and Beyond* serves as a valuable reference, a teaching companion, and a catalyst for further research. Above all, may it inspire readers to approach forensic investigations with scientific curiosity, ethical responsibility, and an unwavering commitment to truth.

About the Editors

Ritesh Kumar Shukla is an Associate Professor in the Department of Forensic Sciences at National Forensic Sciences University, Raipur Campus. He is an accomplished academician and a researcher with extensive expertise in forensic science, particularly in the areas of forensic nanobiotechnology, toxicology, DNA forensics, and diagnostic device development. He obtained his Master's degree in Forensic Science from Dr. B.R. Ambedkar University, Agra, followed by a Ph.D. in Toxicology (Nanomaterial Toxicology) from Jamia Hamdard in 2013. His doctoral research was conducted at CSIR–Indian Institute of Toxicology Research (CSIR-IITR), Lucknow, where he began his research career as a UGC Junior and Senior Research Fellow. Dr. Shukla has held several academic positions, including Associate Professor at Ahmedabad University (2014–2025) and earlier appointments at Guru Ghasidas Vishwavidyalaya, Bilaspur, and the Government Institute of Forensic Science, Nagpur. His research contributions span nanomaterial synthesis, nanotoxicological evaluations, and their applications in forensic investigations. With more than 40 research articles, numerous book chapters, and multiple edited volumes to his credit, he has significantly contributed to advancing forensic nanotechnology and toxicology. His scholarly impact has been recognized through several prestigious awards, including the Forensic Excellence Award and the Young Scientist Medal from the International Association of Advanced Materials (IAAM), Sweden. He has also been listed among the world's top 2% of scientists by Stanford University. Dr. Shukla's current research focuses on integrating nanotechnology with forensic biology and chemistry to develop innovative solutions for evidence analysis, food forensics, and point-of-care diagnostic systems. He continues to mentor students and contribute actively to interdisciplinary research at the interface of nanoscience and forensic applications.

Kamayani Vajpayee is an Assistant Professor at Amity University Rajasthan, Jaipur, with expertise in forensic genetics and forensic DNA analysis. She has recently completed her doctoral research at Ahmedabad University, where her work focused on understanding the mechanisms of PCR inhibition in forensic samples and developing effective mitigation strategies to improve DNA profiling from complex and degraded biological evidence. She holds a Master's degree in Forensic Science from the University of Lucknow and a Bachelor's degree in Biotechnology from Amity University Rajasthan. Her academic training has provided a strong interdisciplinary foundation in molecular biology, forensic serology, and toxicological analysis. Her research contributions focus on PCR inhibition, forensic DNA profiling, and next-generation sequencing, with several publications in reputed international journals and book chapters in the field of forensic science.

Her work has particularly contributed to elucidating the role of inhibitors such as melanin, indigo, and fulvic acid in DNA amplification and improving analytical reliability in forensic casework. She has received multiple recognitions, including awards for research presentations at national and international conferences, reflecting her active engagement in scientific research and dissemination. In addition to research, she is actively involved in teaching, mentoring students, and contributing to academic writing. Her current interests lie in strengthening forensic DNA methodologies and addressing practical challenges in evidence analysis to enhance the accuracy and reliability of forensic investigations.

Hirak Ranjan Dash is an Associate Professor at Centurion University of Technology and Management, with extensive expertise in forensic genetics, DNA fingerprinting, and molecular biology. He has over a decade of experience in teaching, research, and forensic casework, including his role as a Scientific Officer (DNA) at the Forensic Science Laboratory, Madhya Pradesh, where he contributed to the analysis of more than 1200 forensic cases using DNA profiling techniques. He holds a Ph.D. in Life Sciences and has established himself as a leading researcher in forensic DNA analysis, population genetics, and next-generation sequencing. His research focuses on advancing molecular techniques for forensic applications, including PCR-based methodologies, STR analysis, and emerging genomic approaches. Dr. Dash has authored more than 170 research publications, along with several books and book chapters in the field of forensic biology and genetics. His scholarly contributions have earned him significant recognition, including being listed among the top 2% of scientists globally by Stanford University for multiple consecutive years. In addition to his research achievements, he actively contributes to the development of forensic science through teaching, mentoring, and interdisciplinary collaborations. His work continues to impact forensic DNA analysis, with a focus on improving the accuracy, reliability, and applicability of molecular techniques in forensic investigations.

Contributors

Ashish Badiye
Department of Forensic Science
Government Institute of Forensic Science
Nagpur, India
and
Rashtrasant Tukadoji Maharaj Nagpur University
Nagpur, India

Shrijoy Banerjee
UGC-JRF, School of Forensic Sciences
The West Bengal National University of Juridical Sciences
Kolkata, India

Anna Barbaro
Studio Indagini Mediche E Forensi (SIMEF) – Italy
and
Worldwide Association of Women Forensic Experts (WAWFE)
Reggio Calabria, Italy

Amandeep Kaur Bhambara
Department of Forensic Science
Punjabi University
Patiala, India

Chavi Bhardwaj
Division of Biology and DNA
Central Forensic Science Laboratory Hyderabad
Ministry of Home Affairs
Government of India
Hyderabad, India

Dimple Bhatia
Department of Forensic Science
Punjabi University
Patiala, India

Gyaneshwer Chaubey
Cytogenetics Laboratory
Department of Zoology
Banaras Hindu University
Varanasi, India

Tanurup Das
School of Forensic Sciences
The West Bengal National University of Juridical Sciences
Kolkata, India

Tanya Gupta
DNA Division, Central Forensic Science Laboratory (CFSL)
Directorate of Forensic Science Services
Ministry of Home Affairs, Government of India
Hyderabad, India

Bharti Jain
Central Forensic Science Laboratory
Chandigarh, India
and
Institute of Forensic Science & Criminology
Panjab University
Chandigarh, India
and
Department of Forensic Science,
School of Allied Health Sciences
Sharda Unviersity
Greater Noida, UP, India

Rajeev Jain
Central Forensic Science Laboratory
Dakshin Marg, India

E.B. Jimenez
National Institutes of Health, Research Ethics Board
University of the Philippines Manila
Quezon, Philippines

Avantika Johari
School of Forensic Sciences
National Forensic Sciences University
Gandhinagar, India

Neeti Kapoor
Department of Forensic Science
Government Institute of Forensic Science Nagpur
Rashtrasant Tukadoji Maharaj Nagpur University
Nagpur, India

Ajit Kumar
Wildlife Institute of India
Dehradun, India

Roshan Kumar
DNA Division, Central Forensic Science Laboratory (CFSL)
Directorate of Forensic Science Services
Ministry of Home Affairs
Government of India
Chandigarh, India

Satish Kumar
Campus Director, Proferssor, National Forensic Sciences University, Bhopal Campus

Charles E. "Chuck" MacLean
Metro State University
Brooklyn Park, Minnesota

Chandana Basu Mallick
Centre for Genetic Disorders
Institute of Science
Banaras Hindu University
Varanasi, India

Rahul Mishra
Cytogenetics Laboratory Department of Zoology Banaras Hindu University Varanasi, India
Amity University, Raipur, CG, India

Poonam Moon
Karunya Institute of Technology and Sciences
Coimbatore, India

Vidhi Paida
Biological and Life Sciences
School of Arts and Sciences
Ahmedabad University
Ahmedabad, India

Jayapriya Panicker
Division of Biology and DNA
Central Forensic Science Laboratory Hyderabad
Ministry of Home Affairs
Government of India
Hyderabad, India

Chhote Raja Patle
Department of Forensic Science
Medi-Caps University
Indore, India

Rajendra Prasad Pawar
School of Forensic Science
National Forensic Sciences University
Bhopal, India
and
School of Forensic Science,
Uttar Pradesh Institute of Forensic Science,
Lucknow, UP, India

Purvi Pokhariyal
School of Law, Forensic Justice & Policy Studies
National Forensic Sciences University
Gandhinagar, India

Ajay Kumar Rana
Division of Biology and DNA
Central Forensic Science Laboratory Hyderabad
Ministry of Home Affairs
Government of India
Hyderabad, India

Aaliya Sachdev
School of Forensic Science
National Forensic Sciences University
Gandhinagar, India

Girraj Sharma
Jindal Institute of Behavioural Sciences
O.P. Jindal Global University
Sonipat, India

Pankaj Shrivastava
DNA Unit
Regional Forensic Science Laboratory
Jabalpur, India

Rajinder Singh
Department of Forensic Science
Punjabi University
Patiala, India

Ranjeet Kumar Singh
Sherlock Institute of Forensic Science
New Delhi, India

Ankit Srivastava
Centre for Studies and Research in Forensic Sciences
The West Bengal National University of Juridical Sciences
Kolkata, India

Afreen Tarannum
Sherlock Institute of Forensic Science
New Delhi, India

Mukesh Kumar Thakar
Department of Forensic Science
Punjabi University
Patiala, India

M.C.A. De Ungria
Natural Sciences Research Institute
Research Ethics Board
University of the Philippines
Quezon, Philippines

Neelkant Verma
School of Forensic Science
National Forensic Sciences University
Gandhinagar, India

D.S. Yadav
DNA Division, Central Forensic Science Laboratory (CFSL)
Directorate of Forensic Science Services
Ministry of Home Affairs, Government of India
Chandigarh, India

1

Introduction to Forensic Biotechnology

Kamayani Vajpayee, Rajendra Prasad Pawar, and Ritesh K. Shukla

1.1 Introduction

Forensic biotechnology is a subdiscipline of forensic science that applies molecular biology techniques and biotechnological advances to analyze biological evidence in criminal investigations, legal proceedings, and public security (Antil & Joon, 2025). In forensic science, biotechnology has transformed the capacity for collecting, analyzing, and interpreting biological substances, such as biological tissue, DNA, RNA, proteins, and microbes. Forensic biotechnology, which applies techniques such as the polymerase chain reaction (PCR), recombinant DNA technology, and genomic sequencing, provides unparalleled accuracy in individualization, establishment of familial relationships, and detection/screening of biological traces that facilitate the determination of the connection of a suspect to a crime scene or victim (Duster, 2007; Kumar, 2021; Pillay et al., 2007; Shukla, 2016).

Forensic biotechnology goes beyond DNA analysis, encompassing many biotechnological advances. These include body fluid identification (An et al., 2012), establishment of the postmortem interval (PMI) (Franceschetti et al., 2023), microbial forensics for the identification of bioterrorism agents and foodborne toxins (Pattnaik & Jana, 2005; Wani et al., 2022; Yeung, 2016), and toxicological analysis of drugs and toxins (Chung & Choe, 2017; Teotia & Pal, 2009). Forensic biotechnology is significant in environmental forensic science (Murphy & Morrison, 2014), wildlife crime (Kurland et al., 2017) and mass disaster victim identification (Butcher et al., 2014; Lessig & Prinz, 2022). As a representative of its interdisciplinary nature, biotechnology has broadened the scope of forensic science, providing rapid, accurate methods and enabling the analysis of complex biological evidence. Forensic analysis also allows bypassing obstacles such as degraded or minute samples, which were difficult to analyze (Vajpayee, Dash, et al., 2023). Forensic biotechnology is increasingly transforming criminal investigations to make them more efficient, raising important ethical and legal issues of privacy, data protection, and misuse of genetic information (Williams & Wienroth, 2017).

The blending of biotechnology and forensic science is a relatively new development; its roots, however, go back to the earliest breakthrough in the respective field. However, earlier, forensic practice had relied on physical evidence enhanced by fundamental and merely primitive chemical analysis, and biological evidence had been restricted to blood typing and serological analysis. This history includes Watson and Crick's discovery of the double-helical model of DNA in 1953, a milestone and the beginning of the genetic revolution. DNA technologies in forensic science did not start until the 1980s. In 1984, British geneticist Sir Alec Jeffreys made the most crucial discovery of DNA fingerprinting. This technique allows individuals to be recognized by their unique DNA fingerprints. This discovery proved to be a potent and handy

DOI: 10.4324/9781003474296-1

tool for forensic science to link suspects to the murder scene directly. In 1986, the first use of DNA evidence in a forensic setting happened in the United Kingdom when an innocent man was cleared by DNA profiling in conjunction with implicating the actual perpetrator(s) of two slain girls in Leicestershire (Dash et al., 2020).

Advances were made in forensic DNA analysis in the 1990s through PCR, which greatly enhanced the analysis of very degraded or trace amounts of biological materials, resulting in the amplification of DNA to a great extent (Vajpayee, Dash et al., 2023). Short tandem repeat (STR) profiling enhanced forensic DNA discrimination, enabling the establishment of national DNA databases, such as the Combined DNA Index System (CODIS) in the United States. These databases have helped link persistent criminals to many offenses and solve cold cases. As forensic biotechnology has developed, so has its scope of application. Mitochondrial DNA (mtDNA) typing and Y chromosome typing were introduced in the early 2000s to address cases where nuclear DNA was absent or in amounts too minute, e.g., in hair shafts or ancient skeletal remains. Massively parallel sequencing (MPS), or next-generation sequencing (NGS) as it is otherwise called, has revolutionized forensic genomics by providing a broader perspective of genetic variation and the ability to analyze whole genomes from even limited samples (Vajpayee et al., 2020).

Though the use of biotechnology in forensics has accelerated, it is founded on a broad foundation of scientific discovery. The intersection of molecular biology, biochemistry, and forensic science has brought the field into new arenas, from identifying victims of catastrophes to identifying biological weapons. This has improved the accuracy and reliability of forensic evidence, broadened the horizons of forensic investigation, and new technologies will further strengthen the potency of forensic biotechnology.

1.2 Fundamental techniques in forensic biotechnology

1.2.1 Polymerase chain reaction (PCR) and its variants

PCR is the foundation of contemporary forensic DNA analysis, amplifying targeted regions of the DNA molecule from minute and frequently degraded samples. Kary Mullis invented PCR in 1985, transforming forensic biology by allowing scientists to produce enough DNA from trace evidence to analyze previously unanalyzable samples. The reactions of PCR are three primary steps: denaturation, annealing, and extension. Denaturation is done to separate single DNA strands from the double helix due to heating. In the annealing step, the design of specific primers to flank the target region of interest anneals to the single-stranded DNA. The elongation step in PCR involves the interaction of the DNA polymerase enzyme with the primers and adding new nucleotides to yield a new DNA strand, thus amplifying the target region. This process is repeated several times (29–35), and the amount of DNA increases exponentially (Vajpayee, Dash et al., 2023).

PCR is a crucial tool in forensic DNA typing, as it can amplify DNA sequences such as STRs or mtDNA. Several variants of PCR have been modified for DNA analysis:

- *qPCR (Real-Time PCR):* It measures DNA amplification in real time so that the amount of DNA concentration and the index of degradation in the amplification can be known. It is applied in forensic casework to determine the quality and quantity of DNA before further downstream processing (Fraga et al., 2014).
- *Touchdown PCR:* This technique reduces nonspecific amplification by slowly decreasing the annealing temperature, thus improving primer binding specificity. This is

especially effective when working with degraded or mixed DNA samples (Green & Sambrook, 2018).

- *Nested PCR:* This variant of nested PCR uses two primer sets in consecutive PCR rounds to achieve maximum specificity. This method efficiently handles samples of low DNA content or highly degraded DNA (Green & Sambrook, 2019).

1.2.2 Forensic DNA profiling

Forensic DNA analysis has revolutionized the way criminal investigations are conducted, even though it can now identify criminals with hitherto unachievable accuracy. DNA is an excellent molecular marker since each person, barring identical twins, has distinct genetic information. Paternity testing, criminal identification, disaster victim identification, exonerating falsely convicted individuals, and missing person investigations are some of the forensic science applications where forensic DNA profiling is performed (Vajpayee et al., 2020). The accurate collection, preservation, and analysis of DNA samples have been primarily responsible for the efficacy of forensic investigations.

- *Collection and Preservation of DNA Samples:* In forensic DNA analysis, precise and accurate sample collection, preservation, and storage are essential. DNA-containing biological samples can be obtained from bone, hair, skin cells, semen, blood, and saliva. Each sample must be handled carefully to avoid contamination and/or deterioration. Biological evidence is typically collected from the crime scenes, suspects, and victims using sterile tools like swabs, tweezers, and collection tubes. While blood and saliva are swabbed from surfaces, hair or skin cells are carefully lifted using adhesive tapes or forceps. Another critical phase is the preservation of the genetic material. Before being packaged, the biological samples taken from the crime scene are frequently allowed to air dry to inhibit the growth of microorganisms. To avoid cross-contamination or tampering, a stringent chain of custody during collection and storage guarantees that DNA evidence is admissible in court (Dash & Vajpayee, 2021). Chapter 2 has addressed biological sample collection and preservation in detail.

1.2.3 Short tandem repeats (STR) profiling

STR profiling is at the core of forensic DNA analysis that utilizes very highly polymorphic short sequences of DNA, about 2 to 6 bases in length, which are tandemly repeated in the genome. These occur through replication slippage throughout the genome and can be used due to the significant individual-specific variability. The initial steps are isolating DNA from biological materials, quantifying, and assessing its quality. STR loci are amplified using the PCR with fluorescently tagged primers. Multiplex PCR allows for the amplification of several loci in a single response. The amplified products are separated using capillary electrophoresis, producing electropherograms that show unique genetic patterns at different locations. In forensic investigations, STR profiling is frequently used to identify offenders, confirm paternity, locate missing people, and identify victims of natural disasters. Strong discrimination power, compatibility with databases such as CODIS, and the ability to evaluate even deteriorated DNA samples make it incredibly useful. In contrast to whole-genome sequencing, it does have certain disadvantages, including a higher chance of contamination, genetic variations among populations, and a smaller quantity of genetic data (Vajpayee et al., 2020).

1.2.4 Mitochondrial DNA (mtDNA) analysis

mtDNA is maternally inherited, and therefore, it is beneficial for tracing maternal lineages and examining degraded or limited samples. Its compact and high copy number enables successful analysis of skeletal remains, hair, and degraded biological evidence. The high mutation rate of mtDNA produces high variability and is categorized as single-nucleotide polymorphisms (SNPs) and insertion-deletion (InDel) types, facilitating haplogroup assignment. mtDNA analysis is done through DNA extraction, amplifying hypervariable regions (HVR1 and HVR2) by PCR, and sequencing by Sanger or NGS techniques. Results are matched against reference databases for assigning haplogroups. Its applications are mass disaster victim identification, genealogical tracing, and solving cold cases. Although mtDNA provides information on maternal ancestry, individual nonspecificity and susceptibility to contamination remain its limitations. Future improvements in sequencing and bioinformatics will facilitate combining mtDNA with nuclear DNA analysis for more forensic usefulness (Vajpayee et al., 2020).

1.2.5 Y-Chromosome analysis

The patrilineally inherited Y chromosome is essential in male-specific identification and tracking of the paternal lineage. Its non-recombining blocks contain Y-STRs, highly variable repeat sequences among males. Analysis includes DNA extraction, PCR amplification of Y-STR loci, and electrophoresis to resolve amplified products. Loci such as DYS19 and DYS391, known to be common, are selected for their discriminative capabilities. Y chromosome analysis is essential in investigating male suspects, familial lineage research, and population genetics. Nevertheless, since the Y chromosome is passed down in an autosomal mode along the paternal line, its application is limited to identifying males specifically and cannot differentiate between male relatives. This notwithstanding, it remains imperative in sexual assault investigations, paternity disputes, and historical research (Vajpayee et al., 2020).

1.2.6 Next-generation sequencing (NGS) in forensics

NGS (or massively parallel sequencing, or MPS) has been a milestone invention in forensic science, making millions of DNA fragments sequenced simultaneously. The advanced technology offers unprecedented sensitivity, accuracy, and efficiency in genomic analysis. The combination of NGS and bioinformatic tools makes it possible to produce high throughput genomic data, which forensic experts use to probe complex biological samples in a unique detail pattern. Unlike conventional techniques, NGS involves sequencing entire genomes or certain regions with high accuracy, and it differs from techniques that pursue a very narrow panel of genetic markers, such as STRs and SNPs. In particular, degraded or low template DNA samples commonly encountered in forensic casework benefit from this method. With a significant decrease in expense, NGS can produce genetic profiles containing both common and rare variants that improve the precision of individual identity or familial relationship analysis.

One of the key uses of NGS is in resolving complex forensic samples, exceptionally mixed DNA samples. Typically, however, traditional methods have difficulty deconvoluting the DNA profile of multiple contributors with differing proportions. However, NGS is powerful enough to differentiate individual contributions of mixtures, which are encountered in sexual assault kits. Forensic analysts use sophisticated bioinformatics tools to process and interpret these massive datasets and provide more reliable conclusions on who contributed (s). NGS is also critical in typing mtDNA rather than human identification and kinship testing. Forensic experts use high-resolution maternal lineage typing, which is significant for identifying old or deteriorated remains where nuclear DNA is poor, through sequencing of complete mitochondrial genomes. In addition, by being able to sequence SNPs over the genome, paternity tests, and multi-person familial relationship analysis become more stringent even in samples with relatives (Aly & Sabri, 2015).

Likewise, microbial forensics and environmental DNA (eDNA) analysis have been a breakthrough using NGS. This technology enables the detection and identification of pathogens and the determination of the source and possible transmission routes for outbreaks or bioterrorism situations in microbial forensics. eDNA analysis also plays a role in wildlife forensics, helping to monitor illegal wildlife trade and identify species from trace biological samples, helping to show NGS's adaptability for modernistic forensic issues. Although NGS has the potential to revolutionize forensic science, there are substantial challenges when including NGS in forensic labs. This technology is complex. The training requires specialization, investment in infrastructure, and advanced bioinformatics tools for data analysis.

Furthermore, ethical and legal implications on genome data privacy and misuse require the development of strict guidelines for protecting the rights of individuals. These are all important reasons for continuing research and development in forensic science to optimize NGS technologies, increase the performance of NGS data interpretation, and ensure the evidence's reliability in legal settings. Tackling such challenges allows NGS to rewrite levels of forensic examination with reliable and accurate solutions in many different types of investigation contexts (Børsting & Morling, 2015).

1.2.7 Forensic proteomics and metabolomics

In forensic science, forensic proteomics and forensic metabolomics have given rise to transformative approaches that enhance what is known from DNA analysis. These fields concentrate on protein and metabolite fingerprinting to acquire important information regarding an individual's physiological state, environmental exposure, and post-mortem changes. These techniques are especially relevant in solving challenges that DNA methods cannot address and will facilitate a greater appreciation of how DNA evidence is utilized in criminal investigations. In proteomics forensic applications, which is the study of proteins on a large scale, one has large populations of panelist proteins. One example is species identification or tissue type of biological samples, critical for wildlife forensics and cases involving mixed species evidence.

Furthermore, protein degradation's predictable patterns after death can be estimated using proteomic analysis to estimate PMI. It can supply highly reliable timelines even when DNA has degraded beyond utility. Demographic profiling also promises proteomics due to protein variation with age, sex, and genetic ancestry. These variations contribute to the ability of forensic scientists to fill in the details of unknown humans and, as such, can provide the needed clues in cases of missing persons and unidentified remains. Proteomics also has an important role in understanding the trauma or cause of death by detecting specific protein markers of trauma, poisoning, and infection (Martín et al., 2019).

However, metabolomics focuses on studying metabolites, the small molecules that are the terminal products of many biochemical processes. It is of special interest to forensic toxicology because it gives a dynamic snapshot of an individual's physiological state. However, metabolic profiling of different biological samples can distinguish various drugs and poisons and their breakdown products. It is also useful for determining post-mortem biochemical changes, which can help determine PMI and differentiate between natural and unnatural causes of death. Finally, the use of the field is further extended toward the application of stress or trauma markers by having metabolic changes precede death, for example, in conditions such as physical exertion, oxygen deprivation, or exposure to extreme environments.

Additionally, metabolomics can identify environmental exposure sources or dietary habits that may put a victim into a greater context for their lifestyle and act as a potential lead in a criminal investigation. Combined proteomics and metabolomics provide comprehensive biological evidence. For instance, PMI estimation models are improved by combining protein degradation and metabolomic patterns. Likewise, identifying proteins and metabolites that correlate with disease, infection, or death reveals pre-existing conditions and causes of death. In addition, these techniques help determine the age, sex, and ancestry of skeletal remains where DNA is unavailable (Szeremeta et al., 2021).

Nevertheless, forensic proteomics and metabolomics are currently faced with many challenges. These include requirements for highly sensitive techniques to analyze degraded or limited samples, interpreting large and complex datasets, or even just decreasing the time spent inspecting data and writing programs, which is a step in the right direction. Many bioinformatics expertise and tools are needed to make sense of the data. Furthermore, these techniques need strict validation protocols and standardization in forensic science to ensure their reliability and admissibility in courts. Ethical considerations are also raised because the data analysis may reveal sensitive health or lifestyle information about the patient, which could have different meanings depending on the patient (Dawidowska et al., 2021).

Forensic proteomics and metabolomics are expected to revolutionize the investigations as technology evolves. Not only does their power to interpret biological evidence give forensic science a broader scope, but by unlocking the kind of potential to close cases finally, by hands that can deliver justice and never find failure, they create new avenues to solve the most complex type of cases (Akçan et al., 2020; Parker et al., 2021; Szeremeta et al., 2021).

1.3 Emerging technologies in forensic biotechnology

The latest technologies have expanded the scope of forensic biotechnology to such an extent that the novel tools and techniques becoming available are defeating the best forensic analysis. Recently, emerging technologies such as CRISPR, artificial intelligence (AI), and 3D bioprinting not only address issues in forensic science in terms of the accuracy and reliability of forensic processes but also open up new possibilities for forensic services to address modern problems. Combining these emerging technologies revolutionizes evidence processing, interpretation, and management so that black letters are no longer required to describe what these provide to forensic biotechnology.

1.3.1 CRISPR technology

The gene editing technologies that allow the adoption of a revolutionary approach to forensic science are CRISPR (Clustered Regularly Interspaced Short Palindromic Repeats). One such avenue is forensic genetic analysis, in which the precision of CRISPR has opened the doors for targeting and manipulating DNA sequences by editing the DNA sequences that are of interest. For example, CRISPR-based assays ideally offer the opportunity to detect SNPs and other genetic variations at high accuracy and, therefore, have excellent value for human identification, paternity testing, and ancestry analysis. In addition, CRISPR can deal with problems occurring in degraded DNA samples or discrimination between mixed DNA profiles, which frequently happen in complicated forensic cases. Moreover, CRISPR offers a viable outcome and portability, making it possible to perform real-time genetic analysis at the site of occurrence, which is beneficial in forensic investigations in remote or under-resourced contexts (Dash & Arora, 2022).

1.3.2 Artificial intelligence and 3D printing

Forensic biotechnology has emerged as a powerful tool in modern forensic science, increasingly supported by advancements in AI. With the exponential growth of complex biological and analytical data, AI enables efficient processing, analysis, and interpretation at high speed and accuracy. Its ability to identify patterns within large datasets enhances the detection, characterization, and interpretation of forensic evidence, thereby improving the reliability and efficiency of forensic investigations.

AI is used in forensic DNA analysis to increase the interpretation of mixed sample analysis and to predict phenotypic traits from genetic markers such as hair color, skin color, and facial structure. Facial recognition, fingerprint analysis, crime scene reconstruction, and DNA analysis are employed by AI-based devices. Besides these uses of technology in making the forensic process more efficient, these applications also lower human error, enhancing the reliability of forensic investigation results (Sessa et al., 2024).

With the discovery of 3D bioprinting and tissue engineering, age-old problems in forensic science have been offered new solutions. 3D bioprinting allows forensic scientists to print so real that realistic models of human tissues and organs can be completed to recreate trauma patterns, test hypotheses, or recreate a crime scene. However, in contrast, tissue engineering allows the toxicological effect of various toxins, drugs, or environmental exposures on biological tissue in a controlled environment to be studied, making the performance of toxicity tests easier. In addition, these advances are being tested for the authentication of forensic procedures, that is, the analysis of DNA degradation from printed tissue to glean information regarding PMIs. Forensic investigation is improved by the ability to reproduce biological structures with precision, which drives the development of forensic research and teaching (dos Santos Vargas et al., 2021).

1.3.3 Aptamer-based assays

Aptamer-based assays have become essential tools in forensic biotechnology; they offer an inventive method to detect and analyze chemical and biological evidence. Aptamers are short, single-stranded DNA or RNA molecules that adopt a well-defined three-dimensional structure that allows them to bind with high affinity and selectivity to many proteins, small molecules, cells, or whole organisms (Röthlisberger & Hollenstein, 2018).

Forensic science would benefit from making a technological advance through the ability of aptamer-based assays to mimic antibodies with added benefits. The primary application of aptamer-based assays in forensic analysis is in the forensic detection of biological evidence due to the ability to identify different biomarkers. For example, aptamers can be designed to bind to specific proteins or enzymes found in blood, saliva, semen, and other body fluids, so they can be seen when there are complex or degraded samples. This capability is instrumental in crime scene analysis since the samples may be limited, contaminated, or subject to environmental degradation. Aptamers possess high specificity and thus have a low cross-reactivity rate and false positives, improving their study's validity (Gooch et al., 2017).

In addition to detecting body fluids, agent-based assays based upon aptamers have been used to identify chemical adulterants, toxins, and drugs. Forensic toxicologists can use aptamers designed for high-sensitivity drug molecule or toxic compound detection to provide forensic sensitivity for identifying substances in biological samples (Albaba et al., 2015). Due to its rapid and precise nature, which is critically essential in drug-facilitated crime cases, poisoning, or substance abuse, this capability is beneficial. Aptamers are also insensitive to extreme storage conditions and are functional, making them good candidates for field-based testing compared to conventional antibody-based approaches. Other uses for aptamer-based platforms are also being explored to detect forensic interest DNA-binding proteins and for DNA damage analysis. Using the excision of specific DNA lesions as a target, forensic analysts can assay the integrity of DNA samples and, thereby, the quality of samples for effective genetic profiling. Additionally, those aptamers can be applied to multiplex assays where several targets can be assayed in a single run. Such a multiplexing potential greatly improves the efficiency of forensic processes while at all high throughputs (Davydova & Vorobyeva, 2022)

Among the most exciting breakthroughs in using aptamers for forensic assay applications, one of the most interesting is their incorporation into nanotechnology. Researchers have developed highly sensitive, colorimetric, fluorescent, and electrochemical biosensors using aptamers to conjugate nanoparticles. The biosensors are capable of rapid, portable, and on-site forensic evidence detection using minimal equipment. For example, upon detection of a target analyte, aptamers can be conjugated to nanoparticles, resulting in a color change of the nanoparticle conjugates, making this color change an easy and reliable detection technique in the crime scenes (Kim et al., 2016; Zhang et al., 2019).

Despite their many benefits, aptamer-based assays are not devoid of problems in forensic biotechnology. A detailed understanding of the target molecule's structural characteristics and binding properties is required in the aptamer design and selection process, which takes much time. Second, the assay optimization necessary to deal with the extremely high specificity of aptamers rests upon their instability and invariably poor binding efficiency in the presence of the sample matrix. The second significant challenge is the need to standardize and validate aptamer-based assays, strictly ensuring reproducibility and accuracy because such assays cannot be admitted in courtrooms. As a result, aptamer-based assays are likely to become increasingly important with modern scientific tools and biotechnological advances. These allow for rapid and reliable assessment of a multitude of evidence, are highly sensitive and diverse, and can be integrated with leading-edge technology such as nanotechnology. The development of these assays comes as research and development keep pace with the times to supplant forensic practice in such assays, all of which will address some of the most challenging questions in evidence detection and analysis (Gooch et al., 2017; Yoo et al., 2020).

1.4 Applications of forensic biotechnology

1.4.1 Forensic biotechnology in toxicology

Biotechnology has led to revolutionary improvements in characterizing and evaluating toxic substances in forensic toxicology. Biotechnology can circumvent these problems successfully with the help of molecular biology, biochemistry, and emerging technologies to identify drugs, poisons, and toxins in complex biological or environmental matrices. The advances of these technologies in terms of accuracy, sensitivity, and reliability in toxicological analysis have provided robust solutions for problems with traditional techniques. Biotechnology has revolutionized forensic toxicology in key fields such as drug screening, biomarker-based toxin detection, biosensor technologies, and nanotechnology (Singh, 2017).

Biotechnology's recent emergence in toxicology drug screening has provided highly sensitive and specific devices for detecting drugs in biological samples. The conventional methods for detecting the drug have a decreased sensitivity or susceptibility to false positives. Since these drawbacks have been surmounted by biotechnology advances, such as monoclonal antibodies, recombinant enzymes, and high-throughput screening technologies, their use has been skyrocketing. The devices allow for accurate analysis of drug metabolism, enzymatic profiling, and the detection of various substances, including illicit and legitimate pharmaceuticals. These breakthroughs have been proven to be of unparalleled utility in various forensic applications, including workplace, drug screening, post-mortem toxicology, and anti-doping analysis, resulting in total and accurate analysis (Szymański et al., 2011).

The identification of poisons and toxins has been significantly enhanced by using biomarkers. Biomarkers are biological indicators that demonstrate exposure to toxic agents, even if the original compound is not present. Biotechnological tools like proteomics, genomics, and metabolomics make it possible to identify the biomarkers and provide a more precise indication of toxic exposures. For instance, typical patterns of metabolites can indicate exposure to organophosphates, and metallothioneins indicate heavy metal poisoning. Biomarker-based analysis also provides information on mechanisms of toxicity and interindividual variation in susceptibility, proving extremely useful in forensic analysis, especially in delayed or chronic exposure (Amacher, 2010; Heijne et al., 2005).

Biosensors are another significant achievement in forensic toxicology that provide rapid, precise, and on-site detection of toxic compounds. These devices combine biological recognition components, for example, enzymes, antibodies, or nucleic acids, with advanced signal transduction mechanisms and thus provide real-time analysis of complicated samples. Biosensors are incredibly versatile and have been effectively used to detect toxic compounds in biological matrices, such as blood and urine, and environmental matrices such as water and soil. Advances in biosensor technology in recent times, particularly those incorporating nanomaterials and electrochemical or optical detection systems, have further enhanced their performance capabilities and allow them to detect even trace amounts of toxic compounds with excellent specificity and reliability (Keusgen, 2002; Rosendo et al., 2023).

Nanotechnology has become a powerful tool in forensic toxicology, offering new means of detecting and quantitating toxic substances with unprecedented sensitivity. Nanomaterials comprise different nanomaterials (nanoparticles, nanocomposites, and quantum dots) with individual properties that are extremely useful for toxicological purposes. For instance, functionalized nanoparticles may bind selectively to certain toxic substances, facilitating detection signals while minimizing false positives. The high efficiency of such nanotechnology-based approaches is afforded in analyzing complex biological matrices that conventional approaches may struggle with. In support of this, portable analysis instruments with nanotechnology as the design tool have been developed, which allow direct toxicological analysis in different field locations. Nanotechnology offers such revolutionary potential because probability is combined with a potential at the molecular level of detecting toxins (Bülbül et al., 2015; Lodha et al., 2016; Pandya & Shukla, 2018).

Forensic biotechnology web application in toxicology is a conjunction between scientific progress and forensic science that provides an unsurpassed tool to identify, analyze, and understand toxic issues. These developments reflect the centrality of biotechnology in meeting modern toxicologic problems and between research and forensic practice (Lettieri, 2006).

1.4.2 Forensic biotechnology in microbial forensics

Biotechnology has completely changed and revolutionized microbial forensics for microbial identification, analysis, and tracking of microbial evidence in involvement in criminal cases and national security situations. Microbial forensics is a multidisciplinary science combining the convenience of microbiology, molecular biology, and forensic science applied to microorganisms or their DNA signature for investigative and legal purposes. Determination of the origin of microbial agents of bioterrorism attacks, studies regarding the spread of infectious disease outbreaks, and resolution of microbial evidence crimes are all critical sciences. This revolution is at its head with biotechnology's ability to facilitate high-resolution analysis of microbial communities and signatures to facilitate improvement in pathogen detection methods and facilitate the advent of new technologies such as DNA barcoding for more precise microbial forensics (Budowle, Wilson et al., 2005).

In response to the increasing threat of bioterrorism, microbial forensics was developed as a tactic designed to identify the agent used to kill or incapacitate another through the intentional release of a deadly biological agent such as a bacterium, virus, or toxin. This is important because detecting and identifying such pathogens effectively and accurately is essential for preventing and responding to such attacks. In an era of innovation, NGS and quantitative PCR (qPCR) have aided biotechnology with their transformation of the sensitivity and efficiency of pathogen detection (Kuiper, 2016). Forensic scientists can use such methods to detect pathogens at the strain level in different environmental or clinical specimens. Molecular technologies such as CRISPR-based diagnostic systems can now be applied as rapid and precise methods to detect bioterrorism agents. Such developments are vital not only for the identification but also for the origin and potential source of the biological agent used in the attacks, which aids in responding to these attacks effectively by the authorities (Guise, 2024).

Microbial signatures are unique biological identifiers that can help generate valuable evidence in a criminal investigation. Microbial signatures are the genetic, biochemical, or phenotypic characteristics of microorganisms in a given environment, individual, or object. For instance, the microbial mixture of a suspect or object coupled to a specific place (De Alcaraz-Fossoul et al., 2023). Likewise, there are human body microbial communities, referred to as the microbiome, which can also be helpful for forensic investigation. For instance, an individual's skin microbiome differed between individuals and could be used in suspect identification or tracking a person's movements. Forensic analysts can utilize advanced biotechnology tools of metagenomic and microbial genomics to look at microbial fingerprints with a high degree of exactitude, to the point that where traditional evidence falls short, complex cases can be closed (Clarke et al., 2017).

DNA barcoding has proven to be an innovative tool for microbial forensics, delivering a standardized solution to identifying microbial communities or species via specific genetic identifiers (Antil et al., 2023). The protocol involves sequencing conserved microbial DNA regions, such as the 16S rRNA gene of bacteria or the ITS region of fungi, to provide unique genetic signatures or "barcodes." Through DNA barcoding, microbial species can be quickly and accurately identified, even in mixed or degraded samples, and it is a valuable tool for forensic investigations with environmental or biological samples. On top of this, advances in high-throughput sequencing have brought DNA barcoding to entire microbial communities, which means that forensic analysts can explore microbial diversity and dynamics in mixed samples. Such capability helps trace the origins of microbes, trace disease transmission, and link microbial evidence to places or events (Antil et al., 2023).

Biotechnology has revolutionized microbial forensics considerably by offering sophisticated methods of detection, analysis, and interpretation of microbial evidence of unmatched accuracy (Budowle, Schutzer et al., 2005). Biotechnological innovations in this field address bioterrorism concerns and apply microbial signatures in criminal investigations, indicating biotechnology's excellent role in enhancing forensic methods. Using different techniques like DNA barcoding, better genome science, and advanced systems for catching stuff that makes us sick, microbiology forensics is reaching much further today and coming up with nifty new ideas to help the modern science of investigation.

1.4.3 Forensic biotechnology in body fluid identification

Identifying body fluids at crime scenes is a key component of forensic analysis, sometimes providing critical evidence for event reconstruction and linking crime to suspect. Although traditionally essential, conventional body fluid identification techniques are hampered by low specificity, false positives, and the inability to effectively analyze degraded or mixed samples. Biotechnology has revolutionized the process, introducing advanced molecular-based technologies that offer unmatched sensitivity, specificity, and flexibility in body fluid detection (Vajpayee, Desai et al., 2023).

Conventional body fluid identification techniques employ chemical or immunological assays to identify characteristic constituents in fluids like blood, saliva, semen, or vaginal secretions. While these assays, like presumptive colorimetric and confirmatory immunoassays, are simple and pretty standard, they are not always very good at distinguishing closely related substances. It is also tricky to get reliable results with old samples. But there's another way to use biotechnology: by looking at things at a molecular level to counteract those limitations. By focusing on each body fluid's genetic, proteomic, and molecular signature characteristics, biotechnology offers a more consistent platform for fluid identification, even under poor conditions.

RNA-based body fluid identification is a breakthrough in biotechnology. RNA molecules, and messenger RNAs (mRNAs) in particular, have tissue-specific expression patterns and are, therefore, suitable markers to follow the source of biological samples. In the case of blood, for example, HBA (hemoglobin alpha) for blood, STATH (statherin) for saliva, and PRM1 (protamine 1) for sperm are unique. Using reverse transcription PCR (RT-PCR) and real-time quantitative PCR (RQ) techniques lets forensic scientists amplify and find the RNA markers very sensitively. RNA-based methods can also inform the sample's age or whether the liquid originated from a living person, supporting the forensic analysis. Despite RNA's susceptibility to degradation, advances in RNA stabilization and extraction methods have rendered this strategy reliable even under unfavorable forensic circumstances (Lynch & Fleming, 2021).

Another advance in biotechnological methods for identifying body fluids is protein markers and enzyme-based detection techniques. Dependable biomarkers, proteins, and enzymes unique to a given body fluid allow for their identification even if the genetic material is damaged. For the identification of blood, for example, hemoglobin and peroxidase activity are used, whereas seminal plasma proteins like prostate-specific antigen (PSA) and semenogelin are particular markers for sperm. Furthermore, advanced immunoassays, mass spectrometry, and proteomics make these strategies better through biotechnology developments. By these methods, low-abundance proteins can be distinguished, and the profile of the fluid's proteome can be well delineated, therefore raising the accuracy and sensitivity of the identification process (Ornelas-González et al., 2021; Virkler & Lednev, 2009).

New possibilities for forensic science body fluid identification have arisen from recent advancements in aptamer-based sensors. Short-length single-stranded nucleic acids or peptides known as aptamers can specifically and strongly bind to target molecules, including proteins or little metabolites. Compared to antibodies, aptamers are more stable, less costly, and easy to modify and synthesize. Especially in the fast and exact identification of body fluids, aptamer-based sensors are a great forensic tool. Functionalizing aptamers with nanoparticles or immobilizing them onto biosensor platforms has given scientists the means to produce sensitive assays to identify tiny quantities of bodily fluid biomarkers. Multiplexing, which allows for the simultaneous identification of many liquids with one test, is also made possible by the systems. Especially great for field use are aptamer-based technologies thanks to their minor sample volume requirements, mobility, and fast results. In forensic biology, they are a groundbreaking instrument (Chen et al., 2024).

Body fluid identification using biotechnology shows how fundamentally molecular and analytical advances transform forensic science. Employing improved accuracy, specificity, and adaptability, these biotechnological advances are changing the normal of body fluid analysis and raising the dependability of forensic inquiries.

1.4.4 Biotechnology applications in forensic anthropology

By offering fresh tools and techniques to help address fundamental questions about human remains, biotechnology has greatly improved forensic anthropology. Usually based on physical measurements and visual inspection, conventional techniques help to find remains, establish biological profiles, and discern causes of death. Weather damage, incomplete remains, and personal prejudice can restrict these techniques. Biotechnology allows us to conduct thorough molecular analysis that assists us in understanding better skeletal remains, hence more exact profiles and identifications. DNA analysis, ancestry estimation, sex and age identification, and even reconstructing faces using genetic information are essential biotech applications for forensic anthropology (Antil & Joon, 2025; Mastana, 2007).

DNA analysis of bones is one significant way biotechnology has transformed forensic anthropology. With traditional techniques, identifying aged or degraded skeletal remains might be difficult; DNA testing, on the other hand, has dramatically changed the game (Hofreiter et al., 2021). Advances in extracting mtDNA and nuclear DNA have enabled us to use limited samples to identify individuals. Particularly helpful for this is mtDNA since it exists in several copies and is transmitted from mothers. Furthermore, altering the field involves new sequencing methods that allow us to examine a small quantity of DNA. This has greatly aided in solving cold cases, determining victims from large disasters, and assisting missing person inquiries (Kowalczyk et al., 2021).

Forensic anthropology has also advanced from genomic approaches. Skeletal remains were permitted to provide a much more precise estimation of age, sex, and ancestry. Specific genetic markers let us accurately deduce a person's ancestry. Being able today to locate sex markers on the X and Y chromosomes simplifies the determination of biological sex from degraded remains. In addition, epigenetic developments have enabled us to precisely predict biological age through DNA patterns. Genomic technologies benefit complex visual evaluations (Faerman et al., 1995).

Proteomics is another approach that increases the value of gene analysis by offering vital data from bones and other skeletal materials. Because proteins sometimes last longer than DNA in harsh environments, they could be a good substitute in forensic analysis, particularly if the substance is somewhat damaged. By discovering unique protein markers, for example, bone proteomics could assist in identifying gender, roughly estimating age, and determining species. Since recent advancements in mass spectrometry enable us to detect even small quantities of proteins, we may apply them to present forensic cases and ancient artifacts. By looking at bone biomarkers, we have established post-mortem timelines and learned how the environment impacts skeletons, improving forensic studies (Martín et al., 2019; Mickleburgh et al., 2021).

Using genetic data for facial reconstruction is another intriguing angle. The standard approach to reconstructed faces uses craniofacial measurements and creative estimation; this could bias them (Adserias-Garriga et al., 2025). Advances in forensic genomics now enable us to deduce characteristics from genetic markers, including skin color, hair color, eye color, and even face shape. Combining this genome information with computer modeling allows us to produce better depictions of a person's face, which is beneficial when remains are unknown. Though this approach is a new field of study, it offers new chances for identification in historical and contemporary instances (Kayser et al., 2023; Vajpayee & Shukla, 2020).

In essence, biotechnology has changed forensic anthropology. It has offered sophisticated instruments for processing challenging cases, including human remains, and filled in the deficiencies of old techniques. This improvement enables us in forensic and archaeological settings to investigate ideas of identity and human variation and makes forensic analysis more precise and dependable.

1.4.5 Forensic biotechnology in environmental investigations

In this context, forensic biotechnology has expanded to become a highly dynamic technology in the fight against environmental crimes and challenges, offering crime investigation tools against wildlife and illegal trade, eDNA analysis, bioremediation support, and the tracing of ecological pollutants (Moore et al., 2021). They seek answers to basic questions about illegal wildlife trade, habitat loss, environmental pollution, and sources of pollution, which the application of molecular biology tools and biotechnological advances by forensic scientists can achieve. These applications help with law enforcement but also help with conservation biology and environmental management, bridging the gaps between science, policy, and enforcement. The most relevant area of forensic biotechnology in investigating ecological crimes is its role in the fight against wildlife crime—such as poaching and illegal trade (Ogden, 2016).

Because DNA profiling was developed initially for human forensic analysis, it has been modified to identify species (including and excluding subspecies), subspecies, and even individual animals used in wildlife crimes. Forensic scientists can determine the origin and genetic composition of the specimen by doing DNA analysis on animal parts that have been seized, such as horns, tusks, fur, or feathers (Jordan & Mills, 2021). This information is invaluable for identifying poaching hotspots, identifying the species illegally targeted by trade, and assisting in proving prosecution through useable genetic evidence. Wildlife DNA profiling also enables the building of genetic databases, such as the ForCyt, to aid in the global fight against poaching and to protect biodiversity (Ahlers et al., 2017).

Though eDNA analysis has expanded the use of forensic biotechnology in ecological science, it is the genetic material released into its environment by the organism, such as in water, soil, or air. Forensic scientists use eDNA sampling and analysis to detect the presence of particular species in an ecosystem. Observing or catching them is not necessary. The technique has been considered essential for monitoring biodiversity, tracking invasive species, or the presence of protected and endangered species in an area. Additionally, eDNA has proven helpful in environmental forensics in determining the cause of a loss of biodiversity, measuring habitat disturbance, identifying the impacts of human activities, and tracing animal movement (Larson et al., 2020). eDNA analysis is a powerful and non-invasive ethical conservation and legal investigation of environmental crimes (Harper, 2023; Linacre & Tobe, 2013).

Bioremediation based on forensic biotechnology is essential in combating environmental pollution and contamination, the most serious of which are those resulting from polluting and contaminating organisms' activities. It is a process in which biological microorganisms like bacteria, fungi, or plants break down or neutralize soil, air, and water contaminants. Biotechnology innovations have allowed the identification and genetic modification of microorganisms to increase their ability to degrade toxic materials, like oil spills, heavy metals, and industrial chemicals. Forensic scientists use these biotechnological aids to determine the degree of contamination, look for potential sources of pollution, and develop a formula to remediate (Singh et al., 2017). Furthermore, forensic geology, coupled with bioremediation techniques, contributes to tracking the source of pollution within soil, sediment, and mineral content analysis, supporting the legal investigation and its accountability process for environmental crimes.

Biotechnology applications in tracing environmental contamination platforms have completely changed the procedure for measuring and monitoring pollution sources using precision. The recent biotechnology developments, including biosensors and molecular tracing instruments, allow for identifying specific contaminants, for example, heavy metals, pesticides, and industrial chemicals in complex environmental media. Therefore, biosensors, a combination of electronic devices and biological components, are capable of real-time detection of contaminants with high sensitivity and specificity. The instruments help locate sources of pollution, that is, factory effluents or agricultural runoff, and measure long-term contamination trends in ecosystems (Thakur & Kumar, 2022). Additionally, DNA-based methods, such as genetic barcoding and microbial community analysis, can be used by forensic scientists to assess microbial community changes in contaminated environments that reveal the nature and source of the contaminant (Shekhar et al., 2020).

Forensic biotechnology is applied to environmental crime investigation to illustrate its capacity to manage global problems such as wildlife conservation and pollution. In addition to aiding in detecting and evaluating ecological crime, the instruments facilitate sustainable behavior and conservation of natural ecosystems, thus bridging the gap between biotechnology and environmental science.

1.5 Challenges and future directions

The application of tools and technology in biotechnology has transformed investigative science beyond expectations. However, it has also brought about challenges to address. These challenges can broadly be classified as methodological/technical limitations and ethical and legal considerations. To exploit the advancements, these challenges must be addressed. Central to these challenges are instrumentation and methodological limitations. High-end, advanced throughput applications often require sophisticated instruments and costly consumables. Moreover, skilled workers and continuous education and training make acceptance challenging for laboratories.

Furthermore, the demands of the trial call for quick presentation of the reports, which also creates problems, given the slow nature of these methods. Another difficulty is the pushback against using a fresh approach, followed by a thorough verification required for judicial approval. These more sophisticated methods are subject to the sample quality, which affects their performance and effectiveness and presents a challenge in adaptation. A multimodal approach, including capacity-building training and skill development programs and the establishment of standardized processes, would help forensic workflows to incorporate biotechnological advances better (Adams et al., 2013; Budowle, Murch et al., 2005; Kayser et al., 2023).

Also, the ethical and legal aspects are at issue. DNA profiling helps us to delve more into an individual's genetic information as it advances, which damages core rights, including security, data privacy, and consent. Forensic investigations need monitoring and depend on DNA databases (Wallace et al., 2014). Regarding this, particularly during the familial search, data retention and disclosure of sensitive genetic information could also carry significant legal and personal consequences (Clayton et al., 2019). Proper data sharing and biobanking security measures are essential to guarantee governance and accountability. Therefore, one must draft equal and strict legislation to safeguard privacy and individual rights (Budowle, Murch et al., 2005; Nizami et al., 2018).

Furthermore, legal standards and scientific expectations are necessary to adopt these new technologies in forensic investigations. This involves guaranteeing that the methods used meet the criteria of reliability and validity. These have to comply with the legally allowed forensic methods and be admissible in a court of justice. The advanced techniques employed would have to meet evidence standards similar to Daubert or Frye to be sure they are dependable and legally defensible (Mohsin, 2024). Also heightening ethical worries regarding their potential abuse are new technologies, including CRISPR, AI, and machine learning. Therefore, one needs policies guiding the responsible employment of synthetic biology in evidence analysis (Dash & Arora, 2022).

Still, forensic biotechnology offers development guarantees even amid these problems. By enabling swifter and more comprehensive analysis and interpretation, technologies like MPS and AI will help to support advancing forensic biology. For instance, on-field detection kits and lab-on-chip systems let crime scene analysts screen, collect, and analyze biological evidence at the crime scene, lowering analysis time and dependence on a central forensic engine (Yadav, 2025).

The divergence of biotechnology has further allowed for trace evidence investigation in forensic science employing fields like eDNA analysis (Allwood et al., 2020), diatom analysis (Scott et al., 2017), and microbiome analysis (Bennett, 2024). Therefore, combining scientific developments and social–ethical norms requiring cross-disciplinary collaboration will help determine the future of forensic biotechnology. By addressing these complex problems and leveraging new technologies, forensic biotechnology can enhance the precision, reliability, and fairness of forensic science, solidifying its place as a pillar of the modern legal system.

1.6 Conclusion

Contemporary forensic science relies heavily on forensic biotechnology. It has significantly improved the discipline's reliability and accuracy in forensic studies. The combination of biotechnology and forensic science has given criminal investigations sensitive, particular, and practical tools, including DNA methylation techniques, DNA profiling methods, and nanotechnology-assisted protocols. Although these tools and techniques have transformed evidence analysis and individualization beyond expectations, their role in solving cold cases has also been a boon to criminal investigations. As these technologies expand and evolve, they are undoubtedly expected to continue to put forward more accurate and robust solutions to the ever-challenging field of forensic science. Foreseeing the future of forensic biotechnology has great potential. It is now expected that the field of forensic investigations will witness rapid developments soon, aiding professionals with unparalleled capabilities. Thus making the investigation process well-organized, adept, and significant. This multidisciplinary approach will holistically advance evidence collection, preservation, analysis, and interpretation, leading to a convincing trial. Meanwhile, the challenges with implementation, standardization, and public and legal acceptance will remain a stumbling block.

References

Adams, D. E., Mabry, J. P., McCoy, M. R., & Lord, W. D. (2013). Challenges for forensic science: New demands in today's world. *Australian Journal of Forensic Sciences*, *45*(4), 347–355.

Adserias-Garriga, J., Medina-Paz, F., Molina, J., & Zapico, S. C. (2025). Facing the unknown: Integration of skeletal traits, genetic information, and forensic facial approximation. *Genes*, *16*(5), 511.

Ahlers, N., Creecy, J., Frankham, G., Johnson, R., Kotze, A., Linacre, A., McEwing, R., Mwale, M., Rovie-Ryan, J., & Sitam, F. (2017). For Cyt'DNA database of wildlife species. *Forensic Science International: Genetics Supplement Series*, *6*, e466–e468.

Akçan, R., Taştekin, B., Yildirim, M., Aydogan, H. C., & Sağlam, N. (2020). Omics era in forensic medicine: Towards a new age. *Turkish Journal of Medical Sciences*, *50*(5), 1480–1490.

Albaba, D., Soomro, S., & Mohan, C. (2015). Aptamer-based screens of human body fluids for biomarkers. *Microarrays*, *4*(3), 424–431.

Allwood, J. S., Fierer, N., & Dunn, R. R. (2020). The future of environmental DNA in forensic science. *Applied and Environmental Microbiology*, *86*(2), e01504–01519.

Aly, S. M., & Sabri, D. M. (2015). Next generation sequencing (NGS): A golden tool in forensic toolkit. Archiwum Medycyny Sądowej i Kryminologii/*Archives of Forensic Medicine and Criminology*, *65*(4), 260–271.

Amacher, D. E. (2010). The discovery and development of proteomic safety biomarkers for detecting drug-induced liver toxicity. *Toxicology and Applied Pharmacology*, *245*(1), 134–142.

An, J.-H., Shin, K.-J., Yang, W.-I., & Lee, H.-Y. (2012). Body fluid identification in forensics. *BMB Reports*, *45*(10), 545–553.

Antil, S., Abraham, J. S., Sripoorna, S., Maurya, S., Dagar, J., Makhija, S., Bhagat, P., Gupta, R., Sood, U., Lal, R., & Toteja, R. (2023). DNA barcoding, an effective tool for species identification: A review. *Molecular Biology Report*, *50*(1), 761–775. https://doi.org/10.1007/s11033-022-08015-7

Antil, S., & Joon, V. (2025). Biotechnology in forensic science: Advancements and applications. *Journal of Forensic Science and Research*, 9(1), 007–014. https://doi.org/10.29328/journal.jfsr.1001073

Bennett, N. A. (2024). *The Invisible Trail: The Application of The Microbiome in Forensic Investigations*. UCL (University College London).

Børsting, C., & Morling, N. (2015). Next generation sequencing and its applications in forensic genetics. *Forensic Science International: Genetics*, *18*, 78–89.

Budowle, B., Murch, R., & Chakraborty, R. (2005). Microbial forensics: The next forensic challenge. *International Journal of Legal Medicine*, *119*, 317–330.

Budowle, B., Schutzer, S. E., Ascher, M. S., Atlas, R. M., Burans, J. P., Chakraborty, R., Dunn, J. J., Fraser, C. M., Franz, D. R., & Leighton, T. J. (2005). Toward a system of microbial forensics: From sample collection to interpretation of evidence. *Applied and Environmental Microbiology*, *71*(5), 2209–2213.

Budowle, B., Wilson, M. R., Burans, J. P., Breeze, R. G., & Chakraborty, R. (2005). Microbial forensics. In: B. Budowle, R. Breeze, & S. Keim (Eds.), *Microbial Forensics* (pp. 1–25). Elsevier.

Bülbül, G., Hayat, A., & Andreescu, S. (2015). Portable nanoparticle-based sensors for food safety assessment. *Sensors*, *15*(12), 30736–30758.

Butcher, B. A., Bieber, F. R., Budimlija, Z. M., Dennis, S. M., & Desire, M. A. (2014). Identification of missing persons and mass disaster victim identification by DNA. Dragan Primorac and Moses Schanfield (Eds.), In: *Forensic DNA Applications: An Interdisciplinary Perspective* (pp. 277–292). CRC Press.

Chen, L., Yang, G., & Qu, F. (2024). Advances of aptamer-based small-molecules sensors in body fluids detection. *Talanta*, *268*, 125348.

Chung, H., & Choe, S. (2017). Overview of forensic toxicology, yesterday, today and in the future. *Current Pharmaceutical Design*, *23*(36), 5429–5436.

Clarke, T. H., Gomez, A., Singh, H., Nelson, K. E., & Brinkac, L. M. (2017). Integrating the microbiome as a resource in the forensics toolkit. *Forensic Science International: Genetics*, *30*, 141–147. https://doi.org/10.1016/j.fsigen.2017.06.008

Clayton, E. W., Evans, B. J., Hazel, J. W., & Rothstein, M. A. (2019). The law of genetic privacy: Applications, implications, and limitations. *Journal of Law and the Biosciences*, *6*(1), 1–36.

Dash, H. R., & Arora, M. (2022). CRISPR-CasB technology in forensic DNA analysis: Challenges and solutions. *Applied Microbiology and Biotechnology*, *106*(12), 4367–4374.

Dash, H. R., & Vajpayee, K. (2021). Collection, preservation, and transportation of biological evidences. In: H. R. Dash, P. Shricastava, & J. A. Lorente (Eds.), *Handbook of DNA Profiling* (pp. 69–84). Springer.

Dash, H. R., Vajpayee, K., & Agarwal, R. (2020). A glimpse of famous cases in history solved by DNA typing. In: H. R. Dash, P. Shricastava, & J. A. Lorente (Eds.), *Handbook of DNA Profiling* (pp. 1–17). Springer.

Davydova, A., & Vorobyeva, M. (2022). Aptamer-based biosensors for the colorimetric detection of blood biomarkers: Paving the way to clinical laboratory testing. *Biomedicines*, *10*(7), 1606.

Dawidowska, J., Krzyżanowska, M., Markuszewski, M. J., & Kaliszan, M. (2021). The application of metabolomics in forensic science with focus on forensic toxicology and time-of-death estimation. *Metabolites*, *11*(12), 801.

De Alcaraz-Fossoul, J., Wang, Y., Liu, R., Mancenido, M., Marshall, P. A., Núñez, C., Broatch, J., & Ferry, L. (2023). Microbes in fingerprints: A source for dating crime evidence? *Forensic Science International: Genetics*, *65*, 102883. https://doi.org/10.1016/j.fsigen.2023.102883

dos Santos Vargas, B. F., Coutinho, M. A., & Coutinho, F. S. (2021). 3D printing in forensic medicine and crime-solving: Integrative literature review. *Revista de Medicina (São Paulo)*, *100*(1), 62–69.

Duster, T. (2007). The molecular reinscription of race: Unanticipated issues in biotechnology and forensic science. In: S. L. Gilman (Ed.), *Race in Contemporary Medicine* (pp. 134–148). Routledge.

Faerman, M., Filon, D., Kahila, G., Greenblatt, C. L., Smith, P., & Oppenheim, A. (1995). Sex identification of archaeological human remains based on amplification of the X and Y amelogenin alleles. *Gene*, *167*(1–2), 327–332.

Fraga, D., Meulia, T., & Fenster, S. (2014). Real-time PCR. *Current Protocols Essential Laboratory Techniques*, *8*(1), 10.13. 11–10.13. 40.

Franceschetti, L., Amadasi, A., Bugelli, V., Bolsi, G., & Tsokos, M. (2023). Estimation of late post-mortem interval: Where do we stand? A literature review. *Biology*, *12*(6), 783.

Gooch, J., Daniel, B., Parkin, M., & Frascione, N. (2017). Developing aptasensors for forensic analysis. *TrAC Trends in Analytical Chemistry*, *94*, 150–160.

Green, M. R., & Sambrook, J. (2018). Touchdown polymerase chain reaction (PCR). *Cold Spring Harbor Protocols*, *2018*(5), pdb.prot095133.

Green, M. R., & Sambrook, J. (2019). Nested polymerase chain reaction (PCR). *Cold Spring Harbor Protocols*, *2019*(2), pdb.prot095182.

Guise, N. A. (2024). *Subtyping Microbial Species Using CRISPR Arrays and Its Application in a Forensic Context*. University of Otago.

Harper, C. K. (2023). Poaching forensics: Animal victims in the courtroom. *Annual Review of Animal Biosciences*, *11*(1), 269–286.

Heijne, W. H., Kienhuis, A. S., Van Ommen, B., Stierum, R. H., & Groten, J. P. (2005). Systems toxicology: Toxicogenomics, transcriptomics, proteomics and metabolomics applications in toxicology. *Expert Review of Proteomics*, *2*(5), 767–780.

Hofreiter, M., Sneberger, J., Pospisek, M., & Vanek, D. (2021). Progress in forensic bone DNA analysis: Lessons learned from ancient DNA. *Forensic Science International: Genetics*, *54*, 102538.

Jordan, D., & Mills, D. (2021). Past, present, and future of DNA typing for analyzing human and non-human forensic samples. *Frontiers in Ecology and Evolution*, *9*, 646130.

Kayser, M., Branicki, W., Parson, W., & Phillips, C. (2023). Recent advances in forensic DNA phenotyping of appearance, ancestry and age. *Forensic Science International: Genetics*, *65*, 102870.

Keusgen, M. (2002). Biosensors: New approaches in drug discovery. *Naturwissenschaften*, *89*, 433–444.

Kim, Y. S., Raston, N. H. A., & Gu, M. B. (2016). Aptamer-based nanobiosensors. *Biosensors and Bioelectronics*, *76*, 2–19.

Kowalczyk, M., Staniszewski, A., Kamińska, K., Domaradzki, P., & Horecka, B. (2021). Advantages, possibilities, and limitations of mitochondrial DNA analysis in molecular identification. *Folia Biologica (Kraków)*, *69*(3), 101–111.

Kuiper, I. (2016). Microbial forensics: Next-Generation sequencing as catalyst: Using new sequencing technologies to analyze whole microbial communities could become a powerful tool for forensic and criminal investigations. *EMBO Reports*, *17*(8), 1085–1087.

Kumar, A. (2021). *Advanced Forensic Biotechnology*. Akinik Publications.

Kurland, J., Pires, S. F., McFann, S. C., & Moreto, W. D. (2017). Wildlife crime: A conceptual integration, literature review, and methodological critique. *Crime Science*, *6*(1), 15.

Larson, E. R., Graham, B. M., Achury, R., Coon, J. J., Daniels, M. K., Gambrell, D. K., Jonasen, K. L., King, G. D., LaRacuente, N., & Perrin-Stowe, T. I. (2020). From eDNA to citizen science: Emerging tools for the early detection of invasive species. *Frontiers in Ecology and the Environment, 18*(4), 194–202.
Lessig, R., & Prinz, M. (2022). Mass disaster victim identification. *Handbook of Forensic Medicine, 1*, 291–309.
Lettieri, T. (2006). Recent applications of DNA microarray technology to toxicology and ecotoxicology. *Environmental Health Perspectives, 114*(1), 4–9.
Linacre, A., & Tobe, S. (2013). *Wildlife DNA analysis: Applications in Forensic Science*. Wiley-Blackwell.
Lodha, A., Pandya, A., & Shukla, R. (2016). Nanotechnology: An applied and robust approach for forensic investigation. *Forensic Research & Criminology International Journal, 2*(1), 00044.
Lynch, C., & Fleming, R. (2021). RNA-based approaches for body fluid identification in forensic science. *Wiley Interdisciplinary Reviews: Forensic Science, 3*(4), e1407.
Martín, R. D. D., Camacho-Martínez, Z., Hernández, J. R. A., & Valencia-Caballero, L. (2019). Proteomics as a new tool in forensic sciences. *Spanish Journal of Legal Medicine, 45*(3), 114–122.
Mastana, S. (2007). Molecular anthropology: Population and forensic genetic applications. *Anthropology, 3*, 373–383.
Mickleburgh, H. L., Schwalbe, E. C., Bonicelli, A., Mizukami, H., Sellitto, F., Starace, S., Wescott, D. J., Carter, D. O., & Procopio, N. (2021). Human bone proteomes before and after decomposition: Investigating the effects of biological variation and taphonomic alteration on bone protein profiles and the implications for forensic proteomics. *Journal of Proteome Research, 20*(5), 2533–2546.
Mohsin, D. K. (2024). The significance of forensic and scientific evidence and their admissibility in criminal law. *SSRN*. https://doi.org/10.2139/ssrn.4960559
Moore, T., Gaynus, C., Levin, P. S., & Meyer, R. (2021). The intersection of forensic techniques with ecological issues. In: S. C. Underkoffler & H. R. Adams (Eds.), *Wildlife Biodiversity Conservation: Multidisciplinary and Forensic Approaches* (pp. 147–161). Springer.
Murphy, B. L., & Morrison, R. D. (2014). *Introduction to Environmental Forensics*. Elsevier.
Nizami, S. B., Kazmi, S. Z. H., Abid, F., Babar, M. M., Noor, A., Zaidi, N.-u-S. S., Khan, S. U., Hasan, H., Ali, M., & Gul, A. (2018). Omics approaches in forensic biotechnology: Looking for ancestry to offence. In: D. Barh and V. Azevedo (Eds.), *Omics Technologies and Bio-engineering* (pp. 111–129). Elsevier.
Ogden, R. (2016). Forensics in transnational environmental crime. In: *Handbook of Transnational Environmental Crime* (pp. 388–410). Edward Elgar Publishing.
Ornelas-González, A., Ortiz-Martínez, M., González-González, M., & Rito-Palomares, M. (2021). Enzymatic methods for salivary biomarkers detection: Overview and current challenges. *Molecules, 26*(22), 7026.
Pandya, A., & Shukla, R. K. (2018). New perspective of nanotechnology: Role in preventive forensic. *Egyptian Journal of Forensic Sciences, 8*, 1–11.
Parker, G. J., McKiernan, H. E., Legg, K. M., & Goecker, Z. C. (2021). Forensic proteomics. *Forensic Science International: Genetics, 54*, 102529.
Pattnaik, P., & Jana, A. M. (2005). Microbial forensics: Applications in bioterrorism. *Environmental Forensics, 6*(2), 197–204.
Pillay, V. V., Menezes, R. G., Krishnaprasad, R., Pillay, M., Lobo, S. W., Adhikari, D., Vishwanath, P., Bhat, N. B., & Vasudevan, T. K. D. M. (2007). Biotechnology in forensic science: The revolution continues. *Nepal Medical College Journal, 9*(1), 57–62.
Rosendo, L. M., Antunes, M., Simão, A. Y., Brinca, A. T., Catarro, G., Pelixo, R., Martinho, J., Pires, B., Soares, S., & Cascalheira, J. F. (2023). Sensors in the detection of abused substances in forensic contexts: A comprehensive review. *Micromachines, 14*(12), 2249.
Röthlisberger, P., & Hollenstein, M. (2018). Aptamer chemistry. *Advanced Drug Delivery Reviews, 134*, 3–21.
Scott, K. R., Morgan, R. M., Jones, V. J., Dudley, A., Cameron, N., & Bull, P. A. (2017). The value of an empirical approach for assessing diatoms as environmental trace evidence in forensic limnology. *Archaeological and Environmental Forensic Science, 1*(1), 49–78.
Sessa, F., Esposito, M., Cocimano, G., Sablone, S., Karaboue, M. A. A., Chisari, M., Albano, D. G., & Salerno, M. (2024). Artificial intelligence and forensic genetics: Current applications and future perspectives. *Applied Sciences, 14*(5), 2113.
Shekhar, S. K., Godheja, J., & Modi, D. R. (2020). Molecular technologies for assessment of bioremediation and characterization of microbial communities at pollutant-contaminated sites. In: R. N. Bhargava and G. Saxena (Eds.), *Bioremediation of Industrial Waste for Environmental Safety: Volume II: Biological Agents and Methods for Industrial Waste Management* (pp. 437–474). Springer.
Shukla, R. K. (2016). Forensic biotechnology: Application of flow cytometry in legal medicine. *International Journal of Forensic Science, 1*, 1–2.
Singh, M., Srivastava, P. K., Jaiswal, V. K., & Kharwar, R. N. (2017). Biotechnological applications of microbes for the remediation of environmental pollution. *Environmental Protection, 1*, 36.

Singh, Z. (2017). Forensic toxicology: Biological sampling and use of different analytical techniques. *Forensic Research & Criminology International Journal, 4*(4), 117–120.

Szeremeta, M., Pietrowska, K., Niemcunowicz-Janica, A., Kretowski, A., & Ciborowski, M. (2021). Applications of metabolomics in forensic toxicology and forensic medicine. *International Journal of Molecular Sciences, 22*(6), 3010.

Szymański, P., Markowicz, M., & Mikiciuk-Olasik, E. (2011). Adaptation of high-throughput screening in drug discovery—Toxicological screening tests. *International Journal of Molecular Sciences, 13*(1), 427–452.

Teotia, A. K., & Pal, R. (2009). Analysis of drugs and poisons of forensic interest–A review. *Journal of Forensic Medicine and Toxicology, 26*(1), 49–53.

Thakur, A., & Kumar, A. (2022). Recent advances on rapid detection and remediation of environmental pollutants utilizing nanomaterials-based (bio) sensors. *Science of the Total Environment, 834*, 155219.

Vajpayee, K., Dash, H. R., Parekh, P. B., & Shukla, R. K. (2023). PCR inhibitors and facilitators-their role in forensic DNA analysis. *Forensic Science International*, 111773.

Vajpayee, K., Desai, P., Parekh, P. B., Sitwala, H. C., Suri, K., Dash, H. R., & Shukla, R. K. (2023). A novel method for blood detection using fluorescent dye. *Microchemical Journal, 193*, 108987. DOI:10.1016/j.microc.2023.108987

Vajpayee, K., Sagar, D., & Dash, H. R. (2020). Forensic DNA typing: Inception, methodology, and technical advancements. *Forensic DNA Typing: Principles, Applications and Advancements* (pp. 3–26).

Vajpayee, K., & Shukla, R. K. (2020). DNA phenotyping: The technique of the future. In: H. R. Dash, P. Shrivastava, & J. A. Lorente (Eds.), *Handbook of DNA Profiling* (pp. 1–25). Springer Singapore. https://doi.org/10.1007/978-981-15-9364-2_54-1

Virkler, K., & Lednev, I. K. (2009). Analysis of body fluids for forensic purposes: From laboratory testing to non-destructive rapid confirmatory identification at a crime scene. *Forensic Science International, 188*(1–3), 1–17.

Wallace, H. M., Jackson, A., Gruber, J., & Thibedeau, A. (2014). Forensic DNA databases–Ethical and legal standards: A global review. *Egyptian Journal of Forensic Sciences, 4*(3), 57–63.

Wani, A. K., Akhtar, N., Sena, S., & Singh, J. (2022). Microbial forensics: A potential tool for investigation and response to bioterrorism. *Health Sciences Review, 5*, 100068.

Williams, R., & Wienroth, M. (2017). Social and ethical aspects of forensic genetics: A critical review. *Forensic Science Review, 29*(2).

Yadav, B. (2025). Artificial intelligence in forensic science: Navigating ethical frontiers and transformative applications. In: *Generative AI Techniques for Sustainability in Healthcare Security* (pp. 175–194). IGI Global Scientific Publishing.

Yeung, M. (2016). Microbial forensics in food safety. *Microbiology spectrum, 4*(4). https://doi.org/10.1128/microbiolspec.emf-0002-2013

Yoo, H., Jo, H., & Oh, S. S. (2020). Detection and beyond: Challenges and advances in aptamer-based biosensors. *Materials Advances, 1*(8), 2663–2687.

Zhang, Y., Lai, B. S., & Juhas, M. (2019). Recent advances in aptamer discovery and applications. *Molecules, 24*(5), 941.

2

Collection and Preservation of Samples for Forensic DNA Analysis

Anna Barbaro

2.1 Introduction

Forensic DNA analysis is a vital component of the criminal justice system, offering robust and reliable evidence in identifying individuals involved in criminal activities. Biological samples, such as blood, saliva, semen, bones, and teeth, allow linking or excluding individuals to a crime scene. The collection and preservation of biological samples are critical steps that directly impact the reliability and accuracy of DNA analyses. (Lindahl, 1993). Improper handling or preservation can lead to contamination, degradation, and ultimately the loss of genetic information. (Butler, 2011) This chapter discusses best practices, highlighting the scientific principles underpinning these methods and their importance in forensic investigations.

2.2 Crime scene isolation and evidence collection

To preserve the integrity of a crime scene, it is crucial to isolate and protect the area as quickly as possible. This prevents any alteration or contamination of evidence, which is vital for a successful forensic investigation. Evidence collection should only commence after thorough observation, documentation, and inspection of the scene (Singh, 2021).

2.2.1 Scene documentation before evidence collection

Every crime scene is unique, with its own distinct characteristics, as individuals may behave differently during the commission of a criminal act. As a result, crime scene investigation must be conducted by trained specialists with appropriate expertise, adhering to internationally standardized protocols (Houck, 2010).

Before collecting any evidence, it is essential to meticulously document the crime scene. In resume this activity includes (ENFSI, 2022):

- **Marking Evidence**: Each evidence should be marked uniquely, typically with a letter or a number. This ensures that each item is easily identifiable throughout the investigation and analysis process.
- **Photography**: Evidence must be photographed both with and without a metric reference. This provides a visual record of the scene and the precise location and condition of each piece of evidence before it is disturbed.

DOI: 10.4324/9781003474296-2

- **Notes and Sketches**: Detailed notes and sketches should accompany the photographic record. These should include information about the location of evidence, environmental conditions, and any other pertinent observations.

2.3 Collection of biological samples

The first step in a CSI investigation consists in searching samples useful for further analysis. This stage is pivotal to the investigation's success and can be highly challenging, given the often complex and chaotic nature of crime scenes. Different methods are available for crime scene inspection (e.g., quadrant method, spiral method, line method) and biological sample searching (e.g., use of forensic lights or chemical reagents). Recognizing the critical role and complexity of crime scene investigations within the judicial system, the European Network of Forensic Science Institutes (ENFSI) formed the Scene of Crime Working Group, which developed a comprehensive protocol aligned with the ISO/IEC 17020 standards, covering every step of the process (ENFSI, 2022). As such, an experienced investigator who approaches the scene with a systematic evaluation is an invaluable asset (Lee and Ladd, 2001).

Biological samples such as blood, saliva, semen, bones, teeth play a crucial role in linking or excluding individuals from a crime scene. This evidence can be found in various forms, including fluids, stains or trace, on portable or non-portable objects, on porous or non-porous surfaces (Barbaro, 2022).

Each type of biological evidence requires specific collection and handling procedures to ensure its integrity. Improper collection techniques can lead to the destruction or contamination of evidence. Therefore, it is essential to follow strict protocols and prioritize the collection process to prevent degradation or alteration, because biological evidence is very susceptible to environmental factors and contamination (Technical Working Group on Biological Evidence Preservation, 2013).

The Spanish and Portuguese Group of the International Society of Forensic Genetics (GHEP) has developed comprehensive guidelines for the collection and transportation of biological samples useful for DNA analysis (Grupo Español y Portugués de la International Society For Forensic Genetic, 2000). These guidelines emphasize the importance of proper collection techniques, correct packaging, and secure transportation to maintain the integrity of the samples.

Additionally, standards set by the National Institute of Standards and Technology (NIST) and other forensic bodies provide a framework for the accurate collection, preservation and analysis of forensic evidence Technical Working Group on Biological Evidence Preservation, 2013).

By adhering to these guidelines, forensic professionals can ensure that cadaveric remains are properly collected and preserved, providing crucial evidence for forensic DNA analysis.

2.3.1 Crime scene evidence

Biological samples can be found on various surfaces and materials, and different procedures should be used for their collection (Technical Working Group on Biological Evidence Preservation, 2013).

2.3.1.1 Biological stains on transportable objects

When biological stains are present on transportable objects (e.g., small items), the entire object should be collected and placed into an appropriate container for transportation to the laboratory. Each item must be packaged separately to avoid cross-contamination.

Wet evidence, such as items stained with bodily fluids, should be placed in paper bags to allow for air circulation. Alternatively, the samples must be allowed to air-dry prior to packaging. in suitable containers. This prevents the growth of mold or bacteria, which could degrade the biological material.

For transportable objects with soft surfaces (e.g., clothes, bed sheets), it is crucial to avoid folding them while they are still wet. Folding wet evidence on itself can cause the biological material to transfer or smear, potentially compromising the integrity of the sample.

2.3.1.2 Biological stains on non-transportable objects

2.3.1.2.1 Non-porous surfaces When biological material is present on a non-transportable, non-porous surface (e.g., painted wall, floor), the recommended method for evidence collection is swabbing to recover as much of the substance as possible.

In case of dry surfaces, the swab needs to be lightly wetted with sterile distilled water, so to enhance the collection of the biological material. If a large quantity of biological material is present, multiple swabs should be used to ensure thorough collection.

Specialized swabs are available for collecting biological samples designed to maximize sample recovery and are often coated with antimicrobial agents to preserve nucleic acids, eliminating the need for air-drying. They may be used dry.

When the stain is very small, it is advisable to collect all the biological samples on the tip of the swab to maximize recovery. Each swab should be stored in a separate container to avoid cross-contamination, and all swabs should be air-dried before packaging.

In cases of dried stains, scraping with a sterile blade or scalpel may be necessary. This method is stains on hard materials, allowing for efficient collection without contamination. The sterile blade is gently scraped over the stained area, ensuring the biological material is preserved for analysis.

Using the same collection technique, it is recommended to collect material from a clean, unstained area near the evidence that serves as a control sample. This helps in distinguishing potential environmental contaminants from the actual evidence.

2.3.1.2.2 Porous surfaces When the material is found on a non-transportable, porous surface, the best method for evidence collection is to carefully cut out a portion of the stained area using a disposable sterile scalpel or scissors. It is suggested to cut also an unstained area nearby, which functions as a control sample, providing a reference for comparison.

If tape is used for evidence collection, it should be applied repeatedly to the surface to gather as much of the biological sample as possible. This method is particularly useful for collecting trace evidence or fine particles from porous surfaces.

2.3.2 Biological fluids

When dealing with biological evidence in liquid form, such as blood, saliva, or semen, careful collection is essential to preserve the integrity and quantity of the sample for forensic analysis. The selection of the collection technique depends on the volume of the liquid and the surface on which it is found.

For small quantities of liquid or when the liquid is found on a surface, a sterile swab is often the preferred method. The swab should be gently rolled or swiped across the liquid to absorb as much material as possible. This method is especially useful when the liquid is spread thinly across a surface. Swabs should be placed in individual sterile tubes or envelopes designed to protect the sample from contamination and degradation.

If a larger volume of liquid is available, or if the liquid is pooled, a disposable plastic pipette can be used to collect the sample. Pipetting allows for more precise collection of liquid evidence, minimizing the risk of loss or contamination. Once collected, the biological material should be transferred into a sterile, airtight container, such as a tube or vial.

2.3.3 Samples from the victim's body

Various types of biological evidence, such as blood, semen, saliva, skin cells, and other fluids, can be found on the victim's body, including in the mouth, anal, and vaginal regions, under the fingernails, and on clothing. Proper collection of this evidence is crucial for forensic analysis and must be done with care to avoid contamination and degradation (Grupo Español y Portugués de la International Society For Forensic Genetic, 2020).

2.3.3.1 Biological stains

If biological stains are found on the body, they should be collected using a sterile swab dampened with distilled water. The entire stained area should be gently swabbed by pressing lightly to absorb the maximum amount of material.

2.3.3.2 Saliva from bite marks

Saliva residues from bite marks should be collected using a sterile swab lightly moistened with distilled water. The swab should be used to clean the entire area within the bite mark, following the circular pattern left by the teeth. This technique helps ensure the collection of as much biological material as possible from the bite site.

2.3.3.3 Evidence from nails

Evidence such as hairs or fibers can often be found under the victim's fingernails. These should be carefully picked up with tweezers and placed in appropriate containers. The top edge of the nails should also be cut and collected separately for each hand. The nail samples should be stored in individual containers, such as tubes, bags, or they can be carefully wrapped in paper folds and then placed into small paper bags.

2.3.3.4 Hairs and fibers

Hairs and fibers discovered on the victim's body should be gathered using sterile tweezers. They can be stored individually or collectively in paper bags. Another option is to place them on a small sheet of paper, fold it securely, and then place it inside a small paper bag for safekeeping.

2.3.3.5 Sexual assault evidence

In sexual assaults, it is critical that evidence collection, especially from inside the victim's body, be conducted by a qualified personnel (e.g., physician, sexual assault nurse examiner (SANE)). The examination should be performed in a short time after the assault to assess for injuries and sexually transmitted infections.

During sexual assault examinations, the following areas should be swabbed due to potential contact with the assailant:

Oral Cavity: If semen traces are suspected in the oral cavity, sterile swabs should be used to gently collect samples from the teeth, the tongue, and the palate. Oral swabs should be taken immediately because semen tends to degrade quickly in the mouth.

Body Surface: The body surface should be examined for semen stains, saliva, or possible bite marks using a Wood's lamp, which can help visualize biological fluids. Any detected fluids should be collected using sterile swabs.

Pubic Area: Hairs in the pubic area should be collected using a comb or tweezers and placed in paper or plastic bags. Alternatively, the hairs can be collected over a piece of paper, which is then carefully folded and stored in a small paper bag.

Vaginal Cavity, Cervix, Vulvar Region: Two cervical swabs, two vaginal swabs, and one swab from the external genitalia should be collected by swabbing. Then, a vaginal washing should be performed using approximately 10 ml of sterile physiological saline, which should then be collected in a tube.

Anal Samples: Two samples from the anorectal canal and one sample from the anal margin should be taken by using sterile swabs.

2.3.3.6 Clothes

The clothes worn by the victim during the assault should be collected and placed in individually paper bags to prevent contamination and preserve the integrity of any biological sample present. Each item of clothing should be handled with care to avoid cross-contamination, and paper bags are preferred to prevent moisture buildup, which can lead to the degradation of biological materials. By following these protocols, forensic professionals can ensure that evidence from the victim's body is collected in a manner that maximizes its utility in the investigation and subsequent legal proceedings (Grupo Español y Portugués de la International Society For Forensic Genetic, 2020).

2.3.4 Cadaveric remains

The collection of cadaveric remains at a crime scene is determined by the condition and type of remains found. Proper collection and preservation techniques are crucial to maintaining the integrity of the samples for forensic analysis (Barbaro, 2022).

2.3.4.1 Bodies in good condition

Blood Samples: A blood sample of approximately 10 ml should be collected into a tube containing an anticoagulant, such as EDTA, to prevent the blood from clotting, preserving the sample for DNA analysis and other forensic testing.

2.3.4.2 Putrefied bodies

Bones and Teeth: Preferably, long bones, such as the femur, should be collected. Teeth, especially molars, are also valuable for forensic analysis due to their durability and the protection they offer to the DNA inside. Each bone and tooth should be placed individually in paper bags or containers. If residual decomposed tissues are still attached to the bones, it is essential to remove this material before placing the bones into the containers. This prevents further degradation of the skeletal remains and contamination of the sample.

Muscle Tissue: Muscle tissue from the most well-preserved regions of the body should be collected and placed in wide-mouth plastic containers with screw caps. The containers should not contain any fixing liquid, as this could interfere with subsequent DNA extraction.

2.3.4.3 Carbonized bodies

Muscle Tissue: Tissues from deep areas of the body or from the cardiac muscle are used since they are most resistant to heat. Samples should be placed in separate containers.

Bones and Teeth: They are more resistant to fire and heat. Preferably, long bones such as the femur should be collected. It is also useful to collect four teeth, preferably molars and premolars. Samples should be placed in separate containers to avoid cross-contamination.

2.3.5 Reference samples

The collection of reference samples from living persons, whether suspects or victims, requires judicial authorization. Individuals must provide informed consent by signing a document that outlines the purpose of the sample collection, typically for specific forensic tests such as DNA typing. Blood and saliva are the most used reference samples (Grupo Español y Portugués de la International Society For Forensic Genetic, 2020).

2.3.5.1 Blood samples

Blood samples can be collected in two main ways:

- **Vein Puncture**: A blood sample of approximately 5–10 ml is collected via venous puncture. The blood is drawn into a tube containing an anticoagulant (EDTA) to prevent clotting, ensuring the sample remains viable for DNA analysis.
- **Finger Puncture**: A small puncture is made on the fingertip by a sterile needle or surgical lancet. Some blood drops are deposited onto a special filter paper, such as FTA cards (Whatman), which are specifically designed for the collection and preservation of blood. These cards are chemically treated and enable long-term storage of biological material under room conditions without degradation.

2.3.5.2 Saliva samples

Saliva, along with buccal epithelial cells, is another common reference sample. The interior of the mouth is swabbed using a sterile cotton swab. It is recommended to use one swab for the right side and one for the left side of the oral cavity. The swabs should be air-dried at room temperature before being stored in their containers, as moisture can lead to bacterial growth and subsequent DNA degradation. Chemically treated filter paper, such as Whatman FTA cards, is also available for saliva collection, allowing for long-term storage at room temperature.

2.3.5.3 Hair samples

Hair samples can be particularly useful when blood or saliva is not available or in cases of hair comparison analysis. It is suggested to collect at least 10 hairs with roots, using sterile tweezers. Hairs are then placed in folded paper or in paper bags. In sexual assault cases, it is important to gather 10–15 pubic hairs from the victim for examination and comparison purposes.

2.3.5.4 Transfused subjects

In individuals who have recently received a blood transfusion, the collection of reference samples presents specific challenges. After a transfusion, a person's blood may contain DNA from the donor or a mixture of DNA from both the donor and recipient, especially shortly after the procedure. In such cases, saliva or alternatively hairs with roots should be used as the reference samples. These samples are less likely to be contaminated with the donor's DNA, providing a more accurate representation of the individual's genetic profile.

2.4 Biological samples packaging

Proper packaging of collected evidence is crucial to prevent any damage, contamination, degradation, and to maintain the integrity of the sample throughout its transportation and storage. The choice of packaging materials, such as envelopes, cardboard boxes, paper bags, and plastic bags, should be based on the type of evidence, with consideration given to whether the evidence is porous or non-porous. It is essential to use disposable materials for evidence collection to avoid cross-contamination. Once the collection is complete, all used materials, such as gloves, masks, and collection papers, should be disposed of in appropriate containers to prevent any residual contamination at the scene. Each evidence should be distinctly labeled and separately packaged. This practice prevents cross-contamination between evidence and between evidence and reference samples (Dash & Vajpayee, 2022).

Biological samples, which can degrade easily, require special precautions during packaging:

- **Wet Samples**: Wet evidence should be stored in non-porous, waterproof containers to prevent leakage and further contamination. Plastic bags are generally not recommended for long-term storage of wet samples because they favor the growth of bacteria and fungi, which can degrade the evidence. If plastic bags must be used temporarily, breathable plastic bags are preferred. Ideally, wet evidence should be air-dried in a clean environment before packaging to reduce the risk of degradation.

- **Dry Samples**: Dry evidence, such as cigarettes and fibers, can be placed either in paper or in plastic bags. To minimize the loss of biological material, it is crucial to avoid any rubbing or movement of the sample within its container.
- **Liquid Biological Samples**: Plastic bottles or tubes are typically used for the storage of liquids to ensure that they remain sealed and uncontaminated.
- **Hair and Fibers**: These types of evidence can be stored in various containers, including paper bags, envelopes, plastic bags, or vials.

To ensure the correct preservation and transportation of evidence, a three-layer packaging system is often used (Technical Working Group on Biological Evidence Preservation, 2013):

- **Primary Packaging**: This is the container in direct contact with the evidence, ensuring initial protection from contamination or loss.
- **Secondary Packaging**: This is an additional container that holds the primary packaging, providing an extra layer of protection during transportation.
- **Tertiary Packaging**: This container is used to hold multiple secondary packages, facilitating the organized transport of several pieces of evidence at once.

Evidence containers should be securely sealed to prevent tampering, loss, or alteration of the contents. It is necessary to use security bags, tamper-proof tapes, or non-removable seals labeled with the date and the initials of the person sealing the package. These measures ensure that any unauthorized access or tampering is easily detectable.

Every package should be labeled with all essential information required for effective processing and documentation of the evidence collection process (Technical Working Group on Crime Scene Investigation, 2013).

The label should include the following information:

- Case Reference Number
- Summary Description of the Evidence
- Type or Category of Evidence
- Collection Date, Time, and Location
- Name of the individual who collected the evidence

This information is essential for the proper identification and tracking of evidence, ensuring that the label on the package clearly matches the contents inside. Proper documentation also supports the chain of custody, which is critical in legal proceedings to verify the integrity of the evidence (see Section 2.8).

2.5 Biological samples transportation

The transportation of evidence is a critical phase in forensic investigations, ensuring that the integrity, identity, and chain of custody of the evidence are preserved from the point of collection to its arrival at the laboratory or any other intermediate place. Improper transportation can result in the contamination, degradation, or loss of evidence, which could compromise the investigation and subsequent legal proceedings. Once evidence is collected using appropriate tools such as adhesive tapes, tweezers, or swabs, it must be properly packed in suitable containers, such as bags, vials, or boxes. The packaging should be selected based on the type of evidence, considering factors like porosity, fragility, and the potential for contamination. The evidence must be securely sealed and clearly labeled with all relevant information, including the case identifier, description, and the identity of the personnel who collected it (Technical Working Group on Crime Scene Investigation, 2013).

To prevent damage or alteration of the evidence during transportation, the following precautions should be observed:

- **Environmental Control**: Biological evidence is highly prone to deterioration when exposed to environmental conditions like heat, moisture, and sunlight. Therefore, transportation conditions must protect against these factors. For example, biological samples should be kept cool and dry, and exposure to direct sunlight should be minimized. In some cases, refrigerated transport may be necessary to maintain sample viability.
- **Limited and Secured Access**: During transportation, access to the evidence should be strictly limited and secured. Unauthorized access could lead to tampering or contamination, which would invalidate the evidence. Security measures may include the use of tamper-evident seals, locked containers, and secure transport vehicles. Documentation of who has access to the evidence at each stage of transportation is essential for maintaining the chain of custody.
- **Physical Protection**: The packaging must provide adequate protection against physical damage, such as breakage, crushing, or leakage. For fragile items like glass or liquids, additional protective measures, such as padded containers or secondary packaging, may be necessary to prevent damage during transit.

Proper documentation is critical for maintaining the chain of custody, which is the process that ensures the evidence presented in court is exactly the one collected at the crime scene. Each transfer of evidence, whether between officers, to a forensic laboratory, or to another location, must be meticulously documented. The documentation should include:

- **Details of Collection and Transfer**: The circumstances under which the evidence was collected, including date, time, and location, should be recorded. Additionally, each transfer of custody must be documented, noting the individuals involved, the date, time, and location of the transfer, and any changes in the condition of the evidence.
- **Storage Conditions**: Whenever possible, samples should be transported on ice or with cold packs to slow down any enzymatic activity that could degrade the DNA. Before and during transportation, the storage conditions of the evidence must be documented. This includes temperature control, humidity levels, and any other relevant environmental factors. Proper documentation helps to verify that the evidence was stored under appropriate conditions, preserving its integrity for forensic analysis.
- **Signatures and Authentication**: Each individual who handles the evidence during transportation must sign the appropriate documentation, confirming the transfer of custody and the condition of the evidence. These signatures authenticate the chain of custody and provide accountability at each stage of the process.

The choice of transportation method and the location where the evidence will be stored or analyzed must be carefully considered, considering factors such as:

- **Cost and Distance**: The cost and distance of transportation should be balanced with the need to maintain evidence integrity. For example, transporting evidence over long distances may require specialized packaging or temperature control, which could increase costs.

- **Timeframe**: The time required to transport evidence is a critical consideration, particularly for biological samples, which may degrade over time. Expedited shipping or special courier services may be necessary to ensure timely delivery.
- **Compatibility with Evidence**: The transportation method must be compatible with the type of evidence. For example, fragile or hazardous materials may require specialized handling or transport vehicles.

The transportation of evidence must be conducted in accordance with relevant local, national, and international laws and regulations. This is particularly important for certain types of evidence, such as controlled substances, firearms, and hazardous materials, which may require special permits or adherence to specific handling protocols.

In summary, the transportation of evidence is a complex process that requires careful planning, meticulous documentation, and strict adherence to legal and regulatory requirements. By following these guidelines, forensic professionals can ensure that evidence arrives at the laboratory in the same condition it was collected, preserving its integrity for analysis and use in legal proceedings.

2.6 Biological samples preservation and storage

Biological sample preservation is critical for accurate DNA analysis. Although DNA is a relatively stable molecule, its integrity can be compromised under certain conditions, leading to degradation and making it unsuitable for analysis. Main factors influencing DNA preservation are (Butler, 2012):

Temperature: High temperatures accelerate the breakdown of DNA due to increased enzymatic activity. On the other hand, freezing samples at −20°C or lower significantly slows down the degradation process. For long-term preservation, freezing is one of the most effective methods, particularly for liquid samples like blood or tissue. However, repeated freeze—thaw cycles can cause DNA fragmentation, so samples should be stored in a manner that minimizes these cycles.

Humidity: Moist environments promote the growth of bacteria and fungi, both of which can degrade DNA. To prevent this, samples should be thoroughly dried before storage and kept in moisture-proof containers. This is especially important for biological evidence such as swabs and bloodstains, which can retain moisture.

UV Light: Ultraviolet light can cause DNA strand breaks and other forms of damage. Therefore, samples should be stored in dark conditions to prevent exposure to UV light, which can induce thymine dimers and other photoproducts that interfere with DNA analysis.

pH Levels: DNA is most stable in a neutral pH environment. Extremes pH, either acidic or basic, can lead to the hydrolysis of DNA strands. Whenever possible, biological samples should be kept in conditions that maintain a neutral pH to minimize the risk of degradation.

Effective preservation methods that help protect the DNA from factors that may affect its integrity include (Technical Working Group on Biological Evidence Preservation, 2013):

Dry Storage: For short-term preservation, samples such as swabs and bloodstains can be air-dried and stored in paper envelopes or breathable containers. Plastic containers should be avoided as they can trap moisture, leading to DNA degradation.

Table 2.1 Suggested Storing Conditions

Conditions	Temperature	Storage
Room Temperature	Environmental	Very short-term
Controlled	15°C–24°C (60°–75°F)	Short-Term
Refrigerated	2°–8°C (36°–46°F)	Medium Term
Frozen	−10°C (14°F)	Long Term

Source: According to Technical Working Group on Biological Evidence Preservation (2013).

Freezing: Freezing at −20°C or lower is recommended for long-term preservation. This method is particularly important for liquid samples like blood or tissue. Freezing effectively halts the degradation process and maintains the DNA's integrity over time.

Lyophilization: This process involves removing water from the sample through sublimation, leaving it in a stable, dry state. Lyophilized samples can be stored at room temperature for extended periods without significant degradation, making them a suitable option for long-term DNA preservation.

Chemical Preservation: Adding preservatives like ethanol or Tris-EDTA (TE) buffer can stabilize DNA by preventing enzymatic degradation. However, the choice of preservative must be compatible with subsequent DNA analysis methods to ensure the accuracy and reliability of forensic tests.

Evidence must be stored in secure, designated areas with access limited to authorized individuals only, ensuring that items are tightly sealed to prevent any mixture or contamination. Storage in unauthorized locations is strictly prohibited (**Table 2.1**). A comprehensive physical or digital inventory must be maintained, detailing the classification of each piece of evidence, its storage location, the date and time of storage, and the identity of the person responsible for storing it. Additionally, any information regarding individuals who have temporarily handled or reviewed the evidence, including their identity, the reason for access, as well as the dates and times of both removal and return, must be meticulously recorded.

2.7 Contamination issues

Contamination of biological evidence, especially DNA, is a critical issue in forensic science due to the high sensitivity of modern DNA analysis methods. Even minimal contamination can lead to misleading results, potentially compromising the investigation and judicial process. It is essential to understand the main sources of contamination and implement stringent protocols to minimize risks throughout the collection, preservation, and analysis of DNA samples (Technical Working Group on Crime Scene Investigation, 2013). Contamination can occur at multiple stages:

Equipment and PPE: The equipment used in crime scene investigations, such as tools and containers, may become contaminated if not disposable or properly sterilized between uses. Personal protective equipment (PPE), such as protective suits, head covers, masks, gloves, and shoe covers, is crucial in reducing contamination risks, but if not changed between scenes or improperly used, may become sources of contamination.

Evidence Collection: Contamination may arise from the crime scene itself, through the activities of first responders, family members, or onlookers who inadvertently leave their DNA on surfaces. Additionally, improper collection methods, such as touching evidence without gloves or talking, sneezing, or coughing near the evidence, can introduce foreign DNA.

Packaging and Transportation: Inadequate packaging or failure to use sterile, contamination-free containers can lead to cross-contamination between different samples. Biological material from evidence may inadvertently transfer to another if not properly isolated.

Handling and Storage: Poor handling practices, such as reusing pipettes or failing to replace gloves, can introduce contaminants. Improper storage conditions, where samples are not adequately separated or stored in unclean environments, can also lead to cross-contamination.

Laboratory Environment: The laboratory environment presents its own risks, where contamination can occur through contact with the analysts' biological material, such as skin cells, hair, or saliva. Cross-contamination between samples is also a significant risk, especially if the samples are handled or stored inadequately or if equipment is not disposable or sterilized between uses.

To reduce the risk of contamination, several preventive measures must be rigorously applied at different stages, such as:

CSI access: Limiting the number of people at the crime scene and restricting access to the evidence collection area can help reduce the introduction of foreign DNA.

PPE: Wearing personal protective equipment (PPE) at both crime scenes and in laboratories is essential to prevent the contamination of samples with the DNA of personnel.

Evidence Collection: The equipment used for sample collection at crime scene investigations (e.g., tools, containers) should be disposable or must be properly sterilized before use to prevent cross-contamination. Each sample should be placed in a separate, sterile container to prevent cross-contamination.

Storing and Transportation: Samples should be sealed in clearly labeled, contamination-free containers. During transportation, samples should be protected from environmental factors that can degrade DNA, such as heat, moisture, and sunlight. Refrigerated transport may be necessary for certain biological samples (see Section 2.7). Proper labeling and documentation ensure that each sample's identity is maintained throughout the process.

Laboratory Environment: forensic laboratories must implement rigorous protocols for handling, processing, and storing samples, according to international guidelines and standards such as ISO/IEC 17025 (Wilson-Wilde, 2018). This includes using dedicated workstations for different tasks, regular cleaning and decontamination of equipment, and the use of negative controls during the analyses and replicate analyses to detect potential contamination.

Evidence samples and reference samples should be processed in separate areas or at different times to minimize the risk of cross-contamination. Automated workstations can further reduce the risk by limiting human contact with samples.

Laboratory staff should be thoroughly trained in contamination prevention techniques. Staff must wear dedicated PPE for different areas of the lab and follow strict protocols to prevent their DNA from contaminating samples. A DNA elimination database helps quickly identify contamination sources by comparing DNA profiles of the evidence to those of lab personnel.

Despite stringent protocols, contamination can still occur, particularly with "touch DNA," where invisible traces of DNA can be transferred through minimal contact. This form of contamination is particularly insidious because it can be difficult to distinguish between genuine evidence and contamination. The identification and prevention of contamination are crucial to maintaining the integrity of DNA evidence. If contamination is suspected or detected, it must be thoroughly investigated to determine the source and prevent future occurrences. By minimizing contamination risks, forensic professionals can ensure the reliability of DNA evidence, reducing the likelihood of false positives and ensuring that justice is accurately served.

2.8 Chain of custody

The chain of custody encompasses all documentation, such as notes, photographs, and labels, that accompanies evidence. It tracks the complete history of the evidence, identifying every individual who handled it and every location where it was stored, in chronological order, from its collection at the crime scene to its final disposition. This unbroken chain ensures seamless linkage between all phases of evidence management, including collection, transfer, storage, and elimination.

Maintaining the chain of custody is critical for preserving the integrity and identity of the evidence. It ensures proper handling at every stage, from the crime scene to the laboratory, and verifies that the evidence analyzed is the same as that originally recovered. This process confirms both the authenticity of the evidence and the reliability of the results obtained from its analysis (Technical Working Group on Biological Evidence Preservation, 2013). Any alteration in the chain of custody can render crucial evidence inadmissible or unreliable in court proceedings.

To facilitate accurate and efficient documentation, the chain of custody may be recorded using electronic systems, paper records, or a combination of both.

Automated Laboratory Information Management Systems (LIMS) have been developed to streamline the tracking of evidentiary items throughout their examination lifecycle, from collection at the crime scene through analysis and final disposition. These electronic recordings provide precise documentation of all activities associated with each piece of evidence.

Chain of custody records must be retained for varying periods depending on the laws and regulations of each country, even if the evidence has already been lost or destroyed.

This meticulous documentation process is essential for upholding the evidentiary value and ensuring its admissibility and credibility during legal proceedings (Lee et al., 2001).

References

Barbaro, A. (2022). Chapter: Biological Evidence Collection. In: Barbaro, A., Mishra, A. (eds), *Manual of Crime Scene Investigation*. CRC Press, USA.

Butler, J. M. (2011). *Advanced Topics in Forensic DNA Typing: Methodology*. Academic Pr, USA.

Butler, J. M. (ed.) (2012). Chapter 1 - Sample Collection, Storage, and Characterization. In: *Advanced Topics in Forensic DNA Typing: Methodology*. Academic Press, pp. 1–27. ISBN 9780123745132. https://doi.org/10.1016/B978-0-12-374513-2.00001-4

Dash, H. R., Vajpayee, K. (2022). Collection, Preservation, and Transportation of Biological Evidences. In: Dash, H. R., Shrivastava, P., Lorente, J. A. (eds), *Handbook of DNA Profiling*. Springer, Singapore. https://doi.org/10.1007/978-981-16-4318-7_3

ENFSI Scene of Crime Working Group. (2022) ENFSI scenes of crime best practice manual. https://enfsi.eu/wp-content/uploads/2022/02/BPM-SOC-01-v.20220214_final_v2.pdf
Grupo Español y Portugués de la International Society For Forensic Genetic. (2000) Recomendaciones para la recogida y envío de muestras con fines de identificación genética https://ghep-isfg.org/usercontent/content/uploads/medios/Recogida_de_evidencias.pdf
Houck, M. M. (2010). Review of: An introduction to crime scene investigation. *Journal of Forensic Sciences*, 55(5), 1406–1406. https://doi.org/10.1111/j.1556-4029.2010.01481.x
Lee, H. C., Ladd, C. (2001). Preservation and collection of biological evidence. *Croatian Medical Journal*, 42(3), 225–228.
Lee, H. C., Palmbach, T. M., Miller, M. (2001) *Henry Lee's Crime Scene Handbook*. Academic Press, San Diego, CA.
Lindahl, T. (1993). Instability and decay of the primary structure of DNA. *Nature*, 362(6422), 709–15. https://doi.org/10.1038/362709a0
Singh, H. N. (2021). Crime scene investigation. *International Journal of Science and Research (IJSR)*, 10(11), 642–648. https://www.doi.org/10.21275/SR211112005543
Technical Working Group on Biological Evidence Preservation. (2013). *The Biological Evidence Preservation Handbook: Best Practices for Evidence Handlers*, National Institute of Standards and Technology (NIST, NIJ). https://doi.org/10.6028/NIST.IR.7928
Technical Working Group on Crime Scene Investigation. (2013). Crime Scene Investigation: A Guide for Law Enforcement, National Forensic Science Technology Center (NFSTC), USA https://www.nist.gov/system/files/documents/forensics/Crime-Scene-Investigation.pdf
Wilson-Wilde, L. (2018). The international development of forensic science standards — A review. *Forensic Science International*, 288, 1–9. https://doi.org/10.1016/j.forsciint.2018.04.009

3

Insights into Body Fluid Identification for Criminal Investigations

Kamayani Vajpayee and Ritesh K. Shukla

3.1 Introduction

Body fluid identification holds an important place in forensic casework. These include a range of biological samples, such as blood, semen, saliva, bones, and even samples that might contain touch DNA (deoxyribonucleic acid) (Dash & Vajpayee, 2020; Hess & Haas, 2017). As technology has advanced in forensic science, DNA forensics has come to the forefront. It allows for a more precise and accurate analysis of biological evidence from crime scenes, especially body fluids. This has greatly improved the evidentiary value of body fluids in investigations. Identification of body fluids using DNA profiling and other advanced methods helps solve crime cases (Sijen & Harbison, 2021). The comprehensive list of forensically relevant body fluids includes blood (vaginal/menstrual blood or normal blood), saliva, semen, sweat, vaginal fluid, urine, etc., as shown in **Figure 3.1**.

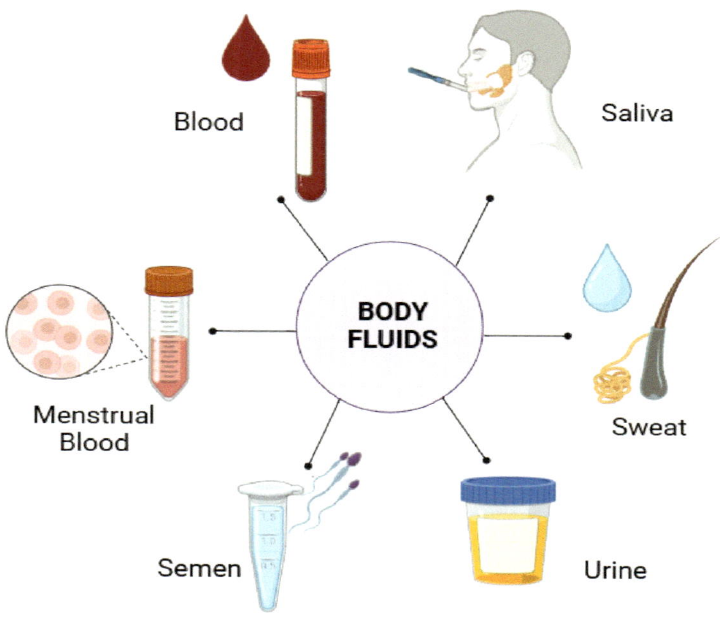

Figure 3.1 *Body fluids most commonly found at the crime scene.*

DOI: 10.4324/9781003474296-3

Composition of Body Fluids

- **Blood** is composed of plasma and cells. Plasma makes up about 55% of blood volume and consists of 90% water, along with proteins such as albumin, globulins, and fibrinogen, as well as electrolytes, nutrients, hormones, and waste products. The cellular component, which constitutes 45% of blood volume, includes red blood cells (erythrocytes), white blood cells (leukocytes), and platelets (thrombocytes).
- **Saliva** is primarily composed of 98% water. It also contains electrolytes like sodium, potassium, calcium, magnesium, bicarbonate, and phosphate. Additionally, saliva includes proteins such as enzymes (amylase, lysozyme), mucins, and immunoglobulins (IgA), along with other components like urea, uric acid, cholesterol, and various antimicrobial agents.
- **Sweat** is mostly water, accounting for 99% of its composition. It also contains electrolytes such as sodium, potassium, calcium, magnesium, and chloride. Other components found in sweat include urea, lactate, ammonia, and trace amounts of substances like glucose and amino acids.
- **Urine** consists of approximately 95% water. The remaining 5% includes solutes such as urea, creatinine, uric acid, and electrolytes like sodium, potassium, and chloride, as well as various metabolic waste products. Urine also contains small amounts of hormones, vitamins, and proteins.
- **Semen** is composed of spermatozoa, which make up 2–5% of its volume, and seminal plasma, which constitutes 95–98%. Seminal plasma contains water, fructose, ascorbic acid, prostaglandins, proteins, enzymes, and various ions, including sodium, potassium, calcium, magnesium, and chloride.
- **Vaginal fluid** is primarily water. Along with proteins like enzymes, immunoglobulins, and mucins, it also contains electrolytes, including sodium, potassium, calcium, and magnesium. Lactic acid, urea, glucose, and other antibacterial agents are additional.

Often, body fluids are present in minimal quantity or are invisible or in mixtures, for example, blood and semen in cases of sexual assault, saliva traces on bitemarks, or cigarette buds. In such cases, the preliminary tests play a vital role in identifying them as biological evidence, but these tests do not directly provide a definite identification due to a lack of sample amount, as well as similarities with other results (false positive). Therefore, for validation, preliminary tests are followed by confirmatory tests that include microscopic and immunological assays, many of which are incompatible with DNA profiling. Identifying these body fluids is very sensitive and specific, and must also be non-destructive, allowing DNA profiling for the sample. The techniques used traditionally are effective, but some are destructive to the DNA, and others might need a larger sample size (Jung et al., 2018). Also, many inhibitors might interact with the biological sample and degrade it for its DNA profiling—all these points to the need for effective body fluid identification tools (Dass et al., 2023).

Expanding on these points, it is crucial to emphasize the importance of both presumptive and confirmatory tests in body fluid identification. Presumptive tests are quick and cost-effective, allowing a forensic scientist to screen larger areas and identify potential body fluid stains. However, they are not conclusive and can sometimes yield false positives, which necessitates the use of confirmatory tests (Sijen & Harbison, 2021). Confirmatory tests, on the other hand, provide definitive identification of body fluids. These tests are highly specific, involving microscopic examination, immunological assays, and advanced molecular techniques. Despite their reliability, some confirmatory tests can be destructive to the sample, potentially hindering subsequent DNA analysis (Jung et al., 2018). Therefore, developing non-destructive and highly sensitive techniques remains a critical area of research in forensic science.

Recent advancements in molecular biology have introduced new technologies for body fluid identification, such as the use of microRNA (miRNA) markers and DNA methylation patterns. These methods offer higher specificity and sensitivity compared to traditional methods. Additionally, technological innovations like portable DNA sequencers and digital PCR provide rapid and accurate results on-site, enhancing the efficiency of forensic investigations (Jung et al., 2018). In Summary, integrating advanced genetic profiling technologies with traditional forensic methods enhances the accuracy and reliability of body fluid identification. This multidisciplinary approach is essential for solving complex criminal cases and ensuring justice (Hamza et al., 2024).

3.2 Types of body fluids and their importance

Several biological samples are found at the crime scene, which are further sent to the forensic laboratory for testing and analysis. Forensic investigations often rely on various biological samples found at the crime scenes. The DNA content in these samples significantly varies (Andréasson et al., 2006). There are different types of samples and different matrices in which they are found, for example, liquid blood from which DNA can be easily extracted and diluted, dried, or contaminated with soil or other materials, where DNA extraction becomes a bit tedious, same for hair samples (Siuta et al., 2023).

Common Body Fluid Samples:

- *Blood:* Liquid blood allows for direct DNA extraction. However, dried or contaminated blood with soil or other materials can complicate DNA extraction.
- *Hair:* Mostly found at the crime scene involving physical altercations particularly. Hair samples are a good source of DNA if the root is present; if not, it becomes challenging to extract DNA.
- *Saliva:* This is often found on bite marks or cigarette butts. The collected samples contain buccal cells, which are a good source of DNA.
- *Semen:* Common in sexual assault cases. The high concentration of sperm cells provides an ample amount of DNA to be extracted.
- *Urine:* This is less commonly analyzed for DNA as it is very challenging to extract DNA from urine samples due to the low concentration of cells, but it can provide genetic material.
- *Sweat:* This is rarely used for DNA analysis due to low cell contents.

It is important to have knowledge about the importance of a sample and the amount of genetic content the sample may contain before analyzing the crime scene for better and ample collection and preservation of the samples. **Table 3.1** provides an insight into how much DNA content a body fluid might contain.

This information helps forensic scientists to ascertain things, such as optimizing the sample collection by prioritizing samples with higher DNA content for reliable results. Ensuring proper preservation of the samples by applying proper techniques for preserving different types of samples to maintain DNA integrity, and improving the analysis techniques by tailoring extraction and analysis methods based on the type and condition of the sample.

There are many factors that affect the quantity and quality of DNA that can be extracted from various body fluids, and these are influenced by multiple factors, including environmental conditions, time since deposition, and the nature of the substrate on which the fluid is found.

Table 3.1 Overview of DNA Content in Common Body Fluids

Type of Body Fluid	DNA Content	References
Blood	30–40 μg/mL	Dash and Vajpayee (2020)
Semen	1.22 ± 0.05 pg/sperm cell	Kanungo et al. (2023)
Saliva	18–433 ng/μL	Livy et al. (2012)
Sweat	15–40 ng/mL (from stain)	Yudianto (2011)
Urine	23.35–40 ng/μL	Hilhorst et al. (2013)
Tears	Yields are typically very low, in the picogram (pg) to low nanogram (ng) range per sample.	Aparna et al. (2021)
Vaginal swab	Concentrations in DNA extracts can be roughly 1 ng/μL to >100 ng/μL, but are highly variable based on sample and elution volume.	Doi et al. (2014)

For example, DNA from blood on a non-porous surface (such as glass) may be easier to extract than from a porous surface (such as cloth), where it may penetrate deeper into the fibers (Heywood et al., 2003). The exposure of biological evidence to sunlight, moisture, or chemicals can degrade DNA, making recovery more difficult (Bhoyar et al., 2024). Understanding these factors is crucial for evidence collection and preservation. Forensic scientists often encounter situations where only trace amounts of biological material are available, or these samples are degraded (Shrivastava et al., 2020). Advanced techniques such as miniSTR analysis and mitochondrial DNA analysis are employed in cases where conventional DNA analysis fails due to sample degradation. These methods are used for analyzing old samples or those exposed to harsh environmental conditions (Livy et al., 2012).

3.3 Collection and preservation of body fluids

Essential stages in forensic investigations involve gathering and preserving bodily fluids from a crime scene. To ensure accurate analysis and trustworthy outcomes, proper treatment preserves the evidence's integrity. Various collection methods are used depending on the type and nature of the body fluid found at any crime scene (Mozayani & Parish-Fisher, 2017). Blood, saliva, semen, and other biological materials are among the bodily fluids for which collection and preservation methods are described in this section (**Figure 3.2**).

For the collection of these samples, various techniques are used, such as:

- Scrapping method: This method is used to collect dry samples from porous surfaces. The first step includes locating the stain and checking if it is dry or not (Pang & Cheung, 2007).
- Double Swabbing method: This is used for dry samples on non-porous surfaces, where samples are collected using a moistened swab, which softens and rehydrates the cells, and then using another swab, the stain is recovered from the surface (Harbison & Fleming, 2016).
- Cutting method: This method is used when the stains are found on large or immovable objects. This involves locating the stain and checking if it is wet or dry. If the stain is wet, let it air dry. Cut the object around where stains are present with sterile scissors (Harbison & Fleming, 2016; Pang & Cheung, 2007).
- Tape-lifting method: Stains, mostly dried ones, which are present on non-absorbent materials, can be recovered using this method. This involves using adhesive tape to

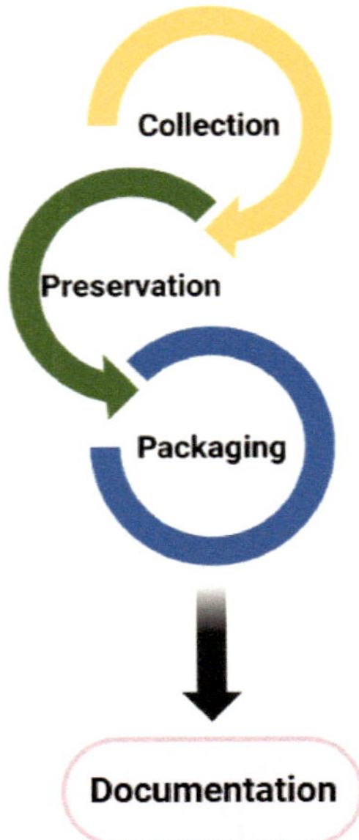

Figure 3.2 *Crime scene to court: Collection, preservation, packaging, and documentation of evidence with precision.*

lift the stain and securing the adhesive side on adhesive paper (Harbison & Fleming, 2016; Pang & Cheung, 2007).

- Picking method: This is used when solid samples are found at the crime scene, for example, bones, hair, etc. Sterile tweezers/forceps are used to pick the sample and carefully transfer it into an envelope or box as required (Vajpayee, Dash, et al., 2023).

3.3.1 Blood samples

- *Collection:*
  For Liquid Blood: A sterile gauze pad or clean cotton cloth should be used to collect liquid blood from surfaces or blood pools. Air dry the sample properly and pack in paper envelopes, and then place it in a non-breakable, air-tight container (s). When present in large amount it can also be collected in ethylenediaminetetraacetic acid (EDTA) vials using sterile syringes.
  Dried Blood Stains: Use a sterile cotton swab moistened with distilled water and gently remove the dried blood stain by rubbing it onto it and let it air dry, then after packing it in a paper envelope.
- *Preservation:* EDTA is an anticoagulant that prevents blood from clotting. Sodium Fluoride is used as a preservative to inhibit glycolysis and bacterial growth. Heparin is another anticoagulant that prevents clotting by inhibiting thrombin and other factors.

3.3.2 Saliva samples

- *Collection:* Use a sterile cotton swab or cotton piece to gather saliva samples. Using the swab, gently massage it into the cheek side of the oral cavity of the suspect's or victim's mouth. Air dry the sample and pack into a paper envelope.
- *Preservation:* Sodium Azide prevents microbial growth in saliva samples.

3.3.3 Semen samples

- *Collection:* Use a sterile cotton swab or cloth piece to collect liquid semen sample or its traces. Air dry it and pack it into a paper envelope. For dried seminal stains, use a sterile moistened swab, gently massage it over the dried seminal stain, and after collecting, let it air dry.
- *Preservation:* Air dry the sample and keep it at room temperature, and if collected in liquid form, then immediately store it at 4°C.

3.3.4 Other body fluids

It is recommended to utilize a sterile container to get samples of other bodily fluids, such as vomit or urine. Refrigerate or freeze the sample as soon as feasible, and ensure the container is leak-proof.

- *For Preservation of Urine:* Thymol is used as a preservative to prevent bacterial growth. Boric acid is another preservative that inhibits bacterial activity in urine samples.
- *For Preservation of Other Body Fluids:* Formalin is used to preserve tissue samples and other body fluids by cross-linking proteins. Phenol is sometimes used to preserve biological samples by denaturing proteins and preventing microbial growth. Chloroform: can be used to preserve samples by creating an anaerobic environment that inhibits microbial activity.

3.4 Packaging

Collect and pack each sample separately to avoid cross-contamination. Use paper bags or envelopes in order to preserve dried samples, and for liquid samples, use air-tight containers. Record collector's initials, the time and date of collection, all details, and information regarding the evidence collected. Do not pack moisture-containing evidence in plastic or paper containers; air dry it first, and do not store it for longer hours to prevent microbial growth, which can contaminate or destroy the collected sample. Dried samples should be kept in a cool, dry place.

3.5 Storage

Refrigerate the sample to prevent its contamination.

Documentation: Keep a thorough chain of custody log for every piece of evidence. Information should be included in this log, such as the date and time of collection and each transfer, information of all individuals who handled the sample, and the purpose of the transfer. Appropriate collection and preservation of body fluids is essential for maintaining the integrity of the evidences in forensic relevance. Following standard procedures and protocols, forensic investigators can ensure to maintain the integrity of the evidence suitable for analysis, ultimately aiding in the investigation of criminal cases.

3.6 Fundamental methods for identifying body fluids

Identifying body fluids at a crime scene is a critical step in forensic investigations, as it can provide valuable evidence linking a suspect to the crime.

The process typically involves two main types of tests: presumptive tests and confirmatory tests (Vajpayee, Dash, et al., 2023) as presented in **Figure 3.3**. Presumptive tests are initial screenings that indicate the possible presence of a specific body fluid, while the confirmatory

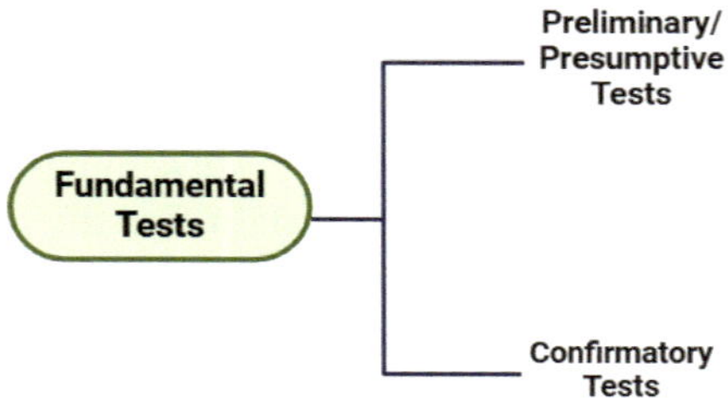

Figure 3.3 *Presumptive and confirmatory tests in body fluids analysis process.*

test provides definitive identification (Vajpayee, Dash, et al., 2023). These methods ensure accurate and reliable results, which are crucial for legal proceedings (C. Wang et al., 2022).

Presumptive tests are quick and cost-effective, allowing forensic scientists to narrow down areas of interest on larger surfaces. However, they are not conclusive and sometimes can yield false positives. Confirmatory tests, on the other hand, are more specific and reliable, conclusively identifying the presence of a particular body fluid. Together, these tests ensure that only relevant samples are analyzed further, optimizing the use of resources and time in forensic laboratories (C. Wang et al., 2022).

3.6.1 Tests for blood

*3.6.1.1 Identification of source of blood (see **Table 3.2**)*

Table 3.2 Identification of the Source of Blood

Source of Blood	Identification Factor
Arterial blood	Bright red in color
Venous blood	Dark in color
Menstrual blood	Endometrial tissues and fibrin content present
Miscarriage/Abortion blood	Alkaline phosphatase and chronic gonadotropins present
Fetal blood	Alpha protein or HbF present

*3.6.1.2 Presumptive tests for blood (see **Table 3.3**)*

Table 3.3 Presumptive Tests for Blood

Test Name	Mechanism	Inference	Disadvantages	First Used By
Benzidine test	This test relies on the peroxidase activity of hemoglobin, which catalyzes the oxidation of benzidine by hydrogen peroxide.	The blue color is observed if blood is present.	Gives false positives as substances with peroxidase-like activity, and some chemical oxidants can produce similar chemical changes • Health Risk: Carcinogenic • Sensitivity to sample condition	Oskar Adler (Chemist), 1904
Phenolphthalein (Kastle–Mayer test)	This test uses a reduced form of phenolphthalein (colorless), which oxidizes in the presence of Hydrogen peroxide and hemoglobin.	Pink color is observed if blood is present.	• Gives false positives with chemical oxidants and environmental contaminants • Not species-specific • Time sensitivity: immediate color change must be observed	Joseph H. Kastle and Oliver March Shedd (Chemists) discovered the technique in 1901, and it was used by Meyer in 1903

(Continued)

Table 3.3 *(Continued)* Presumptive Tests for Blood

Test Name	Mechanism	Inference	Disadvantages	First Used By
Luminol test	Luminol reacts with hemoglobin and hydrogen peroxide, which produces a cyclic peroxide that quickly decomposes to form 3-aminophthalate.	Electrons return to their ground state and emit blue chemiluminescence, indicative of the presence of blood.	• Gives false positives with chemicals and environmental contaminants • Sensitivity to light and duration: short-lived glow and darkness required for visualization • Damage to DNA	Walter Specht (Chemist), 1937
Fluorescein test	The iron in the hemoglobin acts as the catalyst for this reaction. Fluorescein is a fluorescent dye, which, in reaction with hydrogen peroxide and iron, produces a fluorescent compound.	A greenish-yellow glow is observed under UV light if blood is present.	• Gives False positives with chemicals and environmental contaminants • Sensitivity to light and ambient light interference • Potential DNA damage	Robert Cheeseman (Forensic Specialist), 1999
Leucomalachite green (LMG)	The iron in the hemoglobin acts as a catalyzing agent in this reaction. LMG gets oxidized into malachite green (MG) in the presence of hydrogen peroxide and iron.	A blue-green color is observed if blood is present.	• Gives False positives with chemicals and environmental contaminants • Sensitivity to sample condition	Victor Balthazar (Forensic Scientist), 1913

*3.6.1.3 Confirmatory tests for blood (see **Table 3.4**)*

i. Microscopic Examination: Vibert's Reagent is used, which is composed of NaCl, $MgCl_2$, and normal saline. Leishman stain is used as a smearing agent. Human, mammalian, and non-mammalian RBCs features are observed and compared, counting and distinguishing from malaria parasites via smearing. Also, sex determination is done by counting the number of determination is done by counting the number of Davidson bodies present. The method is only useful for fresh or clotted blood samples, as when the sample dries, the cellular components are broken down and are unrecognizable.

ii. Microcrystal Assay:
   a. Teichmann Crystal test (Hematin/Hemin Test): The heme reacts with NaCl and Glacial acetic acid to form hematin chloride (ferriprotoporphyrin chloride). Rhombic-brown colored crystals are formed, which confirm the presence of blood.
   b. Takayama Test (Hemochromogen or Pyridine Hemochromogen Test): The heme reacts with pyridine, sodium hydroxide, glucose, and water and gets crystallized to pyridine-ferroprotoporphyrin (hemochromogen). Feathery-pink crystals are formed, which confirm the presence of blood.
   c. Wagenaar Test: The hemin reacts with HCl and forms acetone-chlor-hemin crystals. Dark brown colored, multi-shaped crystals are observed. Two sets of derivatives are formed, hematin and hemin, so fewer acetone-chlor-hemin crystals are produced, which makes it difficult to interpret the results.

iii. Spectroscopic Examination: Normal saline dilution for fresh and dried blood stains. Dark characteristic bands, in Fraunhofer lines that vary according to hemoglobin and the derivatives: Oxyhemoglobin (2 bands), reduced Hb (1 band), carboxyhemoglobin (2 bands), methemoglobin (4 bands), acid/alkaline hematin (2;1 band respectively), cyan hemochromogen (2 bands), etc. It is a less sensitive, not species-specific, expensive, and sophisticated technique and can be influenced by certain environmental factors that can deteriorate the samples or influence instrument efficiency.

iv. Electrophoresis: Different types of Hb: A1 & A2, C, F, and S move at different rates under the electromagnetic field. Different bands for different Hb are observed. There are higher chances for the sample to get diffused, and the technique is more expensive than microcrystal tests.

v. Immunoelectrophoresis: Due to electric current, the antigen mix separates based on charge and size. Simultaneously, it diffuses with antisera. An antigen-antibody reaction takes place, which results in the formation of precipitin lines. Hb; characteristic pink lines.

vi. RNA-Based Assays: RNA is converted into complementary DNA and amplified by PCR. Blood-specific RNA markers are detected. This technique is highly complex and requires sophisticated equipment and expertise, and there are high contamination risks, sensitivity to sample degradation, and expensive.

vii. Chromatography Method: Ion-exchange chromatography is mainly used to confirm the presence of blood. Separates the Hb from blood or a mixture and purifies it. The technique is inconsistent, limited to ionizable compounds, time-consuming, and expensive.

viii. Serological Tests: This involves the study and diagnostic identification of antibodies in blood and is widely used in medical and forensic fields. These include:

- *Enzyme-Linked Immunosorbent Assay (ELISA)*
  ELISA detects and quantifies antibodies, antigens, proteins, and glycoproteins in samples of biological samples. The methodology involves the immobilization of an antigen on a solid surface and then complexation with an antibody linked to an enzyme.
- *Western Blot*
  This technique is used to detect specific proteins in a sample. The methodology involves protein separation by gel electrophoresis, which is then transferred to a membrane and probed with specific antibodies to the target protein, followed by detection, achieved by enzyme-linked antibodies and chemiluminescence.

Table 3.4 Confirmatory Tests for Blood

Test Name	Mechanism	Inference	Disadvantages	First Used By
Microscopic examination	Examination of intact RBCs and WBCs under a microscope.	Confirms the presence of blood cells.	Limited to fresh or clotted blood; aged stains may not show intact cells.	Mathieu Orfila in the early 19th century
Microcrystal assay a. Teichmann test b. Takayama test c. Wagenaar test	Formation of characteristic crystals when blood reacts with specific chemicals.	Confirms the presence of hemoglobin.	Requires precise conditions; may not work with aged or degraded samples.	a. Ludwig Karol Teichmann, 1853 b. Masaeo Takayama, 1912 c. M. Wagenaar, 1935
Spectroscopic examination	Analysis of the absorption spectrum of hemoglobin.	Confirms the presence of hemoglobin.	Requires specialized equipment; may not be specific to human blood.	Georg Dragendroff, in the late 19th century, and other scientists
Electrophoresis	Separation of blood proteins based on their movement in an electric field.	Identifies specific blood proteins.	It requires specialized equipment and expertise, and there is also the chance of getting diffused easily.	Arne Tiselius, 1937
Immunoelectrophoresis	Combines electrophoresis and immunoassay to detect specific blood proteins.	Identifies specific blood proteins.	It is time-consuming as it takes ~18–24 hours to get the result and is an expensive technique.	Pierre Grabar and Curtis Williams, 1953
RNA-based assays	Detection of specific RNA sequences unique to blood cells.	Confirms presence of blood cells.	Requires specialized equipment and expertise.	First used in the early 2000s
Chromatography method	Separation of blood components based on their interactions with a stationary phase.	Identifies specific blood components.	Requires specialized equipment and expertise.	Mikhail Tswett, 1903

3.6.1.4 Blood typing

Blood typing is a crucial tool in forensic science that provides valuable information and helps link the victim, suspects, and crime scene. There are four main blood types: A, B, AB, and O. According to an international survey, the blood type prevalence in India is as shown in **Figure 3.4**. Also, India comes under the top 10 countries to have the highest prevalence of B+ blood type, along with Thailand, Myanmar, Laos, Bangladesh, etc., with types O and A being the most prevalent globally. This variability can be leveraged to narrow down potential suspects or victims. This helps identify and exclude the individuals as potential suspects and also works as corroborative evidence.

Blood typing is the process that involves testing the sample to determine which antigen is present or absent. Mostly, ABO blood grouping and Rh blood grouping are used in forensic investigations. The basic principle is to mix the blood with antibodies against A and B blood types and observe the agglutination. The agglutination indicates the blood type present in the test sample.

- *ABO Blood Group System*

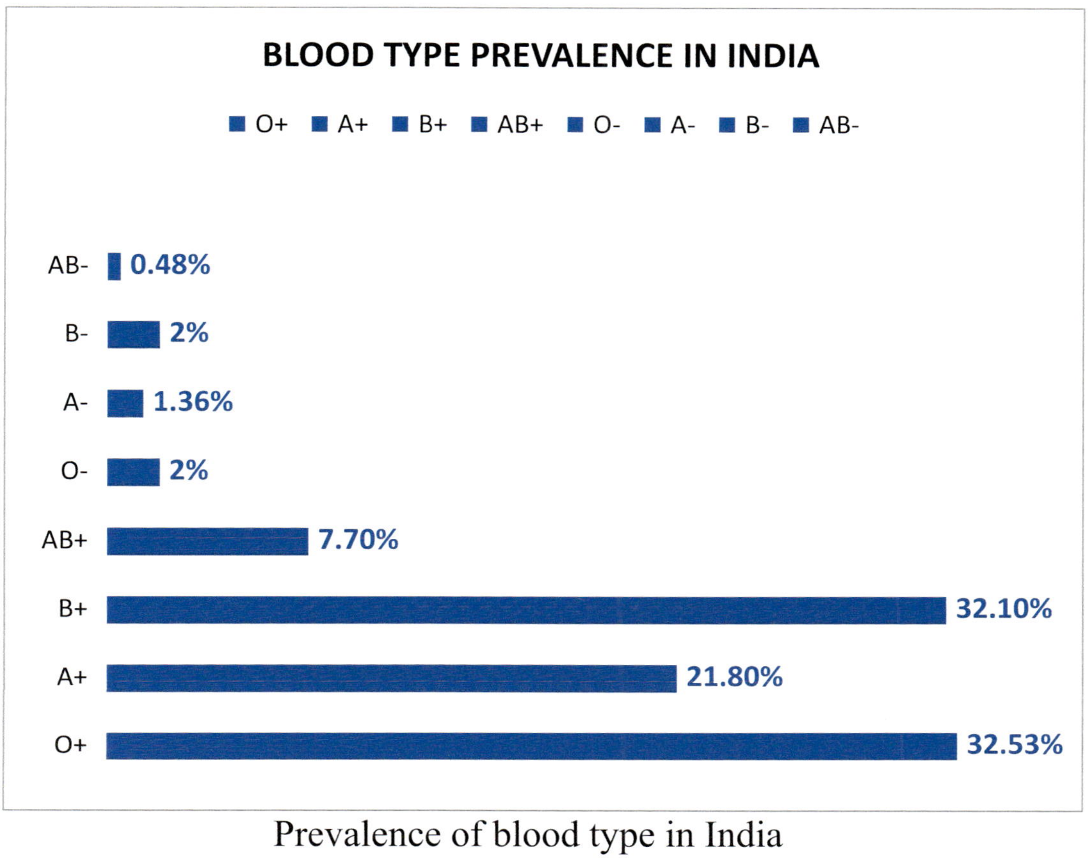

***Figure 3.4** Prevalence of blood type in India.*

The ABO blood grouping system is based on the presence or absence of A and B antigens on the RBC. The test sample is mixed with the antibodies, and agglutination is observed. It is reverse grouping and checks for isoagglutination. If agglutination is observed with anti-A antibodies, the blood type is A; similarly for B, if agglutination is observed in both anti-A and anti-B, the blood type is AB; if no interaction is observed, the blood type is O.

- *Rh Blood Group System*
  This grouping system is based on the presence or absence of the Rh factor, also known as the D-antigen. According to this, blood types can be Rh positive, which means D-antigen is present, and Rh negative, which means D-antigen is absent. Approximately 85% of the population has a positive Rh factor. It is similar to ABO typing, as it involves the addition of anti-D antibodies to the sample to check for the agglutination.

3.6.2 Tests for saliva

Presumptive and confirmatory test for the analysis and examination of saliva samples are represented in the given **Tables 3.5** and **3.6**, respectively.

*3.6.2.1 Presumptive tests for saliva (see **Table 3.5**)*

Table 3.5 Presumptive Tests for Saliva

Test Name	Mechanism	Inference	Disadvantages	First Used By
Starch-iodide test	Amylase in saliva breaks down starch, iodine reacts with the remaining starch	Presence of amylase	Not specific to human saliva, false positives are possible	General forensic use, Early 20th century
Phadebas test	Amylase in saliva hydrolyzes a starch-dye complex, releasing a blue dye	Presence of amylase	Not specific to human saliva, false positives are possible	Pharmacia Diagnostics (Pharmaceutical company), 1970s
Alternate light source (ALS)	Saliva fluoresces under certain wavelengths of light	The possible presence of saliva	Not confirmatory, can be affected by other substances	Forensic investigators, 1980s

*3.6.2.2 Confirmatory tests for saliva (see **Table 3.6**)*

Table 3.6 Confirmatory Tests for Saliva

Test Name	Mechanism	Inference	Disadvantages	First Used By
RSID-saliva test	Uses monoclonal antibodies to detect human salivary amylase	Confirms the presence of human saliva	More expensive requires specific reagents	Independent Forensics, 2008
ABO blood grouping	Detects ABO antigens in saliva if the person is a secretor	Confirms the presence of saliva from a secretor	Only works for secretors, not all individuals are secretors	Karl Landsteiner, 1901
Rapid stain identification	Uses antibodies to detect human salivary α-amylase	Confirms the presence of human saliva	Requires specialized equipment and reagents	Independent Forensics, 2008
DNA analysis	Extracts and analyzes DNA from epithelial cells in saliva	Confirms human origin and can identify individuals	Requires advanced lab equipment, time-consuming, expensive	Alec Jeffreys (DNA profiling), 1984

3.6.3 Tests for semen

Presumptive and confirmatory test for the analysis and examination of semen samples are represented in the given **Tables 3.7** and **3.8**, respectively.

*3.6.3.1 Presumptive tests for semen (see **Table 3.7**)*

Table 3.7 Presumptive Tests for Semen

Test Name	Mechanism	Inference	Disadvantages	First Used By
Acid phosphatase test	Detects acid phosphatase enzyme secreted by the prostate gland	Presence of seminal fluid	Not specific to semen, false positives are possible	Huggins and Hodges, 1954
Alternate light source (ALS)	Semen fluoresces under certain wavelengths of light	Possible presence of semen	Not confirmatory, can be affected by other substances	Forensic investigators, 1980s

*3.6.3.2 Confirmatory tests for semen (see **Table 3.8**)*

Table 3.8 Confirmatory Tests for Semen

Test Name	Mechanism	Inference	Disadvantages	First Used By
Christmas tree Stain	Stains sperm cells with distinctive colors using Nuclear Fast Red and Picroindigocarmine	Presence of sperm cells	Requires microscopic examination, sperm cells may deteriorate	Paul Ehrlich (staining techniques), Early 20th century
Prostate-specific antigen (PSA) test	Detects PSA, a protein produced by the prostate gland	Presence of seminal fluid	Not specific to semen, it can be found in other body fluids	Wang et al., 1979
RSID-semen test	Uses monoclonal antibodies to detect human semen-specific antigen	Presence of human semen	More expensive requires specific reagents	Independent Forensics, 2008
DNA analysis	Extracts and analyzes DNA from sperm cells	Human origin and individual identification	Requires advanced lab equipment, time-consuming, expensive	Alec Jeffreys (DNA profiling), 1984

3.6.4 Tests for urine

Presumptive and confirmatory tests for the analysis and examination of urine samples are represented in the given **Tables 3.9** and **3.10**, respectively.

*3.6.4.1 Presumptive tests for urine (see **Table 3.9**)*

Table 3.9 Presumptive Tests for Urine

Test name	Mechanism	Inference	Disadvantages	First Used By
Dipstick test	Chemical strips react with components in urine, changing the color	Presence of glucose, protein, blood, ketones, etc.	Not specific, as it may give false positives or negatives	Bayer Diagnostics, 1950s
Urease test	Urease enzyme breaks down urea in urine, producing ammonia	Presence of urea	It can be affected by diet or medications	General forensic use, early 20th century
Jaffe reaction	Creatinine in urine reacts with picric acid in an alkaline medium, forming a red complex	Presence of creatinine	Not specific, interference from other substances	Max Jaffe, 1886

3.6.4.2 *Confirmatory tests for urine (see **Table 3.10**)*

TABLE 3.10 Confirmatory Tests for Urine

Test name	Mechanism	Inference	Disadvantages	First Used By
GC-MS (Gas chromatography-mass spectrometry)	Separates and identifies compounds based on mass and charge	Specific identification of drugs, metabolites, etc.	Expensive, requires skilled personnel	Forensic laboratories, 1970s
LC-MS (Liquid chromatography-mass spectrometry)	Separates and identifies compounds in liquid samples	Specific identification of drugs, metabolites, etc.	Expensive, requires skilled personnel	Forensic laboratories, 1980s
High-performance liquid chromatography (HPLC)	Separates compounds based on their interactions with a stationary phase.	Identifies and quantifies specific drugs/ metabolites.	Expensive, requires skilled personnel	Forensic laboratories, mid-1970s. This was particularly after the development of RP (Reverse Phase) chromatography.
Immunoassay	Uses antibodies to detect specific antigens in urine	Presence of specific drugs or hormones. Qualitative detection of drugs/ metabolites.	Can give false positives, requires confirmation	Rosalyn Yalow and Solomon Berson, 1950s

3.7 DNA testing in body fluid identification

DNA is the hereditary material in almost all living organisms. It contains the instructions necessary for the development, functioning, growth, and reproduction of organisms. DNA is composed of molecules called nucleotides, which consist of a phosphate group, a sugar group (deoxyribose), and a nitrogen base. The four nitrogenous bases are Adenine (A), Thymine (T), Guanine (G), and Cytosine (C). These bases pair in specific ways—adenine pairs with thymine (A-T) and guanine pairs with cytosine (G-C)—to form the "rungs" of the DNA double-helix structure.

DNA serves as the blueprint for producing proteins in the cell through a process called transcription and translation, mediated by RNA. It is passed from parents to offspring and plays a key role in inheritance, evolution, and life functioning at the molecular level. DNA plays a crucial role in forensic science by providing a powerful tool for identifying individuals and solving crimes.

Here are some of the key uses of DNA in forensics:

- Identification of Suspects: DNA samples from a crime scene (like blood, hair, or skin cells) can be compared with a suspect's DNA to determine whether they were present at the scene. If the DNA profiles match, strong evidence can link the suspect to the crime.
- Exoneration of Innocent People: DNA testing has been instrumental in overturning wrongful convictions. Many individuals who were wrongfully convicted before the advent of DNA testing have been exonerated through post-conviction DNA testing.
- Identifying Victims: DNA can be used to identify victims of crimes, disasters, or accidents, especially when other identification methods (e.g., fingerprints, facial recognition) are not possible. For instance, DNA is crucial in identifying bodies in mass disasters or cases of severe decomposition.
- Paternity and Family Relationships: In cases involving inheritance disputes, custody battles, or immigration issues, DNA can be used to establish paternity or other family

relationships. By comparing their DNA profiles, forensic scientists can determine whether individuals are biologically related.

- Cold Case Investigations: DNA technology allows investigators to re-examine old cases (often called "cold cases") by analyzing biological evidence that was preserved from crime scenes. New DNA techniques can sometimes yield results that were not possible with earlier methods, leading to the resolution of unsolved cases years later.
- Sexual Assault Cases: The main evidence that is collected in sexual assault or related crimes is DNA, extracted often from samples of biological origin, including body fluids, tissues, and others. The DNA profile can be compared to that of a suspect or maintained in a database for future reference.
- Database Matching (CODIS): Combined DNA Index System; It is a database system used for storing and maintaining DNA profiles in the U.S. Developed by the FBI, these databases can be used to compare DNA profiles obtained from a crime scene to previous offenders or unidentified person/missing person or others. *Crimes committed by a single person can be detected.*
- Missing Persons: DNA testing is a key tool in identifying missing persons by comparing the DNA of unidentified remains with that of the DNA profiles obtained from the relatives of missing persons.
- Contaminant Identification: DNA profiling can also assist in identifying other unidentified biological substances recovered from a crime scene, which helps determine whether a crime was committed and provides context for evidence.
- Bioterrorism: DNA profiling can be used to identify the organisms involved in bio-attacks, such as anthrax, and trace them back to their origin.

3.7.1 DNA isolation and extraction

The process of separating DNA from the rest of the components of cells, such as proteins, lipids, and carbohydrates, to obtain a pure DNA sample for profiling is called DNA extraction. This process is of utmost importance in applications of molecular biology, genetics, forensics, and medical diagnostics. The isolated DNA can then be used for techniques like PCR, sequencing, and genetic analysis.

3.7.1.1 Steps in DNA isolation and extraction

The process generally involves the following key steps:

1. **Cell Lysis**
    - **Goal**: Break open the cells to release their contents, including DNA.
    - **Method**: Cells are treated with a lysis buffer, which often contains detergents or enzymes (e.g., proteinase K) that break down the cell membrane and nuclear envelope. In plant cells, mechanical grinding might be needed to break the tough cell wall.
2. **Removal of Membrane Lipids and Proteins**
    - **Goal**: Dissolve or remove cellular membranes and proteins that are not needed.
    - **Method**: **Detergents** (like SDS) dissolve lipids, and **protease enzymes** help degrade proteins. Often, phenol-chloroform extraction is used to separate proteins and lipids from the DNA.
3. **Precipitation of DNA**
    - **Goal**: Separate the DNA from other cellular debris.

- **Method**: After cell lysis, salt (e.g., sodium chloride) is added to neutralize the DNA's negative charge, which helps it precipitate out. Alcohol (usually ethanol or isopropanol) is added to the solution to cause the DNA to clump together, making it visible as a precipitate. The DNA can then be spooled onto a glass rod or centrifuged to collect it.

4. **Washing the DNA**
   - **Goal**: Remove any remaining contaminants, such as salts and proteins.
   - **Method**: The DNA is often washed with cold alcohol (ethanol or isopropanol) to remove impurities. Repeated washes may be necessary to achieve a high level of purity.
5. **Resuspension of DNA**
   - **Goal**: Store the extracted DNA in a stable form.
   - **Method**: After washing, the purified DNA is dissolved in a buffer (like TE buffer) or distilled water to protect it from degradation by nucleases and to ensure long-term stability.

3.7.2 Types of DNA extraction methods

Different types of DNA extraction protocols are used depending on the sample type (e.g., blood, tissue, bacteria, plants, forensic samples). Common methods include:

1. **Phenol-Chloroform Extraction:**
   - Involves mixing the sample with phenol and chloroform to separate DNA from proteins and lipids. This method is highly effective but involves toxic chemicals.
2. **Silica-Based Methods**:
   - DNA binds to a silica matrix (e.g., spin columns) in the presence of a high salt concentration. This method is common in many commercial DNA extraction kits because it's fast and efficient.
3. **Magnetic Bead-Based Extraction**:
   - Magnetic beads coated with DNA-binding material are used to capture DNA, and then a magnet is applied to pull the beads out of the solution. This method is automated and widely used in high-throughput labs.
4. **Chelex® Extraction**:
   - Chelex resin is used to bind contaminants while DNA is released into the supernatant. This method is frequently used in forensic DNA analysis.
5. **Alkaline Lysis Method**:
   - A quick method where a strong alkaline solution (e.g., NaOH) is used to break open the cells and release DNA. This is often used in plasmid DNA isolation from bacteria.

Extracting DNA from different body fluids requires adapting the general DNA isolation method to accommodate the unique properties of each fluid. Here's an overview of how DNA can be extracted from common body fluids:

1. **Blood**

   Blood is one of the most commonly used sources for DNA extraction, particularly for genetic and forensic analysis. Most DNA in blood is found in the white blood cells (leukocytes) because red blood cells lack nuclei.

   Steps:

   - **Sample Collection**: A small volume of blood (often treated with anticoagulants like EDTA to prevent clotting) is collected.

- **Cell Lysis**: Red blood cells are lysed using a hypotonic solution to remove them, while white blood cells are preserved. This step is essential because red blood cells don't contain DNA.
- **Nuclear Lysis**: White blood cells are lysed using a lysis buffer containing detergents (e.g., SDS) to release the DNA.
- **Protein Removal**: Proteins are degraded using protease (e.g., proteinase K) or separated with phenol-chloroform extraction.
- **DNA Precipitation**: DNA is precipitated using alcohol (ethanol or isopropanol) after the addition of salts.
- **DNA Purification**: Alcohol washings are done with alcohol and then re-suspended in TE buffer for storage or in water for immediate use.

2. **Saliva**
   Saliva is a suitable sample for DNA extraction, commonly used in forensic caseworks, genetic testing and ancestry studies. Epithelial cells in the saliva lining the mouth are the major source of DNA.

   Steps:

   - **Sample Collection**: Saliva is collected using a clean sterile swab or cotton ball, and then is air dried and liquid forms are collected in plastic collection tubes.
   - **Cell Lysis**: The first step of DNA extraction, cell lysis is achieved using a lysis buffer, a solution containing detergents like SDS to disrupt cell membranes and release DNA from the cell.
   - **Protein Removal**: Proteinase-K enzyme is used mainly to digest and disrupt proteins.
   - **DNA Precipitation**: Ethanol or isopropanol is added to precipitate the DNA.
   - **DNA Purification**: washings of the DNA pellet are done with alcohol in order to purify the extracted DNA and then resuspended in water or TE buffer.

3. **Semen**
   Sperm cells contain a high concentration of DNA, making it a valuable sample for forensic analysis, especially in sexual assault cases.

   Steps:

   - **Sample Collection**: Liquid samples generally collected in plastic vials and stored at low temperature.
   - **Cell Lysis**: Strong lysis buffer containing detergents and DTT (dithiothreitol) is used to break down sperm cell membranes, which are more resistant to lysis compared to other cell types.
   - **Protein Removal**: Proteinase K is often used to digest proteins.
   - **DNA Precipitation**: DNA is precipitated using ethanol or isopropanol.
   - **DNA Purification**: The DNA is purified by washing and resuspending in a suitable buffer.

4. **Urine**
   The main source of DNA in urine samples is primarily from epithelial cells from the urinary tract, that usually contain very low amounts of DNA.

Steps:

- **Sample Collection**: Samples are generally collected in plastic vials.
- **Cell Concentration**: Urine samples are often centrifuged to concentrate any cells at the bottom of the tube.
- **Cell Lysis**: The cells are lysed using a buffer with detergents.
- **Protein Removal**: Proteinase K is used to degrade proteins.
- **DNA Precipitation**: Precipitation with ethanol or isopropanol helps to recover the small amounts of DNA.
- **DNA Purification**: The DNA is washed and resuspended.

5. **Sweat** contains even smaller amounts of DNA than urine, typically coming from skin cells. Sweat is sometimes used in forensic cases where other bodily fluids are unavailable.

Steps:

- **Sample Collection**: Sweat is collected on a swab or piece of fabric.
- **Cell Lysis**: Lysis buffer with detergents is applied to break down the cells trapped in the sample.
- **Protein Removal**: Proteinase K or other proteases are used to digest proteins.
- **DNA Precipitation**: Ethanol or isopropanol is used for precipitation.
- **DNA Purification**: The purified DNA is resuspended in buffer or water.
- **Vaginal Fluid and Menstrual Blood** contains DNA from blood and epithelial cells of the vaginal lining. This is used in sexual assault investigations and other forensic analyses.

6. **Vaginal fluid**

Steps:

- Sample Collection: A swab is typically used to collect vaginal fluid samples.
- Cell Lysis: The swab is placed in a lysis buffer containing detergents to break open the cells.
- Protein Removal: Proteinase K is added to digest proteins.
- DNA Precipitation: The DNA is precipitated with alcohol after adding salts.
- DNA Purification: DNA is washed and resuspended for further analysis.

7. **Menstural Blood**

Steps:

The process is similar to other DNA extraction methods but must account for the unique composition of menstrual blood, which contains a mixture of blood and tissue cells.

- The sample is collected using tampons, sanitary pads, menstrual cups, or swabs. Forensic samples may be collected from crime scenes (on clothing or bedding). The sample may be stored at −20°C (frozen) if extraction is not performed immediately to preserve DNA integrity.
- Cell Lysis: The cells (both blood and endometrial cells) are lysed to release DNA. A lysis buffer (e.g., SDS or CTAB) containing detergents and enzymes breaks open cell membranes. Mechanical disruption (vortexing or grinding) may be used if the sample contains clots or tissue fragments.

- Protein Removal: Proteinase K is added to digest proteins and histones bound to the DNA, which is necessary to release DNA from chromatin and remove contaminants. In some protocols, a phenol-chloroform extraction separates DNA from proteins and lipids. The sample is mixed with phenol and chloroform, which causes proteins to precipitate in the organic phase while DNA remains in the aqueous phase.
- DNA Precipitation: Sodium Acetate is added to neutralize or balance the DNA's negative charge. The sample is mixed with cold ethanol or isopropanol, which causes the DNA to precipitate out of the solution. The precipitated DNA forms a visible white or cloudy pellet, which is collected by centrifugation.
- Washing the DNA: The DNA pellet is washed with 70% ethanol, the sample is spun in a centrifuge, and the ethanol is discarded.
- Resuspension of DNA: The purified DNA pellet is dissolved in a suitable buffer (such as TE buffer) or nuclease-free water to preserve the DNA for long-term storage.

Special Considerations for Menstrual Blood DNA Extraction:

1. **Presence of Hemoglobin**: Hemoglobin in menstrual blood can inhibit downstream applications like PCR. Some protocols include additional steps to remove hemoglobin to prevent this issue.
2. **Mixed DNA Sources**: Menstrual blood contains a mixture of blood cells and endometrial cells, so forensic applications must account for potential mixed DNA profiles.
3. **Degraded DNA**: Menstrual blood may contain partially degraded DNA due to the nature of the sample and the presence of nucleases. Proper sample handling (immediate freezing or use of preservative agents) is crucial for maintaining DNA integrity.

In all cases, the general DNA extraction steps (lysis, protein removal, precipitation, and purification) are followed, but each body fluid may require specific modifications due to the unique cellular and protein composition of the sample. The yield and quality of DNA also vary significantly across fluids, with blood and semen typically yielding the highest DNA concentrations.

3.8 Advanced methods for identifying body fluids

Forensic laboratories rely on a combination of traditional and emerging technologies to actually identify body fluids and tissues. Recent advances in biomarkers and technologies are improving the specificity and sensitivity of analyses. Some of the major developments in this area are:

3.8.1 RNA-based methods

RNA is a key molecule in biological processes such as gene expression, including messenger RNA (mRNA) for protein production, miRNA for regulation, and circular RNA (circRNA). miRNA, circRNA, and mRNA are particularly important for the identification of body fluids (Bauer et al., 1999; Lynch & Fleming, 2021). Each body fluid has a distinct mRNA expression profile, which is influenced by the unique cellular origin of those fluids. Bauer et al. (1999) first demonstrated the application of mRNA markers to differentiate between menstrual and peripheral. Various markers have been proposed to detect specific forensically relevant body fluids after this discovery. mRNA markers for different body fluids can be multiplexed together in a single PCR-based assay (Roeder & Haas, 2016). Haas and the research team conducted a series of experiments and successfully demonstrated multiplex reverse transcription endpoint PCR for the detection of blood, saliva, semen, vaginal secretions and menstrual blood, with satisfactory specificity and sensitivity even in the detection of biological stains up to 15 months (Haas et al., 2009). Bamberg and his team developed primers to detect rectal mucosa, which has been overlooked despite its relevance in sexual assault cases. It was the first time that rectal mucosa can be identified using mRNA profiling in a forensic-relevant case (Bamberg et al., 2021). MicroRNAs (miRNAs; non-coding RNAs) that exhibit tissue-specific expression in specific extracellular fluids, including forensically relevant body fluids (Teoh & Das, 2022). Various published research studies have reported promising results identifying differentially expressed miRNAs with great potential as new biomarkers for identifying and distinguishing specific body fluids (Glynn, 2020; Mayes et al., 2018). There are various bodily fluid miRNA markers: miR16 and miR486 for venous blood, miR888 and miR891a for semen, miR214 for semen, miR124a for menstrual blood, vaginal discharge, and miR138-2 for saliva (Chen et al., 2023; O'Leary & Glynn, 2018; Tian et al., 2018).

The combined mRNA and miRNA system offers significant advantages for identifying human body fluids, especially in mixed samples. This approach provides a scientific and accurate reference for the identification of body fluids (Wang & Liu, 2022). RNA rapidly degrades during collection, processing, and testing, affecting reliability. RNA degradation occurs primarily in blood samples, complicating accurate assessment (Park et al., 2014). The rapid degradation of mRNA creates challenges for quantitative experiments.

3.8.2 Epigenetic markers

DNA methylation occurs at CpG dinucleotides-5′-position of the pyrimidine ring of cytosine residue and has been proposed as a new marker for identification of body fluids, as tissue-specific patterns of DNA methylation have been observed in various studies over time (Forat et al., 2016; Mishra et al., 2021). In a study by Forat and his team, in a differential DNA methylation approach, 150 candidate loci related to venous blood, menstrual blood, vaginal fluid, saliva, and sperm were identified (Madi et al., 2012).

Differences in DNA methylation are often insufficient to distinguish between saliva and vaginal secretions, indicating the need for further research to establish specific methylation markers for each body fluid (Tian et al., 2020). Analysis of DNA methylation requires advanced techniques and variability in results due to factors such as age and health complicates standardization. Additionally, the cost of high-throughput sequencing can be prohibitive.

3.8.3 Spectroscopic techniques

Recent spectroscopic advances provide the ability to detect multiple body fluid protein or metabolite markers in a single confirmatory test with high specificity and reliability (Zapata et al., 2015). Recent advancements in spectroscopic techniques such as Raman spectroscopy, Mass Spectroscopy, FTIR, and NMR analysis were found to be accurate techniques for characterization and identification of bodyfluids by generating precise and clear spectral results, which presents its own advantages and limitations.

Raman spectroscopy is an essential tool of utmost reliability for confirming the identity and presence of body fluids, especially in sexual assault cases involving mixtures of semen and vaginal fluid samples (Sikirzhytskaya et al., 2023). Due to its ability to provide detailed molecular information while maintaining the integrity of the sample, makes it an invaluable technique in criminal investigations. NMR spectroscopy allows the identification and characterization of body fluids based on their composition; metabolite markers in them are targeted to detect. NMR offers advantages such as ease of sample preparation and reproducibility of results, an unbiased overview of all metabolite markers, including those which are specifically present in a sample (Speyer & Baleja, 2021). Conventional body fluid identification techniques often rely on specific proteins, which can result in false positive outcomes due to non-specific reactions. Recent advances provide a more confirmatory approach, particularly mass spectrometry: MALDI-MS, that allows detection of multiple protein or metabolite markers (Yang et al., 2013). Jiang and his team demonstrated direct identification of BFs, including blood, semen, saliva, sweat, and urine, by analyzing metabolite markers present in a specific body fluid by using MALDI-MS (Jiang et al., 2019). In their study, they observed that the obtained blood spectra are stable and unaffected by different storage times; however, other fluids have shown less influence of storage time factor. It was also observed that biomarkers can be easily detected in the tested body fluids by increasing the dilution step by step from 1000 to 8000 times; however, the protein signal was difficult to detect beyond 4000 dilutions (Jiang et al., 2019).

Appropriate sample handling and processing are essential to obtain reliable and accurate results, as errors can occur due to contamination or improper handling. The equipment must be regularly calibrated and carefully maintained to ensure accurate and reproducible results.

3.8.4 Biosensors and immunochromatographic methods

Biosensors and immunochromatographic methods are increasingly and effectively utilized in forensic scenarios for the rapid and reliable identification of body fluids. Biosensors are analytical tools that utilize biological macromolecules for specific molecular recognition, translating ligand binding into detectable physical changes (Nawrot et al., 2018). Microfluidic devices utilizing fluorescence employ various materials, dyes, detectors, excitation sources, and filters, allowing for comprehensive analyses of various body fluids in a compact format (Nawrot et al., 2018). Their speed, accuracy, specificity, and low cost make them invaluable in forensic investigations, allowing rapid detection of toxins, body fluids, and explosives without extensive sample preparation (Wani et al., 2023).

Common confirmatory tests include immunochromatographic tests (such as ELISA or ABA hematrace card) to identify body fluids. Immunochromatographic (IC) tests serve as the primary method for forensic human blood identification, highlighted by their sensitivity, specificity, and rapid on-site testing capability. However, destruction of biological stains during testing raises concerns about subsequent DNA typing (Cuttaia et al., 2024). ELISA is a sensitive and specific analytical technique in forensics that uses antigen-antibody interactions to identify biomarkers unique to various body fluids, aiding in their detection and analysis (Alhajj et al., 2024).

In addition to their advantages, these techniques also have limitations. Biosensors for body fluid detection may produce false positives due to limited specificity, exhibit sensitivity to environmental conditions, and encounter stability issues with biological components (Frascione et al., 2013). Similarly, methods like ELISA and chromatographic techniques tend to be destructive, expensive, and non-universal, requiring extensive sample preparation that can be time-consuming.

3.9 Latest developments in body fluid identification

Identification of body fluids plays an important role in forensic investigations, aiding in the reconstruction of crime scenes and linking suspects or victims to specific locations. Traditional methods in the identification of body fluids, including enzyme-based assays, colorimetric testing, and microscopic techniques, which are being widely used, show limitations related to sensitivity and specificity. Recent advances in proteomics, molecular biology, and biotechnology have led to the development of emerging technologies that advance the identification and distinction of body fluids. A comprehensive workflow for the identification of forensically relevant body fluids is represented in **Figure 3.5**.

One of the notable advances in the identification of body fluids is the use of RNA profiling and epigenetic markers (S.-y. Wang et al., 2022). Silva and his team performed experiments for DNA methylation markers such as CG06379435 for blood identification and evolutionary validation in their study (Silva et al., 2016). Chemometrics aids in the efficient analysis and interpretation of complex biological data, increasing accuracy and precision in forensic body fluid identification (Yadav & Sharma, 2024). Tian and his team developed a novel method using with 22 methylation markers and ARMS-PCR and random forest algorithm to detect DNA methylation patterns for body fluids identification, the method achieved high accuracy (0.9966) in the prediction of venous blood, saliva, semen, menstrual blood, and vaginal fluid, demonstrating its robustness in matters of forensic investigation (Tian et al., 2020). DNA methylation-based assays can differentiate different body fluids by targeting specific differentially methylated regions (DMRs). The DNA methylation-based method is

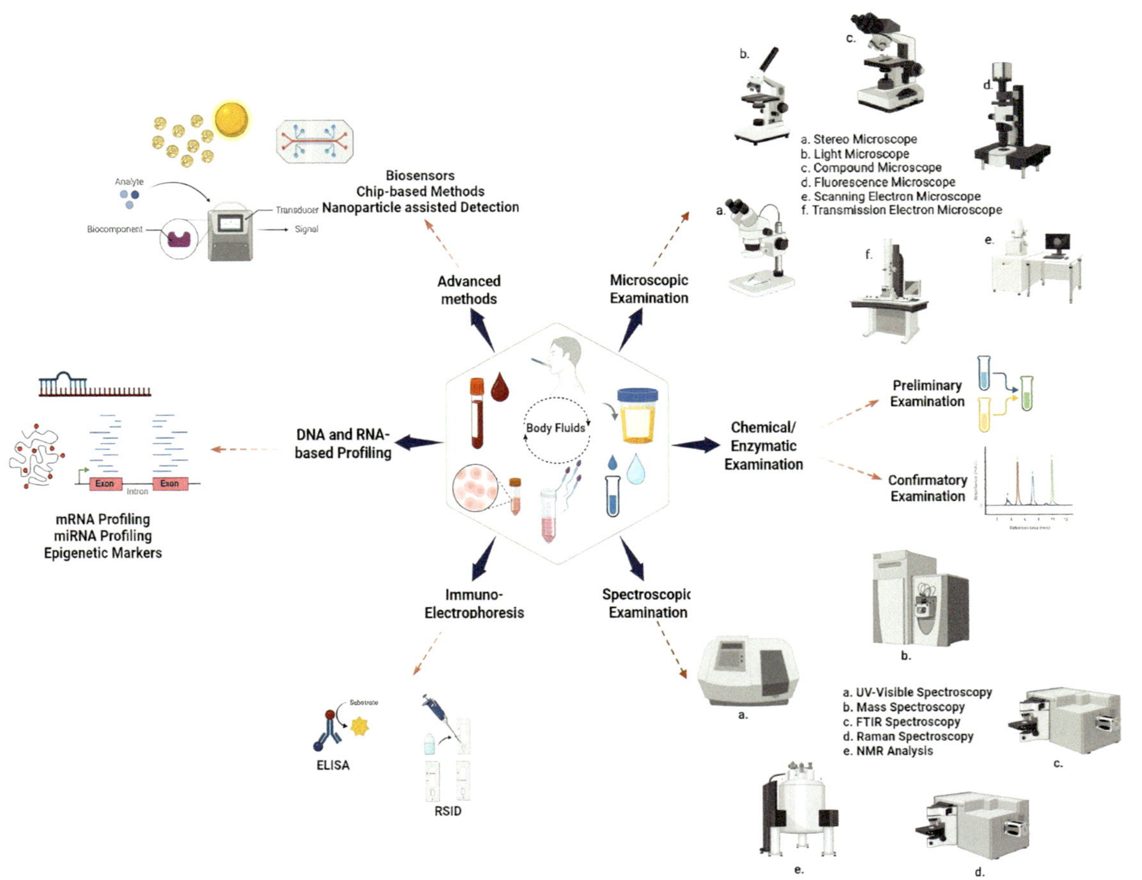

Figure 3.5 *Comprehensive workflow for forensic body fluid identification – Integrating presumptive tests, confirmatory methods, spectroscopic examination, and advanced molecular techniques for accurate analysis.*

advantageous, providing a reliable means of identification, even in degraded samples as DNA is more stable than RNA. Chierto and team demonstrated mRNA profiling assay in identification of the vaginal-specific markers CYP2B7P1, MUC4, and MYOZ1 in vaginal swabs of pre- and post-menopausal women (Chierto et al., 2023). miRNA, being more stable than mRNA, has proven to be an efficient marker for the body fluids identification in forensic relevance (Fujimoto et al., 2019).

Spectroscopic advances allow the detection of multiple body fluid metabolite markers in a single confirmatory test, which presents its own advantages and limitations. In 2010, Sikirzhytski with his team combined Raman spectroscopy with discriminant functional analysis to create a non-destructive, confirmatory, and reliable method for identifying body fluids such as saliva, semen, and blood (Sikirzhytski et al., 2010). In a study, Jiang and team developed a rapid mass spectrometry method for distinguishing body fluids (blood, semen, urine, sweat, and saliva) using microliter samples with no pre-treatment. This method accurately identified molecular biomarkers, achieved high accuracy (90%) in blind tests and took less than 10 minutes to complete the entire process, making it highly suitable for forensic casework (Jiang et al., 2019). In another study, Gowda and team in their study identified 67 new metabolites, a third of which were newer than previously reported metabolites in the human serum metabolome database (Nagana Gowda et al., 2015). Also, scientists investigated the identification and distinction of semen in sexual assault case samples using FTIR spectroscopy. This FTIR analysis validated the identity of semen through specific peaks, in a mixture of body fluids, consisting of semen, saliva, and urine. The findings of this study show that these methods beneficially assist forensic investigations in detecting and distinguishing body fluids, whether present in the sole or in a mixture (Zapata et al., 2015).

In a recent study, researchers identified estimated time since taint and coagulation (TSD) of body fluids using microbial characteristics as biomarkers (Zhang et al., 2024). Recent advances in biotechnology, including nanotechnology-based biosensors, can detect specific biomarkers for specific body fluids with high sensitivity and reliability, such as using gold nanoparticles and quantum dots (Ramesh et al., 2022). Advances in next-generation sequencing (NGS) technology have advantageously enabled the rapid and accurate identification and distinction of body fluids by analyzing their unique genomic and transcriptomic signatures. NGS, especially 16S rRNA sequencing, is widely used for microbiome identification in body fluids (Ji et al., 2023).

Analysis of body fluids when present in mixture forms, such as semen and vaginal fluid in sexual assault or related cases, can be challenging, especially when DNA from multiple individuals is present. Male-specific Y-chromosomal short tandem repeats (Y-STRs) are beneficial in forensic examination, exceptionally in cases of sexual assault or related crimes (Tsuji et al., 2001). The researchers conducted Y-STR analysis and effectively identified the perpetrator from mixed body fluids, including semen, in a case of sexual assault. Typing of three Y-STRs: DYS19, DYS389II, and DYS390 identified male DNA from the male and female fraction mixture in the victim's vagina, vaginal orifice, and anus (Tsuji et al., 2001). The haplotypes matched with those of the suspect, exhibiting the usefulness of Y-STR analysis for mixed samples of body fluids in identifying male contributors. Another team of scientists performed direct amplification of Y-STR profiles from blood and saliva mixture samples collected at 12 different crime scenes. The detection of more than one distinct allele at any Y-STR locus indicates the presence of multiple male contributors, as each male generally has the same Y-STR profile (Dargay & Roy, 2016). Routine DNA typing kits, utilized for Y-STR and autosomal STR analysis in forensic applications, include AmpFℓSTR® Yfiler®, AmpFℓSTR® Identifiler® Plus, PowerPlex® 21 System, and PowerPlex® Y23 System (Biocompare, 2024; Scientific, 2024).

Recent advancements in biosensors and chip-based methods, along with the development of new reagents and nanomaterials, have significantly enhanced the sensitivity and specificity of body fluid detection. Biosensors and microfluidic chips have revolutionized forensic body fluid detection by offering rapid, sensitive, and specific analysis. These technologies integrate biological recognition elements with physical transducers to detect various biomarkers present in body fluids. Wearable biosensors are designed for non-invasive monitoring of body fluids such as sweat, intestinal fluid, and tears (Park et al., 2021). These sensors use electrochemically active materials to detect biomarkers in real-time, providing valuable data for forensic casework. Microfluidic chips, often referred to as lab-on-chip devices, allow for miniaturization and automation of complex biochemical assays. These require a small sample volume and provide quick results. For example, a study demonstrated the use of paper-based microfluidic devices for the colorimetric detection of DNA and saliva using gold nanoparticle (AuNP) composites. This device was able to detect DNA concentrations as low as 10ng/μL, showing its high sensitivity (Bruijns et al., 2016).

The advancements of new reagents and nanomaterials have further improved the detection capabilities of forensic biosensors. Nanomaterials such as gold, silver, and carbon-based nanomaterials have been extensively used to enhance the sensitivity of biosensors. These materials provide a large surface area for biomolecule interaction, leading to improved signal transduction (Bruijns et al., 2016). For example, PMMA-gold nanocomposites have been employed in microfluidic devices for the sensitive detection of DNA (Park et al., 2021). Another example includes the use of graphene oxide-based sensors for the detection of blood proteins, which significantly improved the detection limits compared to traditional methods. The development of specific reagents, such as antibodies and aptamers, has enabled the selective detection of body fluids. These reagents can be immobilized on sensor surfaces to capture target molecules with high affinity and specificity. For example, aptamer-based sensors have been developed for the detection of semen specific antigen (PSA) in forensic samples (Sezgintürk, 2024). Additionally, the use of enzymatic assays and fluorescent probes has facilitated the detection of low-abundance biomarkers in forensic samples. For example, fluorescently labeled antibodies are used for the detection of salivary amylase, which gives specific results even in trace amounts of sample (Sezgintürk, 2024). Additionally, a novel method for blood detection using fluorescein-derived 2′7′-dichlorofluorescein di-acetate (DCFDA) dye has been developed, and this method detects blood and bloodstains aged up to 20 days without interfering with DNA integrity (Vajpayee, Desai, et al., 2023). This method aims to improve the accuracy and efficiency of blood detection compared to traditional reagents like luminol and fluorescein.

3.10 Challenges

Identifying body fluids in forensic science presents several challenges:

1. Degradation and Contamination:
   - Body fluids can degrade over time due to environmental factors (e.g., sunlight, moisture) or bacterial activity.
   - Contamination from other substances (e.g., dirt, chemicals) can interfere with accurate identification.
2. Small Sample Size:
   - Crime scene samples are often limited in quantity, making it challenging to perform multiple tests.
   - Some tests require larger volumes of fluid for reliable results.

3. Complex Mixtures:
    - Real-world samples may contain a mixture of fluids (e.g., blood and saliva), complicating identification.
    - Untangling these mixtures requires advanced techniques.
4. Specificity and Sensitivity:
    - Tests must be specific (identifying a particular fluid) and sensitive (detecting low concentrations).
    - Achieving both simultaneously can be difficult.
5. Aging Effects:
    - Some body fluids change composition over time, affecting detection methods.
    - For example, RNA degradation in blood stains can impact RNA-based tests.
6. Ethical and Legal Considerations:
    - Collecting and analyzing body fluids involves privacy and consent issues.
    - Legal standards for evidence collection vary across jurisdictions.

3.11 Conclusion

The identification of body fluids in forensic investigations plays an essential role in determining the events surrounding a crime scene and establishing the link between the offender and the victim. Body fluids, such as blood, saliva, semen, and urine, contain a wealth of genetic, biochemical, and cellular information that can provide valuable insights for forensic analysis. Over the years, extensive research has been conducted to develop and improve methods for the accurate and reliable identification of body fluids in forensic samples. Traditional methods remain fundamental in the identification of forensically relevant body fluids and even offer some limitations and should be complemented by other advanced biotechnology techniques such as RNA profiling, epigenetic marker profiling, proteomics, and advanced spectroscopic techniques to increase the specificity and sensitivity of forensic body fluid identification. These advanced techniques provide greater accuracy and allow forensic experts to distinguish between fluids that were previously challenging to distinguish. Additionally, recent developments, such as NGS, spectroscopic techniques, and chemometrics, are setting the stage for faster, more efficient, and more reliable methods of body fluid identification. Looking ahead, future directions in body fluid identification are poised to provide exciting developments that enhance forensic investigations that contribute to criminal justice systems. In conclusion, the integration of both conventional and state-of-the-art advanced methods in body fluid identification underpins the continued evolution of forensic science, leading the field toward greater accuracy, reliability, and efficiency in solving crimes.

References

Alhajj, M., Zubair, M., & Farhana, A. (2024). Enzyme linked immunosorbent assay [Updated 2023 Apr 23]. *StatPearls [Internet]*. StatPearls Publishing.

Andréasson, H., Nilsson, M., Budowle, B., Lundberg, H., & Allen, M. (2006). Nuclear and mitochondrial DNA quantification of various forensic materials. *Forensic Science International, 164*(1), 56–64.

Aparna, R., Iyer, R. S., Kumar, N., & Sharma, A. (2021). Forensic DNA profiling of tears stains from commonly encountered substrates. *Forensic Science International, 328*, 111006.

Bamberg, M., Dierig, L., Kulstein, G., Kunz, S. N., Schwender, M., Hadrys, T., & Wiegand, P. (2021). Development and validation of an mRNA-based multiplex body fluid identification workflow and a rectal mucosa marker pilot study. *Forensic Science International: Genetics, 54*, 102542.

Bauer, M., Kraus, A., & Patzelt, D. (1999). Detection of epithelial cells in dried blood stains by reverse transcriptase-polymerase chain reaction. *Journal of Forensic Sciences, 44*(6), 1232–1236.

Bhoyar, L., Mehar, P., & Chavali, K. (2024). An overview of DNA degradation and its implications in forensic caseworks. *Egyptian Journal of Forensic Sciences, 14*(1), 15.

Biocompare. (2024). *PowerPlex® Y23 System*. https://www.biocompare.com/25138-Assay-Kit/7265682-PowerPlex-Y23-System/

Bruijns, B., Van Asten, A., Tiggelaar, R., & Gardeniers, H. (2016). Microfluidic devices for forensic DNA analysis: A review. *Biosensors, 6*(3), 41.

Chen, X., Xu, H., & Zhu, B. (2023). Forensic validation of a combined analysis of mRNA and miRNA markers for precise tissue origin inferences of five kinds of body fluids by RT-qPCR. *Electrophoresis, 44*(21–22), 1714–1724.

Chierto, E., Alessandrini, F., Bini, C., Carnevali, E., Fabbri, M., Fattorini, P., Grignani, P., Scarnicci, F., Tozzo, P., & Verzeletti, A. (2023). An mRNA profiling study of vaginal swabs from pre-and postmenopausal women. *Current Issues in Molecular Biology, 45*(8), 6526–6537.

Cuttaia, C., Di Stefano, B., Sorçaburu Ciglieri, S., Vetrini, R., Previderè, C., & Fattorini, P. (2024). Immunochromatographic detection of human blood: A forensic review. *Separations, 11*(3), 66.

Dargay, A., & Roy, R. (2016). Direct Y-STR amplification of body fluids deposited on commonly found crime scene substrates. *Journal of Forensic and Legal Medicine, 39*, 50–60.

Dash, H. R., & Vajpayee, K. (2020). Collection, preservation, and transportation of biological evidences. *Handbook of DNA Profiling* (pp. 1–16).

Dass, M., Singh, Y., & Ghai, M. (2023). A review on microbial species for forensic body fluid identification in healthy and diseased humans. *Current Microbiology, 80*(9), 299.

Doi, M., Gamo, S., Okiura, T., Nishimukai, H., & Asano, M. (2014). A simple identification method for vaginal secretions using relative quantification of Lactobacillus DNA. *Forensic Science International: Genetics, 12*, 93–99.

Forat, S., Huettel, B., Reinhardt, R., Fimmers, R., Haidl, G., Denschlag, D., & Olek, K. (2016). Methylation markers for the identification of body fluids and tissues from forensic trace evidence. *PloS One, 11*(2), e0147973.

Frascione, N., Gooch, J., & Daniel, B. (2013). Enabling fluorescent biosensors for the forensic identification of body fluids. *Analyst, 138*(24), 7279–7288.

Fujimoto, S., Manabe, S., Morimoto, C., Ozeki, M., Hamano, Y., Hirai, E., Kotani, H., & Tamaki, K. (2019). Distinct spectrum of microRNA expression in forensically relevant body fluids and probabilistic discriminant approach. *Scientific Reports, 9*(1), 14332.

Glynn, C. L. (2020). Potential applications of microRNA profiling to forensic investigations. *RNA, 26*(1), 1–9.

Haas, C., Klesser, B., Maake, C., Bär, W., & Kratzer, A. (2009). mRNA profiling for body fluid identification by reverse transcription endpoint PCR and realtime PCR. *Forensic Science International: Genetics, 3*(2), 80–88.

Hamza, M., Sankhyan, D., Shukla, S., & Pandey, P. (2024). Advances in body fluid identification: MiRNA markers as powerful tool. *International Journal of Legal Medicine, 138*(4), 1223–1232.

Harbison, S., & Fleming, R. (2016). Forensic body fluid identification: State of the art. *Research and Reports in Forensic Medical Science*, 11–23.

Hess, S., & Haas, C. (2017). Recovery of trace DNA on clothing: A comparison of mini-tape lifting and three other forensic evidence collection techniques. *Journal of Forensic Sciences, 62*(1), 187–191.

Heywood, D. M., Skinner, R., & Cornwell, P. A. (2003). Analysis of DNA in hair fibers. *Journal of Cosmetic Science, 54*(1), 21–28.

Hilhorst, M., Theunissen, R., van Rie, H., van Paassen, P., & Tervaert, J. W. (2013). DNA extraction from long-term stored urine. *BMC Nephrology, 14*, 238. https://doi.org/10.1186/1471-2369-14-238

Ji, X., Ni, S., Tian, G., Zhang, L., & Wang, W. (2023). Detection of microorganisms in body fluid samples. Tao Huang, Jialiang Yang, Geng Tian (Eds.), In: *Liquid Biopsies: Methods and Protocols* (pp. 73–88). Springer.

Jiang, Y., Sun, J., Huang, X., Shi, H., Xiong, C., & Nie, Z. (2019). Direct identification of forensic body fluids by MALDI-MS. *Analyst, 144*(23), 7017–7023.

Jung, M., Kristiansen, G., & Dietrich, D. (2018). DNA methylation analysis of free-circulating DNA in body fluids. *DNA Methylation Protocols* (pp. 621–641), Humana New York, NY.

Kanungo, S., Naqvi, S., Mundotiya, N., & Ahmed, U. (2023). A comprehensive study on forensic analysis of semen: Issues related to integration, quantification and quality assessment. *International Journal of Research Publication and Reviews, 6*, 3053–3058.

Livy, A., Lye, S., Jagdish, C. K., Hanis, N., Sharmila, V., Ler, L. W., & Pramod, B. (2012). Evaluation of quality of DNA extracted from buccal swabs for microarray based genotyping. *Indian Journal of Clinical Biochemistry, 27*, 28–33.

Lynch, C., & Fleming, R. (2021). RNA-based approaches for body fluid identification in forensic science. *Wiley Interdisciplinary Reviews: Forensic Science, 3*(4), e1407.

Madi, T., Balamurugan, K., Bombardi, R., Duncan, G., & McCord, B. (2012). The determination of tissue-specific DNA methylation patterns in forensic biofluids using bisulfite modification and pyrosequencing. *Electrophoresis, 33*(12), 1736–1745.

Mayes, C., Seashols-Williams, S., & Hughes-Stamm, S. (2018). A capillary electrophoresis method for identifying forensically relevant body fluids using miRNAs. *Legal Medicine, 30*, 1–4.

Mishra, A., Gondhali, U., & Choudhary, S. (2021). Potential of DNA technique-based body fluid identification. Dash, H.R., Shrivastava, P., Lorente, J.A. (Eds.), In: *Handbook of DNA Profiling* (pp. 1–17). Springer.

Mozayani, A., & Parish-Fisher, C. (2017). *Forensic Evidence Management: From the Crime Scene to the Courtroom*. CRC Press.

Nagana Gowda, G., Gowda, Y. N., & Raftery, D. (2015). Expanding the limits of human blood metabolite quantitation using NMR spectroscopy. *Analytical Chemistry, 87*(1), 706–715.

Nawrot, W., Drzozga, K., Baluta, S., Cabaj, J., & Malecha, K. (2018). A fluorescent biosensors for detection vital body fluid agents. *Sensors, 18*(8), 2357. https://www.mdpi.com/1424-8220/18/8/2357

O'Leary, K. R., & Glynn, C. L. (2018). Investigating the isolation and amplification of microRNAs for forensic body fluid identification. *MicroRNA, 7*(3), 187–194.

Pang, B., & Cheung, B. (2007). Double swab technique for collecting touched evidence. *Legal Medicine, 9*(4), 181–184.

Park, H., Park, W., & Lee, C. H. (2021). Electrochemically active materials and wearable biosensors for the in situ analysis of body fluids for human healthcare. *NPG Asia Materials, 13*(1), 23.

Park, J.-L., Kwon, O.-H., Kim, J. H., Yoo, H.-S., Lee, H.-C., Woo, K.-M., Kim, S.-Y., Lee, S.-H., & Kim, Y. S. (2014). Identification of body fluid-specific DNA methylation markers for use in forensic science. *Forensic Science International: Genetics, 13*, 147–153.

Ramesh, M., Janani, R., Deepa, C., & Rajeshkumar, L. (2022). Nanotechnology-enabled biosensors: A review of fundamentals, design principles, materials, and applications. *Biosensors, 13*(1), 40.

Roeder, A. D., & Haas, C. (2016). Body fluid identification using mRNA profiling. *Forensic DNA Typing Protocols, 1420*, 13–31.

Scientific, T. F. (2024). *Thermo Fisher Scientific. (n.d.). AmpFLSTR™ Identifiler™ Plus PCR Amplification Kit*. Retrieved October 31, from https://www.thermofisher.com/order/catalog/product/4427368

Sezgintürk, M. K. (2024). Microfluidic chip-based biosensing systems. *Microchimica Acta, 191*(5), 275.

Shrivastava, P., Kumawat, R., Kushwaha, P., & Rana, M. (2020). Biological sources of DNA: The target materials for forensic DNA typing. *Handbook of DNA Profiling* (pp. 1–17).

Sijen, T., & Harbison, S. (2021). On the identification of body fluids and tissues: A crucial link in the investigation and solution of crime. *Genes, 12*(11), 1728.

Sikirzhytskaya, A., Sikirzhytski, V., Pérez-Almodóvar, L., & Lednev, I. K. (2023). Raman spectroscopy for the identification of body fluid traces: Semen and vaginal fluid mixture. *Forensic Chemistry, 32*, 100468.

Sikirzhytski, V., Virkler, K., & Lednev, I. K. (2010). Discriminant analysis of Raman spectra for body fluid identification for forensic purposes. *Sensors, 10*(4), 2869–2884.

Silva, D. S., Antunes, J., Balamurugan, K., Duncan, G., Alho, C. S., & McCord, B. (2016). Developmental validation studies of epigenetic DNA methylation markers for the detection of blood, semen and saliva samples. *Forensic Science International: Genetics, 23*, 55–63.

Siuta, J., Dobosz, A., Kawecki, J., & Dobosz, T. (2023). DNA content of various fluids and tissues of the human body. *Genes, 15*(1), 17.

Speyer, C. B., & Baleja, J. D. (2021). Use of nuclear magnetic resonance spectroscopy in diagnosis of inborn errors of metabolism. *Emerging Topics in Life Sciences, 5*(1), 39–48.

Teoh, S. L., & Das, S. (2022). MicroRNAs in various body fluids and their importance in forensic medicine. *Mini Reviews in Medicinal Chemistry, 22*(18), 2332–2343.

Tian, H., Bai, P., Tan, Y., Li, Z., Peng, D., Xiao, X., Zhao, H., Zhou, Y., Liang, W., & Zhang, L. (2020). A new method to detect methylation profiles for forensic body fluid identification combining ARMS-PCR technique and random forest model. *Forensic Science International: Genetics, 49*, 102371.

Tian, H., Lv, M., Li, Z., Peng, D., Tan, Y., Wang, H., Li, Q., Li, F., & Liang, W. (2018). Semen-specific miRNAs: Suitable for the distinction of infertile semen in the body fluid identification? *Forensic Science International: Genetics, 33*, 161–167.

Tsuji, A., Ishiko, A., Ikeda, N., & Yamaguchi, H. (2001). Personal identification using Y-chromosomal short tandem repeats from bodily fluids mixed with semen. *The American Journal of Forensic Medicine and Pathology, 22*(3), 288–291.

Vajpayee, K., Dash, H. R., Parekh, P. B., & Shukla, R. K. (2023). PCR inhibitors and facilitators–Their role in forensic DNA analysis. *Forensic Science International, 349*, 111773.

Vajpayee, K., Desai, P., Parekh, P. B., Sitwala, H. C., Suri, K., Dash, H. R., & Shukla, R. K. (2023). A novel method for blood detection using fluorescent dye. *Microchemical Journal, 193*, 108987.

Wang, C., & Liu, H. (2022). Factors influencing degradation kinetics of mRNAs and half-lives of microRNAs, circRNAs, lncRNAs in blood in vitro using quantitative PCR. *Scientific Reports*, *12*(1), 7259.

Wang, C., Zhao, H., Meng, Q., Sun, H., Xu, X., & Li, W. (2022). Body fluid identification by messenger RNA profiling in sexual assault. *Journal of Forensic Science and Medicine*, *8*(3), 118–122.

Wang, S.-Y., Tao, R.-y, Hou, Y.-P., & Li, C.-T. (2022). Application and Prospect of RNA profiling analysis in forensic body fluid identification. *Journal of Forensic Medicine*, *38*(6), 763.

Wani, A. K., Akhtar, N., Katoch, V., Shukla, S., Kadam, U. S., & Hong, J. C. (2023). Advancing biological investigations using portable sensors for detection of sensitive samples. *Heliyon*, *9*(12), e22679.

Yadav, P. K., & Sharma, S. (2024). Role of chemometrics in forensic identification of body fluids. Praveen Kumar Yadav, Sweety Sharma (Eds.), In: *Advancements in Body Fluid Analysis in Forensics: Current and Emerging Methods* (pp. 279–301). Springer.

Yang, H., Zhou, B., Deng, H., Prinz, M., & Siegel, D. (2013). Body fluid identification by mass spectrometry. *International Journal of Legal Medicine*, *127*, 1065–1077.

Yudianto, A. (2011). DNA isolation from sweat stain in clothes as forensic identification material. *Journal of Universitas Airlangga*, *1*(1), 15–19. http://journal.unair.ac.id/download-fullpapers-01%20Ahmad%20Yudianto.pdf

Zapata, F., de la Ossa, M. F., & García-Ruiz, C. (2015). Emerging spectrometric techniques for the forensic analysis of body fluids. *TrAC Trends in Analytical Chemistry*, *64*, 53–63.

Zhang, J., Yu, D., Wang, T., Gao, N., Shi, L., Wang, Y., Huo, Y., Ji, Z., Li, J., & Zhang, X. (2024). Body fluids should be identified before estimating the time since deposition (TsD) in microbiome-based stain analyses for forensics. *Microbiology spectrum*, *12*(4), e0248023.

4

Introduction to DNA Profiling Techniques

Poonam Moon, Ashish Badiye, and Neeti Kapoor

4.1 Introduction

"DNA profiling" or "DNA typing" is a method for sequencing an individual's DNA. Although inappropriate, the term "DNA fingerprinting" is occasionally used (Gunn, 2009). It was first described by an English geneticist named Alec Jeffreys in 1985 (Butler, 2005, 2009). Previously, identity testing in forensic science utilized the ABO blood group system. Later, new markers for person identification and paternity testing were discovered using serum proteins and red blood cell enzymes. Following that, a human leukocyte antigen system was developed (Butler, 2005, 2009).

Dr. Alec Jeffreys found that in DNA sequence, certain regions of DNA are getting repeated multiple times. During his research, he recognized that the length of these repeating sequences may vary from individual to individual and are informative genetic markers. He then developed a technique for examining the length variation of the repeating sequences and used that for identity testing (Butler, 2005, 2009). He created a radioactive probe with his team, composed of brief sequences that could attach to the repeating sequence and eventually turn up patterns unique to every individual (Butler, 2005, 2009). The development of the polymerase chain reaction (PCR) in 1983 by Kary Mullis paved the revolutionary pathway in biotechnology, making DNA fingerprinting a relatively more straightforward and reproducible technique.

4.2 Forensic DNA profiling

Forensic DNA profiling involves the analysis of human DNA, which is extracted from various biological evidence/sources such as saliva, skin, hair, blood, and sperm, and compares it with the DNA samples obtained from specific individuals. The first case dates back to 1985, where DNA profiling assisted in solving an immigration dispute (Jeffreys et al., 1985).

DNA profiling plays a pivotal role in identifying perpetrators in criminal investigations such as homicides and sexual assaults. Furthermore, it is helpful in civil proceedings, especially in paternity dispute cases. Presently, DNA profiling is a cornerstone of forensic science, aiding in resolving personal identification, paternity dispute cases, tracing ancestry, and identifying the victims of disasters. Over the years, DNA databases have been developed, consisting of the genetic information of criminals, facilitating swift linkage of crimes to offenders.

DOI: 10.4324/9781003474296-4

4.2.1 Forensic DNA sampling and collection

Collecting evidence that can potentially provide DNA is a very meticulous process. To minimize the chances of contamination, the investigating officers must follow a rigorous protocol whenever handling the evidence. Some general protocols include wearing disposable gloves, using single-use instruments to handle one piece of evidence, and avoiding talking, sneezing, or coughing over potential evidence. Some common sources of DNA evidence range from weapons and clothing to personal items like hats, masks, and toothpicks, all of which may harbor skin cells, blood, saliva, or other biological material crucial for forensic analysis. The investigating officers also employ presumptive tests to determine the presence of biological material in the suspected sample. For example, the red-colored fluid may be blood, red wine, or ketchup; thus, presumptive tests aid the investigator's appropriate sample collection (Cătălin et al., 2004).

Packaging is another important aspect while dealing with sample collection. These evidences must be packed in paper bags or envelopes as the plastic bags may retain moisture, potentially damaging the DNA (Shrivastava et al., 2020). The collected biological materials are stored at room temperature and protected from direct sunlight to prevent DNA degradation.

Type of Evidence	Source of DNA
A knife used as a weapon	skin cells, blood cells, or tissue
Hat or mask	skin cells, hair
Facial tissue or cotton swab	mucus, skin cells, blood, earwax
Toothpick, cigarette butt, bottle or postage stamp	saliva
Used condom	semen or vaginal or rectal cells
Bed linen	skin cells, hair, blood, semen
Fingernail or partial fingernail	scraped-off skin cells
The swab of genitals, breasts, and mouth in a case of suspected sexual assault	saliva, semen

4.2.2 Principle of DNA profiling

The principle of DNA profiling is based on the analysis of specific regions of an individual's DNA to create a unique profile that can be used for identification purposes. All humans share 99.9% of the genes which are alike; the rest 0.1% brings about the differences and variations to differentiate one individual from the other. This 0.1% is highly *polymorphic* or variable among all. DNA profiling is based on the knowledge and recognition of genetic markers widely present in an individual's genome and inherited from generation to generation. A genetic marker, also known as a DNA marker, is a gene or DNA sequence that can be used to identify individuals or species. It can be rather long, like minisatellites, or it can be short, like the sequence surrounding a single base-pair change (single-nucleotide polymorphism, or SNP).

We know that humans have 46 chromosomes that inhabit genes and DNA, and these 46 chromosomes are present in 23 pairs. That indicates that we have a copy of each gene on the sister chromatid; this copy of gene is known as *allele*. When both alleles (copies of the gene) are alike, the individual is called *homozygous* for that particular gene; if the alleles are different, then it's called *heterozygous*. These allelic differences come in handy in DNA profiling to distinguish the band patterns on gel-electrophoresis or by southern blotting, as they are unique to an individual and cannot be similar except in the case of identical twins.

4.3 Methods of DNA extraction

DNA is present in cells, but it is not found in pure form. It is associated with proteins, lipids, and many other molecules that can interfere with DNA profiling. To extract DNA from the cell, the cell membrane is broken, and all these molecules are released. This step is called cell lysis. DNA is liberated from the nucleus upon cell lysis, and significant amounts of associated molecules are eliminated. Several methods available for DNA extractions are being implemented routinely (Primorac & Schanfield, 2023).

4.3.1 DNA isolation with organic (phenol-chloroform) solvents

For a long time, organic extraction was the most used technique for DNA extraction because it can give high molecular weight DNA, which was needed for the earlier techniques such as RFLP. Many novel methods have been devised for extraction, which utilize less hazardous substances than phenol (Butler, 2009).

The organic solvent method involves the serial addition of several chemicals, such as sodium dodecyl sulfate (SDS) and proteinase K, which help open the cell membranes and break down proteins that protect DNA molecules while they are in chromosome state, respectively. Later, phenol, chloroform, and isoamyl alcohol mixture are added in the ratio of 25:24:1. Phenol is used to separate proteins from the aqueous mixture, and chloroform is added to increase the density of the organic mix to make it easier to separate the organic layer from the aqueous layer. Isoamyl alcohol is added to the mixture to reduce the foaming. The mixture is then centrifuged to separate the unwanted proteins and cell debris from the aqueous layer. DNA is solubilized in the aqueous phase. The separated proteins will be located at the interphase of the two layers (**Figure 4.1**).

Even though this method yields high molecular weight double-stranded DNA, which can be used for RFLP or PCR-based analysis, it is time-consuming, involves the use of hazardous chemicals, and requires transfer of the sample to multiple tubes, which increases the risk of contamination (Butler, 2009).

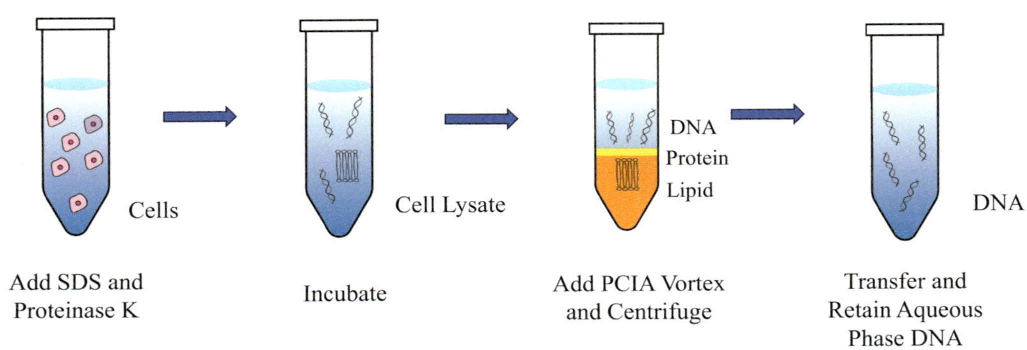

Figure 4.1 *Organic method of DNA extraction. (From* 7 Different DNA Extraction Methods - praxilabs. *(n.d.). Retrieved February 16, 2025, from https://praxilabs.com/en/blog/2022/08/24/7-different-dna-extraction-methods/)*

4.3.2 Chelex extraction method

The forensic DNA community was introduced to Chelex 100 (Bio-Rad) in 1991. This was an alternative and inexpensive method for DNA extraction, making it more popular among forensic scientists. In this method, the chelating resin suspension can be directly added to the sample (e.g., blood, saliva, semen, etc.) and heated. After heating, the proteins and fats are denatured, and the DNA is released into the solution. The Chelex comprises styrene-divinylbenzene copolymers with paired iminodiacetate ions that serve as chelating groups to bind polyvalent metal ions like magnesium. Magnesium is the cofactor of endogenous DNase isolated by the Chelex solution, thus protecting the DNA from degradation. In the next step, the mixture is centrifuged to separate all the molecules associated with Chelex, and the pure DNA remains in the aqueous solution. Magnesium is an essential cofactor for the DNA polymerase used for amplification; hence, it is suggested to avoid transferring the Chelex resin into the DNA amplification solutions (**Figure 4.2**).

The Chelex process is quick, easy, and requires just one extraction tube. But, it denatures the double-stranded DNA and yields only single-stranded DNA, which can only be used for PCR-based techniques and not for RFLP (Butler, 2009; Primorac & Schanfield, 2023).

4.3.3 FTA paper

Another approach for DNA extraction is using the FTA paper. LEE Burgoyne from Flinders University in Australia created this FTA paper for DNA sample storage. FTA originally stands for "Fitzco/Flinder Technology Agreement." In the FTA method, a cellulose-based matrix (filter paper) is treated with a weak base, chelating agent, anionic surfactant or detergent, and uric acid. These compounds protect DNA molecules from nuclease breakdown and the paper against bacterial growth. DNA on the FTA paper remains stable at room temperature for several years.

The procedure consists of applying a spot of blood on the paper and then allowing the stain to dry. When the cells come in contact with the paper, their membrane gets lysed, and the DNA in the WBCs gets trapped in the paper matrix. To separate the DNA, a small punch of the paper containing the bloodstain is removed from the FTA paper and placed in the tube for washing. FTA purification reagents are added to the tube and washed to remove the heme and other inhibitors of the PCR reaction. It can be visually seen that the red color is removed with the

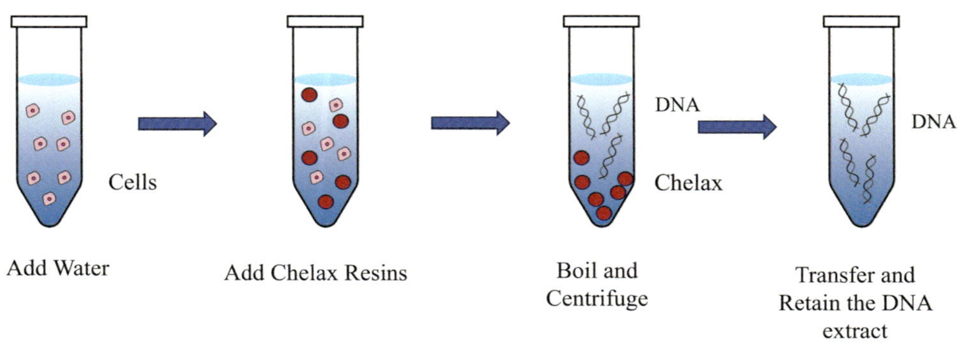

Figure 4.2 *Chelex extraction method. (From* DNA Extraction: Steps and Methods Involved. *(n.d.). Retrieved February 16, 2025, from https://www.sciencedoze.com/2023/12/dna-extraction-steps-and-methods.html#google_vignette)*

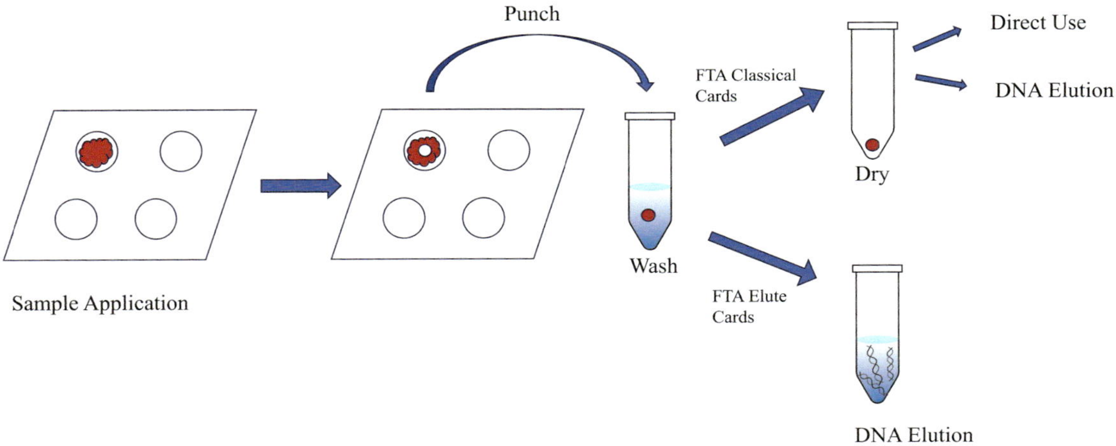

Figure 4.3 *FTA paper-based DNA extraction method. (From Barbosa, C., Nogueira, S., Gadanho, M., & Chaves, S. (2016). DNA extraction: Finding the most suitable method. In N. Cook, M. D'Agostino, & K. C. Thompson (Eds.),* Molecular Microbial Diagnostic Methods: Pathways to Implementation for the Food and Water Industries *(pp. 135–154). https://doi.org/10.1016/B978-0-12-416999-9.00007-1)*

supernatant. At last, the clean punch can be directly added to the PCR reaction (**Figure 4.3**) (Butler, 2009; Primorac & Schanfield, 2023).

4.3.4 Solid-Phase extraction method

Many labs have adapted to solid-phase extraction in an attempt to automate more steps in the DNA analysis process. Solid-phase extraction involves selectively binding DNA to a substrate, such as silica particles, and then releasing the attached DNA molecules following vigorous washes that separate them from proteins and other biological components. Various methods have been developed by using the principle of solid phase extraction. **Table 4.1** shows a comparison of different commercial kits used for DNA extraction which are based on solid-phase extraction methods (Butler, 2009). These includes Qiagen columns, DNA IQ, and PrepFiler.

The solid phase extraction method works on the principle that DNA molecules are reversibly adsorbed on silica (silicon dioxide, SiO_2) in the presence of high concentrations of chaotropic salts. Originally, these chaotropic salts were used to disrupt the three-dimensional structure of proteins. However, its involvement in DNA adsorption to silica remains unclear. Some assume that chaotropic salts' dehydrating actions produce DNA adsorption on silica. Following the dehydration of DNA's phosphodiester backbone, the phosphate residues become available for interaction with the silica surface. Guanidinium salts, such as GuSCN and GuHCl, are popular

Table 4.1 Comparison of Key Features of Different Solid-Phase Extraction Methods

	QIAGEN	Promega	Applied Biosystems
Products	QIAamp	DNA IQ	PrepFiler
Solid-phase	Silica beads	Magnetic silica beads	Magnetic silica beads
Wash performed	Centrifugation or vacuum manifold	Magnet	Magnet
Robotic platform	EZ1, M48, QIACube	Maxwell 16, Tecan	AutoMAte Express, Tecan

Source: Adaptation from Butler (2011).

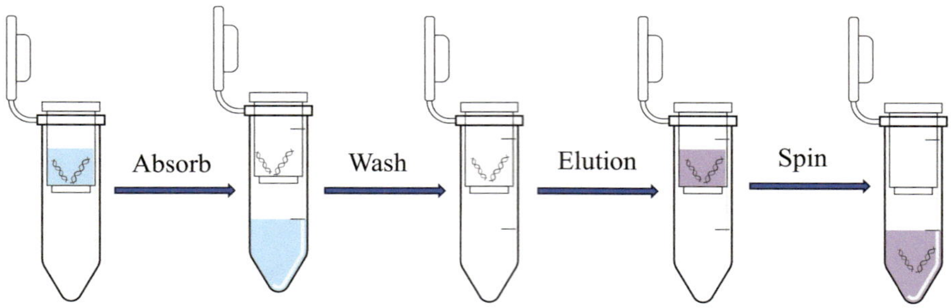

Figure 4.4 *Solid-phase extraction method. (From* CD Bioparticles Launches Silica Spin Column for Nucleic Acid Extraction. *(n.d.). Retrieved February 16, 2025, from https://www.conchovalleyhomepage.com/business/press-releases/ein-presswire/777760049/cd-bioparticles-launches-silica-spin-column-for-nucleic-acid-extraction/)*

chaotropic salts used for DNA extraction. GuSCN is more powerful and promotes cell lysis. The steps involved in the process are as follows (**Figure 4.4**):

- **Cell lysis and Protein Digestion:** Proteinase K is added to break open the cell membrane and to digest the proteins associated with the DNA molecule.
- **DNA Adsorption onto Silica:** This step involves a column with silica as a stationary phase, which adsorbs the DNA in the cell lysate. This happens in the presence of high concentrations of chaotropic salts. In this process, the cell debris and other contaminants that may interfere with the DNA amplification process are not retained in the column. The double-stranded DNA is adsorbed on the silica column.
- **Washing:** In this stage, an ethanol-based wash removes chaotropic chemicals and other impurities from the column. This solution does not elute the adsorbable DNA molecule.
- **Elution of DNA:** DNA elution occurs when DNA is rehydrated using aqueous low salt solutions. The eluted DNA is double-stranded and can be utilized in both RFLP and PCR-based reactions (Butler, 2009).

4.4 DNA quantitation

The quantity and quality of DNA can only be estimated after it has been isolated properly. For majority of PCR-based assays, a specific concentration range of DNA is needed for the optimal results. Therefore, determining the exact amount of DNA in sample is crucial. Excess DNA can cause split or off-scale peaks in the measuring method, and too little DNA may not amplify the DNA, leading to allele "drop-out.". This occurrence is referred as stochastic fluctuation(Butler, 2005). A variety of DNA quantitation methods have been developed over the years to estimate the amount of human or total DNA contained in a sample. Some DNA quantitation techniques have been discussed below, although most forensic DNA quantitation is carried out using commercial kits (Butler, 2005, 2009).

4.4.1 Slot blot assay

Human genomic DNA can be detected in a sample by using slot blot technique. In this method, the DNA sample is first denatured, and then a small volume of the sample is potted onto a nitrocellulose membrane, immobilized. The target sequence is revealed by hybridizing a prime-specific α-satellite DNA sequence at the D17Z1 locus with a tagged 40-nucleotide probe complementary. The slot blot assay can be detected by using three techniques. The first one is using radioisotopes. Radioisotopes might be visualized by exposing the membrane to an X-ray film. The second approach employs an alkaline phosphatase-labeled technique. It can work in conjunction with chemiluminescent detection. The third type is biotinylated probes, which can be used for either colorimetric or chemiluminescent detection. The intensity of the observed signal is proportional to the sample's DNA concentration. By comparing the unknown sample to a set of standards with known concentration, we can determine the quantity of DNA in the given sample (Butler, 2009).

4.4.2 Fluorescent interchelating dye assay

Fluorescent interchelating dye assay is another technique that can be used to estimate small quantities of DNA. This method quantifies the double-stranded DNA using the QuantiTTM PicoGreen reagent, a molecular probe inter-chelating dye. It has a detection limit up to 250 pg. Because this dye is not specific to human DNA, it is typically utilized to quantify known reference samples. This test has been modified for automation, which makes it a high-throughput technique.

The process involves adding the DNA sample to a fluorescent inter-chelating dye solution. The fluorescence is detected using a typical spectrofluorometer with excitation and emission wavelengths. At first, the standard curve is created using samples of known DNA concentrations, and then the tests are subsequently conducted for the unknown samples. Finally, the results of unknown DNA samples are compared to the standard curve (Butler, 2005, 2009).

4.4.3 "Yield" gel method

Yield gel method for DNA quantitation was the first non-spectroscopic technique employed in forensic science. In this method, the purified DNA is added to a gel, often agarose, electrophoresed, and then compared to known-size pieces of high-molecular-weight DNA, which were also tested on the same gel. This method quantifies the entire DNA in a given sample. It was primarily useful for RFLP testing and has minimal utility for PCR-based testing. As a result, its use in contemporary forensic DNA analysis is severely constrained and overshadowed by more recent techniques (Butler, 2005).

4.4.4 End-point PCR

According to the PCR amplification principle, the amount of PCR amplified product correlates with the starting DNA concentration. Thus, it is possible to ascertain a sample's DNA concentration. In the end-point PCR method, the amount of amplified synthesized product is measured at the end of the reaction. Typically, the dyes that intercalate into the double-stranded DNA produce fluorescence. The amount of fluorescence released by dyes like SYBR allows for the detection of the amplified DNA yield. This method detects the quantity of PCR inhibitors in the sample, which makes it a functional test (Butler, 2005, 2009).

4.4.5 Real-time PCR

The primary goal of DNA quantification is to determine the quantity of "amplifiable DNA" in the sample. The presence of coextracted inhibitors, denatured DNA, insufficient DNA, or the combination of all these can make the PCR amplification reactions unsuccessful. Therefore, it is valuable to use a test that can reliably reflect the amount and quality of the DNA template present in an extracted sample when deciding how to progress. Such an evaluation is provided by "real-time" PCR tests. It monitors how a fluorescence signal changes from cycle to cycle because a target sequence is amplified during PCR. A fluorescence reporter monitors the PCR, which increases as the product accumulates with each round of amplification (Butler, 2009).

4.5 Processes of DNA fingerprinting

4.5.1 AFLP

AFLP denotes Amplified Fragment Length Polymorphism. This technique was initially described by Vos et al. (1995), and is a PCR-based DNA fingerprinting method. It allows the analysis of multiple DNA fragments for length differences in a single reaction. The significant advantage of this technique is that it does not require any prior sequence information of the DNA and provides highly reproducible banding patterns. Although the name suggests length polymorphism, the resulting bands on electrophoresis are present-absent polymorphism (**Figure 4.5**).

The technique consists of two or more restriction enzymes that digest/cleave the sample DNA and ligate restriction-site-specific adapters. The reaction thus produced is then

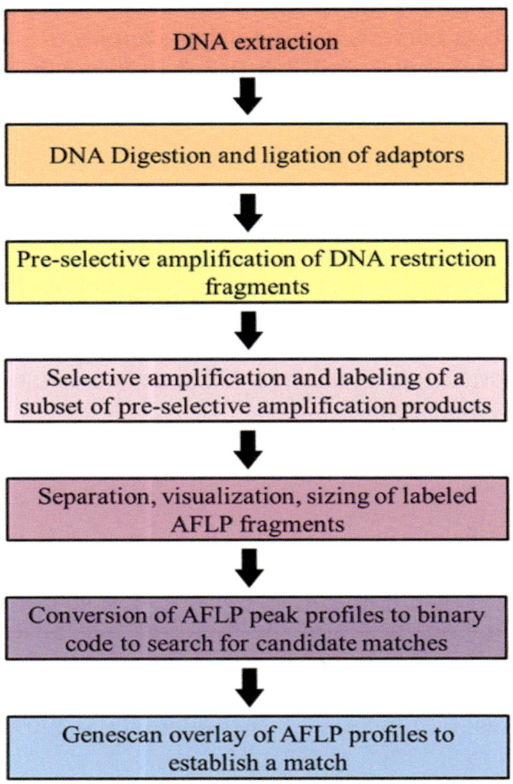

Figure 4.5 *Steps of amplified fragment length polymorphism (AFLP). (From Coyle, H. M., Germano-Presby, J., Ladd, C., Palmbach, T., & Lee, H. C. (2002). Tracking clonal marijuana using amplified fragment length polymorphism (AFLP) analysis: an overview. In* Proceedings of the 13th International Symposium on Human Identification *(pp. 7–13).)*

subjected to PCR with primers designed to complement the earlier selected restriction enzymes.

The procedure consists of the following three steps:

i. Use restriction enzymes to digest whole cellular DNA, then ligate half-site-specific adaptors to all restriction fragments.
ii. Use adapter and restriction site-specific PCR primers to amplify specific segments selectively.
iii. Separating amplicons using electrophoresis on a gel matrix and seeing the resulting band pattern.

AFLP allows the rapid scanning of many different subsets of DNA fragments and is a highly sensitive method for detecting polymorphisms in DNA. This technique has shortcomings with DNA samples where there are low sequence variations.

4.5.2 RFLP

Restriction Fragment Length Polymorphism (RFLP) is one of the oldest and reliable techniques for DNA fingerprinting. It consists of two major components: the restriction endonucleases (REs) (a class of restriction enzymes) and a DNA probe. With the advent of PCR, the tedious process of hybridization by southern blotting is made much easier and is no longer required to analyze the results of RFLP.

The procedure requires the isolation of genomic DNA, followed by PCR, the digestion of genomic DNA with specific REs, and the cleavage of DNA at a specific target site, producing a particular fragmentation pattern and yielding a series of fragments. Gel electrophoresis separates restriction fragments of varied lengths, which are then visualized for analysis. Changes in the DNA sequence linked with an allelic alteration will be visible as a shift in the location of the agarose-gel band. Individuals with various allelic gene variations will have distinct band distribution patterns. The differences in gene fragments in alleles are due to naturally occurring mutations in the genome, causing base substitutions, additions, and deletions. The less related the individuals, the more divergent their DNA. Thus, by employing restriction enzymes to cleave DNA at precise sites and detecting fragments with restriction probes, RFLP enables the comparative analysis of restriction fragments across individual organisms and species (**Figure 4.6**).

The steps involved in PCR-based RFLP are as follows:

i. Sample collection
ii. Isolation of genomic DNA
iii. PCR amplification of a target gene
iv. Restriction endonuclease digestion of the PCR-amplified product
v. Agarose gel-electrophoresis
vi. Visualization of the gel and analysis

4.6 VNTR

A Variable Number Tandem Repeat (or VNTR), also known as minisatellite marker or Hyper Variable Regions (HVRs), is located in a genome where a short nucleotide sequence is organized as a tandem repeat. These markers, located near the Minisatellites, are a type of DNA

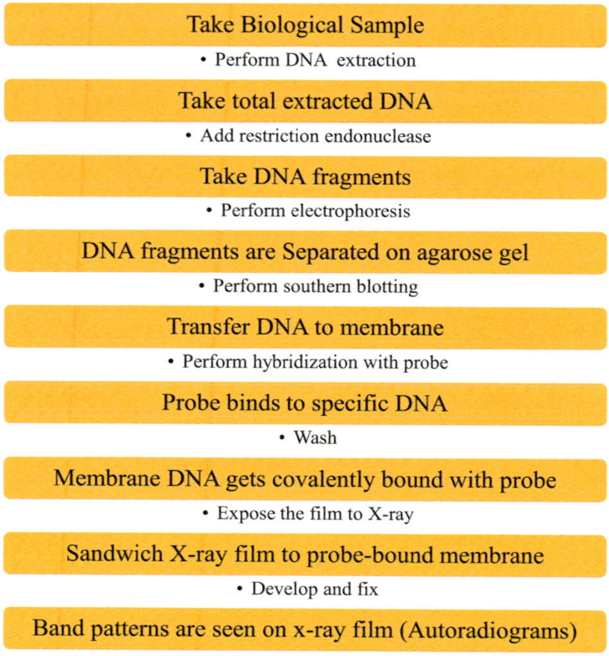

Figure 4.6 *Steps of restriction fragment length polymorphism (RFLP).*

tandem repeat sequence, meaning that the sequences repeat one after another without other sequences or nucleotides in between them. Minisatellites are characterized by a repeat sequence of about 10 to 100 nucleotides, and the number of times the sequence repeats varies from about 5 to 50 times. VNTRs are widely present throughout the genome and in varied lengths (in terms of the number of nucleotide bases) among individuals. These are highly heritable, making them an ideal marker for identifying individual and paternity or kinships. The selection of VNTR loci for forensic analysis considers their compatibility with restriction endonuclease cleavage sites, with many chosen loci being highly polymorphic. The usefulness of VNTR loci in forensic testing is evaluated through their population match probability (Pm), where lower values signify a decreased likelihood of a chance match between individuals (Li, 2025). Due to their high polymorphic content and multi-allelic nature, VNTR is a helpful tool in linkage analysis, forensic identification, paternity testing, and population genetic studies (Pakzad et al., 2014). The limitations include the requirement for a fairly large quantity of DNA, the intricacy of the procedure with degraded DNA, and higher time consumption (**Figure 4.7**).

4.7 STR

Short Tandem Repeats (STRs), also known as microsatellites and sometimes called simple sequence repeats (SSRs), consist of a specific shorter sequence of DNA units or nucleotide bases (A, T, G, C) repeated several times. The repeating sequences, or motifs, in microsatellites can contain up to two to six bases, and each sequence can be repeated about 50 to 100 times. Like VNTRs, microsatellites are highly polymorphic (meaning they exhibit high variation in a

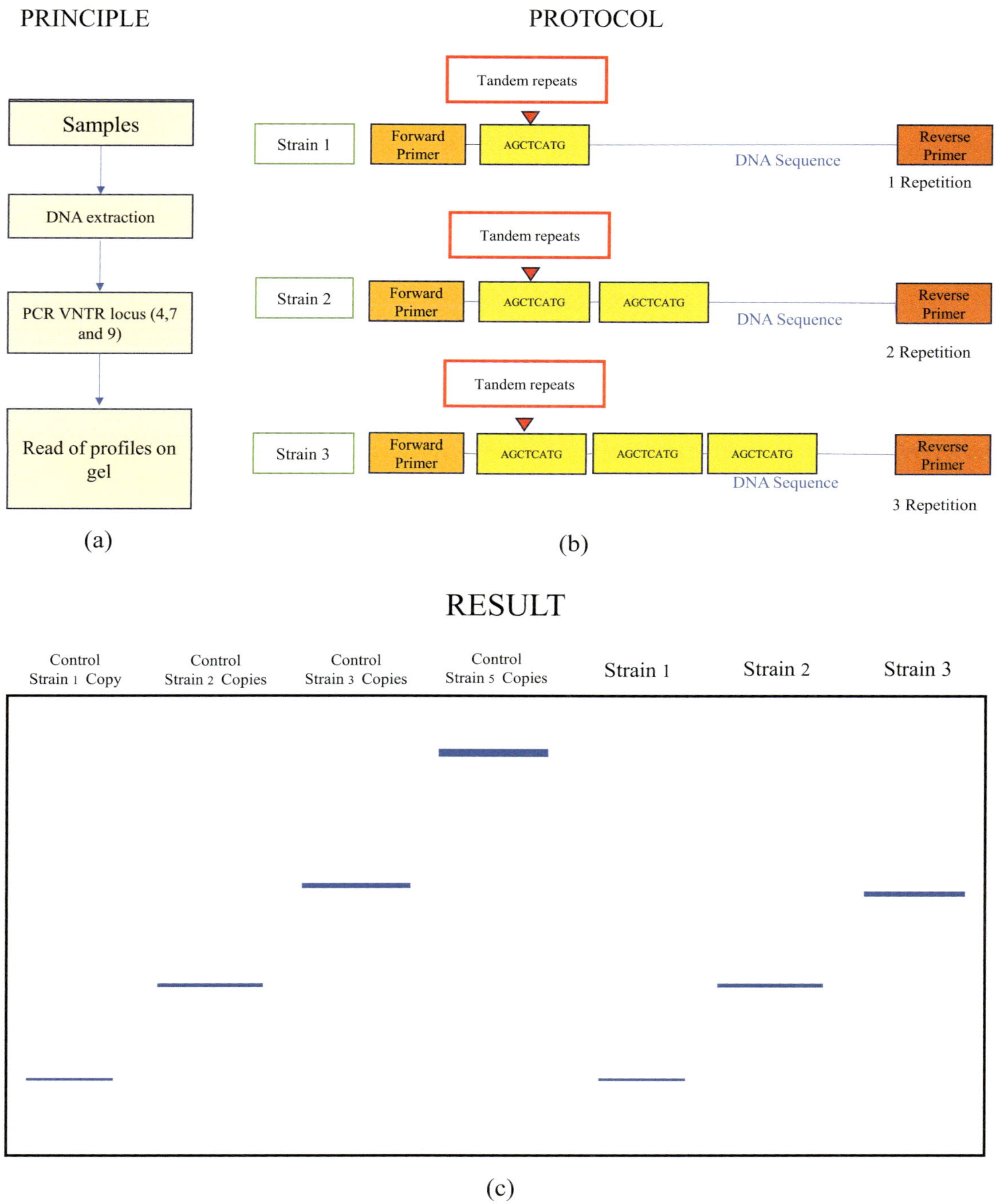

***Figures 4.7** Figures (a), (b), and (c) show the principle, protocol, and result of the VNTR analysis. (From Marquez, A., Djelouadji, Z., Lattard, V., & Kodjo, A. (2017).* Overview of laboratory methods to diagnose leptospirosis and to identify and to type leptospires. *International Microbiology, 20(4), 184–193. https://doi.org/10.2436/20.1501.01.302)*

population) and widely distributed in the genome. The enormous variation conveyed by microsatellite repeats results in repeats being especially prone to DNA replication errors, often through slipped-strand mispairing (Moxon & Wills, 1999). Despite the availability of other DNA markers, such as VNTRs, STR typing has become the preferred method in most forensic laboratories due to its high discriminatory power, ease of analysis, and compatibility with PCR-based methodologies (Keerti & Ninave, 2022).

STRs can be further classified as:

i. *Autosomal STRs*
The term autosomal chromosome refers to the non-sex chromosomes present in all species, i.e., except X and Y; all other 22 chromosome pairs are autosomal chromosomes in humans. Autosomal STRs are mostly present in the non-coding regions. Each individual presents a single copy from both parents; hence, autosomal STRs are most significant in identifying individual, paternity, and kinship. The autosomal STR markers are commonly chosen due to their high discrimination potential; however, they are not ideal for rape cases where more than one male accused is involved (Dixit et al. 2023).

ii. *Y-chromosome linked STRs*
Y-chromosomal genes are specific to human males and thus are of utmost importance in identifying biological son and father disputes and paternity testing in males because this chromosome can only be inherited from father to son in the population. Genetic markers present on these chromosomes are also known as holandric. Y-STRs in forensics become especially crucial when the autosomal STRs are less informative.

iii. *X-chromosome linked STRs*
X-chromosome STRs are known to be more stable than the previously discussed two. Human female individuals possess two X chromosomes, whereas males typically have only one X chromosome and one Y chromosome. X-STRs are extremely useful in missing person identification, cases involving at least one female individual, and cases of deficiency paternity and immigration personnel identification.

Table 4.2 shows some common STR loci characterized for forensic DNA profiling.

Table 4.2 Common STR Loci for Forensic Analysis

SrNo.	Locus	Repeat Motif	Repeat Category	Chromosome Location	Physical Position	Structural Gene
1.	ACTBP2 /SE33	AAAG	Complex	6q14	Chr 6 89.043 Mb	β-actin related pseudogene
2.	CSF1PO	TAGA	Simple	5q33.1	Chr 5 149.436 Mb	c-fms protooncogene, intron 6
3.	D2S1338	[TGCC] [TTCC]	Compound	2q35	Chr 2 218.705 Mb	Anonymous
4.	D3S1358	[TCTG] [TCTA]	Compound	3p21.31	Chr 3 45.557 Mb	Anonymous
5.	D5S818	AGAT	Simple	5q23.2	Chr 5 123.139 Mb	Anonymous

(Continued)

Table 4.2 *(Continued)* Common STR Loci for Forensic Analysis

SrNo.	Locus	Repeat Motif	Repeat Category	Chromosome Location	Physical Position	Structural Gene
6.	D7S820	GATA	Simple	7q21.11	Chr 7 83.433 Mb	Anonymous
7.	D8S1179	[TCTA] [TCTG]	Compound	8q24.13	Chr 8 125.976 Mb	Anonymous
8.	D13S317	TATC	Simple	13q31.1	Chr13 81.620 Mb	Anonymous
9.	D16S539	GATA	Simple	16q24.1	Chr 16 84.944 Mb	Anonymous
10.	D18S51	AGAA	Simple	18q21.33	Chr 18 59.100 Mb	Anonymous
11.	D19S433	AAGG	Simple	19q12	Chr 19 35.109 Mb	Anonymous
12.	D21S11	[TCTA] [TCTG]	Complex	21q21.1	Chr 21 19.476 Mb	Anonymous
13.	FGA	CTTT	Simple	4q31.3	Chr 4 155.866 Mb	α-fibrinogen, intron 3
14.	Penta D	AAAGA	Simple	21q22.3	Chr 21 43.880 Mb	Anonymous
15.	Penta E	AAAGA	Simple	15q26.2	Chr 15 95.175 Mb	Anonymous
16.	TH01	TCAT	Simple	11p15.5	Chr 2 2.149 Mb	Tyrosine hydroxylase, intron 1
17.	TPOX	GAAT	Simple	2p25.3	Chr 2 1.472 Mb	Thyroid peroxidase, intron 10
18.	VWA	[TCTG] [TCTA]	Compound	12p13.31	Chr 12 5.963 Mb	von Willebrand factor, intron 40

Source: Adaptation from *Forensic Biology* Li (2011).

Note: Chromosomal location is based on the cytogenetic map, and physical position is based on the DNA sequence (Mb = megabase). Chr = chromosome.

4.8 Mitochondrial DNA (mt-DNA) analysis in DNA profiling

Mitochondrial DNA is maternal in inheritance; hence is of extreme importance when maternal parentage is disputed or in question. These are highly polymorphic in nature. Every bodily cell contains mitochondria; thus, they are abundant. Identifying ancient discovered entities and even severely old and degraded samples present mitochondrial DNA, making it one of the most important genetic markers. Unlike STR analysis, mtDNA sequence refers to a group of individuals in the same maternal lineage, passed down from mother to child, rather than an individual. MtDNA sequencing is an additional important tool for identifying biological evidence. While its capacity to identify an individual is limited due to the lack of recombination, its high copy number provides substantial advantages when confirming maternal lineage or where nuclear DNA is limited, such as in bone, tooth, and hair studies. Understandably, it is frequently utilized in the analysis of ancient DNA and in disaster victim identification (DVI) (Court, 2021). **Table 4.3** shows the comparison of human nuclear DNA and mitochondrial DNA.

Table 4.3 Comparison of Human Nuclear DNA and Mitochondrial DNA Markers

Characteristics	Nuclear DNA	Mitochondrial DNA (mtDNA)
Size of genome	~3.2 billion bp	~16,569bp
Copies per cell	2 (1 allele from each parent)	Can be 1000
Percent of total DNA	99.75%	0.25% content per cell
Structure	Linear; packaged in chromosomes	Circular
Inherited from	Father and mother	Mother
Chromosomal Pairing	Diploid	Haploid
Generational recombination	Yes	No
Replication repair	Yes	No
Unique	Unique to individual (except identical twins)	Not unique to individual (same as maternal relatives)
Mutation rate	Low	At least 5–10 times nuclear DNA
Reference sequence	Described in 2001 by the Human Genome Project	Described in 1981 by Anderson and coworkers

Source: Adaptation from John M. Butler (2009).

4.9 SNPs

A single-nucleotide polymorphism, abbreviated as SNP (often pronounced as *snip*), is a variation in a single nucleotide that occurs at a specific position in the genome, where each variation is present to some appreciable degree within a population; these are also the most abundant genetic markers available. For example, at a specific base position in the human genome, the C nucleotide may appear in most individuals, but in a minority, the position is occupied by an A. This means that there is an SNP at this specific position, and the two possible nucleotide variations – C or A – are said to be alleles for this position. SNPs can be synonymous or non-synonymous, meaning that they are present in both coding and non-coding regions in the genome. Although most abundantly, SNPs present limitations due to their limited polymorphic nature compared to STRs, requiring many SNPs to avail conclusive results, and the most evident limitation being that only two alleles are present for a given SNP.

In some forensic cases, the biological samples do not yield to the examination even after the STR screening because of too little DNA or denatured samples. In such cases, SNP markers can be used to analyze the genetic data from degraded samples (Sinha & Arjun Rao, 2020). **Table 4.4** shows examples of forensic applications of SNP profiling (**Table 4.5**).

Table 4.4 Forensic Applications of SNP Profiling

Testing	Candidate SNP Loci	Application	References
Identity	Autosomal SNPs	Human identification	Kidd et al. (2006) and Sanchez et al. (2006)
		Human identification via degraded DNA	Budowle (2004)
	mtDNA SNPs	Human identification	Grignani et al. (2006)
	Y chromosome SNPs	Paternity testing	Hammer et al. (2006)
Biogeographical origin	Ancestry informative markers (AIMs)	Ethnic group identification	Frudakis et al. (2003) and Shriver and Kittles (2004)

(Continued)

Table 4.4 *(Continued)* Forensic Applications of SNP Profiling

Testing	Candidate SNP Loci	Application	References
Physical characteristics	MC1R (melanocortin 1 receptor gene)	Hair color identification (investigative lead)	Grimes et al. (2001)
	P (gene has role in pigmentation)	Eye color identification (investigative lead)	(*P Gene as an Inherited Biomarker of Human Eye Color) | Request PDF* (n.d.)
Pathology	KCNH2 (cardiac potassium channel gene)	Determining cause of sudden death from cardiac arrhythmia, long QT syndrome	Lunetta et al. (2002)
	SCN5A (encodes cardiac sodium channel gene)	Determining cause of sudden death from cardiac arrhythmia long QT syndrome	Burke et al. (2005)
Toxicology	CYP2C19, CYP2D6, CYP3A4, CYP2E1 (drug-metabolizing enzyme genes)	Investigation of drug overdose (including death)	Kupiec et al. (2006)

Source: Adaptation from Forensic Biology Li (2011).

Table 4.5 Comprehensive Comparison of DNA Profiling Techniques, Highlighting Their Variations, Benefits, and Drawbacks Spanning the Historical to the Contemporary Era

Analysis Technique	Basis of Differentiation	Advantages	Disadvantages
Restriction Fragment Length Polymorphism (RFLP)	Restriction site sequence and fragment length	• High power of discrimination • Reproducible • No prior sequence information is required • Can differentiate between homozygotes and heterozygotes	• Time-consuming • Partial digests • Need at least 10–25 ng of DNA • Genetic mutations are only identified at restriction-cut sites • It is not ideal for whole genome variation identification • Requires radioisotopes
Short Tandem Repeat (STR)	STR fragment length	• Fast • Highly reproducible • High level of discrimination, codominant alleles • Standardized across forensic laboratories • Uses low DNA amounts for amplification • Database of genetic profiles and allelic frequencies for statistical comparisons	• Mixture deconvolution not easy • PCR artefacts can complicate results • Challenges with highly degraded or low template DNA
Sanger Sequencing	Sequences every base	• Gold standard for sequence analysis • Uses capillary electrophoresis techniques	• Low throughput • Only 500–700 bases sequence at a time • Cannot sequence mixtures without cloning
SNaPshot™	Single base changes	• Detects bi-allelic and multi-allelic SNP markers • Able to distinguish between heterozygotes and homozygotes • Human SNP database for statistical comparisons	• Time-consuming • Need to know SNP sequence in advance to design • Primers • Multiple markers are required for high levels of discrimination
Next-Generation Sequencing (NGS)	Massive parallel sequencing using various technologies	• High throughput • Deconvolve mixture • Sequence entire genomes/metagenomes • Simultaneous detection of STR amplicon lengths and SNPs within the amplicon • Used for any DNA (human, non-human, viral, microbes)	• Massive data output that may be challenging to analyze • Analysis algorithms not standardized • Difficult with some technologies to analyze metagenomes to species level

Source: Adaptation from Alketbi (2023).

4.10 Advantages of DNA profiling

DNA profiling has transformed the field of forensic science by offering such an invaluable technique for person identification and criminal investigation. DNA profiling has numerous advantages in forensic science. DNA profiling focuses on genetic uniqueness, thus helping to identify the person solely through their DNA. DNA profiling has high accuracy in identifying the criminal, as there are very minute chances that two different persons may have the same DNA profile; thus, it has higher evidentiary value (Roman et al., 2008). DNA profiling enhances the ability to link suspects to the crime scene (Panneerchelvam & Norazmi, 2003). This technique can be instrumental in resolving old cases (Kashyap et al., 2004). With time, DNA profiling techniques have shown improvements, being fast and sensitive for identifying forensic samples.

4.11 Disadvantages of DNA profiling

In instances where the amount of DNA is very low, the peaks may be very small or even disappear entirely. At the same time, small blips in the data can be misinterpreted as real peaks. This makes interpretation of results challenging (*DNA Mixtures: A Forensic Science Explainer | NIST*, n.d.). Inherent biasing errors can arise while comparing the sample with the suspect's profile. A place where the victim and assailant both might have touched, a hat which had been worn by 2–3 people, a weapon handled by two or more would result in mixture of DNA samples, and interpretation of mixed DNA samples are challenging and requires advanced laboratory technology and biostatistical software (Barbaro et al., 2004; Hu et al., 2014). There are ethical issues regarding an individual's personal information.

4.12 Application of emerging technologies of DNA typing in forensic investigation

4.12.1 Next-generation sequencing (NGS)

Due to its massive parallel sequencing strategy, NGS technologies can potentially revolutionize DNA sequencing. It facilitates the analysis of very short reads in a cost-effective and time-efficient manner (Yang et al., 2014). The detailed sequence analysis, along with the database, allows for a better interpretation of the mixture of DNA. NSG has been applied to human, animal, and plant biotechnology (Gupta & Gupta, 2013). NSG provides a real-time sequencing platform to solve criminal and civil cases and mass disaster cases (R et al., 2018).

4.12.2 Single-molecule sequencing

The single-molecule sequencing technique facilitates direct sequencing of DNA molecules in real-time. It does not require use of PCR to amplify the DNA sample and thus minimizes human errors. It can be used to analyze DNA mixtures. It can provide significant information regarding the inheritance patterns of genetic variations to reconstruct familial relationships and ancestral lineages. It is highly sensitive in identifying phenotypic characteristics of an individual, genetic diseases, etc.

4.12.3 Microfluidic-based DNA analysis

Microfluidic devices provide a significant advantage while dealing with forensic DNA analysis. These miniaturized devices require a minute quantity of biological samples, quickly producing results. Microfluidic devices have higher sensitivity and precise control over experimental conditions, which enhances the accuracy and reproducibility of DNA typing assays.

4.12.4 CRISPR-based technologies

CRISPR-based technologies are becoming popular in the field of forensic science. It is because CRISPR-based technologies can aid in the development of highly specific and sensitive assays for DNA profiling. Target-specific genetic markers can be made by programming CRISPR-Cas systems. These could be the commonly used genetic markers like SNPs or STRs. CRISPR-based method facilitates multiplexing and simultaneous analysis of multiple loci, enabling efficient processing of forensic samples. CRISPR-based methods can also be used to amplify and detect the targeted DNA sequence, even in degraded samples.

4.13 Conclusion

DNA profiling is a fundamental technique that has a significant role in forensic science. Various DNA profiling techniques have been developed over the years. These techniques facilitate the identification of a person, detection of any genetic disorder, aids in paternity dispute cases, etc. The DNA sequences provide a distinctive genetic profile. Over the years, DNA profiling has become a gold standard in forensic science. With more progression and advanced research continuing in the near future, it will continue to be the heartthrob of forensics.

References

7 Different DNA Extraction Methods - praxilabs. (n.d.). Retrieved February 16, 2025, from https://praxilabs.com/en/blog/2022/08/24/7-different-dna-extraction-methods/

Alketbi, S. K. (2023). The role of DNA in forensic science: A comprehensive review. *International Journal of Science and Research Archive*, *9*(2), 814–829. https://doi.org/10.30574/IJSRA.2023.9.2.0624

Barbaro, A., Cormaci, P., & Barbaro, A. (2004). DNA analysis from mixed biological materials. *Forensic Science International*, *146*(SUPPL.). https://doi.org/10.1016/j.forsciint.2004.09.037

Barbosa, C., Nogueira, S., Gadanho, M., & Chaves, S. (2016). DNA extraction: Finding the most suitable method. In N. Cook, M. D'Agostino, & K. C. Thompson (Eds.), *Molecular Microbial Diagnostic Methods: Pathways to Implementation for the Food and Water Industries* (pp. 135–154). https://doi.org/10.1016/B978-0-12-416999-9.00007-1

Budowle, B. (2004). SNP typing strategies. *Forensic Science International*, *146*(SUPPL.). https://doi.org/10.1016/J.FORSCIINT.2004.09.042

Burke, A., Creighton, W., Mont, E., Li, L., Hogan, S., Kutys, R., Fowler, D., & Virmani, R. (2005). Role of SCN5A Y1102 polymorphism in sudden cardiac death in blacks. *Circulation*, *112*(6), 798–802. https://doi.org/10.1161/CIRCULATIONAHA.104.482760

Butler, J. M. (2005). *Forensic DNA Typing: Biology Technology and Genetics of STR Markers*. 2nd Edition, Elsevier Academic Press, New York.

Butler, J. M. (2009). Fundamentals of forensic DNA typing john. *Angewandte Chemie International Edition*, *6*(11), 951–952.

Butler, J. M. (2011). Advanced topics in forensic DNA typing: Methodology. In *Advanced Topics in Forensic DNA Typing: Methodology*. https://doi.org/10.1016/C2011-0-04189-3

Cătălin, M., Andrei, A., & Mitraşca, O. (2004). Modern Methods of Collection and Preservation of Biological Evidence for Human Identification by DNA Analysis. *Biochemistry Department, University of Medicine and Pharmacy from Timisoara.*, *1*.

CD Bioparticles Launches Silica Spin Column for Nucleic Acid Extraction. (n.d.). Retrieved February 16, 2025, from https://www.conchovalleyhomepage.com/business/press-releases/ein-presswire/777760049/cd-bioparticles-launches-silica-spin-column-for-nucleic-acid-extraction/

Court, D. S. (2021). Mitochondrial DNA in forensic use. *Emerging Topics in Life Sciences*, *5*(3), 415. https://doi.org/10.1042/ETLS20210204

Coyle, H. M., Germano-Presby, J., Ladd, C., Palmbach, T., & Lee, H. C. (2002). Tracking clonal marijuana using amplified fragment length polymorphism (AFLP) analysis: An overview. In *Proceedings of the 13th International Symposium on Human Identification* (pp. 7–13).

Dixit, S., Rana, M., Kushwaha, P., Kumar, A., Dev, K., & Shrivastava, P. (2023). DNA and its significance in forensic science. In P. Shrivastava, J. A. Lorente, A. Srivastava, A. Badiye, & N. Kapoor (Eds.), *Textbook of Forensic Science*. Springer, Singapore. https://doi.org/10.1007/978-981-99-1377-0_19

DNA Extraction: Steps and Methods Involved. (n.d.). Retrieved February 16, 2025, from https://www.sciencedoze.com/2023/12/dna-extraction-steps-and-methods.html#google_vignette

DNA Mixtures: A Forensic Science Explainer | NIST. (n.d.). Retrieved February 22, 2025, from https://www.nist.gov/feature-stories/dna-mixtures-forensic-science-explainer

Forensic Biology Li. (2011).

Frudakis, T., Thomas, M., Gaskin, Z., Venkateswarlu, K., Chandra, K. S., Ginjupalli, S., Gunturi, S., Natrajan, S., Ponnuswamy, V. K., & Ponnuswamy, K. N. (2003). Sequences associated with human iris pigmentation. *Genetics*, *165*(4), 2071–2083. https://doi.org/10.1093/GENETICS/165.4.2071

Grignani, P., Peloso, G., Achilli, A., Turchi, C., Tagliabracci, A., Alù, M., Beduschi, G., Ricci, U., Giunti, L., Robino, C., Gino, S., & Previderè, C. (2006). Subtyping mtDNA haplogroup H by SNaPshot minisequencing and its application in forensic individual identification. *International Journal of Legal Medicine*, *120*(3), 151–156. https://doi.org/10.1007/S00414-005-0059-5

Grimes, E. A., Noake, P. J., Dixon, L., & Urquhart, A. (2001). Sequence polymorphism in the human melanocortin 1 receptor gene as an indicator of the red hair phenotype. *Forensic Science International*, *122*(2–3), 124–129. https://doi.org/10.1016/S0379-0738(01)00480-7

Gupta, A. K., & Gupta, U. D. (2013). Next generation sequencing and its applications. In A. S. Verma & A. Singh (Eds.), *Animal Biotechnology: Models in Discovery and Translation*. https://doi.org/10.1016/B978-0-12-416002-6.00019-5

Hammer, M. F., Chamberlain, V. F., Kearney, V. F., Stover, D., Zhang, G., Karafet, T., Walsh, B., & Redd, A. J. (2006). Population structure of Y chromosome SNP haplogroups in the United States and forensic implications for constructing Y chromosome STR databases. *Forensic Science International*, *164*(1), 45–55. https://doi.org/10.1016/J.FORSCIINT.2005.11.013

Hu, N., Cong, B., Li, S., Ma, C., Fu, L., & Zhang, X. (2014). Current developments in forensic interpretation of mixed DNA samples (review). *Biomedical Reports*, *2*(3). https://doi.org/10.3892/br.2014.232

Jeffreys, A. J., Brookfield, J. F. Y., & Semeonoff, R. (1985). Positive identification of an immigration test-case using human DNA fingerprints. *Nature*, *317*(6040), 818–819. https://doi.org/10.1038/317818A0

Kashyap, V. K., Sitalaximi, T., Chattopadhyay, P., & Trivedi, R. (2004). DNA profiling technologies in forensic analysis. *International Journal of Human Genetics*, *4*(1). https://doi.org/10.1080/09723757.2004.11885864

Keerti, A., & Ninave, S. (2022). DNA fingerprinting: Use of autosomal short tandem repeats in forensic DNA typing. *Cureus*. https://doi.org/10.7759/cureus.30210

Kidd, K. K., Pakstis, A. J., Speed, W. C., Grigorenko, E. L., Kajuna, S. L. B., Karoma, N. J., Kungulilo, S., Kim, J. J., Lu, R. B., Odunsi, A., Okonofua, F., Parnas, J., Schulz, L. O., Zhukova, O. V., & Kidd, J. R. (2006). Developing a SNP panel for forensic identification of individuals. *Forensic Science International*, *164*(1), 20–32. https://doi.org/10.1016/J.FORSCIINT.2005.11.017

Kupiec, T. C., Raj, V., & Vu, N. (2006). Pharmacogenomics for the forensic toxicologist. *Journal of Analytical Toxicology*, *30*(2), 65–72. https://doi.org/10.1093/JAT/30.2.65

Li, R. (2025). *Forensic Biology* (3rd ed.). CRC Press. https://doi.org/10.4324/9781003562726

Lunetta, P., Levo, A., Männikkö, A., Penttilä, A., & Sajantila, A. (2002). Death in bathtub revisited with molecular genetics: A victim with suicidal traits and a LQTS gene mutation. *Forensic Science International*, *130*(2–3), 122–124. https://doi.org/10.1016/S0379-0738(02)00379-1

Marquez, A., Djelouadji, Z., Lattard, V., & Kodjo, A. (2017). Overview of laboratory methods to diagnose leptospirosis and to identify and to type leptospires. *International Microbiology*, *20*(4), 184–193. https://doi.org/10.2436/20.1501.01.302

Moxon, E. R., & Wills, C. (1999). DNA microsatellites: Agents of evolution? *Scientific American*, *280*(1), 94–99.

P gene as an inherited biomarker of human eye color | Request PDF. (n.d.). Retrieved February 23, 2025, from https://www.researchgate.net/publication/11224076_P_gene_as_an_inherited_biomarker_of_human_eye_color

Pakzad, Z., Mozdarani, H., Izadi-Mood, N., & Niromanesh, S. (2014). Variable number tandem repeat (VNTR) genotyping of hydatidiform mole in Iranian patients. *Avicenna Journal of Medical Biotechnology*, *6*(4), 246–253.

Panneerchelvam, S., & Norazmi, M. N. (2003). Forensic DNA profiling and database. *The Malaysian Journal of Medical Sciences: MJMS*, *10*(2), 20. https://pmc.ncbi.nlm.nih.gov/articles/PMC3561883/

Primorac, D., & Schanfield, M. (Eds.). (2023). *Forensic DNA Applications: An Interdisciplinary Perspective*. Boca Raton: CRC Press.

Roman, J. K., Reid, S., Reid, J., Chalfin, A., Adams, W., & Knight, C. (2008). The DNA Field Experiment: Cost-Effectiveness Analysis of the Use of DNA in the Investigation of High-Volume Crimes. *Justice Policy Center: Research Report.*

Sanchez, J. J., Phillips, C., Børsting, C., Balogh, K., Bogus, M., Fondevila, M., Harrison, C. D., Musgrave-Brown, E., Salas, A., Syndercombe-Court, D., Schneider, P. M., Carracedo, A., & Morling, N. (2006). A multiplex assay with 52 single nucleotide polymorphisms for human identification. *Electrophoresis, 27*(9), 1713–1724. https://doi.org/10.1002/ELPS.200500671

Shrivastava, P., Dixit, S., & Dash, H. R. (2020). Guidelines for collection and preservation of samples for forensic DNA testing. In *Forensic DNA Typing: Principles, Applications and Advancements* (pp. 573–583). Springer Singapore, Singapore.

Shriver, M. D., & Kittles, R. A. (2004). Genetic ancestry and the search for personalized genetic histories. *Nature Reviews. Genetics, 5*(8), 611–618. https://doi.org/10.1038/NRG1405

Sinha, M., & Arjun Rao, I. (2020). SNP testing in forensic science. *Forensic DNA Typing: Principles, Applications and Advancements*, 365–376. https://doi.org/10.1007/978-981-15-6655-4_18

Sobiah, R., Syeda, R. H., Zunaira, E., Nageen, Z., Maria, K., Syeda, A. Z., Shahana, S. M., Akifa, M., Abdul, J., & Muhammad, R. K. (2018). Implications of targeted next generation sequencing in forensic science. *Journal of Forensic Research, 9*(2). https://doi.org/10.4172/2157-7145.1000416

Vos, P., Hogers, R., Bleeker, M., Reijans, M., Lee, T. V. D., Hornes, M., & Zabeau, M. (1995). AFLP: A new technique for DNA fingerprinting. *Nucleic Acids Research, 23*(21), 4407–4414.

Yang, Y., Xie, B., & Yan, J. (2014). Application of next-generation sequencing technology in forensic science. *Genomics, Proteomics and Bioinformatics, 12*(5). https://doi.org/10.1016/j.gpb.2014.09.001

5

Next-Generation Sequencing in Forensic Science

Advancements, Limitations, and Applications

Vidhi Paida, Kamayani Vajpayee, and Ritesh K. Shukla

5.1 Introduction

Since the inception of the Sanger sequencing method, advances in DNA sequencing technology have had a tremendous impact on molecular biology and genetics. This method has been successfully used in several large-scale undertakings, including the Human Genome Project and genome programs for numerous species. Nonetheless, the intrinsic disadvantages of classical Sanger sequencing, which include low throughput, high costs, and operational difficulties, have limited its utility in deeper and more complicated genomic investigations (The 100 Genomes Project Consortium, 2015; Jackson, 2016; Kumar et al., 2019; Schook et al., 2015; Tyagi & Bhide, 2020). The recent emergence of next-generation sequencing (NGS) technology has addressed these restrictions, demonstrating increased throughput and cost-effectiveness. As a result, these technologies have been widely applied in an array of life science disciplines, including forensics, agricultural genomics, and ancient DNA study (Di Resta et al., 2018; Esposito et al., 2016; Imam et al., 2018; Satam et al., 2023).

NGS technology refers to high-throughput DNA sequencing methods that are distinct from the Sanger sequencing approach. This approach allows for the parallel sequencing of millions or billions of DNA molecules, significantly increasing throughput and eliminating the need for the fragment-cloning methodology commonly used in Sanger sequencing. It includes second-generation sequencing (SGS) techniques, notably loop array sequencing, which allows for the simultaneous examination of several samples. Furthermore, third-generation sequencing technology is used to determine the base composition of individual DNA molecules. All existing NGS technologies use massively parallel sequencing, which generates thousands of reads at the same time (Dash et al., 2023; Kulski, 2016).

NGS platforms include essential advancements, including ultrahigh-resolution optics, huge parallel chemical processes, and computational techniques for short-read analysis. The cost of sequencing and response times has drastically come down to a few days because of these technical advancements (Ari & Arikan, 2016; Satam et al., 2023).

Regardless of these benefits, academia is currently confronted with the problem of handling the deluge of data generated by these technologies, which could result in deceptive results (Di Resta et al., 2018). These technologies have the potential to transform DNA sequencing, though, and they can be used for tasks like directly sequencing individual RNA or protein molecules and solving problems with data processing and storage (Chen et al., 2021; Hu et al., 2021;

DOI: 10.4324/9781003474296-5

Schwartz & Motorin, 2017; van Dijk et al., 2023). Nowadays, the majority of applications of NGS technology are in the biomedical domains, improving studies in common and rare diseases, genetically diverse disorders, and clinical diseases (Hussen et al., 2022; Liu et al., 2019; Shen et al., 2015).

NGS has been used for whole mitochondrial sequencing in forensic science (Brandhagen et al., 2020). Additionally, in situations where a lack of DNA is a major constraint, NGS has advanced significantly, especially in forensic casework and non-invasive prenatal diagnosis. The low amount and quality of fetal DNA in maternal blood provide difficulties even with the historical knowledge of its detection (Bijarnia-Mahay et al., 2018; Iozzi et al., 2015). With its extensive capabilities, NGS provides a compelling non-invasive prenatal testing option that makes trisomy detection possible. Unfortunately, because there is a lack of information on NGS and the caliber of its output in forensic applications, the progress in prenatal diagnosis does not keep pace with that of forensic genetics (Xu et al., 2020).

The utilization of short reads from NGS outputs and improvements in DNA extraction techniques have been emphasized in historical DNA reviews, increasing the availability of ancient DNA (aDNA) for research (Hofreiter et al., 2015). These studies recommend bespoke single nucleotide polymorphism (SNP) panels for forensic purposes, as well as enriching aDNA for projects. However, the main obstacles to using NGS platforms in the forensic domain continue to be the small sample size and high level of degradation (Alvarez-Cubero et al., 2017).

To provide a reference for future studies and applications in this developing field, the current chapter thus takes a thorough look at the use of NGS technology in forensic science.

5.2 Historical overview

The phrase 'massively parallel sequencing' (MPS), sometimes known as 'next-generation sequencing' (NGS), is used to differentiate more recent developments in DNA sequencing from more traditional techniques (Bruijns et al., 2018). Sanger and associates developed the first DNA sequencing technology in 1977 using a chain-termination technique known as Sanger sequencing (Heather & Chain, 2016). Known as the 'first generation', this sequencing technique became the dominant technology in the early stages of commercial and laboratory sequencing applications (Kchouk et al., 2017; Liu et al., 2012). The Human Genome Project was completed in 2001 with the use of automatic sequencing devices and related software that used capillary sequencing machines and Sanger sequencing technology. The Human Genome Project successfully demonstrated the viability of whole genome sequencing (WGS), but its widespread adoption was hampered by the prohibitively high costs involved (Caulfield et al., 2013; McCook, 2005).

The development of NGS was accelerated by the labour-intensive and expensive nature of WGS (Kulski, 2016). Important turning points were the 454 system, which was first shown by 454 in 2005 and then purchased by Roche in 2007, as well as Solexa's Genome Analyzer (GA), which was introduced in 2006, and SOLiD (Sequencing by Oligo Ligation Detection), which was supplied by Agencourt. The Ion Torrent Personal Genome Machine (PGM) was introduced by Life Technologies in 2011. Illumina systems are at the vanguard of these NGS platforms, which are extensively distributed throughout the world. Together, they exhibit excellent throughput performance, respectable accuracy, and a more cost-effective profile than Sanger sequencing (Liu et al., 2012). All of these second-generation (2G) platforms are based on the basic idea that DNA fragments with certain adaptor sequences at both ends (called libraries) are attached to a surface, whether solid or spherical (Hu et al., 2021; Slatko et al., 2018).

Advancements have been brought about by the shift to third-generation (3G) sequencing technologies, such as PacBio from Pacific Biosciences and HeliScope from Helicos Biosciences, which use single-molecule templates without PCR amplification (Ku & Roukos, 2013). The advantages of this paradigm shift include a lower starting material requirement and a decreased proneness to errors. An enzymatic template replication method in these platforms facilitates the sequencing of individual clonal molecules. The cleavage of fluorescent dye-linked pyrophosphate within a volume-limited viewing window is used to monitor base incorporation as template DNA molecules are collected (Gut, 2013). Interestingly, PacBio is capable of sequencing long areas, including those with 100% GC content or expansions of the CGG trinucleotide repeat, making it especially useful in forensic genetics (Loomis, 2013).

The scope of NGS technologies is rapidly evolving, with the introduction of fourth-generation (4G) systems that entail the direct inspection of specific molecules using single-molecule sequencing (SMS) and nanopore technology. While extreme accuracy might not be required for all applications, it matters most in forensic sciences. In genomics, data quality is critical, and the need for precise data necessitates the use of high-precision single-molecule long-read sequences, especially for investigations requiring dependability of more than 85%. Unlike amplification-based methods, single-molecule techniques provide more uniform coverage and successfully span GC-rich regions by eliminating artefacts such as uneven amplification and GC bias. Additionally, they can navigate repetitive complexes or duplicates, which are difficult for short-read sequencing techniques (Buermans & Den Dunnen, 2014). *In situ* sequencing is still experimental, but it shows promise as a next step in the development of sequencing technology. This method entails carrying out sequencing studies in a particular setting, such as the DNA or RNA of individual cells in each tissue at a specified time (Ku & Roukos, 2013).

5.2.1 First-generation sequencing

Sanger sequencing is a technique used in first-generation sequencing that was invented by Frederick Sanger in 1977. For this accomplishment, Sanger was awarded the 1980 Nobel Prize in Chemistry (Pareek et al., 2011). Like PCR, this technique calls for deoxynucleotide triphosphates (dNTPs), a DNA primer, DNA polymerase, and a single-stranded DNA template. Furthermore, the elongation process is terminated by the addition of di-deoxynucleotide triphosphates (ddNTPs) to the freshly synthesized DNA strand, which is why it is called the 'chain termination method'. In the sequencing reaction, four tubes containing the regular dNTPs are supplemented with only one of the four labelled ddNTPs in a comparatively low concentration. Capillary (gel) electrophoresis and fluorescently labelled ddNTPs are used in this procedure to determine the sequence. Interestingly, compared to many second-generation methods, which usually produce shorter reads, Sanger sequencing produces longer reads (Valencia et al., 2013).

5.2.2 Second-generation sequencing

With the advent of PCR in 1985, sequencing technology advanced significantly, and as substitutes for Sanger sequencing principles became practical, apparatus improved (Chan, 2005). Sanger sequencing of DNA fragments requires bacterial cloning, while new sequencing technologies like 454, SOLiD®, and Ion Torrent™ use a cell-free technique. This change led to parallel analysis techniques that were quicker and less expensive, which increased throughput (Dash et al., 2023). 'Next-generation sequencing' became the term used to characterize these advancements. Other sequencing techniques were also presented, including virtual terminator chemistry and pyrosequencing. The study of small DNA fragments was made possible by the increased throughput, but the precision of each short read was decreased, preventing the sequencing of the entire genome (Grada & Weinbrecht, 2013). Ansorge and associates (Ansorge, 2010) developed a radioactive label-free automated DNA sequencing technique in 1986. They used a fluorescently labelled primer to create a nested set of DNA fragments that were fluorescent as a result of the labelled primer's incorporation. When gel electrophoresis and laser excitation were used together, concentrations as low as 0.1 fmol per band could be detected, and within six hours, 250–300 bases could be read.

The GE Healthcare MegaBACETM system (https://www.artisantg.com/Scientific/70238-1/GE-Healthcare-Amersham-Biosciences-MegaBACE-4500-DNA-Analysis-System), Intelligent Bio-Systems Inc. (a Qiagen® subsidiary) system (https://www.qiagen.com/us), Complete Genomics™ sequencing with nanoballs (https://www.completegenomics.com/technology/), and the Polonator (Church research group, Harvard) (https://arep.med.harvard.edu/Polonator/), created by the Harvard Church research group, are just a few of the SGS devices that have hit the market.

5.2.2.1 Roche 454

Melamede patented the method in 1989, and 454 Life Sciences (later Roche) was the first to develop the sequencing technique known as 454. This pyrosequencing-based technology, known as GS FLX, signalled the introduction of NGS (Bruijns et al., 2018). This technique, also known as sequencing by synthesis (SBS), detects the nucleotide that has been inserted into the new DNA strand by analyzing the signal it emits. A camera records the light emitted by different chemical reactions so that the sequence can be reconstructed. Mismatched dNTPs are broken down by a pyrase, and pyrophosphate is converted to adenosine triphosphate by the enzyme sulfurylase when the right nucleotide is incorporated. Adenosine triphosphate and luciferin are then converted by luciferase into oxyluciferin, which is followed by light emission. Electrophoresis is no longer required for detection and analysis thanks to SBS (Ansorge, 2009; Uhlen & Quake, 2023).

Notwithstanding its benefits, library preparation is necessary prior to sequencing, which entails breaking up the DNA solution and adding adaptors—44-base primer sequences—to both ends of each DNA fragment. One end is ligated with adapter 'A', and the other end is supplemented with adaptor 'B', which has a distinct nucleotide sequence. These adapters are designed to catch bits of DNA on a solid surface. The solid surface for emulsion PCR (em-PCR) is provided by 26 m beads in the 454 sequencers. DNA on the bead inside a water droplet encircled by an oil solution is amplified using Em-PCR. The ability to amplify single DNA molecules without bias by trapping them in lipid microreactors is a crucial component of em-PCR. A fibre-optic slide with recessed 75 pL wells that hold one bead each is loaded with the beads coated with multiple copies of the template DNA after em-PCR. Regretfully, Roche stopped manufacturing the 454 system by the end of 2016.

5.2.2.2 Illumina HiSeq and Illumina MiSeq

After acquiring Solexa in 2007, Illumina unveiled a sequencing method that used reversible dye terminators and sequencing-by-synthesis (Moti et al., 2022). Adapters are placed at the ends of the DNA sequence, just like in the 454 system. But unlike the 454 techniques, which use em-PCR and beads, Solexa uses a solid glass support that is planar (Luo et al., 2012). Adapters 'A' and 'B' have complementary sequences on the inner portion of flow cell lanes. When the complementary sequence on the support and the other end of the target DNA hybridize, a bridge structure is created. Every bridge-amplified cluster has a distinct primed and sequenced DNA template. ddNTPs with detachable blocking groups and unique fluorescent labels are used in the sequencing process. The DNA sequence is ascertained by sequencing each base individually and using a CCD camera to record the fluorescent signal. A quicker camera, polymerases, higher occupancy, monoclonality within wells, and designed flow cells with fixed nanowells all contribute to increased throughput (Bronner & Quail, 2019; Daneshpour et al., 2018). However, because it is difficult to achieve 100% efficiency of base incorporation and cleavage in each cycle, the reversible dye termination SBS approach has a constraint in read duration. The MiSeq® and Solexa Genome Analyzer series were replaced by the HiSeq® series (Jeon et al., 2019; Minoche et al., 2011).

5.2.2.3 Applied biosystems

As time went on, Life Technologies and Invitrogen™ combined to establish Applied Biosystems™ in 2008, and the company also bought a number of businesses that were selling sequencing technologies, such as SOLiD® and Ion Torrent™. Applied Biosystems™ is currently a division of Thermo Fisher Scientific (https://www.thermofisher.com/in/en/home/brands/life-technologies.html).

5.2.2.3.1 SOLiD Sequencing by Oligo Ligation Detection, or SOLiD®, is a technique that uses Polonator technology. It was first presented by Agencourt in 2006, and Applied Biosystems™ later adopted it (Bruijns et al., 2018). McKernan et al. patented the technique in 2006; it is based on sequencing by ligation, where a probe sequence tethered to a fluorophore hybridizes to a DNA segment and is ligated to a nearby oligonucleotide for imaging (Shendure & Ji, 2008). Em-PCR is performed utilizing 1 m small magnetic beads and octamers of oligonucleotides with specific fluorescent labels on the fourth and fifth bases, covering all conceivable complementary DNA strand variants. It is possible to sequence particular sites by cleaving the ligated octamer after the 5th base in each cycle (Liu et al., 2012).

5.2.2.3.2 Ion torrent After Jonathan Rothberg left 454 Life Science in late 2010, Ion Torrent™ unveiled the Ion Personal Genome Machine (PGM) (Bruijns et al., 2018). In this SBS approach, Em-PCR is performed on beads in the first step, and the beads are then maintained in place in wells on a chip/picowell plate. Unlike fluorescence-based methods, the detection mechanism relies on pH change upon nucleotide addition. Proton release upon pyrophosphate cleavage, detected by monitoring potential, allows localization and retention of proton release. Sequencing of homopolymeric sections in the template DNA is made possible by the signal, which is proportionate to the released protons. Millions to billions of sequencing reactions are measured concurrently by a complementary metal-oxide semiconductor (CMOS) sensor array chip, whose sensor surface is near the bottom of the solid plate (Bustillo et al., 2013). Compared to fluorescence-based methods, this potential-based measurement is quicker, more affordable, and possible with smaller devices (Rothberg et al., 2011).

5.2.3 Third-generation sequencing

Third-generation sequencing techniques, also known as next-next-next generation (NNN), utilize the real-time reading of individual molecules, whereas SGS methods depend on amplification (Schadt et al., 2010). Compared to previous sequencing technologies, these developments produce longer reads and considerably faster processes (Lee et al., 2016). Technologies for SMS fall into three categories: SBS approaches (like PacBio®), nanopore sequencing (like Oxford Nanopores), and methods that use sophisticated microscopy to directly see individual DNA molecules (like VisiGen). Third-generation sequencing began with the first SMS technology, created by Quake's group and subsequently introduced by Helicos in 2007. Systems that use SMS, such as Oxford Nanopore and PacBio®, are described in more depth below. Starlight, Heliscope (later Life Technologies/Invitrogen™), and VisiGen are further third-generation real-time sequencing technologies (Gut, 2013; Lavezzo et al., 2016; Ozsolak, 2012; Uhlen & Quake, 2023; Xiao & Zhou, 2020).

5.2.3.1 PacBio

The technology used by Pacific Biosciences® (PacBioR) is Single Molecule Real-Time® (SMRT®). PacBio®, the first individual DNA molecule sequencer with real-time detection capabilities, was introduced in late 2010. This methodology is appropriate for long-read sequencing, in contrast to earlier disclosed methods. It makes use of real-time sequence sequencing, Single Base Synthesis (SBS), and fluorescent labelling (Korlach, 2013; McCarthy, 2010). Zero-mode waveguides (ZMWs) are created by the instrument's metal-coated glass wells, which have diameters of less than 100 nanometres. An immobilized polymerase is kept in the bottom of these minuscule wells, which have a volume of the order of zeptolitres. The integrated nucleotide's real-time fluorescence is captured by a high-resolution camera. ZMWs' design makes it possible to illuminate the lowest 30 nm, which makes it easier to detect nucleotide incorporation by the polymerase (Carneiro et al., 2012; Weirather et al., 2017).

5.2.3.2 Nanopore based systems

Like other third-generation methods, nanopore-based technologies allow the reading of a single DNA molecule without the requirement for costly fluorescent labelling or amplification. The ion current changes that occur when a charged molecule, like DNA, travels through a nanoscale pore in a membrane provide the basis for the detection principle. A voltage drives the DNA strand through the pore, creating a fluctuating ionic current, and this membrane divides two chambers that are filled with a conductive electrolyte. Theoretically, DNA strand sequencing is made easier by the four various current levels produced by the bases. In 2014, Oxford Nanopore Technologies introduced the MinION™, a nanopore device that weighs under 100 grams and connects to a computer via USB (Healy, 2007; Singer et al., 2010).

5.3 Applications of NGS in different fields

A revolutionary age in scientific study has begun with the introduction of NGS, which has transformed genetic analysis in a variety of fields. Numerous applications of NGS are covered in the section that follows forensic genomics, environmental genomics, pharmacogenomics, evolutionary genomics, medical research, cancer genomics, microbial genomics, epigenomics, metagenomics, neurogenomics, agricultural genomics, and environmental genomics. A quick discussion of the various subtleties and significant effects of NGS in each area highlights the revolutionary contributions this technology has made to deciphering the complexity of genetic information.

5.3.1 Genomics and next-generation sequencing

The introduction of NGS technologies has brought about a radical change in the scientific field of genomics, which is devoted to the thorough investigation of complete genomes. This significant advancement has changed the field of genetic study by giving scientists access to previously unheard-of instruments and techniques for delving into the complicated workings of the genome. NGS is a technological wonder that has ushered genomics into a new age thanks to its high-throughput capabilities and quick sequencing. NGS's capacity to make it easier to interpret genetic variants inside the genome is one of its most significant contributions to genomics. The size and scope of genomic investigations were constrained by the time-consuming and frequently arduous nature of traditional sequencing techniques. These restrictions have been removed by NGS, which can sequence millions of DNA fragments at once, giving researchers the capacity to do extensive genomic analysis with previously unheard-of efficiency. Comprehensive surveys of genetic variants across several populations have been made possible by this scalability, which has shown the intricate web of genomic diversity (Stapley et al., 2010; Zhang et al., 2011).

Another crucial area of genetic research that has seen a revolutionary change because of NGS is the identification of mutations. From single-nucleotide variants to intricate structural changes, genetic modifications may be precisely detected thanks to the great resolution and accuracy of NGS techniques (Werner, 2010). Studies to determine the genetic causes of several diseases, such as cancer, neurological diseases, and uncommon genetic problems, have benefited greatly from this capability. Today, scientists can identify certain genetic abnormalities linked to illnesses, providing vital information about the mechanisms underlying these conditions and possible treatment options. Furthermore, NGS has been essential in the identification of new genes, which has increased our knowledge of the genomic environment (Mardis, 2008).

Gene discovery activities have advanced due to the rapid and cost-effective sequencing of whole genomes. NGS makes it easier to identify regulatory elements, non-coding RNA molecules, and previously unidentified genes that add to the complexity of genetic regulation. Our understanding of gene function, pathways, and the underlying genomic architecture driving different biological processes has expanded as a result (Guffanti et al., 2014; Weikard et al., 2013).

Beyond individual gene research, NGS has had a profoundly revolutionary impact on our understanding of population genetics (Fumagalli et al., 2014), evolution (Tan et al., 2019), and hereditary disorders (Boycott et al., 2013). NGS-powered large-scale genomic research has produced thorough lists of genetic variants in a variety of populations. Studies on population dynamics, migratory trends, and human evolution have been aided by this abundance of genetic data.

NGS emerges as a game changer in *cancer research*, revealing the complex genomic changes that drive cancer formation. Cancer is fundamentally a genome illness, defined by a slew of genetic mutations that drive uncontrolled cell proliferation. NGS enables researchers to delve deeply into the genomic landscape of cancer, resulting in a complete catalogue of mutations that underpin various forms. This granular understanding serves as the foundation for designing targeted medicines to combat specific genetic abnormalities that drive cancer growth. Tumour heterogeneity, or the occurrence of many cell types inside a single tumour, presents a substantial problem in cancer treatment. NGS excels at unravelling this complexity by analyzing tumour genomic heterogeneity. Researchers can discover subclones with diverse genetic profiles, telling doctors about the wide range of potential therapeutic targets inside a single tumour. NGS makes it easier to find possible biomarkers for cancer that have predictive, prognostic, and diagnostic significance. These biomarkers, which can be anything from particular genetic abnormalities to patterns of gene expression, help doctors customize treatment plans for each patient. With the help of NGS-derived biomarkers, precision oncology seeks to match patients with the best treatments according to their individual genetic profiles. Personalized cancer treatments are the result of NGS discoveries in cancer genomics. Treatment effectiveness is increased, and side effects are reduced when medicines are customized according to the unique genetic composition of a patient's tumour. NGS-guided precision oncology is a paradigm shift that shifts the focus from one-size-fits-all methods to patient-centred, personalized cancer care (Petric et al., 2015; Serratì et al., 2016; Shyr & Liu, 2013; Soto et al., 2016; Tyulyandina et al., 2018).

5.3.2 Application of NGS in microbial science

Understanding the microbial world has undergone a revolution thanks to the combination of NGS and microbial genomes. The tedious and time-consuming nature of sequencing methods has historically made studying microbial genomes extremely difficult. NGS overcomes these limitations and provides a quick and affordable way to sequence the genomes of fungi, viruses, and bacteria. This discovery has ushered in a new era of investigation in microbial genomics by enabling scientists to quickly decipher the genetic codes of various microbes. Determining the genetic foundations of infectious diseases has become made possible thanks in large part to NGS. Researchers can learn more about virulence factors, infection mechanisms, and possible intervention targets by sequencing the genomes of diseases. NGS-driven microbial genomics advances our knowledge of pathogen evolution and host interaction, guiding the development of control and treatment options for infectious diseases (Koonin et al., 2021). Public health is at risk worldwide due to the emergence of antibiotic resistance. An effective method for determining the genetic foundation of antibiotic resistance in microbial populations is NGS. Researchers can find resistance genes and comprehend the processes that make some microbes resistant to antibiotics by examining the genomes of resistant strains. This information is essential for developing plans to stop the spread of infections that are resistant to antibiotics (Saracino et al., 2021; Tuan et al., 2019). By revealing the intricate relationships among microbial populations in various settings, NGS broadens its scope into microbial ecology. NGS makes it possible to conduct metagenomic research, which entails the simultaneous sequencing of DNA from many organisms found in a sample. The ecological dynamics, functional potential, and biodiversity of microbial communities in soil, water, air, and the human body are revealed by this method (De Mandal et al., 2015).

5.3.3 Application of NGS on forensic genomics

With the introduction of NGS, the area of forensic genomics, which lies at the nexus of genetics and forensic science, has undergone a paradigm shift. Conventional profiling techniques are challenged by forensic investigations, which frequently come across materials with low DNA quantities or damaged DNA. In these situations, NGS proves to be an effective method because it has a better sensitivity and can produce thorough genetic profiles from little or tainted DNA samples. When it comes to DNA degradation, the conventional method of capillary electrophoresis (CE), which is based on short tandem repeat (STR) analysis, has drawbacks (Aly & Sabri, 2015; Børsting & Morling, 2015).

A more thorough and precise representation of the genetic information contained in a sample can be obtained by parallel sequencing several loci, thanks to NGS's high-throughput sequencing capabilities (Iozzi et al., 2015). This is especially important in situations where conventional approaches might produce outcomes that are insufficient or ambiguous. By helping to overcome the drawbacks of traditional methods, NGS's improved resolution enables forensic scientists to extract useful information from difficult materials (Iozzi et al., 2015).

NGS's capacity to improve DNA profiling's discriminatory strength is one of its main benefits in forensic genomics. Even while they work well, traditional techniques could not have the level of detail required to distinguish between people with similar genetic profiles. NGS improves the discriminatory potential by analyzing a wider range of genetic markers and sequencing longer segments of DNA. NGS's discriminatory capacity is especially useful in situations when traditional approaches would not be able to produce definitive results, such as those involving familial links. Forensic experts can more precisely differentiate between people thanks to the higher resolution, which lowers the possibility of false positives or misidentifications. The foundation of forensic investigations is accurate individual identification, and NGS plays a major role in accomplishing this objective. NGS gives forensic professionals a powerful toolkit for matching and differentiating people based on their distinct genetic signals by obtaining a more complete genetic fingerprint. This has significant ramifications for missing persons cases, criminal investigations, and clearing falsely accused people (Imam et al., 2018; Iozzi et al., 2015).

5.4 Forensic applications of NGS

The use of NGS technologies for the 'de novo' characterization of complete genomes, as demonstrated by the entire mitochondrial DNA (mtDNA) sequence of Neophema chrysogaster, as well as the resequencing of human genomes (such as the Human Genome Project, which involved 1092 genomes from 14 populations and used a combination of low-coverage whole-genome and exome sequencing), has been thoroughly validated (Matullo et al., 2013; Miller et al., 2013). This broad scope shows prospective applications of NGS in forensic sciences, addressing the fundamental technological constraints of conventional technologies such as CE and the obstacles faced by deteriorated or limited forensic materials (Lyons et al., 2013). Recent research has included NGS data collected from human hair, showing notable results in metagenomic analyses. A total of 42 DNA extracts from human scalp and pubic hairs generated 79,766 reads, resulting in 39,814 post-control and abundance filtering. This information could be useful in situations involving sexual assault victims and offenders, especially when other associative evidence is limited (Tridico et al., 2014).

Indels and copy number variations (CNVs) have not gotten much attention, even though repetitions make up over half of the human genome and STRs only make up 15%. This is because, in forensic genetics, pyro-sequencers are the main platforms with sufficient read length for sequencing the key STR loci. To date, the pyrosequencing technique has been used to generate the majority of forensic literature utilizing NGS STR data. Nevertheless, a few studies have used semiconductor sequencing and SBS, with libraries created directly via adaptor ligation or PCR (Mason-Suares et al., 2016).

5.4.1 STR and SNP analysis

DNA profiling with STRs and SNPs is a common approach in modern forensic investigation (Vajpayee et al., 2023). STR typing is currently the established gold standard technique, although DNA sequencing has mostly focused on examining the hypervariable portion of mtDNA (Davis et al., 2015). This targeted technique aids in understanding the origin and relationships among DNA samples with limited numbers (Berglund et al., 2011). STR analysis has become widely accepted in human forensic cases (Dash, Rawat, et al., 2021; Dash, Vajpayee, Agarwal, et al., 2021; Irwin et al., 2016). Several STR database systems, including the Combined DNA Index System (CODIS) in the United States, have been developed. CODIS uses a standardized set of 13 highly polymorphic, genetically unlinked STR loci in non-coding areas (Ge et al., 2014). The usual investigation of these STR alleles comprises multiplexed PCR followed by CE for separation (Vajpayee et al., 2020), with traditional laboratory procedures needing 8–10 hours to provide a DNA profile.

Rapid PCR, direct PCR, microfluidic chip-based amplification and detection, and integrated microfluidic devices have been the main methods employed recently to speed up the creation of STR profiles. By offering alternate methods, these developments hope to facilitate automated DNA profile creation in the shortest amount of time. Booking stations, field sites like airports and border crossings, kinship determination, and catastrophe victim investigation are among possible uses for this automated approach (Romsos & Vallone, 2015).

It has been shown that relying only on regular STR typing can induce mistakes in kinship analysis, even if conventional STR typing is generally effective for a variety of applications (Bodner et al., 2016; Tsai et al., 2024). Furthermore, resolving dye artefacts and differentiating stutter results from alleles of a secondary participant in mixes present difficulties (Butler et al., 2018). Bi-allelic SNPs are less polymorphic than multi-allelic STRs, which raises additional challenges when using autosomal SNPs for human individualization in forensic settings. Using a greater number of SNPs in combination with different genotyping technologies may compensate for the fact that SNPs might not be as meaningful when evaluating DNA mixes from numerous individuals (Jin et al., 2022).

One of the many improvements that large-scale sequencing offers to forensic analyses is the use of more meaningful markers that are obtained from this type of sequencing (Berglund et al., 2011). Compared to Sanger capillary sequencing, NGS technologies offer a new method for creating DNA sequence data, enabling the sequencing of millions of individual DNA strands (such as those found in DNA mixtures) at a much lower cost per nucleotide and with a higher throughput. When STR amplicons are sequenced using NGS, their discriminative strength as a genetic marker for individuals is increased, allowing full profiles to be generated with a little 62 pg input (Zeng et al., 2015). There is evidence that nucleotide changes and deletions or insertions that are not detectable by CE analysis take place in the flanking area or repeat unit of STRs (Rockenbauer et al., 2014).

On the other hand, NGS technologies reveal sequence variations inside the core unit of STRs in addition to the number of repeats. Furthermore, NGS can produce distinct sequences for each allele found in a mixture of STR amplicons, increasing the statistical power of STR analyses and ultimately lowering the quantity of typed loci needed for case resolution. By offering additional characterization for alleles that could seem identical in CE analysis, NGS can also help resolve DNA mixes in criminal investigations. Separating people with identical allele lengths is essential, particularly in forensic or kinship research, since almost half of the STRs in the European Standard Set and CODIS set are simple repeats with little sequence diversity (Rockenbauer et al., 2014).

Presently, Thermo Fisher Scientific has developed a Human Identification STR called HID-STR 10-plex panel, including amelogenin, D3S1358, D16S539, CSF1PO, D5S818, D8S1179, D7S820, TH01, vWA, and TPOX. This panel is specifically designed for NGS, and the Ion TorrentTM software supports the data analysis. Successful experiments with this panel have yielded full profiles for all samples in actual crime and identification cases, where CE assays previously only provided partial profiles. However, challenges related to stutters in the data persist, and efforts to address these issues are ongoing, including the development of enhanced software (Fordyce et al., 2015)

The inclusion of new complex STRs is essential for the best use of NGS. STRs, SNPs, Insertion/Deletion markers (InDels), and mtDNA markers are among the relevant forensic markers that can be included in a comprehensive multiplex that can be created using NGS (Dash, Vajpayee, Srivastava, et al., 2021; Irwin et al., 2011). Notably, several businesses have released forensic-oriented panels using the recognized markers, simplifying sample preparation for researchers who would otherwise need to create their own primers and panels.

In the realm of forensic DNA phenotyping, recent genome-wide association studies have identified SNPs that can predict physical characteristics and ethnic backgrounds. This information helps with individual identification, which is then confirmed by reference samples, and is crucial in the resolution of criminal cases. Furthermore, one of the most important goals in identifying unknown individuals is to predict individual-specific appearance using DNA-derived facial morphology (Irwin et al., 2011; Rockenbauer et al., 2014). With few genetic markers, such predictions are unlikely to be possible, hence sophisticated technologies like NGS are required for the parallel genotyping of many SNPs. These technologies are especially useful when there is little to no DNA degradation (Kumar et al., 2012).

MPS is an outstanding approach that has enabled the discovery of forensic SNP markers (Seo et al., 2013). Recently, 160 SNPs linked to phenotype, ancestry, and human identification were discovered using TruSeq™ Forensic Amplicon, a modified version of the TruSeqTM ChIP technique. The process prepares the libraries with just 1 ng of input DNA and requires less effort (Warshauer et al., 2015). Affymetrix has created a novel genome-wide SNP array that includes 906,600 SNPs in the autosomal DNA, mitochondrial, Y, and X chromosomes (Sehrawat et al., 2011; Vermeulen, 2020). By making it possible to generate large amounts of data at a reasonable cost—up to one billion short reads every instrument run—NGS offers a substantial advancement (Metzker, 2010). Nuclear microsatellite markers have been the subject of recent NGS applications, showing that a tiny portion of a genome may be randomly sequenced to provide a high density of putative microsatellite loci at low cost and with fast processing times. According to Van Dijk et al. (2014), these polymorphic microsatellite loci are widely used in forensic applications like genetic identification and population genetics (Van Dijk et al., 2014).

The enormous amount of data makes manual genotype calling or NGS data analysis impossible. Creating a forensic NGS tool for sequencing data analysis is a significant task. To guarantee dependability, the software program must go through extensive validation prior to implementation in real-world case work. Although software solutions for their kits have been introduced by Illumina and Thermo Fisher Scientific, they are not sophisticated enough for forensic genetics (Alvarez-Cubero et al., 2017). In addition to identifying two identical STR alleles with distinct sequences, the perfect software program should also give them names that adhere to the ISFG's suggested nomenclature (Børsting & Morling, 2015; Metzker, 2008).

5.4.1.1 Lineage markers

Since mtDNA is more abundant in cells than nuclear DNA (nuDNA), it is more likely to yield useful DNA information in forensic materials when nuDNA typing yields unsatisfactory findings (Parson et al., 2013). Furthermore, mtDNA, which is mostly derived from the hypervariable area, provides information about maternal biogeographic heritage (Larmuseau et al., 2015; Westen et al., 2015). As the only significant non-coding area in the mtGenome with a random match probability (RMP) of roughly 1 in 30, hypervariable segment 1 found in the mitochondrial regulatory region has historically been the focus. The RMP rose to 1 in 120 individuals when the complete mtDNA regulatory area was examined, which was beneficial in certain situations (Aizpurua-Iraola et al., 2023). To improve the informative value of mtDNA analysis, it is reasonable and desirable to expand the study to encompass the complete mt genome (Brandhagen et al., 2020). By analyzing individual polymorphisms in the coding region that Sanger sequencing is unable to read, whole mitochondrial genome sequencing from samples that share a control region enables differentiation (Holland et al., 2011; Melton et al., 2012). NGS technique requires very small amounts of sample and can provide entire mtDNA genome sequences at a lower cost than previous methods (Miller et al., 2013).

Sanger capillary sequencing of the entire mtDNA was the sole method available for identifying mtDNA variations up until recently (Alkanaq et al., 2019). However, this approach is costly and time-consuming (Davis et al., 2015). Another significant restriction on this type of research is the availability of adequate evidence. Recent advancements in digital imaging systems, bioinformatics, and microfluidics have overcome the limitations of NGS technology. Furthermore, the NGS's vast parallel sequencing capability is revolutionizing the creation of genetic data and may make it possible to economically create entire mtDNA genome sequence profiles from degraded samples. In order to determine the rarity of mtGenome haplotypes, the use of mtDNA coding region data in standard forensic casework is anticipated to rely more on the availability of large, high-quality complete mtGenome population databases than on sample size (Herrnstadt et al., 2002; Legati et al., 2021).

As mtDNA is present in hundreds of copies per cell in numerous organs, it contains a significant amount of genetic information (Tang & Huang, 2010). Both homoplasmic and heteroplasmic mutations are highly prevalent as a result. Certain punctual mutations may be helpful in the development of new haplogroups or in aiding with personal identification, even though this variation might not add much new information to forensic research. Massive parallel mtDNA sequencing can be done using a variety of platforms, including 454, SOLiD, Illumina, Affymetrix re-sequencing chip, and IonTorrent (Ye et al., 2014). Understanding the structure of the mitochondrial genome, which includes regions with high GC content, homopolymeric stretches, and STRs, is necessary to recognize the limitations of various sequencing approaches. Platform selection may affect multiplexing capabilities and coverage depth (Zhang et al., 2012). According to recent studies, the detection sensitivity of NGS for single-nucleotide changes can be compared to that of Sanger sequencing, particularly for results from the 454 systems. In addition to four heteroplasmic variants previously discovered by Sanger sequencing, more than 400 single-nucleotide alterations have been discovered with high concordance (98%) (Alvarez-Cubero et al., 2017; Zaragoza et al., 2010; Zavodna et al., 2014). Research on mtDNA has demonstrated the advantages of using NGS technology, particularly the Illumina GAIIx platform, in synthesizing short reads (suitable for degraded samples), lowering error rates, and providing broad coverage due to the large number of reads synthesized (Zaragoza et al., 2010). Using the Ion Torrent Personal Genome Machine, 64 mtGenomes were used to compare various biological sources and phylogenetic backgrounds in order to evaluate the performance and dependability of contemporary NGS, with Sanger sequencing serving as a reference technique (Zaragoza et al., 2010). The study's conclusions demonstrated the importance of pre-alignment data filtering methods, aligners, and parameters in determining the final NGS haplotypes. Thus, alignment software needs to be improved in order to increase the use of NGS in mtDNA forensic biology (Zaragoza et al., 2010).

Conversely, when the minor allele surpasses 20%, heteroplasmy can be examined using Sanger techniques. However, the Illumina GAII can detect heteroplasmy precisely at a rate of one in ten thousand copies of the mitochondrial genome (He et al., 2010). Consequently, the potential benefits of NGS technology for studying whole mtDNA sequences are demonstrated (Yang et al., 2014). There are numerous NGS methods in use today, primarily for mtDNA. As part of ongoing efforts to enhance the quality of NGS data, two additional procedures are being added: primer extension capture and preliminary DNA repair. These methods ensure consistent read coverage across the whole mtGenome by using primers for target enrichment. Primers extension capture and DNA repair are used in forensic applications to enhance the quantity and quality of NGS data. These modifications are especially made to manage large volumes of high-quality DNA, particularly in the analysis of ancient DNA (Loreille et al., 2011). Additionally, entire mtDNA sequencing is made more successful by the use of barcoding and DNA libraries, especially when working with severely fragmented samples for individualization (Templeton et al., 2013). Regions downstream of length heteroplasmy homopolymers present difficulties in Sanger sequencing of mtDNA because they are not reliably sequenced, which results in downstream sequences that are difficult to interpret because several length molecules are being read at once.

NGS has provided numerous new insights into Y-chromosomal research, chief among them the identification of previously unidentified Y-SNPs. The construction of Y-chromosomal phylogenies and haplogroup nomenclatures necessary for anthropological and forensic studies depends on these Y-SNPs. Because they uncover new classifications that incorporate Y-SNPs and evolutionary lineages from all available, verified, or not-so-much as a small set of Y-SNPs arranged in a condensed Y-chromosome tree that includes major global Y-haplogroups, these findings have implications for Y-chromosomal phylogeny trees. To incorporate new Y-SNPs and create international agreement, extensive efforts in the domains of anthropology and forensics are still needed. Advances in NGS analysis have led to the discovery of new evolutionary lineages, necessitating an update to the Y-chromosomal tree (Larmuseau et al., 2015). Y phylogeny and Y-SNPs found by Karafet are being used in the majority of Y-chromosomal research (Karafet et al., 2008). Complete Genomics used the complete genomics analysis (CGA) platform to enable SNP calls for 35 male whole genomes with high sequencing coverage (Drmanac et al., 2010). The Personal Genome Project (PGP) and the Singapore Sequencing Malay Project (SSMP) later made use of the same infrastructure. Currently, the SSMP has made 46 Malays publicly available, whereas the PGP makes 40 male genomes available (Wong, 2013). By sequencing more than 1000 human genomes, the 1000 Genomes Project hopes to eventually compile an extensive database on genetic variation in humans. A collection of 526 SNP profiles from phase 1 data were released in 2012 (The 100 Genomes Project Consortium, 2015). The Y-chromosomal phylogenetic tree has been revised for forensic reasons by combining all of the information from multiple studies (Van Geystelen et al., 2013).

5.4.2 MicroRNA assessment

One noteworthy feature of microRNAs, which have just lately entered the forensic field, is their small size. Their tiny size helps explain their tissue-specific or highly tissue-divergent expression as well as their resistance to destruction. MicroRNAs are therefore useful for activities including species identification, forensic body fluid identification, and post-mortem interval (PMI) inference analysis (Song et al., 2023; Yang et al., 2014).

5.4.3 Forensic epigenetics

Epigenetic indicators have been shown to have potential uses in forensic science in recent studies. In particular, there is evidence that these markers can accurately predict tissue types, distinguish between monozygotic (MZ) twins, and determine the age of a DNA contributor. Although the majority of research has used large DNA samples, encouraging results have also been seen with small initial DNA concentrations (100 pg), which is crucial for forensic applications (Yang et al., 2014).

5.4.4 Microbiome analysis

The investigation of the oral microbiota is an intriguing and useful application of NGS technology. By recognizing species at the molecular level and figuring out the relative quantity of bacteria, this technique can reveal a microbiome profile. Since these profiles are frequently individual-specific, they are useful for genetic research. Additionally, an individual's oral microbiota can provide insight into their geographic location and recent behaviours (Guo et al., 2016; Nilendu, 2023). NGS is expected to soon emerge as a significant area of research in forensic genetics, despite its limited use at this time. It could serve as a suitable alternative to PCRCE analysis (Cláudia-Ferreira et al., 2023).

Necrobiome or thanato-microbiome analysis employing NGS technology, specifically pyrosequencing, is a relatively new addition to the toolkit in the field of forensic research (Tozzo et al., 2022). The population of organisms participating in a corpse's decomposition is known as the necrobiome. It is possible to identify important bacterial taxa within these communities and determine the decomposition stage and succession by analyzing the microbiota profile of human remains (Javan et al., 2016). Studies have shown a relationship between the evaluation of carrion decomposition and bacterial populations, indicating that the necrobiome may be a useful tool for estimating the PMI. It serves as a microbial clock indicator in this way (Guo et al., 2016). The genetic development of DNA bacteria, including pathogens and decomposition microbiota, might be more thoroughly investigated with the use of a paleoforensic science adaptation of this technique (Irfan et al., 2024; West, 2019).

5.5 Issues with legitimacy, acceptance, and dependability

In the field of forensic investigation today, CE analysis of STRs mostly concentrates on noncoding DNA regions (Vajpayee et al., 2020). But with the introduction of more recent MPS programs, it is now possible to examine coding areas linked to ancestry and phenotype. Privacy problems may arise since these areas may contain information that could be medically relevant (Dash, Vajpayee, Srivastava, et al., 2021). States in the United States have the authority to regulate such issues; states like Indiana, New Mexico, Rhode Island, and Wyoming specifically forbid the phenotyping of DNA stored in CODIS, whereas Texas expressly allows it.

Significantly, the move to sequencing the whole mitochondrial genome rather than just the hypervariable section entails examining genes that might have information about disease. The United States is currently concerned about this issue (Greig et al., 2005). Europe, on the other hand, permits the use of more meaningful whole mitochondrial genome data for individual identification in court (Mavraki et al., 2023). According to a recent evaluation of 33 European laboratories, integrating MPS into forensic DNA analysis presents logistical, legal, and ethical issues (Hastings et al., 2012). These difficulties include a lack of financing and law for MPS forensic applications, incompatibility with current databases and population data for statistical analysis, and the absence of defined nomenclature and reporting standards.

Even while cloud computing is becoming a viable option for data storage, privacy and security issues still exist. Although MPS has the potential to revolutionize the criminal justice system, its implementation might result in difficult and divisive admission hearings (Machado et al., 2024). The 2017 McNair case marked the first such hearing, which was distinctive as it evaluated the technology as well as its use in differentiating between MZ twins. In this case, Dwayne McNair and his twin brother, Anwar Thomas, shared a DNA match based on normal STR profiles, and WGS was used to distinguish between them. In spite of Eurofins Scientific's assertion that WGS can identify identical twins, the presiding judge rejected the DNA evidence at the admission hearing, stating that more thorough testing and crime scene simulation were required before such evidence could be admitted (Erlich et al., 2020). In spite of this setback, additional evidence led to Dwayne McNair's final conviction for the rape (Erlich et al., 2020; Gerth, 2017).

The admissibility of MPS technology was not specifically addressed in the admission hearing, which focused on a particular use of WGS by MPS (Erlich et al., 2020; Gerth, 2017). As a result, the scientific issues raised by the MPS examination of evidentiary samples are still under discussion in later sessions. However, it is important to recognize that the fundamental ideas and techniques of polymerase chain reaction (PCR) and DNA sequencing (Roth, 2020) have been understood and approved by the scientific community for many years (Hotis, 1997). Over the past 30 years, these methods have been rigorously tested, evaluated, and validated by scientists using casework samples. More than 20 years have passed since manual techniques for sequencing mtDNA were accepted as early as 1996 (Budowle et al., 2003).

It is important to acknowledge that the scientific foundations of MPS are well established, even though more testing, validation, and implementation of this technique in forensic DNA applications inside the United States are necessary (Erlich et al., 2020). For more than ten years, the forensic community has been actively involved in MPS testing. According to the Daubert and Frye admissibility requirements, privacy concerns have been the main focus of some people's criticisms of MPS in forensic applications rather than issues of dependability or widespread acceptability. In addition to requiring affirmative informed consent, these critics contend that permitting the state to search criminal DNA evidence against public genomic databases is extremely intrusive and may reveal secondary private genetic information about people and their family members (Chandra & Sharma, 2023; Erlich et al., 2020).

5.6 Advantages and limitations

Comprehensive datasets have been methodically assembled into DNA databases ever since genetic technology was introduced into the forensic field. Notable databases such as CODIS and the UK National DNA Database (NDNAD) have become essential, and they are complemented by national repositories that are kept up to date by police and civil guards. Professionals, including attorneys, police officers, physicians, and individuals, are now skilled at using and handling this data, exhibiting a general understanding of how to analyze it. However, considering the emergence of a unique data format and the need for specialized training for researchers across several fields, integrating NGS into forensic science presents a substantial challenge. Furthermore, switching from traditional genetic analysis to NGS produces different kinds of data, which calls for changes to the law.

Certain kits, such as the AmpFLSTR® Identifiler® Direct PCR Amplification Kit (produced by Applied Biosystems, Foster City, CA), have been designed to simplify and speed up the process for database and single-source reference samples. Without the requirement for DNA extraction or purification procedures, these kits enable immediate amplification of the 16 loci included in the Identifiler® Kit. This technical development makes it possible to quickly gather a large amount of information in a short amount of time. On the other hand, NGS does not yet demonstrate the same degree of data-generating efficiency.

One significant benefit of NGS is its ability to economically generate large datasets; as evidenced by the work of Zaragoza and team (Zaragoza et al., 2010), several instruments can generate over one billion short reads each run. Despite its revolutionary potential, NGS's data generation speed is currently slower than that of well-established genetic analysis techniques. This disparity highlights the complex issues and difficulties involved in using NGS in forensic procedures.

Recent developments in the field of forensics, namely in population genetics, have introduced new genetic markers that overcome some of the major limitations of traditional and earlier DNA profiling techniques (Dash, Rawat, et al., 2021). In situations like mass disasters or occasions where forensic specimens and samples are compromised and degraded, the use of NGS technologies is expected to be crucial for DNA human typing (Dash et al., 2023). NGS provides the opportunity for simultaneous analysis of mtDNA, X and Y chromosomal markers, and standard autosomal DNA (short tandem repeats, or STRs, and single nucleotide polymorphisms, or SNPs) (Irwin et al., 2011). Additionally, NGS offers comprehensive genomic and phenotypic data that may have applications in forensic settings, such as physical evidence.

The usefulness of NGS technologies in routine forensic genetics casework has not yet been proven, despite its bright future. The application of these cutting-edge methods in real forensic casework is preceded by financial difficulties, strict forensic genetics requirements, and the necessity of multi-centric standardization exercises (Prewett, 2022). As is customary with traditional sequencing methods, there is yet no official forensic evaluation of the data quality produced by NGS technologies (Aalbers et al., 2023). The need for specific bioinformatics analysis tools designed for NGS data and its forensic applications should also be highlighted.

Although NGS has the potential to significantly alter genetic identity genotyping, its use in the forensic industry is expected to develop gradually. One significant benefit of NGS is its capacity to produce a wide range of data quickly (Ballard et al., 2020). Significant challenges must be overcome, too, including exorbitant expenses, a need for intensive training, and the complex administration of extensively pre-processed data. Because of the current state of software applications, businesses, forensic researchers, and scientists must work together to overcome this difficulty.

The successful integration of NGS into forensic sciences depends on the creation, expansion, and validation of software solutions. Furthermore, a major obstacle to inter-laboratory comparability of results is the lack of consistency in process and data analysis among forensic genetic laboratories involved in SGS STR typing (Ballard et al., 2020; Fordyce et al., 2015). In order to successfully integrate NGS technology into standard forensic procedures, it will be essential to address these issues.

5.7 Future prospects

Beyond the existing field of DNA analysis, NGS in forensic research has the potential to make revolutionary strides in the future. Increased sensitivity provided by NGS technologies makes it possible to detect even minute amounts of DNA, which is very useful when examining damaged or mixed DNA samples. This could improve the accuracy of individual identification by giving forensic investigators a more thorough DNA profile. Furthermore, NGS outperforms conventional techniques by enabling the simultaneous investigation of several genomic regions. Whole-genome sequencing provides forensic experts with detailed genetic information that allows them to comprehend biological materials more deeply. NGS is essential for the detailed analysis of mtDNA, which improves the identification of maternal lineage. This opens up new possibilities for identification in forensic investigations and is extremely helpful in situations where nuDNA analysis may be difficult. Forensic scientists can examine phenotypic and ancestral features encoded in DNA by utilizing NGS. This contributes to a comprehensive strategy in forensic identification by forecasting physical traits and offering information about people's geographic origins.

Furthermore, it expands forensic databases beyond conventional markers and enables a more thorough investigation of population genetics. A more inclusive forensic environment is promoted by the use of more varied genetic markers, which improve the precision of ancestry determination and individual identification. Continuous improvements in data processing techniques and bioinformatics make it easier to analyze large NGS datasets, which leads to more effective forensic result interpretation. This facilitates prompt decision-making and speeds up investigative procedures. The multidisciplinary uses of the tools in forensic science promote cooperation with other scientific fields, including epidemiology, anthropology, and microbiology. By taking a comprehensive approach to solving complex situations, this expands the scope of forensic investigations.

To handle privacy, permission, and the admissibility of complicated genomic data, ethical and legal frameworks must change as NGS becomes more widely used in forensics. To overcome these obstacles, strong legal and ethical frameworks are necessary. Forensic scientists must complete extensive training programs in order to incorporate NGS into standard forensic workflows. In order to improve reliability and repeatability, standardization initiatives are essential to ensuring consistency in procedures, data interpretation, and reporting across forensic laboratories. It is anticipated that continued technological development and growing use would lower NGS prices, opening up the technique to a wider variety of forensic labs. Improved pricing will make sophisticated genomic analysis tools more accessible to all.

Therefore, NGS has the potential to completely transform forensic science in the future by providing hitherto unheard-of capabilities for genetic analysis, identification, and investigative insight. In order to fully utilize NGS for the benefit of forensic investigations and justice systems around the world, interdisciplinary cooperation and ethical considerations will be crucial as technology develops.

References

Aalbers, S. E., Khan, A. T., & Weir, B. S. (2023). Perceptions of forensic scientists on statistical models, sequence data, and ethical implications for DNA evidence evaluations: A qualitative assessment. *Forensic Science International: Synergy*, *6*, 100335.

Aizpurua-Iraola, J., Rasal, R., Prieto, L., Comas, D., Bonet, N., Casals, F., Calafell, F., & Vásquez, P. (2023). Population analysis of complete mitogenomes for 334 samples from El Salvador. *Forensic Science International: Genetics*, *66*, 102906. https://doi.org/10.1016/j.fsigen.2023.102906.

Alkanaq, A. N., Hamanaka, K., Sekiguchi, F., Taguri, M., Takata, A., Miyake, N., Miyatake, S., Mizuguchi, T., & Matsumoto, N. (2019). Comparison of mitochondrial DNA variants detection using short-and long-read sequencing. *Journal of Human Genetics*, *64*(11), 1107–1116.

Alvarez-Cubero, M. J., Saiz, M., Martínez-García, B., Sayalero, S. M., Entrala, C., Lorente, J. A., & Martinez-Gonzalez, L. J. (2017). Next generation sequencing: An application in forensic sciences? *Annals of Human Biology*, *44*(7), 581–592.

Aly, S. M., & Sabri, D. M. (2015). Next generation sequencing (NGS): A golden tool in forensic toolkit. *Archiwum Medycyny Sądowej i Kryminologii/Archives of Forensic Medicine and Criminology, 65*(4), 260–271.

Ansorge, W. J. (2009). Next-generation DNA sequencing techniques. *New Biotechnology, 25*(4), 195–203.

Ansorge, W. J. (2010). Novel next-generation DNA sequencing techniques for ultra high-throughput applications in bio-medicine. In G. Patrinos & W. Ansorge (Eds) *Molecular Diagnostics* (pp. 365–378). Elsevier.

Ari, Ş, & Arikan, M. (2016). Next-generation sequencing: Advantages, disadvantages, and future. In K. R. Hakeem, H. Tombuloglu, & G. Tombuloglu (Eds.) *Plant Omics: Trends and Applications* (pp. 109–135). Springer.

Ballard, D., Winkler-Galicki, J., & Wesoły, J. (2020). Massive parallel sequencing in forensics: Advantages, issues, technicalities, and prospects. *International Journal of Legal Medicine, 134*(4), 1291–1303.

Berglund, E. C., Kiialainen, A., & Syvänen, A.-C. (2011). Next-generation sequencing technologies and applications for human genetic history and forensics. *Investigative Genetics, 2*(1), 1–15.

Bijarnia-Mahay, S., Arora, V., Puri, R. D., Lall, M., Saxena, R., Verma, J., Baijal, A., Dimri, N., Sharma, N., & Kohli, S. (2018). The changing scenario in prenatal diagnosis of genetic disorders: Genetics to genomics. *Current Medicine Research and Practice, 8*(6), 203–208.

Bodner, M., Bastisch, I., Butler, J. M., Fimmers, R., Gill, P., Gusmão, L., Morling, N., Phillips, C., Prinz, M., & Schneider, P. M. (2016). Recommendations of the DNA commission of the International Society for Forensic Genetics (ISFG) on quality control of autosomal short tandem repeat allele frequency databasing (STRidER). *Forensic Science International: Genetics, 24*, 97–102.

Børsting, C., & Morling, N. (2015). Next generation sequencing and its applications in forensic genetics. *Forensic Science International: Genetics, 18*, 78–89.

Boycott, K. M., Vanstone, M. R., Bulman, D. E., & MacKenzie, A. E. (2013). Rare-disease genetics in the era of next-generation sequencing: Discovery to translation. *Nature Reviews Genetics, 14*(10), 681–691.

Brandhagen, M. D., Just, R. S., & Irwin, J. A. (2020). Validation of NGS for mitochondrial DNA casework at the FBI laboratory. *Forensic Science International: Genetics, 44*, 102151.

Bronner, I. F., & Quail, M. A. (2019). Best practices for Illumina library preparation. *Current Protocols in Human Genetics, 102*(1), e86.

Bruijns, B., Tiggelaar, R., & Gardeniers, H. (2018). Massively parallel sequencing techniques for forensics: A review. *Electrophoresis, 39*(21), 2642–2654.

Budowle, B., Allard, M. W., Wilson, M. R., & Chakraborty, R. (2003). Forensics and mitochondrial DNA: Applications, debates, and foundations. *Annual Review of Genomics and Human Genetics, 4*(1), 119–141.

Buermans, H., & Den Dunnen, J. (2014). Next generation sequencing technology: Advances and applications. *Biochimica et Biophysica Acta (BBA)-Molecular Basis of Disease, 1842*(10), 1932–1941.

Bustillo, J., Fife, K., Merriman B., and Rothberg, J. Development of the Ion Torrent CMOS chip for DNA sequencing, *2013 IEEE International Electron Devices Meeting*, Washington, DC, USA, 2013, pp. 8.1.1–8.1.4, https://doi.org/10.1109/IEDM.2013.6724584.

Butler, J. M., Kline, M. C., & Coble, M. D. (2018). NIST interlaboratory studies involving DNA mixtures (MIX05 and MIX13): Variation observed and lessons learned. *Forensic Science International: Genetics, 37*, 81–94.

Carneiro, M. O., Russ, C., Ross, M. G., Gabriel, S. B., Nusbaum, C., & DePristo, M. A. (2012). Pacific biosciences sequencing technology for genotyping and variation discovery in human data. *BMC Genomics, 13*(1), 1–7.

Caulfield, T., Evans, J., McGuire, A., McCabe, C., Bubela, T., Cook-Deegan, R., Fishman, J., Hogarth, S., Miller, F. A., & Ravitsky, V. (2013). Reflections on the cost of "low-cost" whole genome sequencing: Framing the health policy debate. *PLoS Biology, 11*(11), e1001699.

Chan, E. Y. (2005). Advances in sequencing technology. *Mutation Research/Fundamental and Molecular Mechanisms of Mutagenesis, 573*(1–2), 13–40.

Chandra, G., & Sharma, R. (2023). Admissibility of forensic evidence in investigations: A comparative study between India, UK & USA. *Journal of Legal Studies & Research, 9*(5), 154–168.

Chen, Y. S., Yang, W. L., Zhao, Y. L., & Yang, Y. G. (2021). Dynamic transcriptomic m5C and its regulatory role in RNA processing. *Wiley Interdisciplinary Reviews: RNA, 12*(4), e1639.

Cláudia-Ferreira, A., Barbosa, D. J., Saegeman, V., Fernández-Rodríguez, A., Dinis-Oliveira, R. J., Freitas, A. R., Forensic, E. S. G., & Microbiology, P.-M. (2023). The future is now: Unraveling the expanding potential of human (Necro) microbiome in forensic investigations. *Microorganisms, 11*(10), 2509.

Daneshpour, M., Masjoodi, S., Ghafari, N., Raesi, M., & Fallah, M. (2018). Performance of Illumina next generation sequencing. *Pathobiology Research, 21*(3), 153–161.

Dash, H. R., Elkins, K. M., & Al-Snan, N. R. (2023). *Next Generation Sequencing (NGS) Technology in DNA Analysis*. Elsevier.

Dash, H. R., Rawat, N., Vajpayee, K., Shrivastava, P., & Das, S. (2021). Useful autosomal STR marker sets for forensic and paternity applications in the Central Indian population. *Annals of Human Biology*, *48*(1), 37–48.
Dash, H. R., Vajpayee, K., Agarwal, R., Gang, A., Shukla, R., & Srivastava, A. (2021). Evaluation of diallelic STR markers with inter-population allelic database for their usefulness in paternity trios in the Central Indian population. *Annals of Human Biology*, *48*(7–8), 605–613.
Dash, H. R., Vajpayee, K., Srivastava, A., & Das, S. (2021). Prevalence and characterisation of size and sequence-based microvariant alleles at nine autosomal STR markers in the Central Indian population. *Annals of Human Biology*, *48*(7–8), 614–620.
Davis, C., Peters, D., Warshauer, D., King, J., & Budowle, B. (2015). Sequencing the hypervariable regions of human mitochondrial DNA using massively parallel sequencing: Enhanced data acquisition for DNA samples encountered in forensic testing. *Legal Medicine*, *17*(2), 123–127.
De Mandal, S., Panda, A., Bisht, S., & Kumar, N. (2015). Microbial ecology in the era of next generation sequencing. *Journal of Next Generation Sequencing & Applications*, *1*(2). https://doi.org/10.4172/2469-9853.S1-001
Di Resta, C., Galbiati, S., Carrera, P., & Ferrari, M. (2018). Next-generation sequencing approach for the diagnosis of human diseases: Open challenges and new opportunities. *Ejifcc*, *29*(1), 4.
Drmanac, R. (2010). Complete human genome sequencing for large-scale human disease studies [Abstract]. *Journal of Biomolecular Techniques*, *21*(3 Suppl), S27.
Erlich, H., Calloway, C., & Lee, S. B. (2020). Recent developments in forensic DNA technology. In H. Erlich, E. Stover, & T. J. White (Eds.), *Silent witness: Forensic DNA Analysis in Criminal Investigations and Humanitarian Disasters* (pp. 105–127). Oxford University Press.
Esposito, A., Colantuono, C., Ruggieri, V., & Chiusano, M. L. (2016). Bioinformatics for agriculture in the next-generation sequencing era. *Chemical and Biological Technologies in Agriculture*, *3*, 1–12.
Fordyce, S. L., Mogensen, H. S., Børsting, C., Lagacé, R. E., Chang, C.-W., Rajagopalan, N., & Morling, N. (2015). Second-generation sequencing of forensic STRs using the Ion Torrent™ HID STR 10-plex and the Ion PGM™. *Forensic Science International: Genetics*, *14*, 132–140.
Fumagalli, M., Vieira, F. G., Linderoth, T., & Nielsen, R. (2014). ngsTools: Methods for population genetics analyses from next-generation sequencing data. *Bioinformatics*, *30*(10), 1486–1487.
Ge, J., Sun, H., Li, H., Liu, C., Yan, J., & Budowle, B. (2014). Future directions of forensic DNA databases. *Croatian Medical Journal*, *55*(2), 163.
Gerth, O. M. (2017). Identical twins as a misnomer: How advancing technology protects the interest of justice in the courtroom. *Georgetown Journal of Legal Ethics*, *30*, 783.
Grada, A., & Weinbrecht, K. (2013). Next-generation sequencing: Methodology and application. *Journal of Investigative Dermatology*, *133*(8), 1–4.
Greig, T. W., Moore, M. K., Woodley, C. M., & Quattro, J. M. (2005). Mitochondrial gene sequences useful for species identification of western North Atlantic Ocean sharks. *Fishery Bulletin*, *103*(3), 516–523.
Guffanti, A., Simchovitz, A., & Soreq, H. (2014). Emerging bioinformatics approaches for analysis of NGS-derived coding and non-coding RNAs in neurodegenerative diseases. *Frontiers in Cellular Neuroscience*, *8*, 89.
Guo, J., Fu, X., Liao, H., Hu, Z., Long, L., Yan, W., Ding, Y., Zha, L., Guo, Y., & Yan, J. (2016). Potential use of bacterial community succession for estimating post-mortem interval as revealed by high-throughput sequencing. *Scientific Reports*, *6*(1), 24197.
Gut, I. G. (2013). New sequencing technologies. *Clinical and Translational Oncology*, *15*(11), 879–881. https://doi.org/10.1007/s12094-013-1073-6
Hastings, R., De Wert, G., Fowler, B., Krawczak, M., Vermeulen, E., Bakker, E., Borry, P., Dondorp, W., Nijsingh, N., & Barton, D. (2012). The changing landscape of genetic testing and its impact on clinical and laboratory services and research in Europe. *European Journal of Human Genetics*, *20*(9), 911–916.
He, Y., Wu, J., Dressman, D. C., Iacobuzio-Donahue, C., Markowitz, S. D., Velculescu, V. E., Diaz, L. A., Jr, Kinzler, K. W., Vogelstein, B., & Papadopoulos, N. (2010). Heteroplasmic mitochondrial DNA mutations in normal and tumour cells. *Nature*, *464*(7288), 610–614. https://doi.org/10.1038/nature08802
Healy, K. (2007). Nanopore-based single-molecule DNA analysis. *Nanomedicine*, *2*(4), 459–481. https://doi.org/10.2217/17435889.2.4.459
Heather, J. M., & Chain, B. (2016). The sequence of sequencers: The history of sequencing DNA. *Genomics*, *107*(1), 1–8.
Herrnstadt, C., Elson, J. L., Fahy, E., Preston, G., Turnbull, D. M., Anderson, C., Ghosh, S. S., Olefsky, J. M., Beal, M. F., & Davis, R. E. (2002). Reduced-median-network analysis of complete mitochondrial DNA coding-region sequences for the major African, Asian, and European haplogroups. *The American Journal of Human Genetics*, *70*(5), 1152–1171.

Hofreiter, M., Paijmans, J. L., Goodchild, H., Speller, C. F., Barlow, A., Fortes, G. G., Thomas, J. A., Ludwig, A., & Collins, M. J. (2015). The future of ancient DNA: Technical advances and conceptual shifts. *BioEssays*, *37*(3), 284–293.
Holland, M. M., McQuillan, M. R., & O'Hanlon, K. A. (2011). Second generation sequencing allows for mtDNA mixture deconvolution and high resolution detection of heteroplasmy. *Croatian Medical Journal*, *52*(3), 299–313.
Hotis, K. K. (1997). The admissibility of PCR-based DNA evidence: State v. Lyons. *Jurimetrics*, *37*(4), 495–506.
Hu, T., Chitnis, N., Monos, D., & Dinh, A. (2021). Next-generation sequencing technologies: An overview. *Human Immunology*, *82*(11), 801–811.
Hussen, B. M., Abdullah, S. T., Salihi, A., Sabir, D. K., Sidiq, K. R., Rasul, M. F., Hidayat, H. J., Ghafouri-Fard, S., Taheri, M., & Jamali, E. (2022). The emerging roles of NGS in clinical oncology and personalized medicine. *Pathology-Research and Practice*, *230*, 153760.
Imam, J., Shrivastava, P., Dixit, S., & Shrivastava, A. (2018). Future of DNA fingerprinting: Application of NGS in forensic science. In H. R. Dash, P. Shrivastava, B. K. Mohapatra, & S. K. Jain (Eds.): *DNA Fingerprinting: Advancements and Future Endeavors* (pp. 259–275). Springer
Iozzi, S., Carboni, I., Contini, E., Pescucci, C., Frusconi, S., Nutini, A., Torricelli, F., & Ricci, U. (2015). Forensic genetics in NGS era: New frontiers for massively parallel typing. *Forensic Science International: Genetics Supplement Series*, *5*, e418–e419.
Irfan, M., Khan, K., & Khan, N. (2024). Application of NGS in maternal genome analysis in ancient human remains. In P. Shrivastava, H. R. Dash, & S. K. Jain (Eds.) *Next Generation Sequencing (NGS) Technology in DNA Analysis* (pp. 195–217). Elsevier.
Irwin, J. A., Just, R. S., & Parson, W. (2016). Massively parallel mitochondrial DNA sequencing in forensic genetics: Principles and opportunities. In A. Amorim & B. Budowle (Eds.): *Handbook of Forensic Genetics: Biodiversity and Heredity in Civil and Criminal Investigation* (Vol. 2, pp. 293–322). World Scientific. https://doi.org/10.1142/9781786340788_0012
Irwin, J. A., Parson, W., Coble, M. D., & Just, R. S. (2011). mtGenome reference population databases and the future of forensic mtDNA analysis. *Forensic Science International: Genetics*, *5*(3), 222–225.
Jackson, S. A. (2016). Rice: The first crop genome. *Rice*, *9*(1), 14.
Javan, G. T., Finley, S. J., Can, I., Wilkinson, J. E., Hanson, J. D., & Tarone, A. M. (2016). Human thanatomicrobiome succession and time since death. *Scientific Reports*, *6*(1), 29598.
Jeon, S. A., Park, J. L., Kim, J.-H., Kim, J. H., Kim, Y. S., Kim, J. C., & Kim, S.-Y. (2019). Comparison of the MGISEQ-2000 and Illumina HiSeq 4000 sequencing platforms for RNA sequencing. *Genomics & Informatics*, *17*(3), e32. https://doi.org/10.5808/GI.2019.17.3.e32
Jin, X. Y., Liu, Y. F., Cui, W., Chen, C., Zhang, X. R., Huang, J., & Zhu, B. F. (2022). Development a multiplex panel of AISNPs, multi-allelic InDels, microhaplotypes, and Y-SNP/InDel loci for multiple forensic purposes via the NGS. *Electrophoresis*, *43*(4), 632–644.
Karafet, T. M., Mendez, F. L., Meilerman, M. B., Underhill, P. A., Zegura, S. L., & Hammer, M. F. (2008). New binary polymorphisms reshape and increase resolution of the human Y chromosomal haplogroup tree. *Genome Research*, *18*(5), 830–838.
Kchouk, M., Gibrat, J. F., & Elloumi, M. (2017). Generations of sequencing technologies: From first to next generation. *Biology and medicine*, *9*(3).
Koonin, E. V., Makarova, K. S., & Wolf, Y. I. (2021). Evolution of microbial genomics: Conceptual shifts over a quarter century. *Trends in Microbiology*, *29*(7), 582–592. https://doi.org/10.1016/j.tim.2021.01.005.
Korlach, J. (2013). Understanding Accuracy in SMRT® Sequencing. https://www.pacb.com/wp-content/uploads/2015/09/Perspective_UnderstandingAccuracySMRTSequencing1.pdf
Ku, C. S., Pawitan, Y., Wu, M., Roukos, D. H., & Cooper, D. N. (2013). The evolution of high-throughput sequencing technologies: From sanger to single-molecule sequencing. In: W. Wu & H. Choudhry (Eds) *Next Generation Sequencing in Cancer Research*. Springer, New York, NY. https://doi.org/10.1007/978-1-4614-7645-0_1
Kulski, J. K. (2016). Next-generation sequencing—an overview of the history, tools, and "Omic" applications. *Next Generation Sequencing-Advances, Applications and Challenges*, *10*, 61964.
Kumar, K. R., Cowley, M. J., & Davis, R. L. (2019). Next-generation sequencing and emerging technologies. *Seminars in thrombosis and hemostasis. Seminars in Thrombosis and Hemostasis*, *45*(7), 661–673. https://doi.org/10.1055/s-0039-1688446
Kumar, S., Banks, T. W., & Cloutier, S. (2012). SNP discovery through next-generation sequencing and its applications. *International Journal of Plant Genomics*, *2012*. https://doi.org/10.1155/2012/831460
Larmuseau, M. H., Van Geystelen, A., Kayser, M., van Oven, M., & Decorte, R. (2015). Towards a consensus Y-chromosomal phylogeny and Y-SNP set in forensics in the next-generation sequencing era. *Forensic Science International: Genetics*, *15*, 39–42.

Lavezzo, E., Barzon, L., Toppo, S., & Palu, G. (2016). Third generation sequencing technologies applied to diagnostic microbiology: Benefits and challenges in applications and data analysis. *Expert Review of Molecular Diagnostics*, *16*(9), 1011–1023.
Lee, H., Gurtowski, J., Yoo, S., Nattestad, M., Marcus, S., Goodwin, S., Richard McCombie, W., & Schatz, M. C. (2016). Third-generation sequencing and the future of genomics. *bioRxiv*, 048603. https://doi.org/10.1101/048603
Legati, A., Zanetti, N., Nasca, A., Peron, C., Lamperti, C., Lamantea, E., & Ghezzi, D. (2021). Current and new next-generation sequencing approaches to study mitochondrial DNA. *The Journal of Molecular Diagnostics*, *23*(6), 732–741.
Liu, L., Li, Y., Li, S., Hu, N., He, Y., Pong, R., Lin, D., Lu, L., & Law, M. (2012). Comparison of next-generation sequencing systems. *Journal of Biomedicine and Biotechnology*, *2012*. 251364. https://doi.org/10.1155/2012/251364
Liu, Z., Zhu, L., Roberts, R., & Tong, W. (2019). Toward clinical implementation of next-generation sequencing-based genetic testing in rare diseases: Where are we? *Trends in Genetics*, *35*(11), 852–867.
Loomis, E. W. (2013). *Molecular Consequences of Expanded CGG Repeats in the Human FMR1 Gene: R-Loops, Methylation, and DNA Sequencing*. University of California.
Loreille, O., Koshinsky, H., Fofanov, V., & Irwin, J. (2011). Application of next generation sequencing technologies to the identification of highly degraded unknown soldiers' remains. *Forensic Science International: Genetics Supplement Series*, *3*(1), e540–e541.
Luo, C, Tsementzi, D, Kyrpides, N, Read, T, Konstantinidis, K. T. (2012) Correction: Direct comparisons of Illumina vs. Roche 454 sequencing technologies on the same microbial community DNA sample. *PLoS ONE*, *7*(3). https://10.1371/annotation/64ba358f-a483-46c2-b224-eaa5b9a33939.
Lyons, E. A., Scheible, M. K., Sturk-Andreaggi, K., Irwin, J. A., & Just, R. S. (2013). A high-throughput Sanger strategy for human mitochondrial genome sequencing. *BMC Genomics*, *14*. https://doi.org/10.1186/1471-2164-14-881
Machado, H., Neiva, L., Granja, R., & Silva, S. (2024). Ethical, legal, and social challenges of next-generation sequencing technologies (NGS) in forensic criminal identification. In P. Shrivastava, H. R. Dash, & S. K. Jain (Eds.) *Next Generation Sequencing (NGS) Technology in DNA Analysis* (pp. 551–569). Elsevier.
Mardis, E. R. (2008). The impact of next-generation sequencing technology on genetics. *Trends in Genetics*, *24*(3), 133–141.
Mason-Suares, H., Landry, L., & Lebo, M. S. (2016). Detecting copy number variation via next generation technology. *Current Genetic Medicine Reports*, *4*, 74–85.
Matullo, G., Di Gaetano, C., & Guarrera, S. (2013). Next generation sequencing and rare genetic variants: From human population studies to medical genetics. *Environmental and Molecular Mutagenesis*, *54*(7), 518–532. https://doi.org/10.1002/em.21799
Mavraki, E., Labrum, R., Sergeant, K., Alston, C. L., Woodward, C., Smith, C., Knowles, C. V., Patel, Y., Hodsdon, P., & Baines, J. P. (2023). Genetic testing for mitochondrial disease: The United Kingdom best practice guidelines. *European Journal of Human Genetics*, *31*(2), 148–163.
McCarthy, A. (2010). Third generation DNA sequencing: Pacific biosciences' single molecule Real Time Technology. *Chemistry & Biology*, *17*(7), 675–676. https://doi.org/10.1016/j.chembiol.2010.07.004
McCook, A. (2005). The automated DNA sequencer: Where would the human genome project, bioinformatics, and life sciences in general, be without it? *The Scientist*, *19*(16), 15–20.
Melton, T., Holland, C., & Holland, M. (2012). Forensic mitochondria DNA analysis: Current practice and future potential. *Forensic Science Review*, *24*(2), 101.
Metzker, M. L. (2008). *Advances in Next-Generation DNA Sequencing Technologies*. CRC Press.
Metzker, M. L. (2010). Sequencing technologies—the next generation. *Nature Reviews Genetics*, *11*(1), 31–46.
Miller, A. D., Good, R. T., Coleman, R. A., Lancaster, M. L., & Weeks, A. R. (2013). Microsatellite loci and the complete mitochondrial DNA sequence characterized through next generation sequencing and de novo genome assembly for the critically endangered orange-bellied parrot, Neophema chrysogaster. *Molecular Biology Reports*, *40*, 35–42.
Minoche, A. E., Dohm, J. C., & Himmelbauer, H. (2011). Evaluation of genomic high-throughput sequencing data generated on Illumina HiSeq and genome analyzer systems. *Genome Biology*, *12*(11), 1–15.
Moti, T. B., Guyassa, C., Dima, C., & Home, M. (2022). Illumina sequencing technology review. *Image*, *10*(3), 25–31.
Nilendu, D. (2023). Toward Oral thanatomicrobiology—An overview of the forensic implications of oral microflora. *Academic Forensic Pathology*. https://doi.org/10.1177/19253621231176411
Ozsolak F. (2012). Third-generation sequencing techniques and applications to drug discovery. *Expert Opinion on Drug Discovery*, *7*(3), 231–243. https://doi.org/10.1517/17460441.2012.660145
Pareek, C. S., Smoczynski, R., & Tretyn, A. (2011). Sequencing technologies and genome sequencing. *Journal of Applied Genetics*, *52*, 413–435.

Parson, W., Strobl, C., Huber, G., Zimmermann, B., Gomes, S. M., Souto, L., Fendt, L., Delport, R., Langit, R., & Wootton, S. (2013). Evaluation of next generation mtGenome sequencing using the Ion torrent personal genome machine (PGM). *Forensic Science International: Genetics*, *7*(5), 543–549.

Petric, R. C., Pop, L.-A., Jurj, A., Raduly, L., Dumitrascu, D., Dragos, N., & Neagoe, I. B. (2015). Next generation sequencing applications for breast cancer research. Clujul Medical, *88*(3), 278.

Prewett, T. (2022). *A Path Towards General Acceptance: The Feasibility of Massively Parallel Sequence Technology, Nomenclature, and Statistical Analysis in Forensic Casework*. University of Central Oklahoma.

Rockenbauer, E., Hansen, S., Mikkelsen, M., Børsting, C., & Morling, N. (2014). Characterization of mutations and sequence variants in the D21S11 locus by next generation sequencing. *Forensic Science International: Genetics*, *8*(1), 68–72.

Romsos, E. L., & Vallone, P. M. (2015). Rapid PCR of STR markers: Applications to human identification. *Forensic Science International: Genetics*, *18*, 90–99.

Roth, A. (2020). Admissibility of DNA evidence in court. In H. Erlich, E. Stover, & T. J. White (Eds.) *Silent Witness: Forensic DNA Evidence in Criminal Investigations and Humanitarian Disasters* (pp. 291–322). Oxford University Press. https://doi.org/10.1093/oso/9780190909444.003.0014

Rothberg, J. M., Hinz, W., Rearick, T. M., Schultz, J., Mileski, W., Davey, M., Leamon, J. H., Johnson, K., Milgrew, M. J., & Edwards, M. (2011). An integrated semiconductor device enabling non-optical genome sequencing. *Nature*, *475*(7356), 348–352.

Saracino, I. M., Pavoni, M., Zullo, A., Fiorini, G., Lazzarotto, T., Borghi, C., & Vaira, D. (2021). Next generation sequencing for the prediction of the antibiotic resistance in helicobacter pylori: A literature review. *Antibiotics*, *10*(4), 437.

Satam, H., Joshi, K., Mangrolia, U., Waghoo, S., Zaidi, G., Rawool, S., Thakare, R. P., Banday, S., Mishra, A. K., & Das, G. (2023). Next-generation sequencing technology: Current trends and advancements. *Biology*, *12*(7), 997.

Schadt, E. E., Turner, S., & Kasarskis, A. (2010). A window into third-generation sequencing. *Human Molecular Genetics*, *19*(R2), R227–R240.

Schook, L. B., Collares, T. V., Darfour-Oduro, K. A., De, A. K., Rund, L. A., Schachtschneider, K. M., & Seixas, F. K. (2015). Unraveling the swine genome: Implications for human health. *Annual Review of Animal Biosciences*, *3*(1), 219–244.

Schwartz, S., & Motorin, Y. (2017). Next-generation sequencing technologies for detection of modified nucleotides in RNAs. *RNA Biology*, *14*(9), 1124–1137.

Sehrawat, B., Sridharan, M., Ghosh, S., Robson, P., Cass, C. E., Mackey, J. R., Greiner, R., & Damaraju, S. (2011). Potential novel candidate polymorphisms identified in genome-wide association study for breast cancer susceptibility. *Human Genetics*, *130*, 529–537.

Seo, S. B., King, J. L., Warshauer, D. H., Davis, C. P., Ge, J., & Budowle, B. (2013). Single nucleotide polymorphism typing with massively parallel sequencing for human identification. *International Journal of Legal Medicine*, *127*, 1079–1086.

Serratì, S., De Summa, S., Pilato, B., Petriella, D., Lacalamita, R., Tommasi, S., & Pinto, R. (2016). Next-generation sequencing: Advances and applications in cancer diagnosis. *Onco Targets and Therapy*, *9*, 7355–7365. https://doi.org/10.2147/OTT.S99807

Shen, T., Pajaro-Van de Stadt, S. H., Yeat, N. C., & Lin, J. C.-H. (2015). Clinical applications of next generation sequencing in cancer: From panels, to exomes, to genomes. *Frontiers in Genetics*, *6*, 215.

Shendure, J., & Ji, H. (2008). Next-generation DNA sequencing. *Nature Biotechnology*, *26*(10), 1135–1145. https://doi.org/10.1038/nbt1486.

Shyr, D., & Liu, Q. (2013). Next generation sequencing in cancer research and clinical application. *Biological Procedures Online*, *15*, 1–11.

Singer, A., Wanunu, M., Morrison, W., Kuhn, H., Frank-Kamenetskii, M., & Meller, A. (2010). Nanopore-based sequence-specific detection of duplex DNA for genomic profiling. *Biophysical Journal*, *98*(3), 598a–599a.

Slatko, B. E., Gardner, A. F., & Ausubel, F. M. (2018). Overview of next-generation sequencing technologies. *Current Protocols in Molecular Biology*, *122*(1), e59.

Song, B., Qian, J., & Fu, J. (2023). Research progress and potential application of microRNA and other non-coding RNAs in forensic medicine. *International Journal of Legal Medicine*, *138*(3), 1–22.

Soto, J., Rodriguez-Antolin, C., Vallespín, E., de Castro Carpeño, J., & Ibanez de Caceres, I. (2016). The impact of next-generation sequencing on the DNA methylation-based translational cancer research. *Translational Research*, *169*, 1–18. https://doi.org/10.1016/j.trsl.2015.11.001

Stapley, J., Reger, J., Feulner, P. G., Smadja, C., Galindo, J., Ekblom, R., Bennison, C., Ball, A. D., Beckerman, A. P., & Slate, J. (2010). Adaptation genomics: The next generation. *Trends in Ecology & Evolution*, *25*(12), 705–712.

Tan, M. P., Wong, L. L., Razali, S. A., Afiqah-Aleng, N., Mohd Nor, S. A., Sung, Y. Y., Van de Peer, Y., Sorgeloos, P., & Danish-Daniel, M. (2019). Applications of next-generation sequencing technologies and computational tools in molecular evolution and aquatic animals conservation studies: A short review. *Evolutionary Bioinformatics, 15,* 1176934319892284.

Tang, S., & Huang, T. (2010). Characterization of mitochondrial DNA heteroplasmy using a parallel sequencing system. *BioTechniques, 48*(4), 287–296.

Templeton, J. E., Brotherton, P. M., Llamas, B., Soubrier, J., Haak, W., Cooper, A., & Austin, J. J. (2013). DNA capture and next-generation sequencing can recover whole mitochondrial genomes from highly degraded samples for human identification. *Investigative Genetics, 4*(1), 1–13.

The 100 Genomes Project Consortium. (2015). A global reference for human genetic variation. Nature, 526(7571), 68.

Tozzo, P., Amico, I., Delicati, A., Toselli, F., & Caenazzo, L. (2022). Post-mortem interval and microbiome analysis through 16S rRNA analysis: A systematic review. *Diagnostics, 12*(11), 2641.

Tridico, S. R., Murray, D. C., Addison, J., Kirkbride, K. P., & Bunce, M. (2014). Metagenomic analyses of bacteria on human hairs: A qualitative assessment for applications in forensic science. *Investigative Genetics, 5*(1), 1–13.

Tsai, Y.-T., Hrytsenko, Y., Elgart, M., Tahir, U., Chen, Z.-Z., Wilson, J. G., Gerszten, R., & Sofer, T. (2024). A parametric bootstrap approach for computing confidence intervals for genetic correlations with application to genetically-determined protein-protein networks. *HGG Advances, 5*(3). https://doi.org/10.1016/j.xhgg.2024.100304

Tuan, V. P., Narith, D., Tshibangu-Kabamba, E., Dung, H. D. Q., Viet, P. T., Sokomoth, S., Binh, T. T., Sokhem, S., Tri, T. D., & Ngov, S. (2019). A next-generation sequencing-based approach to identify genetic determinants of antibiotic resistance in Cambodian Helicobacter pylori clinical isolates. *Journal of Clinical Medicine, 8*(6), 858.

Tyagi, P., & Bhide, M. (2020). History of DNA sequencing. *Folia Veterinaria, 64*(2), 66–73.

Tyulyandina, A., Gorbunova, V., Khokhlova, S., Kolomiets, L., Filipenko, M., Imyanitov, E., Demidova, I., Moliaka, Y., Cherdyntseva, N., & Vodolajskiy, D. (2018). Profile of BRCA1/BRCA2 mutations in Russian ovarian cancer population detected by NGS and MLPA analysis: Interim results of OVATAR study. *Cancer Research, 78*(13_Supplement), 1241–1241.

Uhlen, M., & Quake, S. R. (2023). Sequential sequencing by synthesis and the next-generation sequencing revolution. *Trends in Biotechnology. 41*(12), 1565–1572. https://doi.org/10.1016/j.tibtech.2023.06.007

Vajpayee, K., Dash, H. R., Parekh, P. B., & Shukla, R. K. (2023). PCR inhibitors and facilitators – their role in forensic DNA analysis. *Forensic Science International. 349,* 111773. https://doi.org/10.1016/j.forsciint.2023

Vajpayee, K., Sagar, D., & Dash, H. R. (2020). Forensic DNA typing: Inception, methodology, and technical advancements. In P. Shrivastava, H. R. Dash, J. A. Lorente, & J. Imam (Eds.) *Forensic DNA Typing: Principles, Applications and Advancements* (pp. 3–26). Springer.

Valencia, C. A., Pervaiz, M. A., Husami, A., Qian, Y., Zhang, K., Valencia, C. A., Pervaiz, M. A., Husami, A., Qian, Y., & Zhang, K. (Eds.) (2013). Sanger sequencing principles, history, and landmarks. In *Next Generation Sequencing Technologies in Medical Genetics* (pp. 3–11). Springer

Van Dijk, E. L., Auger, H., Jaszczyszyn, Y., & Thermes, C. (2014). Ten years of next-generation sequencing technology. *Trends in Genetics, 30*(9), 418–426.

van Dijk, E. L., Naquin, D., Gorrichon, K., Jaszczyszyn, Y., Ouazahrou, R., Thermes, C., & Hernandez, C. (2023). Genomics in the long-read sequencing era. *Trends in Genetics. 39*(9), 649–671. https://doi.org/10.1016/j.tig.2023.04.006

Van Geystelen, A., Decorte, R., & Larmuseau, M. H. (2013). AMY-tree: An algorithm to use whole genome SNP calling for Y chromosomal phylogenetic applications. *BMC Genomics, 14,* 1–12.

Vermeulen, M. C. (2020). *Improving and Estimating Y Chromosome Loss in Blood and Brain Tissues Using High-Throughput Sequencing.* University of British Columbia.

Warshauer, D. H., Davis, C. P., Holt, C., Han, Y., Walichiewicz, P., Richardson, T., Stephens, K., Jager, A., King, J., & Budowle, B. (2015). Massively parallel sequencing of forensically relevant single nucleotide polymorphisms using TruSeq™ forensic amplicon. *International Journal of Legal Medicine, 129,* 31–36.

Weikard, R., Hadlich, F., & Kuehn, C. (2013). Identification of novel transcripts and noncoding RNAs in bovine skin by deep next generation sequencing. *BMC Genomics, 14,* 1–15.

Weirather, J. L., de Cesare, M., Wang, Y., Piazza, P., Sebastiano, V., Wang, X.-J., Buck, D., & Au, K. F. (2017). Comprehensive comparison of pacific biosciences and Oxford nanopore technologies and their applications to transcriptome analysis. *F1000Research, 6,* 100. https://doi.org/10.12688/f1000research.10571.2

Werner, T. (2010). Next generation sequencing in functional genomics. *Briefings in Bioinformatics, 11*(5), 499–511.

West, F. L. (2019).*From ancient DNA to forensic genetics: Validation studies in degraded DNA*. [Doctoral dissertation, University of Tennessee]. TRACE: Tennessee Research and Creative Exchange. https://trace.tennessee.edu/utk_graddiss/5396

Westen, A. A., Kraaijenbrink, T., Clarisse, L., Grol, L. J., Willemse, P., Zuniga, S. B., de Medina, E. A. R., Schouten, R., van der Gaag, K. J., & Weiler, N. E. (2015). Analysis of 36 Y-STR marker units including a concordance study among 2085 Dutch males. *Forensic Science International: Genetics, 14*, 174–181.

Wong, L.-J. C. (2013). *Next Generation Sequencing*. Springer.

Xiao, T., & Zhou, W. (2020). The third generation sequencing: The advanced approach to genetic diseases. *Translational Pediatrics, 9*(2), 163.

Xu, J., Chen, M., Liu, Q. Y., Hu, S. Q., Li, L. R., Li, J., & Ma, R. M. (2020). Detecting trisomy in products of conception from first-trimester spontaneous miscarriages by next-generation sequencing (NGS). *Medicine, 99*(5), e18731. https://doi.org/10.1097/MD.0000000000018731

Yang, Y., Xie, B., & Yan, J. (2014). Application of next-generation sequencing technology in forensic science. *Genomics, Proteomics & Bioinformatics, 12*(5), 190–197.

Ye, F., Samuels, D. C., Clark, T., & Guo, Y. (2014). High-throughput sequencing in mitochondrial DNA research. *Mitochondrion, 17*, 157–163.

Zaragoza, M. V., Fass, J., Diegoli, M., Lin, D., & Arbustini, E. (2010). Mitochondrial DNA variant discovery and evaluation in human cardiomyopathies through next-generation sequencing. *PLoS One, 5*(8), e12295.

Zavodna, M., Bagshaw, A., Brauning, R., & Gemmell, N. J. (2014). The accuracy, feasibility and challenges of sequencing short tandem repeats using next-generation sequencing platforms. *PLoS One, 9*(12), e113862.

Zeng, X., King, J. L., Stoljarova, M., Warshauer, D. H., LaRue, B. L., Sajantila, A., Patel, J., Storts, D. R., & Budowle, B. (2015). High sensitivity multiplex short tandem repeat loci analyses with massively parallel sequencing. *Forensic Science International: Genetics, 16*, 38–47.

Zhang, J., Chiodini, R., Badr, A., & Zhang, G. (2011). The impact of next-generation sequencing on genomics. *Journal of Genetics and Genomics, 38*(3), 95–109.

Zhang, W., Cui, H., & Wong, L.-J. C. (2012). Comprehensive one-step molecular analyses of mitochondrial genome by massively parallel sequencing. *Clinical Chemistry, 58*(9), 1322–1331.

6

From Data to Interpretation

Statistical Software in Forensic DNA Typing

Ranjeet Kumar Singh and Afreen Tarannum

6.1 Overview of forensic DNA analysis

DNA profiling (DNA Fingerprinting) has emerged as a keystone, revolutionizing the way evidence is analyzed and interpreted. The goal of DNA profiling is individualization of biological sample encountered at scene of crime. In forensic science, the concept of class and individual characteristics is used to differentiate evidence. Class characteristics are features shared by a group of individuals, whereas individual characteristics are unique and can be linked to a specific individual. Similarly, in case of biological sample analysis, the blood group is considered as a class characteristic. For instance, 44.29%, 34.08%, 16.8%, and 5.21% of indigenous populations globally have blood types "O," "A," "B," and "AB," respectively. Therefore, if we found a blood sample of B blood group at scene of crime, then it is common to 1,344,000,000 individuals, considering present world population to be 8 billion.

In forensic DNA analysis, it is a well-known fact 99.9% of human DNA is shared across all humans, but for forensic DNA analysis expert focuses on 0.1% of the non-coding region of human genome that varies. Out of approximately 3 billion nucleotide base pairs in the human genome, around 3 million nucleotide base pairs are of interest to forensic examiners. Forensic DNA analysis examines a set of 20 expanded CODIS short tandem repeat (STR) markers, despite the fact that STRs occur approximately every 2000 base pairs throughout the human genome. With over half a million STRs characterized, which make up about 3% of the human genome, there remains a chance of random matches in the population. To address this, a portion of the STR DNA profile is essential to determine the likelihood of a match and provide a reliable assessment in forensic contexts. Statistical interpretation plays a vital role in forensic DNA analysis due to intricacies involved in DNA profile matching of paternity disputes or mutilated body's cases identification. Mixed DNA profiles cases, such as from sexual assaults or touch DNA, the presence of multiple alleles makes interpretation challenging. DNA Experts' analysis is required to differentiate between major and minor contributors, and several statistical approaches are employed to ensure accurate attribution of DNA profiles in mixed samples. Hence, the true strength of DNA profiling lies in statistical discrimination, aiding precise identification.

In the Forensic Science Laboratory, DNA experts compare the biological samples found at scene of crime and match against a known reference sample for DNA profiling. This helps the investigating officer in establishing link between suspects and crime scenes for solving crimes. If two DNA profiles differ, reference person cannot be the source of crime scene DNA. However, if DNA profiles match, it indicates a potential match. While matching crime scene DNA profile to reference profile, it's essential to identify the alleles (peaks) in the profile.

DOI: 10.4324/9781003474296-6

According to Gill et al. (2006), three kinds of alleles appear in a crime stain profile i.e. *"Alleles that are unmistakable; Alleles that may be masked by artifacts; Alleles that have dropped out completely and cannot be detected."* Hence, the significance of a match is established by a statistical analysis (Dash et al., 2023; Taupin, 2014).

6.2 Role of statistical software in DNA analysis

Darrell Huff, in his book *How to Lie with Statistics,* mentioned H.G. Wells' quote, *"Statistical thinking will one day be as necessary for efficient citizenship as the ability to read and write."* Statistics has played a crucial role in forensic science and law by dealing with variability and uncertainty within these two disciplines. James Curran, Department of Statistics, University of Auckland in New Zealand, concluded that *"Statistical interpretation is a vital part of a modern forensic scientist's toolbox. It is incumbent upon us, as a community, to make sure that we have the best tool set available and that everyone knows how to use it."* Experts in field of forensic science use statistical methods to interpret evidence collected from scene of crime and submit their findings and interpretations to courts for legal proceedings. Experts employ a variety of methods for analysis, ranging from basic techniques like percentages, means, and standard deviations to more advanced methods such as hypothesis testing, regression, calibration, cluster analysis, logistic regression, and Fisher's exact test. For DNA analysis, statistical methods range from elementary to more complex methods such as least-square deconvolution. Fung et al. (2006) highlighted several reasons why statistics has attracted (much) more attention in DNA forensics than in any other forensic science areas. *"First, DNA profiling is generally scientifically unambiguous and very powerful. Since the DNA evidence is repeatable, statistical evaluation would then be possible and in most situations objective. Second, when there is a match to the DNA evidence, people would like to know how likely there is a random match. Third, extremely small probabilities are commonly encountered in DNA profiling, and people are curious about their derivations and interpretations (note: these probabilities are sometimes interpreted incorrectly, e.g., prosecutor's fallacy). Fourth, many forensic scientists are not that familiar with statistics, particularly on different approaches of the subject. Fifth, some problems such as kinship determinations and DNA mixtures need complex statistical analysis."*

The question may arise: "When is statistical interpretation required?" The answer is that statistical interpretation is required when DNA analysis cannot definitively exclude an individual as a contributor to a crime scene sample, statistical methods are used to evaluate the likelihood of their involvement. DNA profiles obtained from scenes of crime are used to establish a link between suspects and crime scene. Since the complete DNA databases are not available worldwide, it can be challenging to determine the true source. Suspect's DNA profile is compared with DNA profile found at crime scene using population data statistics to determine if there's a match or not. This process includes calculating the probability of random match, which is important for concluding whether DNA profile matches crime evidence or not. Population statistics help to measure uncertainty and provide the likelihood of a random match, estimated using frequency of a specific DNA profile within a population. This frequency data is derived from a small, random sample of individuals from a particular geographic area. Conclusively, statistics has become an essential tool in modern forensic science, particularly in DNA analysis. By quantifying uncertainty and providing objective measures of likelihood, statistical methods bridge the gap between scientific evidence and legal interpretation (Butler, 2015; Fung & Hu, 2008; Fung et al., 2006; Shrivastava et al., 2020).

6.3 Fundamental aspects of statistics

DNA evidence is powerful and influential, but to understand the real importance of DNA in a case, it needs to be carefully measured. Statistical methods serve as the tool for analyzing and interpreting genetic data and determining the evidential value of DNA profiles. This section of the chapter highlights three fundamental aspects of statistics that are essential in forensic DNA analysis i.e. Probability, Likelihood Ratios, and the Bayesian Framework.

6.3.1 Probability

Absence of direct witnesses often poses significant challenges in determining the true offender of the serious crime in cases of sexual assault and murder. In such scenarios, DNA evidence becomes an essential investigative tool for connecting offender to crime scenes without any direct witnesses. The result of DNA sample from a crime scene matches an offender's profile in terms of probabilities instead of certainty. The language of probabilities is employed to quantify this uncertainty, providing numerical values that reflect the strength of the DNA match. For example, a DNA expert states that there is a one in a million chance that DNA found at crime scene could match to someone other than suspect. These probability statements are crucial because they help the judicial system understand the weight of DNA evidence while recognizing and identifying that no forensic method is entirely free of doubt. Hence, the use of probability in DNA analysis ensures that findings are communicated with appropriate accuracy and vigilance.

Probability is defined as *"the number of times an event happens divided by the number of opportunities for it to happen (i.e. the number of trials)."* In simple words, it can be said that whenever we are not sure about the event outcome, we can talk about the possibilities of certain outcomes. Understanding probability can be challenging because we are often in the mindset that something either happened or it did not. Probability, however, exists on a scale ranging from zero (impossible) to one (certain), where zero indicates that an event will not occur, whereas one indicates that the event has occurred. In reality, events in science and life fall somewhere between these extremes. This simply means that decisions frequently must be made without complete certainty.

From the forensic perspective, profile probability is the fundamental and essential concept in the field of DNA profiling, which enumerates probability of evidentiary DNA profile occurring in a population. It is essentially a statistical hypothesis test used to determine probability of evidentiary DNA profile (denoted by E) under a given hypothesis (Ho). Profile probability denoted with symbols Pr and given by the notation or formula:

$$\mathrm{Pr} = \frac{E}{Ho}$$

In cases involving DNA evidence from a single contributor, the hypothesis is made that the DNA profile is from a person who is not related to suspect. Under this approximation, probability $\mathrm{Pr}(E|Ho)$ can be calculated by multiplying the frequencies of each individual genetic marker (allele) in the profile. This probability equates to frequency with which specific DNA profile appears in general population, meaning that if hypothesis Ho suggests that DNA sample originates from a random person unrelated to suspect, probability calculated would reflect how common that DNA profile is among the population at large. This approach is vital in forensic science because it provides a statistical basis for measuring and evaluating the strength of DNA evidence in relation to a particular hypothesis. By comparing frequency of profile in population with evidence at hand, forensic experts can better understand probability that evidence points to suspect or to someone else. This statistical framework plays a significant role in DNA evidence interpretation in legal contexts, helping to ensure and guarantee that conclusions drawn from DNA analysis are grounded in a rigorous, probabilistic understanding (Butler, 2010; Evett & Weir, 1998; Gill et al., 2006; National Research Council, 1996).

6.3.2 Likelihood ratios

In forensic DNA typing, comparison of DNA profiles between known sample (K) and questioned sample (Q) from a crime scene is a fundamental process to evaluate the significance of any potential match. While traditional methods like calculating Random Match Probability (RMP) offer understanding and insights, a more statistically rigorous approach involves using Likelihood Ratio (LR). Likelihood Ratios, denoted as LR, is a crucial concept in forensic DNA typing that provides a statistical measure to quantify the strength of DNA evidence by comparing probability of observed DNA profile under two hypotheses.

Hypothesis Hp: Prosecution Hypothesis, suspect is source of DNA collected from a crime scene.

Hypothesis Hd: Defendant Hypothesis, unrelated individual is source of DNA.

Likelihood ratio (LR) is calculated by:

$$\mathrm{LR} = \frac{Hp}{Hd}$$

Prosecution's hypothesis suggests that suspect is indeed person who committed crime. Under this hypothesis, if we assume suspect commits the crime and there is 100% probability that suspect has left DNA at crime scene, then Hp = 1. This confirms that suspect is source of DNA found at the scene of crime, hence likelihood of obtaining observed DNA profile is certain. Defendant hypothesis suggests that someone other than suspect is source of DNA found at crime scene, which means that TR typing is heterozygous. In such case, the probability is calculated based on allele frequencies in the relevant population and would be Hd = 2pq, where p and q are occurrences of allele 1 and 2 for a locus in a relevant population. These frequencies reflect how common the observed alleles are in the population. If STR typing is homozygous, then probability of defendant hypothesis would be Hd = p2. Therefore, equation for heterozygous would become:

$$\mathrm{LR} = \frac{Hp}{Hd} = \frac{1}{2pq}$$

Therefore, equation for homozygous would become:

$$\mathrm{LR} = \frac{Hp}{Hd} = \frac{1}{p2}$$

Likelihood Interpretation would be:

LR > 1: Likelihood ratio greater than 1 supports prosecution hypothesis, signifying that DNA evidence is more likely that suspect is source of DNA.

LR < 1: Likelihood ratio less than 1 supports defendant hypothesis, signifying that DNA evidence is more likely that someone else is source of DNA.

LR = 1: Likelihood ratio equal to 1 indicates that DNA evidence does not favor either hypothesis

Table 6.1 Strength of the Evidence According to Likelihood Ratio Result

Likelihood Ratio	Corresponding Evidence
1 to 10	Limited support
10 to 100	Moderated support
100 to 1000	Moderated strong support
1000 to 10,000	Strong support
10,000 or greater	Very strong support

LR provides a framework for evaluating DNA evidence found at scene of crime, helping to clarify strength of evidence in supporting the involvement of a suspect. The strength of LR can be interpreted numerically as presented in **Table 6.1**, given by Butler.

Conclusively, LR is a powerful and statistically sound approach in forensic DNA typing, offering a quantitative measure to assess strength of DNA evidence by comparing two competing hypotheses that can help in making informed decisions about significance of a match (Alamoudi et al., 2018; Butler, 2010; Butler, 2015; Gill et al., 2006).

6.3.3 Bayesian framework

When dealing with the evidence found at scene of crime, challenges of interpreting multiple stains, suspects, or combining different types of evidence have led many to seek substitutes to frequentist approach. Bayesian approach (also known as logical approach) is one of the prominent alternatives. This approach has been an essential in paternity testing since the 1930s. However, in later stages of the 20th century, it made inroads into many other fields of forensic science. This evolution marked a broadening of their application, highlighting their growing importance and versatility in forensic investigations. Bayesian statistical methods are becoming more popular in area of applied and fundamental research. It provides a comprehensible method for appraising beliefs in light of new evidence. It is named after the Reverend Thomas Bayes, who first provided the foundation for this approach in the 18th century. In 2001, Finkelstein and Levin stated that "*Bayes's theorem is a fundamental tool of inductive inference.*"

Given by:

$$\text{Posterior Odds} = \text{LR} \times \text{Prior Odds}$$

Prior odds are established through non-DNA evidence, which can at times be influenced by subjective factors. In contrast, LR is grounded in scientific DNA evidence, which provides a more objective basis for evaluation. Despite this scientific foundation, LR must still be calculated within a Bayesian framework to integrate both the prior odds and DNA evidence efficiently and effectively. This approach allows for a rigorous valuation of evidence while acknowledging initial uncertainty that non-DNA evidence might introduce (Buckleton et al., 2005; Butler, 2015).

6.4 Understanding of mixed profile

Mixed DNA profile contain genetic material from multiple individuals, making it more challenging to analyze than a single-source DNA profile, as the presence of more than three peaks at a given STR locus indicates mixed profile. In mixed profile interpretation cases, it's not enough for the person of interest to just share some alleles with profile. Instead, their DNA must be a strong match to entire mixed profile. First step of statistical analysis is to analyze the mixed DNA profile and determine number of contributors, which is a subjective and complex process. Typically, number of contributors is inferred from locus with the highest number of alleles, although the presence of homozygotes (which display two or fewer peaks) can make this challenging. In many sexual assault cases, forensic examiners have a prior knowledge about number of contributors to mixed DNA profile. In such type of cases, it's important to carefully examine DNA evidence for multiple peaks to avoid any error. Modern techniques might erroneously oversee low levels of DNA from the consensual partners in approximately 3% of cases, potentially leading to incorrect conclusions. CPI stand for combined probability of inclusion, a broadly used statistical method to evaluate mixed DNA profiles. CPI represents the population proportion that could potentially contribute to observed mixture. Calculation of CPI involves three steps: assessing the profile, determining inclusion or exclusion of the person of interest, and performing statistical calculations. Another approach is LR calculation, which compares probability of two competing hypotheses. Numerous software programs are now available to assist with interpreting mixed profiles and other forensic DNA applications (Dash et al., 2023).

6.5 Statistical software

Statistical software plays a significant role in forensic DNA profiling by converting raw data of genetics into interpretable evidence. Forensic DNA profiling includes the complex analysis of markers (genetic markers) to connect the links between an individual and crime scene. The statistical software advances the reliability and accuracy of such analyses. Software tools are applied to bring out the match probabilities, assess LR of DNA samples, and to handle large genetic databases. These tools are important to ensure the accuracy of conclusions of forensic and maintain the probity of legal activities.

6.5.1 STRmix

STRmix™ is a software tool designed for forensic DNA analysis that utilizes a continuous probabilistic genotyping technique to evaluate complex DNA profiles. It basically addresses challenges of interpreting DNA mixtures, which can be particularly difficult when multiple individuals contribute to a sample. By using sophisticated statistical methods, STRmix™ provides forensic experts with a powerful means of estimating likelihood of different contributors to a DNA sample, enhancing the accuracy and reliability of forensic evidence. STRmix™ can analyze samples with or without reference. STRmix™ utilizes a well-defined mathematical procedure known as Markov Chain Monte Carlo (MCMC) to randomly assign probabilities to measure the genotype combinations for a certain contributor, which is a crucial feature of the program. This method efficiently explore vast number of genotypes and determine most likely contributors to the DNA mixture. Importantly, results of earlier reports remain valid even after STRmix™ implementation, ensuring the integration of STRmix™ into forensic practice and enhances interpretation process without undermining the integrity of prior findings.

STRmix™ offers several characteristics that enhance its efficacy in forensic DNA typing. It provides faster interpretation of the complex DNA profiles, significantly improving proficiency of casework. The software is accessible, and its results are very precise, which facilitates clear communication in court. STRmix™ includes analytical thresholds and models for stutter and peak height, as well as drop-out and drop-in, which help in accurately interpreting complex DNA profiles. Furthermore, it allows for integration of DNA profiles from different kits within the same elucidation, ensuring a comprehensive and cohesive analysis.

6.5.1.1 STRmix™ working

STRmix™ is a software used to analyze DNA evidence, particularly in cases of complex DNA samples from multiple individuals. It works by combining biological modeling and mathematical processes to interpret complex DNA profiles. Using statistical methods, software generates millions of hypothetical DNA profiles and compares them to evidence, identifying combinations that best match observed DNA. This allows STRmix™ to determine likelihood of different DNA contributors being present in a sample. By using statistical methods and considering factors like allele peak heights and stutter, STRmix provides a more accurate and informative analysis of DNA evidence. It uses Bayesian algorithms to assess likelihood of a specific individual's DNA being part of a complex sample versus it coming from other, unknown contributors. This basically allows forensic experts to provide more accurate and interpretable results in criminal investigations, even when dealing with low-level or highly complex DNA samples. In court, the reliability of STRmix™ is supported by empirical studies, ensuring that its mathematical framework is readily accessible to DNA analysts.

6.5.1.2 Limitations

STRmix has its own limitations, one primary constraint is its dependency on quality and quantity of DNA sample. Degraded or low-concentration DNA samples can reduce result accuracy and precision. Furthermore, STRmix also struggle with extremely complex DNA mixtures, where multiple individuals are present in equal proportions, leading to potential doubts in interpretation. The software relies heavily on reference databases and models used; if these are outdated or incomplete, it can impact reliability and consistency of conclusions. Finally, while STRmix provides a robust statistical framework, its outputs still require careful interpretation by forensic experts, as software does not eliminate need for professional judgment in forensic analyses (Bright et al., 2019; Shrivastava et al., 2020)

6.5.2 LRmix

LRmix released on 15th December by Hinda Haned and Jeroen, assist forensic experts in interpreting complex DNA profiles by calculating LRs, which quantify weight of evidence. It was partially funded by a grant from the Netherlands Genomics Initiative/Netherlands Organization for Scientific Research (NWO) through Forensic Genomics Consortium Netherlands. It is a free, open-source (GPLv3) software for interpreting forensic DNA profile, with an emphasis on complex DNA mixes. LRmix studio is written in Java and can determine probative value of any (autosomal STR-based) forensic DNA profile. This model clearly accounts for ambiguity in DNA profiles due to allelic drop-out and drop-in occurrences. The program estimates these numbers based on available data and utilizes them to calculate LRs. LRmix Studio is also beneficial for interpreting SNP mixtures (Tasselli et al., 2015).

6.5.2.1 Features of LRmix

The current version of LRmix can calculate LRs for the DNA profiles categorized with the autosomal STR kits and has undergone extensive testing and validation for scenarios involving up to 4 unknown contributors, with a maximum of 5 contributors in total. The software operates under assumption that hypothesized contributors in both prosecution and defense hypotheses are unrelated to each other, though they may be related to an unidentified contributor under defense hypothesis. The LRmix compare any number of replicates from specific DNA sample to any number of reference profiles; it has been thoroughly tested and validated for scenarios involving no more than 3 reference profiles and up to 5 replicates. The software implements model described in Haned et al. (2012), and its optimal use requires familiarity with related literature, as well as provided tutorial and training materials. It is important to note that uncommon or untested situations may lead to unreliable results.

6.5.2.2 Procedure

This software is compatible for any platform having Java version more than or equal to 8. The initial step in using program is uploading sample profiles. The LRmix Studio supports reading files in LRmix and Genemapper® IDX formats (with/without peak heights). Software is not case-sensitive; it is sensitive to the spaces inside maker names. After importing the sample profile, the user can now add profile of interest. The software has profile summary that helps user to visualize alleles contained in sample and reference profiles. The software also has analysis tab where user can explain hypothesis of prosecution and defense. The sensitivity analysis (SA) depicts log 10 LRs, as well as individual likelihoods of prosecution and defense hypotheses. Sensitivity analysis evaluates propositions defined by a user on Analysis tab. Preceding step also defines drop-out criteria for known and unknown contributors. The analysis carried out by user can be generated in PDF format by automatic report generator tab. This software also provides comment section to user if they want to make addition in the report but it is optional (Haned, 2013). Gill and Haned (2013) suggested non-contributor tests to assess robustness of case-specific LR.

6.5.2.3 Limitations

The current version of LRmix Studio is limited in its capabilities, as it cannot analyze sample profiles with missing data or deconvolute mixtures due to the absence of specific peak height information. It is assumed that the postulated contributors for Hp and Hd are unrelated to each other and to any unknowns. Additionally, untested situations may lead to incorrect results, and the software does not handle missing data in reference profiles. Mixtures also cannot be deconvoluted, and propositions involving individuals who have not been profiled cannot be analyzed.

6.5.3 TrueAllele™

Twenty years ago, Cybergenetics developed the TrueAllele probabilistic genotyping approach to address the challenges posed by DNA mixtures. To produce LR match statistics, TrueAllele employs Bayesian modeling to evaluate genotype combinations from multiple contributors, assign genotype probabilities, and compare genotypes (Perlin & Szabady, 2001). TrueAllele and similar software systems are now widely used in the forensic laboratories to resolve the complex DNA mixtures (Tvedebrink et al., 2012). Initial validation studies for TrueAllele focused on DNA samples from single sources (Kadash et al., 2004). Later peer-reviewed validations extended the analysis to mixtures of DNA from two, three, four, or even five contributors (Ballantyne et al., 2013; Greenspoon et al., 2015). The primary parameters used in these validations were repeatability, specificity, and sensitivity. Critical factors for

TrueAllele's effectiveness included low-template DNA, collaborative data analysis, and the quantity of data. These studies examined both the laboratory-generated and casework DNA mixture data.

TrueAllele utilizes a hierarchical Bayesian probability model to account for variance, artifacts, and genotype alleles in STR data. It computes uncertainty and parameter values by solving Bayesian equations through statistical sampling using MCMC methods. The joint posterior probability provides marginal distribution for the contributor genotypes, mixture weights, and other explanatory factors. TrueAllele starts with DNA mixture data and distinguishes contributing genotypes up to a likelihood, as it requires no data selection or separation procedure. The system analyzes hundreds of possible values for each variable, resulting in comprehensive inferences. Variations in pipetting or amplification can cause observed mixture proportions to deviate from planned proportions in laboratory data (Perlin et al., 2015). TrueAllele can calculate a more accurate mixture weight by conditioning on all known contributor genotypes, leading to more precise results.

TrueAllele calculates an LR match statistics by comparing a separated unidentified genotype to reference genotype, relative to population genotype. This comparison measures the "weight of evidence" between two genotypes, expressed in log (LR) ban units. LR serves as final common metric in the forensic DNA analysis, consolidating identity information from entire STR process into a single value that can be reported or presented in the court of law. Key validation metrics can be statistically assessed using additive log (LR) values. It also allows the input of known genotypes that remain constant during the analysis, reducing number of unknown genotypes in mixture. Providing more information enhances the probability distributions for remaining genotype. For instance, in sexual assault case involving two individuals, inputting victim's known genotype can help identify the unknown assailant's genotype more effectively.

6.5.3.1 Features

TrueAllele® casework product offers a comprehensive approach to DNA analysis by utilizing all available data, including low-level peaks, without imposing any thresholds. It reports a match statistic error rate directly from the evidence, providing an accurate assessment of the match's reliability. The system is capable of resolving all contributors to a DNA mixture, including those with small contributions, and can analyze complex mixtures without need for reference samples. TrueAllele® supports an unlimited number of contributors, free from artificial limitations. Its reporting process is straightforward, offering a simple ratio of match probabilities. Additionally, the system effectively separates mixed DNA data into individual genotypes, allowing for evidence-to-evidence comparison and the generation of match statistics. Even in cases of low-level data, TrueAllele® identifies statistical uncertainties directly from the evidence, ensuring a robust and reliable analysis (Cybergenetics, 2022)

6.5.3.2 TrueAllele system

Cybergenetics TrueAllele Casework system is designed to deconvolute DNA mixtures into their individual genotypes. Developed using MATLAB, this software employs the MCMC sampling to address a hierarchical probability model. The system features a Visual User Interface (VUIer) client that connects via a network to server that manages database and utilizes parallel processor to run interpretation algorithms. The DNA identification process is divided into

2 phases. First, TrueAllele system infers genotype from evidence data, an objective process that does not incorporate any reference genotype comparisons. Once genotypes are inferred, they can be compared to reference genotypes to calculate LR relative to population. By isolating mixture data into distinct genotypes, TrueAllele simplifies the interpretation and reporting of results (Perlin et al., 2014).

6.5.3.3 Limitation

TrueAllele, efficacy can be restrained by several factors. Complex mixtures, especially those involving multiple contributors or degraded DNA, can challenge TrueAllele's accuracy. Furthermore, software may struggle with low-template DNA samples due to inherent noise and uncertainty associated with limited DNA material. Allelic dropout, a phenomenon where an allele fails to amplify in PCR, and stochastic effects can introduce errors into the analysis, even with TrueAllele's robustness. Moreover, while TrueAllele provides a probabilistic framework, accurate interpretation requires experienced analysts to prevent misinterpretations and ensure reliable conclusions. Finally, concerns have arisen regarding the extent of independent validation for TrueAllele, potentially affecting its overall reliability and accuracy (Perlin et al., 2014).

6.5.4 LikeLTD

LikeLTD (Likelihoods for Low-Template DNA profiles) is an open-source R software designed to evaluate DNA profiles with low amounts of degraded DNA. It employs a probabilistic model to calculate LRs for hypotheses (Hp and Hd) to identify potential DNA donors. Unlike some other programs, LikeLTD does not directly use heights of EPG (electropherogram) peaks, but instead classifies these peaks as either allelic, non-allelic, or uncertain. Peaks classified as uncertain are treated as though presence or absence of allele is unknown when computing likelihoods, simplifying the analysis.

In mixed-source DNA profiles, peak heights can provide crucial information about quantity of DNA from different contributors, but variability in peak heights due to degradation and methodological differences can make this data challenging to use. LikeLTD mitigates this by focusing on reference profiles and control sample profiles (CSPs) to label allelic positions as present, doubtful, or missing across replicates. This approach helps forensic scientists determine nature of peaks, such as distinguishing between allelic and stutter peaks, by combining the presence/uncertain/absence indicators across alleles, loci, and replicates.

The software improves performance by setting intercept to zero, which reduces the number of parameters that need to be estimated. LikeLTD's drop-in model represents contaminant DNA as a contribution to projected peak height for each allele, proportional to allele's population percentage. The program assumes that predicted peak height decreases exponentially with length of DNA fragment, which aligns with behavior observed in DNA deterioration.

The probability distributions used for modeling peak heights vary among different programs. For example, most models utilize a gamma distribution, whereas STRmix employs a lognormal distribution and TrueAllele uses a truncated normal distribution. Unlike some models that do not account for DNA degradation, LikeLTD incorporates this factor into its calculations. Notably, LikeLTD and EuroForMix are among the few truly open-source programs available for this type of forensic analysis.

6.5.4.1 *Limitation*

A key limitation of LikeLTD software is its reliance on accurate modeling of stochastic phenomena, which can be challenging with low-level or degraded DNA samples. The software's performance may be impacted by inherent variability in peak heights, which can be sensitive to specific profiling protocols used and may not always be reliably informative. Moreover, while likeLTD can handle profiles with up to two unprofiled contributors, absence of replicate analyses can introduce extra uncertainty and potentially lead to reduced evidential weight, especially when sample size is already very small. LikeLTD software plays a significant role in analyzing complex DNA profiles, but its effectiveness can be limited by inconsistency in DNA sample quality and profiling conditions, making careful interpretation and validation crucial for reliable forensic analysis (Balding, 2013; Steele et al., 2016).

6.5.5 Lab Retriever

Lab Retriever software is designed to enhance the analysis of forensic DNA data, originally derived from an open-source R-code program introduced by David Balding. The key objectives is to provide a user-friendly interface, optimize computational efficiency, and ultimately facilitate more accurate and faster forensic DNA analysis. Balding created a visually appealing and intuitive graphical user interface (GUI), using combination of JavaScript, CSS, and HTML, for typical forensic DNA analyst who is not computer programmer. To accelerate execution on back end, program was re-coded using C++.

LR computed by Lab Retriever is used to assess probability of an evidence profile matching a suspect versus a random individual. Consider the following hypotheses:

H0: Suspect is the source of evidence profile
H1: Random individual is the source of evidence profile

Lab Retriever can be written for calculating Likelihood:

$$\mathrm{LR} = \frac{\mathrm{P}(\mathrm{E} \mid \mathrm{s})}{\sum_{\text{all } j} \mathrm{P}(\mathrm{E} \mid \mathrm{j})\,\mathrm{P}(\mathrm{j})}$$

Here,

P(E|s) = Probability that evidence (E) would be found if suspected person (s) is source of DNA.
P(E|j) = Probability of evidence if random person (j) is source.
P(j) = Probability of finding specific DNA profile (j) in population.

The method calculates ratio of P(E|s) to P(E|j), known as LR. A higher LR indicates a stronger likelihood that suspected person is source of DNA. The denominator (P(E|j)) is calculated by considering all possible DNA profiles in population and their corresponding probabilities. This can be computationally intensive, especially with multiple unknown contributors. The text mentions a dynamic programming algorithm in Lab Retriever that speeds up this calculation.

6.5.5.1 Program parameter

Lab Retriever is a software that can analyze DNA collected from crime scenes by Globalfiler® and PowerPlex® Fusion kits, widely used in forensic DNA analysis. The software automatically recognizes which DNA markers are present in evidence and only shows results for those markers. It uses allele frequency data from different backgrounds people such as Caucasian, African-American, and Hispanic in the United States for its calculation. This data was collected by the National Institute of Standards and Technology (NIST) and is available on STRbase website. One important factor is chance that DNA sample might be incomplete or missing information, called P(DO). Estimates the chance that DNA profile might be missing information due to low DNA levels, for accurately weighing evidence when deciding if a particular person contributed DNA to profile. The P(DO) value, which should be based on a specific logistic regression model for sample and lab, is entered into P(DO) field. To estimate this parameter for given sample, use a system and lab-specific logistic regression model. Likelihood can also be calculated in Lab Retriever by adjusting probability of drop-out (P(DO)) to see how different values affect the results. P(DO) is assumed to be the same across all loci in the current model, though this simplifies real biological conditions. Research shows that using a unvarying P(DO) gives results similar to those obtained with varying P(DO) across loci. For profiles with multiple contributors, the same P(DO) is used for everyone. Probability of drop-in (P(DI)), which accounts for contamination, is set by default to 0.01, though this value can be adjusted based on lab data. Lab Retriever also allows you to test hypotheses about relationships between contributors by accounting for shared alleles. Users can specify hypotheses for both numerator and denominator of LR calculation and handle profiles with up to four unknown contributors. Software divides contributors into Suspected, Assumed, and Unknown categories, with specific rules for including or excluding alleles. Co-ancestry coefficient (FST) is set to 0.01 by default but can be adjusted by the user.

6.5.5.2 Limitation

Lab Retriever inability includes its incompetence to automatically incorporate peak height information or other parameters. It relies on user-defined assumptions and manual adjustments, which can lead to potential inaccuracies or incomplete analyses, especially in complex DNA profiles where peak height and additional parameters are critical. Moreover, Lab Retriever's inability to fully integrate peak height data or account for allelic drop-out automatically can impact the precision of the LR calculations. Comparative studies have also shown that Lab Retriever might produce smaller LRs compared to other programs, particularly in cases of significant divergence between major and minor profile components. Additionally, inadequate training and comparison against other software tools pose challenges in evaluating its performance comprehensively (Inman et al., 2015; Kelly et al., 2012; Scientific Collaboration, Innovation & Education Group, 2014).

6.5.6 ArmedXpert

ArmedXpert is a software designed to automate complex process of DNA interpretation, particularly in cases involving mixed DNA profiles. Tom Overson, a scientist at DFSC, originally developed ArmedXpert to simplify organization of DNA profiles. He had worked with FBI's CODIS database and sought a more efficient method for managing DNA data. The software's commercial development by NicheVision Forensics has further expanded its application, making it a vital tool in forensic investigations worldwide. In 2010, NicheVision was awarded by the United States Army Criminal Investigation Laboratory (USACIL) exclusive distribution rights to ArmedXpert™, a full case DNA management tool using a binary approach for mixture deconvolution." In 2015, "NicheVision, along with the United States Army Medical Research and Materiel Command, received the Federal Laboratory Consortium's (FLC) 'Excellence in Technology Transfer' award for the outstanding commercial success of ArmedXpert™. ArmedXpert™ currently has more than 50 installations worldwide, including the United States, Mexico, Europe and Africa.

ArmedXpert™ is a user-friendly software that streamlines the process of analyzing DNA mixtures, reducing the time required from hours to minutes. It can handle mixtures with up to three contributors and can be tailored to meet specific needs of each laboratory. Software helps forensic scientists to analyze DNA mixtures, but it's not a replacement for experts. It calculates RMP for DNA mixtures in three different ways: unrestricted RMP, restricted RMP, or as a mixture of both handled on a locus-by-locus basis. If there's a suspicion that some DNA alleles are missing (dropout) at a specific location, software can adjust RMP calculations accordingly. It also allows for combining different RMP approaches within the same DNA profile to produce the most accurate statistics. These statistical techniques are used for two or three contributors and are all aligned with the 2010 SWGDAM criteria in order to prevent duplicate counting, formulae suitably exclude combinations. ArmedXpert™ software analyzes DNA samples and helps experts to understand and interpret DNA profiles. An analyst has to decide how to use data during interpretation, and sometimes more than one analysis may be needed to fully understand the results.

6.5.6.1 Features

ArmedXpert™ is a software for DNA analysis, offers variety of features for both single-person and 2 or 3 individual DNA samples. For single-person samples, it calculates RMP statistic, providing a measure of likelihood of a random match occurring. For complex DNA samples from 2 or 3 individuals, ArmedXpert™ assists in interpreting mixtures, separating individual DNA profiles, filtering out stutter artifacts, calculating individual contributions based on peak height ratios, identifying main DNA profile, and performing statistical analyses. Additionally, software allows for use of known DNA references, such as from someone who wore item (owner), to enhance accuracy of result interpretation.

6.5.6.2 Limitation

ArmedXpert™ is a valuable software for forensic DNA analysis, but it does not provide expert opinions or interpretations, only serves as an aid for forensic scientists. Moreover, accuracy of its results is contingent on quality of DNA data and user's ability to interpret findings correctly (ArmedXpert™, 2019; ArmedXpert™ - NicheVision Forensics, LLC, 2020)

6.5.7 DNA·VIEW

DNA·VIEW stands as universally accepted standard for DNA identification, highly esteemed in both paternity and forensic laboratories. Since 1988, forensic laboratories have trusted DNA·VIEW for its reliability and advanced features in field of genetic analysis. Dr. Charles Brenner developed and constantly supports DNA·VIEW, well-regarded for its advanced software features and its authoritative standing in the scientific community. The DNA·VIEW software design has been influenced by extensive interaction and feedback from its users, combining real-world needs with advanced technology. DNA·VIEW is more than a software program; it signifies a comprehensive amalgam that includes forensic mathematical expertise and support from a worldwide user community. This holistic approach ensures that DNA·VIEW remains at forefront of DNA analysis, providing unmatched accuracy and support in genetic identification.

DNA·VIEW is designed for various DNA analysis, such as mixture analysis, symbolic kinship program, paternity testing, and disaster victim identification. Mixture Solution™ computer program that allows analysis of many-contributor mixtures, accounting for drop-out and drop-in (fully continuous), and stutter. It can effectively handle different combinations of contributors and references, providing accurate and precise results in a fraction of time compared to other methods. It offers a more effectual and comprehensive analysis approach compared to traditional MCMC methods by employing mathematical techniques. Symbolic Kinship Program™ analyzes irregular kinship cases, including issues like reconstruction, incest, inheritance, missing bodies, and twin zygosity, with enhanced accuracy through Y-haplotypes. Additionally, its Kinship Simulation feature allows for prediction of potential efficacy of DNA in resolving kinship problems. This program has been applied in different contexts, such as immigration casework in multiple countries and in identifying victims of different events such as the Croatian and Bosnian wars, Argentine orphans, the SwissAir crash, and the World Trade Center attacks. Paternity Case Analysis – Routine Version recognizes allele sharing and effectively handles motherless cases, maternity cases, missing alleles (nulls), and mutations. It calculates paternity index (PI), exclusion probability (A), probability of paternity (W), and avuncular index. Additionally, software can interface to PATER paternity report program (not included) or to custom reports, making it suitable for handling a high volume of casework. Crime case analysis calculates DNA profile and Y-haplotype matching odds, and handles mixture analysis using LR or exclusion. In September 2001, DNA·VIEW turned out to be an essential tool for victim identification after the World Trade Center attacks. Screening module and Kinship program are essential tools for establishing DNA identifications, particularly those that link to living relatives. The DNA·VIEW has also aided in identifying victims from disasters like Thailand and Japanese tsunamis, Hurricane Katrina, and mass identification efforts for disappeared children from El Salvador, as well as war victim from countries including Argentina, Bosnia, Guatemala, Mexico, Kuwait, and Iraq.

6.5.7.1 Limitations

DNA·VIEW program is a DOS-based application, compatible with 32-bit Windows versions, making it incompatible for Mac systems. Its search functionality for databases of convicted offenders or missing persons is limited to some extent. Whereas it offers a strong search capability for large databases, this feature is an additional cost and is precisely designed for searching rather than maintaining offender lists. Additionally, software's search capabilities are restricted to around 100,000 offenders when using the "Screen disaster matches" command. For managing offender lists and searching for direct matches, software can only handle a limited number of offenders. This process involves setting up appropriate DNA·VIEW databases, regularly updating them, and using the "List and search database profiles" facility. The integration with LIMS is not fully comprehensive, and there is limited documentation available for newer, more advanced integration approaches (Brenner, 2019; DNA•VIEW™, 2024).

6.5.8 DNAMIX

DNAMIX is a software designed to calculate LRs as they pertain to mixed DNA samples. The latest version, DNAMIX v.3, builds upon John Storey's earlier versions, DNAMIX v.1 and v.2. The methods and formulas presented by Curran et al. (1999) rely on the assumption that the proportions of all alleles in mixture are independent, applying concepts from Hardy-Weinberg equilibrium and linkage equilibrium. While these assumptions hold that individuals are completely independent, they overlook the slight dependencies that may exist among individuals within the same population (Dash et al., 2023).

6.5.9 EuroForMix

EuroForMix is an open source software designed to analyze STR-DNA profiles. It uses a continuous model to calculate weight of evidence for DNA profiles that may be from a mixture of contributors. It can infer mixture proportions, peak height properties, stutter proportion, and degradation, and include models for allele drop-out, allele drop-in, and subpopulation structure. Further, it offers two inference approaches: maximum likelihood estimation and Bayesian inference, which allow users to incorporate their own prior information. While limited to four unknown contributors due to computational time constraints, EuroForMix provides a valuable platform for comparing different solutions and serves as a reference point for other software in the field (Bleka et al., 2016).

6.5.10 DNAmixtures

Therese Graversen developed the R-package DNAmixtures, a powerful statistical tool for analyzing DNA samples from multiple contributors. It depends on the standard computations framework for Bayesian network models offered by Hugin, accessed through interface provided by R package RHugin. DNAmixtures does not feature a GUI, it aims to provide an intuitive and user-friendly experience without compromising flexibility for users with a basic familiarity with R. It can be installed in Windows, Linux, or macOS, provided that the following three software programs installed i.e. R (statistical software); Hugin (Bayesian network software); and RHugin (API that enables calling Hugin from within R) (Cowell et al., 2013).

Statistical software has revolutionized forensic DNA profiling by enhancing the accuracy and reliability of DNA analyses, crucial for linking individuals to crime scenes. Software discussed above employs sophisticated algorithms and probabilistic models to address the complexities of interpreting DNA mixtures, accounting for factors such as allelic dropout, peak heights, and contributor identification. These tools not only improve the efficiency of forensic analyses but also facilitate clear communication of results in legal contexts, thereby upholding the integrity of forensic evidence. Forensim, FST, DNAxs, Ceeslt, Kongoh, NOCIt, etc. are several other software's work on the mixtures of DNA sample and calculate LR value and interpret STR Data.

6.6 Validation of statistical software

Validation in forensic science is a critical topic and when using statistical software for DNA profiling, the question arises "How to ensure that complex computer program gives the 'correct answer'?" Gill quoted, *"Since all models are wrong the scientist cannot obtain a 'correct' one by excessive elaboration. On the contrary following William of Occam he should seek an economical description of natural phenomena. Just as the ability to devise simple but evocative models is the signature of the great scientist so over-elaboration and over-parameterization is often the mark of mediocrity."* In other words, there is no single "correct" answer, and the goal is to find a balance between model simplicity and accuracy. A model ensures it is reliable and suited for its intended purpose (Gill et al., 2020).

Model validation is crucial to ensure that a model is suitable for its intended purpose. This process involves two main steps: conceptual validation and operational validation. Conceptual validation evaluates whether the model's mathematical formulation and its foundational assumptions are fundamentally sound. On the other hand, operational validation is the process of determining whether *"the model's output behavior meets the accuracy required for its intended purpose across its applicable domain."* Typically, a "computerized model" is used to validate operational processes. In simple terms, the model's operational validity cannot be validated until a computer implementation is present that can execute a profile and produce a result. Operational validity is checked using user-defined criteria, which can be approved or refused. These can be estimated for LR-based models. Model and software validation are typically performed simultaneously, because it is a computerized version of model that facilitates model validation process. Software validation examines program algorithms to ensure that it adhere to mathematical concepts outlined in the model follow. The following are the steps suggested for effective software validation:

- *Statistical Requirement:* This step includes outlining theoretical framework behind the model intended for computerized implementation.
- *Conduct Analytical Verification:* This involves performing analytical calculations, such as LRs for simple cases (for instance, single-source and two-person mixtures), and comparing these results to software output. Depending on model's complexity, analytical verification may or may not be feasible. This situation has been referred to as "complexity paradox," indicating that as models increase in complexity, verifying their various components becomes more challenging. In such cases, software output can be compared to results from alternative software that implements a similar model.
- *Compare to Parallel Implementations:* Comparing software to alternative programs that utilize either similar/different probabilistic models can aid in verifying the software's behavior. These comparisons leverage the "convergence principle," which suggests that differences in modeling between software that correspond to one unit on log10 scale are considered negligible.
- *Code Verification through Inspection and Recoding:* This involves verifying code through visual examination and recoding. The Open-source software offers unrestricted access to its code, facilitating this verification process.

6.7 Legal and ethical considerations

Statistical software in DNA typing introduces significant legal and ethical implications to maintain the integrity of justice system. The primary legal implication of using statistical software is the need for accuracy in interpreting DNA evidence. Misinterpretation or inaccurate conclusions can have serious consequences, including wrongful convictions or unfair acquittals. For instance, if a software program inaccurately assesses probability of a DNA match, it could lead to a perpetrator being wrongfully convictions based on flawed evidence. Therefore, it is imperative that forensic scientists use validated and reliable software to minimize risk of errors. Continuous validation of these software programs against known datasets is important to ensure they perform accurately and consistently in various scenarios. Further, transparency is also crucial when integrating statistical analysis into the field of forensic. Legal professionals and judges must be able to understand the methods used to analyze DNA evidence during the trial procedure. Courts increasingly rely on DNA evidence, making it imperative that the legal system incorporates guidelines and standards for admissibility of statistical analyses.

Moreover, ethical considerations are related to confidentiality and handling of DNA data must be emphasized. The collection, storage, and analysis of DNA samples should adhere to strict protocols to protect individuals' privacy rights. Additionally, the use of DNA evidence in legal proceedings raises questions about the potential for profiling and discrimination, necessitating careful consideration of how DNA data is utilized in law enforcement practices. The integration of statistical software in forensic DNA typing presents significant legal and ethical challenges that must be addressed to maintain the integrity of the justice system.

6.7.1 Case study

6.7.1.1 Case study 1 – Federal criminal trail

In a federal criminal prosecution, a plastic bag containing smuggled goods was a key piece of evidence linking the defendant to crime. However, the defense questioned the involvement of defendant's by hiring a private laboratory to conduct DNA analysis on bag. After collecting the swab, laboratory analyzed it by using a Globalfiler PCR amplification kit and an ABI 3130 Genetic Analyzer. A low-level partial STR-DNA profile was obtained, revealed a mixture of DNA from at least two individuals. The primary contributor was eliminated based on the electropherograms, even though the defendant's partial DNA profile was appropriate with at least two contributors—including one male. After this lab calculated the LRs using STRMix software, they came to the conclusion that the DNA combination had a 24 times higher probability of coming from two unidentified sources than from defendant. In parallel, Cybergenetics employed TrueAllele to analyze the same data, which showed much stronger support for defendant's non-contribution, with LRs several million times greater. Both software indicated that the DNA mixture was far more likely to have originated from two unidentified individuals than from the defendant and an unknown contributor (Thompson, 2023).

6.7.1.2 Case study 2 – Mysterious death of a young boy

A young Caucasian boy was killed by a train in 2012. The victim's family believed he was killed, but investigators concluded it was a suicide. Family of boy provided boy's relatively new coat, which he was wearing on the day of accident, for forensic examination to identify any genetic material from others. Coat examination revealed a mixed DNA profile under the armpit, like someone had lifted the victim and pushed them under the train. Utilizing LRmix Studio for further analysis indicated the involvement of multiple individuals in the victim's demise. Moreover, the family provided two cigarette butts discarded by victim's friend, who was alleged to have been associated to incident. These new findings encouraged authorities to reopen the case for further investigation (Tasselli et al., 2015).

6.8 Conclusion

Forensic DNA analysis has revolutionized field of forensic science, offering a powerful tool to link accused to scene of crime with a high degree of probability. Statistical interpretation plays a significant role in DNA profiling, as it provides a framework for quantifying uncertainty and assessing the strength of a match. Particularly in complex cases, including mixed DNA profiles, probabilities, LRs, and Bayesian methods, empower forensic experts to make informed decisions about DNA evidence. Statistical tools allow the judicial system to interpret weight of DNA evidence and support legal process by providing reliable and scientifically grounded conclusions. Forensic science and statistical methods together strengthen the credibility of

DNA profiling, making it a crucial component to criminal justice system. Tools like STRmix™, LRmix, TrueAllele, LikeLTD, Lab Retriever, ArmedXpert, and DNA•VIEW offer unique features to enhance accuracy and reliability of DNA analysis. While these software tools are very helpful in forensic investigations and court trials, it is important to remember that they are, after all, just tools. Hence, careful interpretation by forensic experts is required to ensure accuracy and reliability of DNA evidence. Advancements in these statistical software are revolutionizing field of forensic genetics, enhancing judicial process, and contributing to the integrity of forensic science. As technology continues to evolve, the development and validation of these tools will further strengthen their role in DNA analysis, ensuring justice is served based on scientific evidence.

References

Alamoudi, E., Mehmood, R., Albeshri, A., & Gojobori, T. (2018). DNA profiling methods and tools: A review. In: *Smart societies, infrastructure, technologies and applications: First international conference, SCITA 2017, Jeddah, Saudi Arabia, November 27–29, 2017, Proceedings 1* (pp. 216–231). Springer International Publishing. https://doi.org/10.1007/978-3-319-94180-6_22

ArmedXpertTM - NicheVision Forensics, LLC. (2020, March 3). *NicheVision Forensics, LLC.* Retrieved September 2, 2024, from https://nichevision.com/armedxpert/

Balding, D. J. (2013). Evaluation of mixed-source, low-template DNA profiles in forensic science. *Proceedings of the National Academy of Sciences, 110*(30), 12241–12246. https://doi.org/10.1073/pnas.1219739110

Ballantyne, J., Hanson, E. K., & Perlin, M. W. (2013). DNA mixture genotyping by probabilistic computer interpretation of binomially-sampled laser captured cell populations: Combining quantitative data for greater identification information. *Science & Justice, 53*(2), 103–114. https://doi.org/10.1016/j.scijus.2012.04.004

Bleka, Ø., Storvik, G., & Gill, P. (2016). EuroForMix: An open source software based on a continuous model to evaluate STR DNA profiles from a mixture of contributors with artefacts. *Forensic Science International: Genetics, 21*, 35–44. https://doi.org/10.1016/j.fsigen.2015.11.008

Brenner, C. (2019). *DNAVIEW user's manual.* Retrieved October 5, 2024, from https://dna-view.com/downloads/documents/manuals/DNAVIEW%202019%20US.pdf

Bright, J. A., Cheng, K., Kerr, Z., McGovern, C., Kelly, H., Moretti, T. R., & Buckleton, J. (2019). STRmix™ collaborative exercise on DNA mixture interpretation. *Forensic Science International: Genetics, 40*, 1–8. https://doi.org/10.1016/j.fsigen.2019.01.006

Buckleton, J. S., Triggs, C. M., & Walsh, S. J. (2005). *Forensic DNA evidence interpretation.* CRC Press.

Butler, J. M. (2010). *Fundamentals of Forensic DNA Typing.* Academic Press/Elsevier.

Butler, J. M. (2015). *Advanced Topics in Forensic DNA Typing: Interpretation.* Academic Press.

Cowell, R. G., Graversen, T., Lauritzen, S. L., & Mortera, J. (2013). Analysis of forensic DNA mixtures with artefacts. *Journal of the Royal Statistical Society Series C (Applied Statistics), 64*(1), 1–48. https://doi.org/10.1111/rssc.12071

Curran, J. M., Triggs, C. M., Buckleton, J., & Weir, B. S. (1999). Interpreting DNA mixtures in structured populations. *Journal of Forensic Sciences, 44*(5), 987–995. https://doi.org/10.1520/jfs12028j

Cybergenetics. (2022, July 15). *TrueAllele® Casework Product | CyberGenetics.* Retrieved September 2, 2024, from https://www.cybgen.com/products/casework/

Dash, H. R., Elkins, K. M., & Al-Snan, N. R. (2023). *Advancements in Forensic DNA Analysis.* Springer Verlag.

DNAVIEW™. (2024). Product information. Retrieved October 5, 2024, from https://dna-view.com/dnaview.htm

Evett, I. W., & Weir, B. S. (1998). *Interpreting DNA Evidence: Statistical Genetics for Forensic Scientists.* Sinauer Associates.

Finkelstein, M. O., & Levin, B. (2001). Statistics for lawyers (2nd ed.). Springer. https://doi.org/10.1007/b97319

Fung, W. K., & Hu, Y. (2008). *Statistical DNA Forensics: Theory, Methods and Computation.* Wiley.

Fung, W. K., Hu, Y., & Chung, Y. (2006). On statistical analysis of forensic DNA: Theory, methods and computer programs. *Forensic Science International, 162*(1–3), 17–23. https://doi.org/10.1016/j.forsciint.2006.06.025

Gill, P., Bleka, Ø., Hansson, O., Benschop, C., & Haned, H. (2020). *Forensic practitioner's guide to the interpretation of complex DNA profiles.* Elsevier.

Gill, P., Brenner, C. H., Buckleton, J. S., Carracedo, A., Krawczak, M., Mayr, W. R., Morling, N., Prinz, M., Schneider, P. M., & Weir, B. S. (2006). DNA commission of the international society of forensic genetics: Recommendations on the interpretation of mixtures. *Forensic Science International, 160*(2–3), 90–101. https://doi.org/10.1016/j.forsciint.2006.04.009

Gill, P., & Haned, H. (2013). A new methodological framework to interpret complex DNA profiles using likelihood ratios. *Forensic Science International: Genetics*, *7*(2), 251–263. https://doi.org/10.1016/j.fsigen.2012.11.002

Greenspoon, S. A., Schiermeier-Wood, L., & Jenkins, B. C. (2015). Establishing the limits of TrueAllele® casework: A validation study. *Journal of Forensic Sciences*, *60*(5), 1263–1276. https://doi.org/10.1111/1556-4029.12810

Haned, H. (2011). Forensim: An open-source initiative for the evaluation of statistical methods in forensic genetics. *Forensic Science International: Genetics*, *5*(4), 265–268.

Haned, H. (2013). *LRMix tutorial, version 4.1*. Forensim. Retrieved October 5, 2024, from https://forensim.r-forge.r-project.org/misc/LRmix.pdf

Inman, K., Rudin, N., Cheng, K., Robinson, C., Kirschner, A., Inman-Semerau, L., & Lohmueller, K. E. (2015). Lab retriever: A software tool for calculating likelihood ratios incorporating a probability of drop-out for forensic DNA profiles. *BMC Bioinformatics*, *16*, 1–10. https://doi.org/10.1186/s12859-015-0740-8

Kadash, K., Kozlowski, B. E., Biega, L. A., & Duceman, B. W. (2004). Validation study of the TrueAllele® automated data review system. *Journal of Forensic Sciences*, *49*(4), JFS2003336. https://doi.org/10.1520/jfs2003336

Kelly, H., Bright, J. A., Curran, J., & Buckleton, J. (2012). The interpretation of low level DNA mixtures. *Forensic Science International: Genetics*, *6*(2), 191–197. https://doi.org/10.1016/j.fsigen.2011.04.013

National Research Council. (1996). *Evaluation of Forensic DNA Evidence*. National Academies Press.

Perlin, M., & Szabady, B. (2001). Linear mixture analysis: A mathematical approach to resolving mixed DNA samples. *Journal of Forensic Sciences*, *46*(6), 1372–1378. https://doi.org/10.1520/jfs15158j

Perlin, M. W., Dormer, K., Hornyak, J., Schiermeier-Wood, L., & Greenspoon, S. (2014). TrueAllele casework on Virginia DNA mixture evidence: Computer and manual interpretation in 72 reported criminal cases. *PLoS One*, *9*(3), e92837. https://doi.org/10.1371/journal.pone.0092837

Perlin, M. W., Hornyak, J. M., Sugimoto, G., & Miller, K. W. (2015). TrueAllele® genotype identification on DNA mixtures containing up to five unknown contributors. *Journal of Forensic Sciences*, *60*(4), 857–868. https://doi.org/10.1111/1556-4029.12788

Sharma, V., Young, B., Armogida, L., Khan, A., & Wurmbach, E. (2022). Evaluation of ArmedXpert software tools, MixtureAce and Mixture Interpretation, to analyze MPS-STR data. *Forensic Science International: Genetics*, *56*. https://doi.org/10.1016/j.fsigen.2021.102603

Scientific Collaboration, Innovation & Education Group. (2014). *Lab retriever*. Retrieved September 2, 2024, from https://scieg.org/wp-content/uploads/2017/07/Lab_Retriever_manual_2.2.pdf

Shrivastava, P., Dash, H. R., Lorente, J. A., & Imam, J. (2020). *Forensic DNA typing: Principles, applications and advancements*. Springer.

Steele, C. D., Greenhalgh, M., & Balding, D. J. (2016). Evaluation of low-template DNA profiles using peak heights. *Statistical Applications in Genetics and Molecular Biology*, *15*(5), 431–445. https://doi.org/10.1515/sagmb-2016-0038

Tasselli, G., Lancia, M., Tommolini, F., Margiotta, G., Massetti, S., & Carnevali, E. (2015). Complex DNA mixture analysis: Report of two cases. *Forensic Science International: Genetics Supplement Series*, *5*, e664–e666. https://doi.org/10.1016/j.fsigss.2015.10.015

Taupin, J. M. (2014). *Introduction to Forensic DNA Evidence for Criminal Justice Professionals*. CRC Press, Taylor & Francis Group.

Thompson, W. C. (2023). Uncertainty in probabilistic genotyping of low template DNA: A case study comparing Strmix™ and TrueAllele™. *Journal of Forensic Sciences*, *68*(3), 1049–1063. https://doi.org/10.1111/1556-4029.15225

Tvedebrink, T., Eriksen, P. S., Mogensen, H. S., & Morling, N. (2012). Identifying contributors of DNA mixtures by means of quantitative information of STR typing. *Journal of Computational Biology*, *19*(7), 887–902. https://doi.org/10.1089/cmb.2010.0055

7

Challenges and Limitations of Forensic DNA Profiling Technique

Shrijoy Banerjee, Tanurup Das, and Ankit Srivastava

7.1 Introduction

DNA profiling technique has transformed the area of criminal investigation since its discovery in the 1980s. It is a crucial tool in crime solving because of its precise ability to connect biological evidence to a suspect or victim. In 1985, Sir Alec Jeffreys published the first description of DNA profiling, sometimes known as fingerprinting. He discovered that nitrogenous base sequences repeatedly appeared next to one another in specific DNA locations. He also found that the quantity of repeated parts in a sample can vary from person to person. Dr. Jeffreys made it possible to do human identity checks by devising a method to analyse the length variation of these DNA repeat sequences. These DNA repeat regions were called variable number of tandem repeats, or VNTRs. Because it included cutting the DNA areas surrounding the VNTRs with a restriction enzyme, Dr. Jeffreys' method of examining the VNTRs was known as restriction fragment length polymorphism (RFLP). First applied to an English immigration case, the RFLP approach was later utilised to solve a double homicide case. Since then, DNA profiling techniques for human identification testing have proliferated. Most crime labs have extensive expertise in using various biological samples to perform human DNA identification tests and presenting the results to juries and judges in court. The potential for individual identification has significantly increased with the discovery of techniques to separate DNA from almost all biological specimens. Analysis of certain DNA areas has become a potent tool for the criminal justice system thanks to improvements and adjustments made to the primary polymerase chain reaction (PCR) method and Multiplex PCR-based short tandem repeats (STRs).

In order to characterise DNA recovered from forensic evidence, several multiplex PCR-based kits have been used for the PCR to amplify DNA sequences known as STRs. Many STR loci, along with a section of X and the Y chromosome, the Amelogenin gene, can be amplified simultaneously with these multiplex kits. The Genetic Analysers use capillary gel electrophoresis to segregate the amplified fragments based on size, and laser excitation is used to identify them. A charged-coupled device (CCD) camera captures the resulting emission spectra. Peaks in scan numbers and relative fluorescence units (RFUs) graphically represent the resultant data. We refer to this display as an electropherogram (Butler, 2011). DNA analysis is now a crucial component of forensic science for identity and relationship testing in cases like missing persons, mass disasters, complex paternity analysis, parentage testing with only one available parent, sexual assault, homicide, etc., because of its exceptional sensitivity and discriminating power.

DOI: 10.4324/9781003474296-7

Nevertheless, there are drawbacks and restrictions to using DNA profiling. DNA analysis must overcome technological, interpretive, ethical, and legal obstacles despite its scientific rigour to guarantee its efficacy and dependability in forensic applications. The limitations of DNA profiling in forensic investigations are examined in this chapter, along with these difficulties (Butler, 2005).

7.2 Challenges for forensic DNA profiling technique

Forensic DNA labs frequently need more DNA samples. For instance, the biological evidence may have been subjected to a severe environment during a missing person investigation for days, months, or even years. Murder victims are typically taken to isolated areas where they remain until their remains are located. Instead of being kept in a freezer away from harsh chemicals that could break down the DNA, the molecules may have been kept in wet forests or direct sunlight. In any event, the DNA molecules that are recovered from a crime scene come from an unclean environment, usually from molecular biology labs. The potential recovery of a small amount of biological material is equally important. Therefore, accurate sample analysis is crucial since a forensic scientist may only collect enough evidence for one analysis attempt. This chapter examines the forensic challenges of analysing STRs. These difficulties, which include managing deteriorating DNA samples, avoiding contamination, getting beyond PCR inhibition, and evaluating mixes, are typical in forensic cases, especially those involving sexual assault. We also discuss using low-copy number DNA profiling to retrieve genetic information from materials with limited cell availability for analysis (Butler, 2011).

7.2.1 DNA degradation

Adverse environmental circumstances have the potential to destroy or harm DNA. DNA molecules break down into smaller bits randomly when exposed to the environment. Water, oxygen, ultraviolet light, and DNA-digesting nuclease enzymes are all enemies of intact DNA molecules. These conditions are frequent and may result in fragmented DNA, depending on the specific environmental conditions of a biological sample. It would have been challenging, if not impossible, to analyse severely degraded DNA samples using older technologies like RFLP because relatively high molecular mass (also known as molecular weight) DNA molecules had to be intact in order to detect large VNTRalleles. Smaller STR alleles are more helpful in recovering older, damaged DNA samples. An ethidium-bromide-stained agarose "yield gel" can be used to evaluate the quality of a DNA sample. A high molecular weight, high-quality genomic DNA typically exhibits a highly tight band of approximately 20,000 bp compared to an appropriate relative molecular mass marker.

Conversely, degraded DNA appears as a smear much more diminutive than 20,000 bp. Modern PCR techniques, such as multiplex STR typing, are successful because they can measure and detect minuscule amounts of DNA by amplifying them to a detectable level. In contrast to RFLP, which may have required 100 ng or more of DNA a few years ago, multiplex PCR amplification of STR alleles may currently analyse less than 1 ng of DNA. However, this sensitivity to small amounts of DNA also makes it difficult to prevent contamination by the police officer or crime scene specialist who collects the biological evidence. For PCR amplification, the DNA template must stay intact where the two primers bind and between the primers for full extension to occur. Since the primer extension will halt at the template break, PCR cannot proceed if there is no intact DNA strand surrounding the STR repeat region to use as a template strand. The template breaks more frequently, and fewer and fewer DNA molecules have the full length needed for PCR amplification as a DNA sample degrades (Butler, 2011).

7.2.2 DNA contamination

If caution is not used, the sensitivity of PCR may become an issue. Validated laboratory procedures, such as those of the DNA analyst, must be followed to prevent contamination from greater DNA concentrations. However, it is crucial to remember that contamination will almost always favour the defendant and produce an "exclusion" or "inconclusive" finding. The unintentional transfer of DNA is implied by contamination. When performing PCR, there are three possible sources of contamination: contaminating a sample with amplified DNA from a prior PCR reaction, contaminating a sample with genomic DNA from the environment, and contaminating samples during preparation (Lygo et al., 1994). The initial source of contamination is mostly determined by the care with which the evidence-collecting team gathers samples at the crime scene. "Substrate controls" can only monitor environmental pollution in a restricted sense (Gill, 1997). The latter two sources of contamination can be lessened or even eliminated using appropriate laboratory procedures and designated work areas. The risk of laboratory contamination is reduced by "negative controls," which look for contamination in PCR chemicals and tubes. A negative control is a blank sample run through the entire process alongside the forensic case evidence. The same volume of purified water is used as the DNA template in the other samples in a negative control PCR process. If detectable PCR findings are discovered in the negative control, those sources of contamination should be identified and eliminated before proceeding. After thoroughly examining the possible sources of PCR contamination, Henry Lee and associates found that PCR contamination was never found under the usual circumstances in casework analysis. They examined four broad PCR aspects—aerosolisation, handling the PCR results, PCR amplification setup, and DNA storage—where contamination may occur. To accomplish this, the AmpliType PM and DQA1 systems were detected by reverse dot blot (the same PCR principles apply as with STR markers detected by fluorescence). Detectable contamination only appeared in these tests when severe deviations were from basic prevention measures. They concluded that pollution could not be caused by careless actions alone (Scherczinger et al., 1999). In order to detect any contamination caused by an individual in the lab, the genotypes of laboratory staff are usually kept for comparison. Before determining that a created DNA profile accurately reflects the evidence and is not the result of contamination from laboratory personnel, this "staff elimination database" is frequently searched. Mixing two DNA types (the analyst's and the sample's) would only occur in the event of a serious laboratory error. During the mixture analysis, these kinds of errors would probably be eliminated by comparing the results with the genotypes of laboratory staff. Some expert system software has been created to help identify possible cross-contamination in a batch of samples. Recently, the Forensic Science Service pointed out that infrequent contamination of DNA testing supplies, including the tiny tubes used for PCR amplification, can introduce unnecessary DNA profiles. Over three years, the FSS found 11 casework-contaminating DNA profiles in over a million samples. In order to determine the overall batch quality, quality control testing of a few chosen tubes was analysed once these contaminated profiles were found in negative controls. Ten of the eleven casework-contaminating DNA profiles were fully matched to anonymous samples submitted by 300 tube maker workers. By introducing extra anti-contamination procedures in the production process due to communication with the tube manufacturer, the number of contamination occurrences attributed to their manufacturing workers was dramatically reduced. Potential sources of contaminated DNA can be found using elimination databases and negative control log information for forensic laboratory staff, manufacturing employees of DNA consumable suppliers, and police officers handling crime scene evidence. Additionally, as fewer personnel come into close touch with materials in the lab, the growing use of robotics for process automation lowers the possibility of contamination. In a recent research titled "Development of a Simulation Model to Assess the Impact of Contamination in Casework Using STRs," Peter Gill and Amanda Kirkham examined issues related to the influence of contamination on casework reporting requirements. Overall contamination in an operable DNA unit was predicted using negative controls. Negative controls performed with a batch of samples cannot assure that the corresponding batch of extracted casework material is contaminant-free, since PCR contamination can be tube-specific. Because contaminating DNA material can be preferentially amplified over shallow levels of original material present from the casework sample or may conceal the perpetrator's profile in a resulting combination, this study indicates that false exclusion is the most likely outcome of a contamination incident. Although a meticulous forensic DNA laboratory may not be affected by this contamination possibility very often, it may have important implications for cases being reviewed from the past, such as the Innocence Project. Assume, for instance, that ungloved police officers or evidence custodians handled biological evidence from a case that was 20 years old (before they knew about the sensitivity of contemporary DNA testing). In that instance, the collecting officer's contamination may obscure the DNA of the real offender. As a result, the DNA of the police officer or the person in charge of the evidence would be found during a DNA test, not the actual criminal. The incarcerated person may be wrongfully deemed "innocent" in the absence of additional proof since his DNA profile was not discovered in the initial crime scene evidence. The significance of seeing DNA evidence as an investigative tool in the context of a case rather than the exclusive, conclusive evidence of guilt or innocence is highlighted by this example. Physical separation of all pre-PCR and post-PCR amplification activities is necessary to reduce contamination issues during laboratory operations. Although it is most likely impossible to prevent laboratory contamination, it can be proactively evaluated using staff removal databases and negative controls (Gill & Kirkham, 2004).

7.2.3 Presence of inhibitors

When working with materials from difficult conditions, inhibitors pose a significant obstacle to acquiring accurate DNA profiles. Inhibitors cause partial or absent profiles by interfering with DNA extraction, amplification, or analysis (Butler, 2011). The substrates on which the sample is located might be biological materials or environmental pollutants among the sources of inhibitors. Blood haemoglobin, bile salts, and plant-based polysaccharides are examples of common biological, whereas soil and humic compounds are environmental inhibitors. Furthermore, forensic samples frequently interact with several inhibitors because of their exposure to different environmental factors like heat, moisture, or chemicals. Exhibits collected from crime scenes, including clothing, could be subjected to chemicals like detergents or cleaning solutions that prevent DNA extraction or PCR amplification by acting as inhibitors. Precisely, it has been demonstrated that materials such as tannins in plants or indigo dye in denim can impact the effectiveness of DNA profiling. By disrupting several stages of the DNA profiling technique, inhibitors impact DNA analysis (Alaeddini, 2012).

Inhibitors can attach to the DNA or disrupt the DNA polymerase (the enzyme that amplifies DNA during the PCR). This obstructs the amplification of the target DNA sequences, resulting in non-amplifiable DNA profiles. In forensic casework, where the DNA amount is already low or damaged, the impact of inhibitors might be very problematic. It might be challenging to draw firm conclusions since the presence of trace levels of inhibitors can result in incomplete or low-quality DNA profiles. In order to enhance DNA recovery and amplification in the presence of inhibitors, forensic experts have explored several methods to minimize their impact. When inhibitors are suspected, diluting the sample can lower the inhibitor concentration and enable the target DNA to be amplified successfully. However, this technique might also dilute the DNA, reducing the analysis's sensitivity. Commercial kits are available to eliminate inhibitors from forensic samples. These kits increase the likelihood of obtaining a complete DNA profile by removing materials that can obstruct PCR amplification using filtration or chemical treatments (Vajpayee et al., 2023). Early in the DNA amplification process, real-time PCR techniques can assist in identifying the presence of inhibitors. Reactions with internal PCR controls are used to track amplification effectiveness and determine the effect of inhibitors (Butler, 2011).

7.2.4 Mixed DNA samples

When two or more people contribute to the sample under test, mixtures are created. With much practice and careful instruction, mixtures can be easier to identify and understand. The capacity to see small components in the DNA profile of mixed samples has significantly increased over what was possible with RFLP approaches just a few years ago, thanks to advancements in detection technologies, such as PCR sensitivity combined with fluorescence measures. Similarly, a closer look has been given to the theoretical elements of statistical computations for mixture interpretation (Curran et al., 1999). With more loci and genetic markers and a high prevalence of heterozygotes, the likelihood of a mixture being discovered increases. Multiple DNA sources can be detected in a single sample depending on the ratio of DNA from each source, the particular genotype combinations, and the total amount of amplified DNA. Specific mixes will be more complex to find than others. There is a higher likelihood of detecting changes between the two components of a mixture when using highly polymorphic STR markers, which have more potential alleles. For instance, TPOX only has 15 known alleles, whereas D18S51 has 51 potential alleles, making D18S51 a more effective marker for mixture detection.

Similarly, the likelihood of observing many components in a mixture increases with the number of markers analysed (e.g., using a multiplex STR amplification). The detection of all contributors to a mixed sample depends on the amount of each component in the mixture. For instance, it will be far simpler to identify two DNA sources if their amounts are comparable than if one is present in a small percentage of the other.

For mixture ratios below the 5% threshold or 1:20, the minor component of a mixture is typically undetectable. To get a trustworthy genotyping result, the AmpFlSTR kits advise that the minor component be present in quantities greater than 35 pg. Multiple people's mixed DNA profiles can be found in crime scene evidence. Interpreting these combinations can be difficult, and the individual components may need to be separated based on the alleles in the profile. PCR inhibitors or damaged DNA can occasionally result in incomplete profiles where entire loci have disappeared. Multiple pieces of DNA evidence are used in many forensic cases; not all will be mixed. Therefore, rather than using complex mixtures, it is better to look for extra samples that are simpler to understand if they can be examined. Suppose a thorough differential extraction is carried out in a sexual assault case, and the sperm fraction can be separated from the victim's DNA. In that case, fortunately, mixes do not represent the majority of cases. For instance, one forensic laboratory handled 2424 samples in 1547 criminal cases over four years, but only 163 displayed a mixed profile, or 6.7% (Torres et al., 2003).

7.3 Legal and ethical challenges

The uncertainty of partial DNA profiles can have profound legal and ethical implications. In criminal cases, where the stakes are high, relying on partial DNA evidence can lead to wrongful convictions or acquittals if the limitations are not adequately understood and communicated. Courts must carefully consider the limitations of partial DNA profiles and the associated uncertainty when evaluating DNA evidence. Forensic experts must be cautious when presenting partial DNA profile results, clearly conveying the limitations and uncertainties. Judges and jurors may be misled by exaggerated claims of the strength of partial DNA evidence, which could lead to injustices. Family law is changing, and different jurisdictions require consistent rules. While family searching is illegal or uncontrolled in some nations and US states, it is regulated and subject to stringent standards in others. Because law enforcement agencies may have different thresholds for commencing familial searches, the lack of clear and consistent legal rules exacerbates privacy problems (Suter, 2009).

The ethical discussion concerns balancing people's right to privacy and the public interest in criminal investigations. Proponents contend that familial searching is an effective tool for law enforcement and can deter future offences, especially when dealing with serial offenders. However, the technique's detractors contend that it violates the privacy rights of people whose genetic information has not been used for investigative purposes. Important legal and ethical questions are also brought up by the limitations of STR analysis, especially when using DNA evidence in court. Although juries may require assistance in comprehending the limitations and difficulties connected with STR analysis, particularly in situations involving partial profiles or blends, DNA evidence is frequently considered extremely reliable. Forensic specialists must clarify these restrictions to prevent the court from exaggerating or misinterpreting DNA findings. Complex DNA profiles, including those produced from mixed or low-template samples, are increasingly being interpreted using probabilistic algorithms. Reliance on these models, however, adds uncertainty since, especially in low-quality DNA cases, little adjustments to the input data might result in significant alterations in the output (Buckleton et al., 2018).

7.3.1 Admissibility of DNA evidence

Since DNA evidence can individualise people with absolute certainty, it has become a crucial tool for forensic science in criminal trials. DNA evidence has been pivotal in convicting offenders and clearing charges against the innocent. However, it is still difficult to admit DNA evidence in court. Since, scientific validity and the interpretation of the evidence greatly determine the acceptance of DNA evidence in court. Legal requirements that guarantee the validity of the evidence offered in court govern the admissibility of forensic evidence, including DNA, in many conventional legal systems. The Frye and Daubert standards are the two that are most frequently used.

The Frye Criteria: The Frye test started in the 1923 case *Frye v. United States*. It said that for scientific evidence to be admitted in court, it must be "generally accepted" by the relevant scientific community (Frye v. United States, 1923). Under this standard, DNA evidence would be admissible if the scientific techniques used to obtain and analyse the DNA are widely accepted as reliable by forensic experts. While the Frye standard remains influential, it has been criticised for its reliance on the scientific community's consensus, which may exclude emerging but valid scientific techniques (Masten, 2001).

The Daubert Standard: In 1993, the U.S. Supreme Court case *Daubert v. Merrell Dow Pharmaceuticals* established a new standard for the admissibility of scientific evidence. The Daubert standard requires that scientific evidence meet criteria related to the methodology's validity, including whether the technique has been tested, subjected to peer review, has known error rates, and is widely accepted (Daubert v. Merrell Dow Pharmaceuticals, 1993). Courts apply the Daubert standard that acts as gatekeepers, ensuring that the DNA evidence is based on reliable and scientifically valid methods before being presented to the jury (Fisher, 2012). This standard is used in many federal courts of the United States. It is considered more flexible than the Frye standard, as it allows new scientific techniques to be evaluated on their merits rather than solely on acceptance.

7.3.2 Standards of reliability and validity

For DNA evidence to be admissible, it must fulfil the standards of reliability and validity. Courts typically assess a few factors while determining the credibility of DNA evidence, like Scientific Methodology, which includes DNA testing methods, such as STR analysis, that must be scientifically sound and reproducible. This involves DNA extraction, amplification through PCR, and comparison of DNA profiles (Butler, 2011). Another factor that is taken into account is the Chain of Custody. Maintaining the integrity of the DNA sample is critical for its admissibility in court. The chain of custody denotes the documentation of every person who handled the DNA evidence from the crime scene to the courtroom. Any break in this chain can result in questions about the credibility of the sample, leading to the exclusion of the evidence.

Statistical interpretation is also a vital factor that courts consider. Statistical interpretations of DNA evidence can introduce complexities. For example, mixed DNA samples or partial profiles may produce inconclusive results. Experts must clearly and accurately represent the statistical probabilities associated with DNA matches, as misinterpretation of probabilities can mislead the court proceeding (Buckleton et al., 2018).

7.3.3 Familial searching and privacy concerns

A forensic method called "familial searching" looks through DNA databases to find a person's near biological relatives whose DNA profile is unknown. Familial searches seek partial matches between profiles in the database and DNA from crime scenes, unlike typical DNA searches that seek an exact match. Researchers may be able to identify close relatives like parents, siblings, or kids thanks to these partial matches, which imply a genetic tie (Haimes, 2006).

When a close relative's DNA is already in a law enforcement database, but the offenders are not, the procedure is crucial. Investigators can employ other techniques to narrow down the suspect after a possible family match is discovered. High-profile crimes have been resolved using this technique, such as the "Grim Sleeper" case in Los Angeles, when detectives used the suspect's son's DNA in a DNA database to identify him.

Familial searching presents serious privacy issues despite its value in law enforcement. In addition to an individual's identifying information, DNA also carries information about their relatives, many of whom might not have given their permission for their genetic data to be utilised in criminal investigations. The use of family DNA searches brings up numerous significant privacy concerns:

Non-Consensual Inclusion in Investigations: One of the main issues is that people who have never provided their DNA to a database may be the subject of an investigation by law enforcement because they share a genetic heritage with someone who has. Family members are examined as a result, without their knowledge or consent (Haimes, 2006). Furthermore, this calls into question the morality of putting innocent people through police investigations based only on their biological ties to a suspect (Suter, 2009).

Genetic Monitoring: Because more and more people may be indirectly involved in criminal investigations due to the extensive use of familial searches, genetic surveillance may become increasingly common. The likelihood of law enforcement discovering partial matches rises with the size of DNA databases, potentially resulting in the profiling of larger groups of people, including those who have never engaged in criminal conduct. Critics contend that this extension compromises personal autonomy and privacy by establishing genetic surveillance.

Disproportionate Impact on Minority Groups: Due to their over-representation in criminal databases, minority groups may be disproportionately impacted by familial searching. People from minority groups are more likely to be implicated in familial searches since law enforcement databases often have more profiles from these groups, which raises questions about racial bias in the technique's employment (Suter, 2009). This might result in the unjust targeting of particular populations and worsen already-existing disparities in the criminal justice system.

Possibility of Misidentification: DNA matches within the family are only sometimes reliable. A partial match shows the likelihood of a relationship but does not confirm that the person is related to the suspect. This ambiguity raises the possibility of incorrect identification and false charges, mainly if other research methods must be thoroughly used to verify the family relationship (Haimes, 2006).

7.4 Limitations of current forensic DNA technology

7.4.1 Low-quantity DNA samples

As little DNA in a single collected cell can produce STR typing results, the PCR is vulnerable (Findlay et al., 1997). Attempts to retrieve DNA profiles from touch evidence that could be useful in a case have been prompted by this capability. However, unless steps are made to show that allele calls are reproducible, this low-level DNA analysis can occasionally push the boundaries of a reliable result. This chapter will discuss methods for increasing DNA sensitivity and initiatives to make low-level DNA testing more reliable. We should first review several frequently used words to refer to low-level DNA testing. A low copy number, sometimes called LCN, is most likely the most prevalent. In light of minimal amounts of DNA being analysed, some labs have referred to LCN as low template DNA (LT-DNA or LTDNA) in recent years. While LTDNA intends to concentrate on the sample, LCN usually uses a higher sensitivity approach. This method is also called DNALQQ (analysis of DNA of poor quality and quantity). Although "low-level DNA" is the main word used throughout this chapter, these other terms are generally interchangeable. Despite this, there has yet to be a clear distinction between low-level DNA and traditional STR typing (Gill & Buckleton, 2010), and various definitions have been discussed.

The definitions fall into several categories: A methodology that typically involves increasing the number of PCR cycles (e.g., 34 cycles instead of 28 cycles) to improve assay sensitivity; (2) those that refer to the amount of DNA testing in the PCR reaction (e.g., 100pg or 200pg) based on some quantitation assay, and (3) DNA profile appearance that shows an increased imbalance of observed alleles (e.g., heterozygote peak height ratios below 60%). It is acknowledged in all of these definitional techniques that data dependability is reduced when smaller amounts of DNA are analysed; therefore, more steps must be taken to increase the likelihood of receiving results that accurately reflect the sample under investigation. A DNA profile from less-than-ideal biological materials can be obtained using LTDNA, described as "an ultra-sensitive technique" (Caddy et al., 2008). A more precise definition was created in January 2009 by a committee based in the United States: "Any DNA typing results produced from a limited quantity and quality DNA template using conditions that have demonstrated increased stochastic effects are defined as LCN, or Low-Level DNA analyses, based upon a laboratory's internal validation." Increased stutter, increased intra-locus peak height imbalance, and allelic drop-ins or drop-outs are examples of stochastic effects that can be seen in DNA samples exposed to these circumstances. LCN settings may include nested PCR, post-amplification purification, reduced reaction volume, injection enhancement by higher voltage or time, and additional amplification cycles to assess limited quantity and quality DNA templates below the stochastic thresholds. Here, performance concerns are prioritised over defining a specific DNA concentration, which could yield less accurate results. Numerous labs have shown they can extract DNA profiles from minimal material. Typically, low-level DNA testing looks at fewer than 100 pg of input DNA, or roughly 15 diploid cells (Gill et al., 2000). Others have increased the threshold for defining LCN to 200 pg, or roughly 30 diploid cells (Budowle et al., 2009; Caddy et al., 2008).

7.4.2 Partial DNA profiles

Since DNA profiling is a highly effective method of accurately identifying people, it has completely changed the field of forensic research. DNA evidence, however, is only sometimes flawless or comprehensive. Due to sample degradation, low DNA concentrations, or contamination during collection or processing, only partial DNA profiles are often acquired in forensic situations. In addition to adding confusion to the conclusions derived from the evidence, partial DNA profiles can make forensic investigations more difficult. When not all loci (genetic markers) are amplified or readable in the study, the DNA profile is said to be incomplete. Forensic DNA analysis frequently uses STRs, and a comprehensive DNA profile would include information from all the necessary loci.

However, incomplete profiles need information at one or more loci, which limits how much may be compared to a reference profile (Butler, 2011). Partial DNA profiles may be obtained from forensic evidence for several reasons. Environmental conditions such as heat, moisture, and UV light can break down DNA molecules, making it challenging to produce a complete profile. Low DNA quantity may further lead to a partial DNA profile. Incomplete results may arise from the PCR, inhibited by the presence of inhibitors in the sample (Butler, 2011). Foreign DNA may be introduced by improper sample handling or contamination, which could affect the profiling procedure (Gill et al., 2006). A comprehensive DNA profile can accurately identify between people because it has a solid discriminating potential. However, the likelihood of a coincidental match rises when loci are absent from a partial profile, particularly in populations with high genetic similarity. This can cause the comparison procedure to have inaccurate inclusions or exclusions (Evett et al., 1998). The "allele dropout" condition occurs when an allele does not amplify and is hence absent from the DNA profile. Allele dropout is more likely to occur in partial profiles, especially in low-template DNA samples. This can produce false results, particularly in combinations containing DNA from several sources (Gill et al., 2006). Partial DNA profiles become even more trouble causing when working with mixed samples from several people because it can be challenging to determine which alleles belong to which person. This raises the possibility of inaccurate linkages and data misinterpretation (Buckleton et al., 2018).

7.4.3 Limitations of STR analysis

STR analysis is a popular technique in forensic DNA profiling that compares particular areas of a person's DNA where brief base pair patterns repeat tandemly. STR analysis has emerged as a fundamental component of forensic research because of its great discriminatory power and precision in matching DNA samples from crime scenes to possible suspects. However, STR analysis has its limits even after its wide application. The quality and amount of DNA, the challenges of mixed DNA samples, and the impact of low-template DNA contribute to these restrictions. The quantity and quality of DNA for testing determine how well STR analysis works. When DNA is exposed to environmental elements like heat, moisture, or ultraviolet light, it degrades and breaks into constituent units. As a result, some alleles may not amplify during the PCR, leading to incomplete STR profiles (Butler, 2011). This is problematic, particularly in forensic investigations that use DNA from aged biological samples or touched surfaces. Another drawback of STR analysis is the challenge of interpreting mixed DNA samples that arise when DNA from several people is present in a sample. Mixed DNA profiles are typical in cases of sexual assault with multiple perpetrators that require more sophisticated analytical methods and tools (Buckleton et al., 2018).

The STR analysis statistically interprets the probability that a DNA profile matches a specific person. Because STR loci are highly polymorphic—that is, there are numerous potential alleles at each locus—STR analysis has a high degree of discriminating between people. However, trustworthy population genetics data are necessary for this approach to be accurate. The likelihood of a coincidental match can differ according to the population under study, which is one drawback of STR analysis. The databases' representativeness may impact the reliability of the match probability to determine the frequency of particular STR profiles, which are frequently based on particular demographic groupings (Butler, 2011). The probability of accidental matches rises in populations with lower genetic diversity, which lowers the discriminatory strength of STR analysis. Moreover, STR databases do not always consider population substructures, which might affect allele frequencies. In these situations, the statistical estimates used to determine the likelihood of a match may exaggerate the evidence's strength, which could skew how DNA evidence is interpreted (Evett & Weir, 1998).

When nuclear DNA is insufficient or unavailable for conventional STR analysis, forensic genetics methods such as mitochondrial DNA (mtDNA) and Y-STR (Y-chromosomal short tandem repeat) analysis can be beneficial. While mtDNA is used to trace maternal ancestry, Y-STR analysis is frequently employed in male-specific lineage tracing cases. Although both approaches have intrinsic limits impacting their accuracy, resolution, and applicability, they offer crucial insights in forensic investigations. Forensic specialists must know these restrictions to evaluate the findings and appropriately explain their importance in court.

7.4.3.1 Y-STR analysis and its limitations

The Y chromosome transferred from father to son with little recombination is the focus of the Y-STR study. Because of this, Y-STRs help determine male contributions in mixed DNA samples and track paternal lineage, particularly in situations of sexual assault (Butler, 2009). Notwithstanding these benefits, several restrictions limit the applicability and efficacy of Y-STR analysis in forensic settings. Absence of personal discrimination: The absence of discriminating power is one of Y-STR analysis's most significant drawbacks. All male relatives within the same paternal lineage will have identical or almost identical Y-STR profiles because the Y chromosome is inherited as a single unit and largely stays constant over generations (Kayser, 2017). Y-STR analysis's utility in identifying individuals is thus limited because it cannot differentiate between closely related males, such as siblings or father and sons. In forensic investigations involving gang or group assaults, this might provide serious difficulties if paternal relatives are involved. Diverse Y-haplogroups exist in various populations, and Y-STR profiles show notable regional and ethnic heterogeneity. This variance complicates the interpretation of the results, even though it can occasionally help identify the geographic origin of an unknown individual. The reliability of population frequency estimations may be impacted by the irregularities in Y-STR databases, which could result in mistakes when determining match probabilities (Kayser, 2017).

Furthermore, this population substructure may result in biased conclusions in specific communities where particular Y-STR haplotypes are over-represented. Emphasis on the Y chromosome also leaves out female suspects, which can be a significant drawback when both male and female offenders are involved.

7.4.3.2 Mitochondrial DNA (mtDNA) analysis and its limitations

Mitochondrial DNA is inherited maternally, meaning it is passed from mothers to all their offspring, regardless of gender. MtDNA is helpful in forensic situations involving severely degraded evidence, such as skeletal remains or hair without roots, because it mutates more slowly and is found in higher copies than nuclear DNA (Parson & Dür, 2007). However, despite its advantages, mtDNA analysis comes with several limitations. Limited discriminatory power, similar to Y-STR, mtDNA analysis lacks the individual discriminatory power of autosomal STR analysis. Since all maternal relatives share the same mtDNA, this method cannot distinguish between siblings, cousins, or even distant maternal relatives (Parson & Dür, 2007). When individual identification is needed, this restriction lessens the value of mtDNA.

Furthermore, when utilised alone for forensic identification, the significant degree of maternal inheritance renders it less helpful. Heteroplasmy, in which a person has numerous mitochondrial DNA sequences, is a significant problem in mtDNA analysis because forensic scientists may come across distinct mtDNA sequences depending on the sample source. This variability makes it more challenging to analyse mtDNA profiles (Budowle et al., 2003). One of the advantages and disadvantages of mtDNA is its slow mutation rate, which restricts its capacity to differentiate between closely related individuals, even while it aids in its long-term preservation and makes it useful for historical or ancient DNA research (Amorim et al., 2019). The large copy number of mtDNA in cells raises the possibility of contamination, even though it is beneficial for recovering damaged samples. Due to the abundance of mtDNA, even minute contamination levels can hamper the analysis results (Budowle et al., 2003). To determine how rare a particular mtDNA haplotype is, the interpretation of mtDNA results mainly depends on population databases. However, compared to autosomal STR databases, mtDNA databases are smaller and less sophisticated, and their ability to represent populations worldwide has to be improved (Amorim et al., 2019). This can make it challenging to assess the significance of an mtDNA match, especially when rare haplotypes or under-represented groups are involved.

7.5 Emerging challenges

7.5.1 Synthetic biology and fabricated DNA samples

The use of evidence in criminal investigations has changed dramatically due to advances in DNA analysis as a result of advances in biotechnology. However, these improvements have raised several challenges too. The development of "synthetic biology" is raising concerns regarding the integrity of DNA evidence in forensic investigations. This deals with designing and constructing biological organisms or systems. The confidence in DNA as an indisputable evidence in court is questioned by the possibility of fabricated DNA samples in which synthetic or modified genetic materials could be introduced. This article examines the technological potential, possible risks, and legal perspectives of combining synthetic biology and DNA evidence fabrication.

Here, "fabrication" usually indicates the use of synthetic biology to manipulate DNA samples. The production or modification of genetic material to mislead forensic investigation is seen as a challenge for the near future. The fabrication or modification of DNA could have severe repercussions for criminal justice in a court of law where DNA evidence is regarded as highly trustworthy. The capacity to synthesize DNA questions the notion that DNA evidence is infallible and increases the risk of wrong convictions.

7.5.2 Impacts of advancing technology

It is now possible to produce artificial DNA sequences with remarkably higher accuracy, all because of advanced DNA synthesis technology. Scientists can now code a particular DNA sequence in the lab. Synthetic DNA could be planted at crime sites to deceive the investigation since these sequences could be made to match a person's genetic built (Frumkin et al., 2010). Theoretically, a forensic sample may be genetically modified to either diverge from or match the DNA profile of a particular person.

References

Alaeddini, R. (2012). Forensic implications of PCR inhibition—a review. *Forensic Science International: Genetics*, *6*(3), 297–305.

Amorim, A., Fernandes, T., & Taveira, N. (2019). Mitochondrial DNA in human identification: A review. *PeerJ*, *7*. https://doi.org/10.7717/peerj.7314

Buckleton, J. S., Bright, J. A., & Taylor, D. (Eds.). (2016). *Forensic DNA evidence interpretation* (2nd ed.). CRC Press. https://doi.org/10.1201/9781315367125

Budowle, B., Allard, M. W., Wilson, M. R., & Chakraborty, R. (2003). Forensics and mitochondrial DNA: Applications, debates, and foundations. *Annual review of genomics and human genetics*, *4*(1), 119–141.

Budowle, B., Eisenberg, A. J., & Daal, A. V. (2009). Validity of low copy number typing and applications to forensic science. *Croatian Medical Journal*, 50(3), 207–217. https://doi.org/10.3325/cmj.2009.50.207

Butler, J. M. (2005). *Forensic DNA Typing: Biology, Technology, and Genetics of STR Markers* (Vol. 660). Elsevier Academic Press.

Butler, J. M. (2009). *Fundamentals of Forensic DNA Typing*. Academic Press. https://doi.org/10.1016/C2009-0-01945-X

Butler, J. M. (2011). *Advanced Topics in Forensic DNA Typing: Methodology*. Academic Press. https://doi.org/10.1016/C2011-0-04189-3

Caddy, B., Taylor, G. R., & Linacre, A. M. T. (2008). A review of the science of low template DNA analysis. UK Home Office Report. https://assets.publishing.service.gov.uk/media/5a7a4f86e5274a34770e5c1d/Review_of_Low_Template_DNA_1.pdf

Curran, J. M., Triggs, C. M., Buckleton, J., & Weir, B. S. (1999). Interpreting DNA mixtures in structured populations. *Journal of Forensic Sciences*, 44(5), 987–995. https://doi.org/10.1520/JFS12028J

Daubert v. Merrell Dow Pharmaceuticals, 509 U.S. 579 (1993). https://supreme.justia.com/cases/federal/us/509/579/

Evett, I. W., Gill, P. D., & Lambert, J. A. (1998). Taking account of peak areas when interpreting mixed DNA profiles. *Journal of Forensic Sciences*, 43(1), 62–69. https://doi.org/10.1520/JFS16091J

Evett, I. W., & Weir, B. S. (1998). *Interpreting DNA Evidence: Statistical Genetics for Forensic Scientists*. Sinauer Associates.

Findlay, I., Taylor, A., Quirke, P., Frazier, R., & Urquhart, A. (1997). DNA fingerprinting from single cells. *Nature*, 389(6651), 555–556. https://doi.org/10.1038/39225

Fisher, B. A. (2012). *Techniques of Crime Scene Investigation* (8th ed.). CRC Press.

Frye v. United States, 293 F. 1013 (D.C. Cir. 1923). https://nys-fjc.ca2.uscourts.gov/programs/10-4-18%20Frye%20v%20US.pdf

Frumkin, D., Wasserstrom, A., Davidson, A., & Grafit, A. (2010). Authentication of forensic DNA samples. *Forensic Science International: Genetics*, *4*(2), 95–103.

Gill, P. (1997). The utility of `substrate controls' in relation to `contamination. *Forensic Science International*, 85(2), 105–111. https://doi.org/10.1016/S0379-0738(96)02085-3

Gill, P., Brenner, C. H., Buckleton, J. S., Carracedo, A., Krawczak, M., Mayr, W. R., & Weir, B. S. (2006). DNA commission of the International Society of Forensic Genetics: Recommendations on the interpretation of mixtures. *Forensic Science International*, 160(2–3), 90–101. https://doi.org/10.1016/j.forsciint.2006.04.009

Gill, P., & Buckleton, J. (2010). A universal strategy to interpret DNA profiles that does not require a definition of low-copy-number. *Forensic Science International: Genetics*, 4(4), 221–227. https://doi.org/10.1016/j.fsigen.2009.09.008

Gill, P., & Kirkham, A. (2004). Development of a simulation model to assess the impact of contamination in casework using STRs. *Journal of Forensic Sciences*, 49(3). https://doi.org/10.1520/JFS2003366

Gill, P., Whitaker, J., Flaxman, C., Brown, N., & Buckleton, J. (2000). An investigation of the rigor of interpretation rules for STRs derived from less than 100 pg of DNA. *Forensic Science International*, 112(1), 17–40. https://doi.org/10.1016/S0379-0738(00)00158-4

Haimes, E. (2006). Social and ethical issues in the use of familial searching in forensic investigations: Insights from family and kinship studies. *The Journal of Law, Medicine & Ethics*, *34*(2), 263–276.

Kayser, M. (2017). Forensic use of Y-chromosome DNA: A general overview. Human Genetics, 136(5), 621–635. https://doi.org/10.1007/s00439-017-1776-9

Lygo, J. E., Johnson, P. E., Holdaway, D. J., Woodroffe, S., Kimpton, C. P., Gill, P., & Clayton, T. M. (1994). The validation of short tandem repeat (STR) loci for use in forensic casework. *International Journal of Legal Medicine*, 107, 77–89. https://doi.org/10.1007/BF01225493

Masten, A. S. (2001). Ordinary magic: Resilience processes in development. *American Psychologist*, *56*(3), 227.

Parson, W., & Dür, A. (2007). EMPOP—A forensic mtDNA database. *Forensic Science International: Genetics*, 1(2), 88–92. https://doi.org/10.1016/j.fsigen.2007.01.018

Scherczinger, C. A., Ladd, C., Bourke, M. T., Adamowicz, M. S., Johannes, P. M., Scherczinger, R., & Lee, H. C. (1999). A systematic analysis of PCR contamination. *Journal of Forensic Sciences*, 44(5), 1042–1045. https://doi.org/10.1520/JFS12038J

Suter, S. M. (2009). All in the family: Privacy and DNA familial searching. *Harvard Journal of Law & Technology*, *23*(2), 309–399.

Torres, Y., Flores, I., Prieto, V., López-Soto, M., Farfán, M. J., Carracedo, A., & Sanz, P. (2003). DNA mixtures in forensic casework: A 4-year retrospective study. *Forensic Science International*, 134(2–3), 180–186. https://doi.org/10.1016/s0379-0738(03)00161-0

Vajpayee, K., Dash, H. R., Parekh, P. B., & Shukla, R. K. (2023). PCR inhibitors and facilitators—Their role in forensic DNA analysis. *Forensic Science International*, *349*. https://doi.org/10.1016/j.forsciint.2023.111773

8

Ethical and Normative Implications of Forensic DNA Technology

M.C.A. De Ungria and E.B. Jimenez

8.1 Introduction

DNA fingerprinting, or DNA testing, is a highly effective tool for human identification. However, it is important to consider its limitations, such as sample degradation and the potential for environmental and human contamination, which can result in partial or mixed genotypes that necessitate more complex analyses. Using the uniqueness of an individual's DNA, scientists have developed various protocols to enhance the accuracy of identifying the human source of DNA in samples collected from crime scenes. The identification process involves generating a person's DNA profile using multiple DNA markers characterized by different levels of variability, inheritance patterns, size, quantity, and complexity.

During forensic investigations, samples collected from a crime scene are used to assess the presence and level of involvement of persons of interest. The availability of DNA profiles from these samples has provided valuable genetic investigative leads because this information addresses the fundamental questions of who may be involved in a crime and who was present at the scene. To properly reconstruct the events leading to the crime, it is essential to corroborate genetic leads with other evidence. Since its introduction in the mid-1980s, DNA technology has significantly aided forensic investigations and has played a vital role alongside other forensic methods. The Year 2025 marks 40 years since the initial discovery of DNA fingerprinting, now more commonly known as DNA profiling. This occasion presents a timely opportunity to reflect on the evolution of DNA technology and examine the ongoing ethical challenges in various sectors, some of which remain unresolved.

This chapter examines the ethical and normative issues related to the rapid growth of DNA technology. It emphasizes concerns about public access to whole genome sequences (WGS) and genetic data (Caulfield et al., 2008), the increasing use of human DNA databases (Machado & Granja, 2020), and the global trend toward open science in order to accelerate scientific progress (Lunshof et al., 2008; Pisani et al., 2016). This push for open sharing of genetic data, including whole genome data, should be pursued with care, considering the cultural values or the vulnerability of individuals from whom samples were collected (Hill, 2017). The rapid expansion of the use of genomic data for forensic applications presents significant privacy risks to individuals and communities, many of whom do not have the resources or technical expertise to protect themselves from possible harm (Shringarpure & Bustamante, 2015). Moreover, the effect of over-representation of particular ethnic, marginalized, displaced groups, or rural populations in national databases (Piquado et al., 2019) must be evaluated. Mechanisms to mitigate risks should be developed before fully adopting kinship-based data searches. Institutions conducting forensic DNA testing must recognize the importance of

DOI: 10.4324/9781003474296-8

aligning forensic DNA practices with international human rights standards and ethical principles. As forensic DNA technologies continue to expand in scope and reach, establishing transparent, proportional, and rights-based frameworks that safeguard individual liberties while supporting valid investigative objectives to help maintain public security is crucial. In conclusion, we present recommendations for initiating conversations at local, national, and international levels about the critical need for DNA laws and policies that protect individuals, secure DNA information, and prevent the misuse of genetic and forensic data.

8.2 Forensic DNA technology

Before DNA profiling was discovered, forensic investigations relied heavily on traditional methods such as fingerprint analysis, blood typing, hair and fiber comparisons, and eyewitness testimony (National Research Council, 2009). Although these techniques provided valuable leads, they lacked the specificity and reliability that DNA profiling later introduced to human identification (Cooper, 2016).

8.2.1 Discovery of DNA fingerprinting

It all began with Sir Alec Jeffreys' studies on polymorphic minisatellite regions of DNA, which revealed highly variable patterns among individuals, much like the public's appreciation for the uniqueness of actual fingerprints (Jeffreys et al., 1985). The original technique, restriction fragment length polymorphism (RFLP), utilized restriction enzymes to cut DNA at specific recognition sites throughout the human genome, producing a series of fragments of varying lengths. When these fragments were separated by gel electrophoresis and visualized using hybridization with specific probes, they created distinct banding patterns that could differentiate individuals. This method became known as "DNA fingerprinting," an important breakthrough in human identification that provided law enforcement agencies with objective leads in their investigations.

This technique was first utilized in a real-life case involving a disputed maternity claim that was referred to Sir Alec Jeffreys' laboratory. A British nurse, Christiana Sarbah, who returned to the United Kingdom after working in Ghana, was accompanied by a 13-year-old boy she claimed to be her son. Immigration authorities questioned the biological relationship and requested DNA testing to verify the claim. Because the alleged father was unavailable, his genetic profile was reconstructed using the DNA profiles of the couple's three undisputed children. The results confirmed that the boy was indeed their biological son. The UK Home Office accepted the findings, marking one of the earliest non-criminal applications of DNA fingerprinting and highlighting its potential in immigration and kinship cases (Jobling, 2013).

However, the case of Colin Pitchfork is best known as the first criminal investigation in which DNA fingerprinting played a decisive role. This case involved the sexual assault and murder of two teenage girls in Leicestershire, England, in 1983 and 1986. During the investigation, the police initiated a local mass DNA screening, requesting that all males within a defined area voluntarily provide blood samples for DNA analysis. Five thousand samples were collected from local men, and only those that passed the initial screening underwent DNA testing. The DNA profiles obtained from biological evidence at both crime scenes matched, confirming that the same individual was responsible for both crimes. Interestingly, this DNA profile did not match Richard Buckland, who had previously confessed to one of the murders. The turning point came when it was discovered that Colin Pitchfork evaded the mass screening by convincing a friend, Ian Kelly, to submit a sample on his behalf. His DNA sample was tested, and the evidence matched the DNA profiles from both victims (Goodwin, 2015).

8.2.2 Types of DNA markers

Since the introduction of RFLP analysis, several DNA markers have been discovered, and methods have been developed that enhance DNA-based testing for identifying persons of interest, aiding in the prosecution of actual perpetrators, and exonerating those wrongfully accused. Along with the emergence of DNA fingerprinting, the polymerase chain reaction (PCR) also surfaced as an innovative tool in molecular biology. Discovered by Dr. Kary Mullis in 1983, PCR enabled the exponential amplification of specific DNA sequences *in vitro*, allowing scientists to generate millions of copies of targeted DNA regions from small, degraded, or environmentally compromised samples (Mullis & Faloona, 1987). This innovation addressed the significant limitations of RFLP analysis, which required relatively large amounts of intact, high-molecular-weight DNA. Forensic samples collected at crime scenes are often environmentally compromised and contain only trace amounts of DNA.

The development of thermostable DNA polymerases, particularly *Taq* polymerase, isolated from *Thermus aquaticus*, marked a significant milestone that enabled the widespread use of PCR in various laboratory settings (Chien et al., 1976), including forensic DNA analysis. By the mid-1990s, PCR became the primary protocol used in forensic DNA labs. This technique allowed scientists to replicate DNA *in vitro* from various types of evidence, many of which contained only trace amounts. This capability was essential for forensic investigations, provided that the crime scene samples were properly collected and managed (Federal Bureau of Investigations, 1994). The adoption of PCR-based techniques marked a significant advancement in the field, enabling faster, more sensitive, and more comprehensive DNA analysis by utilizing various DNA markers and requiring much less DNA than earlier protocols, such as RFLP.

8.2.2.1 Short tandem repeat (STR) DNA markers

Short tandem repeat (STR) DNA markers are characterized by short sequences of 2 to 7 base pairs (bp) and are found throughout human chromosomes. Many STRs are among the most informative and polymorphic genetic markers in the human genome (Edwards et al., 1991; Gettings et al., 2015). The number of repeat units at each STR locus varies between individuals, contributing to the overall combined power of discrimination during DNA profile comparisons.

STRs selected for forensic applications are chosen from non-coding regions of the human genome. These loci are preferred because they are not influenced by selective pressure and do not affect phenotypic traits or contain potentially medically relevant genetic information, thereby preserving individual privacy. Their genetic neutrality promotes greater allele diversity, as indicated by fragment length and sequence variations. STR alleles of the same length may vary in the sequence of the internal core repeat or flanking regions. These variations, known as "isoalleles," can be detected using massively parallel sequencing (MPS) technology, further enhancing our ability to distinguish DNA among individuals (Gettings et al., 2015).

Forensic STRs are often found on autosomes (autosomal STRs or aSTRs), making these autosomal STRs ideal for individual identification and most routine parentage testing (Butler, 2014; Gettings et al., 2015). However, STR markers on the sex chromosomes, such as those on the Y chromosome (YSTRs), are inherited strictly along the paternal line (Kayser, 2017; Purps et al.,

2014). They are useful in cases involving male contributors, such as during DNA mixture deconvolution of samples collected in sexual assault investigations (Roewer, 2019). STRs on the X chromosome (XSTRs), on the other hand, demonstrate a unique inheritance pattern that helps resolve complex kinship analyses involving paternity tests of daughters from different partners (Gusmão et al., 2025; Nothnagel et al., 2012). Given the variation in the inheritance patterns of aSTRs, YSTRs, and XSTRs, the selection of STR testing strategies is determined on a case-by-case basis.

8.2.2.2 Single-nucleotide polymorphism (SNP) DNA markers

Single-nucleotide polymorphisms (SNPs) are DNA sequence variations that occur at a single nucleotide position within the genome. SNPs appear approximately once every 300 base pairs, making SNPs the most common form of human genetic variation (Budowle & van Daal, 2008). Moreover, unlike STRs, individual SNPs exhibit a much lower mutation rate and are typically bi-allelic, which means they have only two allele variants per locus (Butler et al., 2007). As a result, SNPs are less informative per locus for individualization purposes. However, their small amplicon size makes them suitable for analyzing highly degraded or trace DNA samples, such as those obtained from abandoned crime scenes, disaster victim remains, or aged biological material.

Moreover, SNPs used in forensic science are categorized into several distinct types. First, identity SNPs (iSNPs) (Pakstis et al., 2007) are selected for their high heterozygosity and serve in personal identification, similar to the role of STRs in forensic investigations. Second, ancestry informative SNPs (aiSNPs or AIMS) (Kidd et al., 2011; Phillips et al., 2014) infer an individual's biogeographic ancestry, enabling estimations of continental or regional origins that can provide important investigative leads. Third, phenotype-informative SNPs (piSNPs) are used to predict externally visible traits such as eye, hair, and skin color (Kayser, 2015). These predictions derived from forensic phenotyping offer a "genetic face" that can assist in identifying unknown individuals when no matches are found in forensic databases. In addition, Tillmar and colleagues reported the use of a novel set of 3931 autosomal SNPs for extended kinship analysis (Tillmar et al., 2021).

In addition to nuclear SNPs, Y-chromosomal SNPs (YSNPs) (Ralf et al., 2025) and mitochondrial SNPs (mtDNA SNPs) (Holland & Parsons, 1999) provide insights into paternal and maternal lineages. These lineage markers are helpful in genealogical investigations to identify persons of interest when none have been specified and to locate missing persons when reference samples from direct relatives are unavailable. The ForenSeq Intelligence Kit produced by Verogen makes use of 9867 Kinship SNPs, 22 piSNPs, 106 XSNPs, and 85 YSNPs while retaining the 56 AIMS marker panel and the 94 iSNPs for kinship analysis (www.verogen.com).

8.2.3 Types of DNA tests

Due to their varying lengths, STR markers are typically analyzed using fragment length separation methods such as capillary electrophoresis (CE). This technique uses commercially available multiplex kits for the simultaneous analysis of multiple markers, ranging from 21 to 27. A summary of the various kits available for aSTR is provided in Gettings et al. (2015), YSTR in Shrivastava (2019), and XTR in Gomes and Arroyo-Pardo (2021).

The introduction of MPS technology, also known as next generation sequencing (NGS), provides both length and sequence-level information for STRs, thereby increasing discriminatory power by distinguishing isoalleles that were not detectable with CE (Berry, 2024; De Knijff,

2019). Furthermore, MPS technology has expanded the number and types of DNA markers that can be analyzed in a single assay (Al-Snan, 2024). For instance, the ForenSeq™ DNA Signature Prep Kit with Primer Mix B, developed by Illumina (now Verogen), allows for the simultaneous analysis of 27aSTR, 24YSTR, 7XSTR, 94 iSNPs, 22piSNPs, and 56 AIMS (Silvia et al., 2017), which significantly increased the amount of genetic information that could be generated in one assay.

Notably, MPS-generated DNA evidence played a critical role in a criminal case in the Netherlands, where DNA samples from a 28-year-old woman who was sexually assaulted in 2015 led to the first criminal conviction based on MPS data (Verogen, 2020). This milestone underscores the growing significance of MPS in forensic investigations, particularly in complex cases requiring high-resolution genetic profiling.

Verogen also released a fully SNP-based MPS assay that can be used for forensic investigative genetic genealogical (FIGG) testing. The ForenSeq Intelligence Kit replaced all the STR markers with 9867 Kinship SNPs, 22 piSNPs, 106 XSNPs, and 85 YSNPs while retaining the 56 AIMS marker panel and the 94 iSNPs.

8.2.4 Types of genetic databases

The evaluation of DNA test results and their statistical significance relies on databases established for specific purposes. Since the types of human samples in each database align with their intended functions, management and use of these databases must remain consistent with their original purpose (**Figure 8.1**).

8.2.4.1 Population reference databases

A population reference database, also known as a population database, consists of DNA profiles from individuals representing the local population in the area where a crime occurred. It is used to estimate the frequency of observed genotypes, forming the statistical basis for interpreting DNA matches. These frequency estimates help determine whether a DNA match is likely to have occurred by chance or is statistically significant, whether in direct comparisons between crime scene evidence and a person of interest or in kinship testing to assess biological relationships. Notably, countries characterized by extensive genomic diversity, due to refugee and displaced populations, migrant populations, diaspora communities, ethnic minorities, and Indigenous peoples (IPs), require a more strategic and inclusive approach to population sampling.

In Southeast Asia, academic institutions and research agencies have created valuable population reference databases by gathering samples from persons from diverse backgrounds, including those belonging to regional mainstream populations, Indigenous peoples, and ethnic communities (Abdulla et al., 2009). Countries such as Brunei Darussalam (Cheong et al., 2008), Indonesia (Venables et al., 2016), Malaysia (Rashid et al., 2020), the Philippines (Rodriguez et al., 2015; Salvador et al., 2019), Singapore (Yong et al., 2004), Thailand (Boonderm et al., 2020), and Vietnam (Huong et al., 2024) have reported establishing and continuously expanding population databases aimed at genomic research and the statistical assessments required to gauge the significance of matching DNA evidence (**Figure 8.1**). To improve the precision of these evaluations, it is crucial to broaden the population databases and incorporate the extensive genetic diversity present within each country's inhabitants, especially in regions such as Southeast Asia, where local populations are still underrepresented in global databases (Fatumo et al., 2022). International initiatives, such as the Y-Chromosome Haplotype

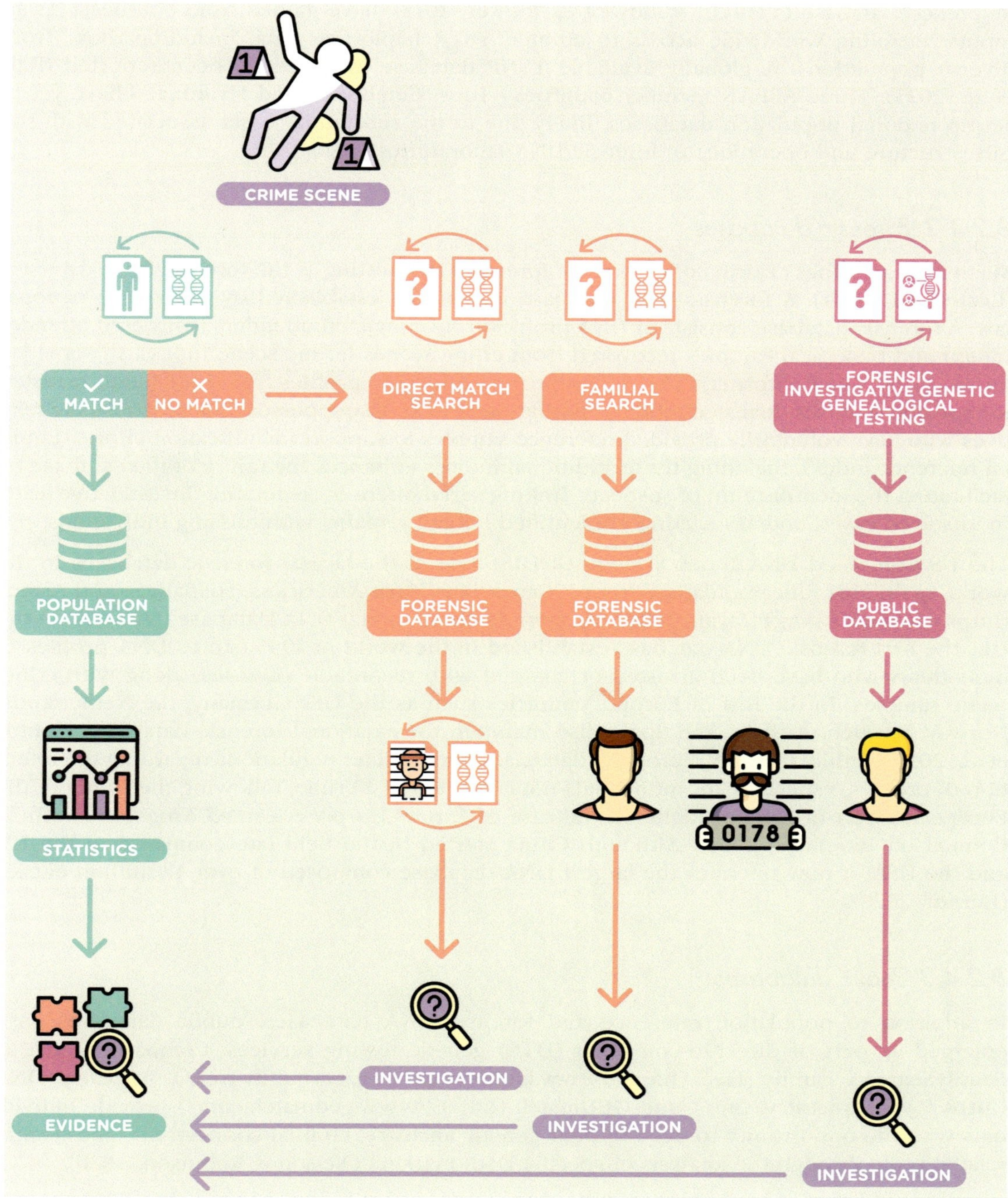

Figure 8.1 *DNA-based searching across different DNA databases. This figure illustrates the various approaches to DNA-based searching in forensic investigations, highlighting different types of databases and their respective search methods. It includes a population reference database, which stores samples mainly used for internal validation and statistical evaluation of matching DNA profiles; criminal DNA databases, which maintain offender profiles for direct comparisons with new crime scene profiles and familial searching; and public databases for forensic investigative genetic genealogy testing, which facilitate indirect identification through genetic relatives.*

Reference Database (YHRD) (Willuweit & Roewer, 2015), have gained from Southeast Asian inputs, enabling worldwide access to curated YSTR haplotype data, including those from diverse populations. A globally available YSNP database has recently been reported (Ralf et al., 2025). Three ASEAN member countries—Laos, Cambodia, and Myanmar—have yet to set up national population databases, likely due to the substantial costs associated with the infrastructure and operations of forensic DNA laboratories.

8.2.4.2 Forensic databases

Meanwhile, another critical component of forensic DNA testing is the forensic DNA database (Corte-Real, 2004). A forensic DNA database is typically established by enacting a national law. A forensic database consists of DNA profiles from convicted offenders (convicted offender index) and biological samples recovered from crime scenes (crime scene index). In some US jurisdictions, like California, the database may also include profiles from individuals arrested for specific offenses (arrestee index), missing persons (missing person index), and their relatives who have voluntarily provided reference samples to support identification efforts (familial reference index). Including these additional indices enhances the utility of the database by facilitating the identification of suspects, linking serial offenses, generating investigative leads in unsolved cases, and associating unidentified human remains with missing individuals.

The US Combined DNA Index System (CODIS), one of the largest forensic databases in the world, maintains four population groups: Caucasians, Afro-Americans, Hispanics, and Asians (https://www.fbi.gov/) (Myall, 2020). Meanwhile, the National DNA Database (NDNAD) in the UK, the first forensic DNA database established in the world in 1995, stores DNA profiles of individuals who have been arrested or charged with recordable offenses, along with crime scene samples. In the rest of Europe, countries such as the UK, Germany, the Netherlands, Norway, Sweden, Austria, and Spain also maintain their national forensic databases (Santos et al., 2013). Unlike the USA, European databases have stricter regulations regarding the retention of profiles, especially for individuals not convicted of a crime, following the ruling of the European Court of Human Rights in the case of *S. and Marper v. United Kingdom* (Nuffield Council on Bioethics, 2007). Although China started in the field late compared to the UK and the USA, it now operates the largest DNA database composed of over 17 million entries (Bernot, 2020).

8.2.4.3 Public databases

In addition to population reference and forensic DNA databases, public databases have emerged as part of direct-to-consumer (DTC) genetic testing services. Companies such as FamilySearch's Family Tree (https://www.familysearch.org/en/family-tree/), Ancestry DNA (https://www.ancestry.com/), and GEDmatch (https://www.gedmatch.com/) provide individuals with the opportunity to explore their genetic ancestry, familial connections, and health-related traits through the analysis of specific DNA markers (Nelson & Robinson, 2020).

8.2.5 Evolution in the use of forensic DNA technology

Since its initial development in 1985, "DNA fingerprinting" or DNA profiling has undergone significant advancements (**Table 8.1**). Driven by rapid progress in molecular biology and the expansion of analytical tools that enable forensic DNA scientists to study larger regions of the genome, along with the completion of the human genome sequence (Daly et al., 2001; Nurk et al., 2022; Subramanian et al., 2001), forensic DNA analysis is evolving toward the field of

Table 8.1 Comparison of Forensic DNA Investigative Methods

Method	Main Purpose	Data Used	Identifies Individual	Typical Database Used	Investigative Application	Limitations	Ethical/Legal Concerns
1. Traditional Forensic DNA Database Search	Identify direct matches from known offender or arrestee	Autosomal STR profiles	Yes	National forensic databases (e.g., CODIS in the US)	Routine identification of suspects or linking to old cases	No match if suspect is not in the database	Inclusion of arrestees in the search
2. Forensic DNA Phenotyping (FDP)	Predict externally visible characteristics (EVCs)	Phenotype-informative SNPs (piSNPs)	No	Research or reference EVC panel	Visual ID for investigative leads	Probabilistic; not identity-specific	Risk of reinforcing stereotypes; regulation gaps
3. Biogeographical Ancestry (BGA) Testing	Estimate ancestry linked to geographic origins	Ancestry-informative SNPs (aiSNPs), mtDNA	No	Research or reference AIM panels	Infers continental/ regional origin of unknown individuals	Low resolution in admixed populations	Risk of ethnic profiling
4. Familial Searching (FS)	Identify close relatives of unknown DNA source	Autosomal STR profiles	Indirectly	Forensic databases with familial search function	Identifies potential relatives of perpetrator (parent, sibling, etc.)	Limited to close relatives	False leads; privacy of relatives of those in the database
5. Forensic Investigative Genetic Genealogy (FIGG)	Identify unknown individuals through distant relatives	Genome-wide SNP profiles (DTC-style)	Yes (via genealogical reconstruction)	Open-source or DTC genealogy databases (e.g., GEDmatch)	Solving cold cases; identifying human remains	Labor-intensive	Consent for secondary use; unclear oversight, data misuse

forensic genomics. This field utilizes genomic sequences from biological samples of interest, including those that will eventually serve as DNA evidence in court, to identify the human sources of crime scene samples.

8.2.5.1 Predictive applications of DNA testing

Forensic DNA phenotyping and biogeographical ancestry testing signal a transition from purely identity-based forensic methods to predictive applications of DNA. The early generation of DNA markers used for population and forensic DNA databases consisted of various STRs in noncoding regions of the human genome. These markers were often labeled as "junk DNA." This characterization justified the establishment and expansion of forensic DNA databases, from the state to the national level, without rigorous ethical scrutiny or public debate. The prevailing notion was that since STRs did not contain sensitive phenotypic or medical information, their use posed minimal risk to individual privacy while significantly enhancing law enforcement's ability to ensure community safety.

However, creative efforts to increase the amount and type of genetic information that could be extracted from a biological sample have led to the development of forensic DNA phenotyping (Kayser, 2015). This method enables the prediction of externally visible characteristics (EVCs), such as the eye color, hair color, and skin pigmentation of the human source of the biological sample. By targeting specific phenotype-informative SNPs (piSNPs), forensic DNA phenotyping can create a "biological person of interest." Although forensic DNA phenotyping does not enable individual identification, it helps provide a physical description. Notably, the effectiveness of the current forensic DNA phenotyping panels for the available EVCs is limited in Southeast Asian populations, where most individuals display similar phenotypic traits such as black hair, brown eyes, and medium to dark brown skin.

Biogeographical ancestry (BGA) analysis complements forensic DNA phenotyping by estimating an individual's ancestral origins using ancestry-informative markers (AIMs)—a subset of SNPs selected for their ability to distinguish between major continental or regional population groups. BGA may also incorporate mitochondrial DNA (mtDNA) analysis to trace maternal lineage and Y-SNP analysis for paternal lineage identification (Phillips, 2015; Shriver et al., 2003). However, it does not identify individuals directly and serves only as an investigative tool in cases where a person of interest is unavailable.

Since BGA relies heavily on population reference databases, its discriminatory power can be limited in regions such as Southeast Asia, where these databases are less extensive compared to those in the USA and Europe. Additionally, the existing set of 56 AIMs used in ForenSeq FGx System (Verogen) has limitations in differentiating populations within the Asia-Pacific region. To address this, researchers developed the Multiplex for the Asia-Pacific (MAPlex), a MPS assay designed to enhance ancestry differentiation among East, Southeast, and Near Oceanian populations (Phillips et al., 2019). However, its initial validation primarily involved mainstream populations, and further research, including ethnic minorities and Indigenous groups, will be essential to improving its accuracy in predicting ancestry from human samples.

8.2.5.2 Kinship-based database searches

When a crime scene DNA profile is available, but no direct match is found in the national forensic database, familial DNA searching and forensic investigative genetic genealogical testing (FIGG) may be employed as investigative strategies (**Figure 8.1**). Both methods

Table 8.2 Methods for Kinship-Based Database Searches

Aspect	Familial Searching	Forensic Investigative Genetic Genealogy (FIGG)
Purpose	Identify close biological relatives (e.g., parent, sibling, child) of a suspect via partial DNA matches in government forensic DNA databases	Identify distant biological relatives (e.g., 2nd-4th cousins) of a suspect via shared DNA in public databases
Data Source	National criminal DNA databases (e.g., US CODIS)	Public genealogy databases (e.g., GEDmatch)
DNA Markers	STRs	SNPs
Analysis Method	Searches for partial STR profile matches	Compares SNP data to build genetic family trees
Investigative Steps	Match → Background check → Family tree construction → Covert DNA collection (e.g., from discarded items) → Confirmatory STR test	Match → Build extended family tree using genealogical records → Narrow down to person of interest → Collect DNA covertly or via warrant → Confirmatory testing
Privacy Concerns	Limited, due to use of government database	Significant, due to law enforcement use of public data originally submitted for personal use
Ethical Issues	Risk of implicating innocent family members; ethnic disparities if database is not representative	Consent and privacy; potential for re-identification of users who did not intend their data to be used in criminal investigations
Legal Oversight	Often regulated and limited to serious crimes; varies by country or place	Still emerging; lacks formal legislation or clear ethical guidelines in many areas

leverage genetic relationships to generate investigative leads but differ in scope, data sources, and ethical concerns. **Table 8.2** provides a more detailed comparison of both procedures.

First, family DNA searching employs reduced stringency search parameters and kinship matching algorithms to identify first-degree biological relatives—such as parents, children, or siblings—of the individual linked to the crime scene DNA (Greely et al., 2006). This technique is used in the USA (Greely et al., 2006) and the UK (Jobling & Gill, 2004) to help identify suspects in serious crimes when traditional DNA searches have been unsuccessful. After detecting a partial match, investigators conduct a thorough background investigation of the individual in the forensic database. This includes reviewing criminal records, residential history, and known family members. Based on this analysis and any additional information, new persons of interest who are relatives of the database individual are identified. Law enforcement initiates covert surveillance to gather samples of discarded DNA from biological materials left by the persons of interest in public areas, such as cigarette butts, tissues, coffee cups, or water bottles. Under current US legal standards, individuals are generally not deemed to have a reasonable expectation of privacy in garbage or discarded materials, and collecting this material for DNA testing does not require a warrant (Skene, 2014). The DNA profile obtained from the discarded sample is compared to the original crime scene profile. If the profiles match and the likelihood of a match is high, law enforcement will seek a court order to acquire a confirmatory DNA sample from the person of interest (Krimsky & Simoncelli, 2011).

When family DNA searching does not identify a first-degree relative or a viable person of interest from the forensic database, investigators may turn to FIGG, which makes use of a wider SNP DNA panel specifically developed for extended kinship analysis (Greytak et al., 2019). In this process, the SNP DNA profile of crime scene sample is generated using targeted sequencing or more recently, MPS technology is used to search public or open-source genetic genealogy databases like Family Tree, Ancestry DNA, and GEDmatch.

Public databases primarily consist of voluntarily submitted profiles from individuals looking to explore their ancestry or family connections (**Figure 8.1**). FIGG allows for identifying distant biological relatives, such as second, third, or fourth cousins, who share smaller segments of DNA with the unknown suspect. By analyzing the shared DNA segments and employing traditional genealogical research methods, such as family trees created from historical records, investigators can deduce the suspect's identity or significantly narrow the pool of potential suspects.

8.3 Ethical and normative considerations

The advancement of forensic technology has benefited the criminal justice system by providing a robust and accurate tool for processing crime scenes, thereby promoting the public good. The integration of DNA technology into forensic investigations has enhanced the legal system's ability to identify individuals, solve crimes, and exonerate the wrongfully accused (Laporte, 2018; Reilly, 2001). However, the use of DNA in forensics raises various ethical and normative concerns that relevant government agencies, professional forensic organizations, and international institutions must carefully evaluate. These concerns stem from the personal nature of genetic information and the broader legal, social, and moral frameworks within which DNA technology is utilized for forensic purposes.

8.3.1 Personal autonomy, informed consent, data privacy, access, and protection

These ethical and legal concepts are profoundly interconnected and form the basis for the responsible governance of forensic DNA technologies and databasing, especially regarding the inclusion of historically marginalized populations, ethnic minorities, and Indigenous peoples. Together, they set the standards for recognizing and protecting individual rights in collecting, storing, and using genetic information.

Personal autonomy refers to an individual's right to make informed and voluntary decisions about their body and identity, including whether to provide genetic material or consent to its use. It dictates that individuals own their DNA and must have complete control over all activities performed with it. Personal autonomy is the moral and legal foundation for obtaining informed consent, which is required before collecting DNA for research, forensics, or medical purposes (Iphofen, 2020).

Informed consent operationalizes this autonomy (freedom to choose or to decide) while protecting individuals whose autonomy may be diminished (Budowle & Sajantila, 2023). It is generally regarded as the cornerstone of ethical norms, ensuring that individuals understand potential risks (and benefits) and can make decisions with genuine options available. DNA testing accesses personal genetic information to reveal a person's identity. The collection, retention, and use of this data without explicit consent from the individual whose DNA is being tested can violate their autonomy. Notably, regardless of the level of suspicion surrounding their involvement in a crime, individuals who have not been convicted must continue to uphold the principle of autonomy, meaning they should maintain control over how their DNA profile is utilized. Informed consent is central to the ethical use of genetic data, ensuring individuals understand: (1) the nature of the data being collected, including DNA, health information, and family history; (2) the rationale for the collection, encompassing purposes like research and forensic investigations; (3) how the material will be used, for instance, matched against a crime scene profile or stored in a database for future use; (4) the potential risks of participation, such as re-identification despite assurances of anonymization, surveillance if profiles match future database searches, and discrimination, particularly concerning incidental findings from genomic analysis; and (5) their rights to withdraw samples in the future or limit the scope of use of their samples, including secondary use. For Indigenous peoples, individual consent must also reflect collective rights articulated through Free, Prior, and Informed Consent (FPIC), based on international principles like the United Nations Declaration on the Rights of Indigenous Peoples (UNDRIP) (UN General Assembly, 2007).

Once consent is granted and data is collected, data privacy laws and guidelines protect individuals from unauthorized access, sharing, or misuse of their genetic information (Ursin, 2008). Privacy safeguards ensure that autonomy remains respected after consent, particularly concerning long-term storage, database inclusion, and secondary use scenarios. Privacy is widely recognized as a fundamental right, allowing individuals to maintain ongoing control and access to their personal information. As a principle, an individual's genetic information requires protection from unauthorized access. Even for convicted criminals, the right to privacy may be diminished or restricted, but it is not entirely eliminated.

Meanwhile, data access and protection serve as legal safeguards. Data access defines who can use or view genetic data, including law enforcement, researchers, private companies, and foreign governments. Data protection encompasses the legal, technical, and ethical measures that prevent misuse, breaches, or exploitation. Examples of data protection include encryption, retention limits, de-identification, and judicial oversight. These mechanisms uphold privacy, ensure ongoing respect for consent, and safeguard personal and collective autonomy from violations.

8.3.1.1 Forensic database operations: direct matching and familial DNA searching

These considerations raise immediate questions about DNA searches conducted on forensic and public databases using crime scene DNA profiles without individuals' consent (**Figure 8.1**). One could argue that direct, high-stringency matching of the crime scene DNA profile against forensic databases to identify persons of interest and link other crimes is essential for forensic investigations. Additionally, it can be argued that retaining the DNA profiles of convicted offenders in the forensic database relies on the principle of proportionality—that is, the infringement of privacy is deemed acceptable given the seriousness of the crime and the need to protect public safety (National Research Council, 2014). From a rights-based ethical framework (https://socialprotection-humanrights.org/), this could be viewed as a permissible limitation of human rights under defined circumstances.

However, the principle of proportionality no longer applies to arrestees whose DNA profiles are retained in the forensic database, even without a conviction. Legal practices vary across jurisdictions when addressing this issue. In the USA, state laws differ regarding the inclusion and retention policies of forensic DNA databases. California retains DNA from all felony arrestees, regardless of the case outcome. If the case does not go to trial or if a person is acquitted, they must submit a request to have their genotypes removed from the forensic database. On the contrary, Texas includes DNA in the forensic database only after a conviction is obtained, which eliminates the need to remove DNA profiles from individuals who have been acquitted or whose cases did not go to trial (Piquado et al., 2019). Including arrestee DNA profiles in forensic databases, like those in California, is viewed by many human rights advocates as a violation of civil liberties. Internationally, several European legal systems do not include arrestee DNA profiles in their national forensic databases following the 2008 ruling of the European Court of Human Rights in *S. and Marper v. United Kingdom*. After careful consideration of the petition of Mr. S and Mr. Marper, this court required the UK to revise its policy of indefinitely storing all arrestee DNA profiles in the NDNAD. This case involved two individuals, Mr. S, who was arrested as a child in 2001 for attempted robbery, and Mr. Marper, who was charged with harassing his wife. Both had their fingerprints and biological samples collected and stored despite Mr. S's acquittal and the dropping of charges against Mr. Marper. The police rejected all requests to delete these two profiles from the NDNAD. Ultimately, the European Court of Human Rights ruled in favor of the individuals, stating that the retention of fingerprints, biological samples, and DNA profiles breached Article 8 of the European Convention on Human Rights, which guarantees respect for private and family life (Tuazon, 2021). Additionally, the UK passed the Protection of Freedoms Act 2012, which required the removal of DNA profiles of individuals who have not been convicted after a proportionate retention schedule has elapsed.

Because of the potential harm caused by erroneous matches, the high-stringency DNA database search parameters must be clearly defined in laboratory operations. Notably, the number and type of STR markers used for routine DNA profiling have increased to improve identification accuracy and decrease the likelihood of adventitious matches, even with very close relatives. For example, the USA has expanded its CODIS markers from the original 13 aSTRs to 20aSTRs in 2012 (Hares, 2012). European countries utilize the European Standard Set (ESS), which consists of 12 aSTR markers and newer recommendations to incorporate YSTRs into routine DNA profiling (Ralf et al., 2024). The forensic database should only contain high-quality reference DNA profiles. Direct DNA matching searches necessitate a complete match with either a potential person's reference genotype or a DNA profile from a different crime scene. Setting these standards is essential to ensure the reliability of database search results and to prevent unintended matches, which could result in incorrect investigations or wrongful convictions due to misleading DNA evidence.

The second major issue in database searching for forensic applications is the absence of informed consent from the family members of convicted offenders present in forensic databases (Greely et al., 2006). These searches included persons resulting solely from a family member's criminal actions. Therefore, jurisdictions allowing family DNA searches must enforce stringent regulatory measures, including case-by-case authorizations, restricting usage to violent offenses, and necessitating judicial or policy-level monitoring (Piquado et al., 2019). When considering arrestee DNA profiles, it was clear that retaining these no longer adheres to the principle of proportionality. Notably, in the USA, this method is not standardized, including those defined to be "unsolvable" cases. A survey conducted on local familial DNA search practices revealed that the primary reason for its lack of use is the absence of clear norms or guidelines for data utilization and analysis (Piquado et al., 2019).

Another concern is that familial DNA searching disproportionately affects "racial" and ethnic minorities who are overrepresented in the US CODIS database (Murphy, 2020). The resulting imbalance increases the likelihood of identifying African Americans as persons of interest, further contributing to perceived racial bias in many forensic investigations. While offering valuable leads, forensic database searching can disproportionately target marginalized communities, perpetuate structural biases, and involve individuals in criminal investigations based solely on genetic relatedness rather than direct evidence (Kennett, 2019; Murphy, 2013).

This issue of racial bias is particularly significant as an ethical and normative concern in Asia, characterized by extensive genetic diversity due to its many ethnic groups, population subgroups, and Indigenous peoples (Wall, 2019). When establishing databases in these countries, many of which are still defining the norms of operations and formulating the laws and policies to govern forensic DNA technology, examining the genetic ancestral background of individuals included in the forensic database is vital for determining whether any database search, particularly familial DNA searching, may be biased toward certain groups. This raises profound concerns regarding structural discrimination, the lack of presumption of innocence, and the collective harm to ethnic minorities and Indigenous peoples. Furthermore, obtaining consent and understanding its implications—including consent for all future use of genetic information—is not standardized due to differences in legal systems and cultural practices (Sessa et al., 2024).

Historically, genetic data from Indigenous populations have been collected without meaningful consultation or benefit-sharing. Including such legacy samples in forensic research or databasing, especially when facilitated through international collaborations, raises issues of biocolonialism. These acts infringe upon the rights of Indigenous peoples to control access to their genetic resources, as enshrined in the UNDRIP and articulated through the principle of FPIC. Significant ethical challenges have been reported regarding the invitation extended to some Philippine Indigenous peoples to provide saliva samples for a population reference database intended solely for research (Rodriguez, 2022). The voluntary consent of these groups to participate in any database that law enforcement would use is likely to be considered unacceptable

8.3.1.2 Forensic database operations: DNA legislation

The proper practice of obtaining informed consent includes full disclosure of relevant information, understanding such information, and a voluntary decision integrated into the procedures used by forensic laboratories to collect and generate DNA profiles that are later included in forensic DNA databases. In Cyprus, France, Germany, Luxembourg, Portugal, the Republic of Ireland, and Spain, even if coercive sampling is allowed under legal mandates or regulations, informed consent is still obtained for samples collected during criminal proceedings (Machado & Silva, 2015). A critical question arises in countries where a national DNA law has not yet been enacted: Can a country reliably depend on the results of direct matching and familial searches from a forensic DNA database established and operated without a national legislative mandate, independent oversight, or public accountability? Such normative oversight is essential to ensure the integrity of DNA analysis workflows, which involve comparing crime scene DNA profiles with those of persons of interest and the reliability of database search outputs aimed at producing investigative leads, such as in familial DNA searching. The lack of external audits and independent oversight is especially concerning in situations involving vulnerable or historically marginalized populations. In fact, in countries like the USA with already established laws and procedures for handling DNA evidence, it took a while to review more than 800 rape cases and identify those that involved mishandling of DNA evidence by a lab technician in New York City's medical examiner's office (Goldstein, 2013).

In Southeast Asia, the Philippines, Brunei, Thailand, and Indonesia have reported establishing and operating forensic DNA databases. However, these countries have not enacted formal national laws to authorize their establishment or provide mechanisms for regulation and sustainable financing. The gap in regulatory oversight raises concerns about governance and transparency in managing DNA evidence. In contrast, several Southeast Asian nations have successfully enacted legislation to regulate forensic DNA practices. Malaysia passed the DNA Identification Act of 2009, establishing clear procedures for collecting, analyzing, and storing DNA profiles (Hashom et al., 2020). Singapore utilizes the Comprehensive DNA Database under the Registration of Criminals Act, which governs DNA collection and database management for criminal justice purposes (https://sso.agc.gov.sg/SL/RCA1949-S358-2023). Vietnam enacted Law No. 26/2023/QH15, which addresses DNA collection, database operations, and privacy protections. In the Philippines, which has experienced over 20 years since the first bill was proposed, the country adopted an alternative approach to address the delayed passage of a DNA law by institutionalizing judicial oversight. In 2007, the Supreme Court of the Philippines promulgated the Rule on DNA Evidence, providing a framework for the admissibility and evaluation of DNA evidence in court proceedings (De Ungria et al., 2008). While the Rule does not serve as a substitute for a DNA law, it has provided interim guidance on using forensic DNA in all Philippine courts. The Rule highlights the judiciary's recognition of the technology's potential to address the nation's issues with unresolved cases, significant trial delays resulting in overcrowded jails, and a heightened rate of wrongful convictions (De Ungria & Jose, 2018).

8.3.2 Potential misuse of genetic data

The significant risk of potential harm arising from the misuse of genetic data highlights the need for precise regulation that aligns with the norms and expectations of governments responsible for ensuring national security and individual liberties. Such regulations should aim at solutions as well as reconciliation with the needs of the parties, involving facilitation of responsiveness to normative expectations. The responsiveness of regulation is particularly critical when the issue intersects with the rights of vulnerable populations, such as marginalized groups, ethnic minorities, Indigenous peoples, diaspora communities, and migrant populations, especially when operating without strong legal and ethical protections. The likelihood of harm has been documented in some secondary forensic uses of genetic data and adverse outcomes of technological determinism and misuse in legal proceedings.

8.3.2.1 Secondary use of non-forensic genetic data

The effectiveness of any forensic DNA database is shaped by its size, scope, and data composition, including sample diversity, DNA marker selection, and the nature of stored genetic information—whether length-based or sequence-based data. Traditionally, population databases used for allele frequency calculations in forensic matching have also contributed to broader genetic research, exploring human diversity, adaptation, health disparities, and migration patterns. However, the introduction of MPS, which enables the simultaneous analysis of hundreds of DNA markers, has blurred the lines between forensic and non-forensic genetic data.

Expanding forensic genetics into phenotypic and ancestral prediction introduces further ethical challenges, including potential misuse and the reinforcement of racial or ethnic profiling. The validity of specific analytic methods and the data produced, especially those used to infer "race," now referred to as ancestry from DNA sequences, requires careful consideration (Cho & Sankar, 2004). The application of population genetics in forensic investigations can collectively affect marginalized communities, particularly Indigenous groups, who may view genetic data collection through the lens of historical exploitation and systemic mistrust (Baker & Rattray-Te Mana, 2024). Without culturally sensitive engagement and group-rights safeguards, forensic practices risk violating collective identities and ethical principles (Baker & Rattray-Te Mana, 2024). The ethical concerns surrounding genetic surveillance and targeted collection have been particularly salient in cases involving vulnerable populations, such as Uyghur minorities in China (Schiermeier, 2021). Ensuring rigorous ethics review and oversight is essential for protecting both individual and community rights, reinforcing responsible research standards and forensic justice (De Ungria & Jimenez, 2022).

The emergence of advanced investigative methods, such as FIGG, raises the risk of state overreach and misuse (**Figure 8.1**). The issue of informed consent becomes more complicated when public databases are involved. While many consumers have embraced direct-to-consumer DNA testing, various groups have raised ethical concerns regarding law enforcement's secondary use of genetic data without the individuals' knowledge or consent (Tuazon et al., 2024). These platforms were initially designed for personal use, but law enforcement agencies have increasingly accessed public databases, particularly open-source platforms, to assist in criminal investigations. While companies like AncestryDNA maintain policies restricting law enforcement access to their databases without legal compulsion, others, like GEDmatch, have permitted such use if individuals submitting their DNA profiles also consent to having their profiles searched against those from law enforcement.

The first criminal conviction using FIGG in the USA is that of the Golden State Killer in 2020 (Levenson, 2020). Joseph Angelo DeAngelo Jr. was a former police officer who admitted to committing crimes against 87 victims from 1975 to 1986. His criminal portfolio included 13 murders, 51 sexual assault and over 120 burglaries in California from 1976 to 1986. Investigators were able to link him to these crimes using DNA from one of the crime scene samples collected in 1980 and searched against the GEDMatch public database (gedmatch.com).

While FIGG has successfully aided in solving more cold cases, its use has raised privacy, consent, and ethical concerns (Brown, 2019). Law enforcement's use of public databases has ignited significant debates about privacy, consent, and ethical boundaries. Consequently, FIGG often faces stricter protocols, with many jurisdictions requiring judicial authorization, transparency, and proportionality assessments before it can be utilized in active criminal investigations. With advances in molecular biology technologies, data mining, and access to metadata, it is crucial to evaluate whether the past informed consent process and associated risks are consistent with these increased capabilities (Budowle & Sajantila, 2023).

In some jurisdictions, DNA samples collected during medical procedures, immigration processing, or through research participation can be repurposed for forensic or law enforcement use without appropriate consent. For example, in Denmark and Sweden, newborn blood samples in biobanks for screening phenylketonuria and other diseases have been used for DNA tests to identify suspected criminals without prior consent (Machado & Silva, 2015). Moreover, large-scale sequence data and techniques used to analyze genetic differences between humans for medical purposes are now applied in forensic investigations to identify persons of interest.

8.3.2.2 Technological determinism and the risk of misuse in legal proceedings

While harm from secondary use is more widespread or easily identifiable, the potential likelihood of harm from adverse outcomes of technological determinism can be overlooked, especially its risk of misuse in legal proceedings. Courts may focus on DNA evidence without fully acknowledging the limitations of forensic analyses, particularly in cases involving complex DNA mixtures or low-template samples. Without proper regulation, probabilistic genotyping software, phenotypic predictions, and ancestry inference lacking adequate internal validations can lead to misinterpretation, ethnic profiling, or wrongful convictions. These risks increase when individuals are implicated based on uncertain or speculative genetic inferences.

The ethical and normative reflection ensures that the application of forensic DNA technology aligns with broader principles such as justice, equity, transparency, and accountability. These considerations are crucial for preventing wrongful convictions, promoting public trust in forensic institutions, and ensuring that forensic science serves the interests of justice, not just investigatory efficiency (Krimsky & Simoncelli, 2011).

8.4 Recommendations

When the Forensic Genetics Policy Initiative (FGPI) was launched, a global consensus on human rights standards for DNA databases emphasized: (1) the need for legal bases to manage biospecimens and to remove DNA profiles of individuals whose cases did not proceed to trial, as well as those acquitted at trial; (2) standards for the justified use of DNA in court; (3) appropriate safeguards for DNA collection from persons of interest; and (4) standards on data access, use, sharing across borders, and data protection (Forensic Genetics Policy Initiative, 2017). More than a decade later, the initiative has successfully engaged global stakeholders and mapped best practice expectations regarding forensic investigations involving genetic evidence. While these do not significantly deviate from the FGPI's original objectives, these recommendations aim to contribute to this decade-long discussion by sharing specific issues and experiences documented in recent years; an effort to chart progress toward achieving a balance where the requirements of the criminal justice system are met without compromising individual human rights.

8.4.1 Promoting equitable access to forensic DNA technology

Collecting and analyzing large numbers of individual samples is costly, as is maintaining extensive databases (Machado & Silva, 2015). The prohibitive costs of technology, including significant startup expenses, widen the gap between developed and low-income developing countries. Problems often arise from insufficient physical infrastructure and institutional support, as well as a lack of human resources, limited funding, and the absence of established pipelines necessary for long-term infrastructure development. This includes specialized biobank facilities and dedicated data centers for the computational and bioinformatics tasks needed to operate forensic DNA laboratories (Forensic Genetics Policy Initiative, 2017). Therefore, countries that are just beginning may benefit from outsourcing services through satellite forensic collaborations and data-sharing between nations. Although this cross-country initiative faces challenges, such as turnaround times, it can promote the sharing of resources, including personnel and equipment across collaborating laboratories (De La Chica et al., 2025). Supported by consultation, a regulated DNA database can be streamlined to include a defined set of essential categories that balance crime detection and prevention with protecting individuals' rights to data privacy and democratized access to forensic DNA technology through cost-effective use of resources (Forensic Genetics Policy Initiative, 2017). Likewise, mechanisms to reduce operational costs and the price of molecular reagents and supplies through local tax subsidies should be explored. In Asia, members of the DNA working group of the Asia Forensic Science Network (AFSN) have discussed collaborations and resource sharing since its initial establishment in Singapore in 2008. Member nations with more resources have sought to assist others through collaborations and partnerships.

8.4.2 Increasing the critical mass of DNA, ethics, and legal experts

A sufficient number of forensic experts, including DNA scientists, data analysts, database managers, ethicists, and lawyers, is essential for managing and overseeing a forensic genetics investigation system. Many regions face limited access to trained personnel due to the relative novelty of forensic DNA science in those areas (De La Chica et al., 2025). Addressing this gap requires government investment in establishing and enhancing forensic science education, particularly in DNA forensics, and active collaboration programs with countries with functional and robust forensic science systems. In countries lacking such systems, the curriculum must be developed by including best practices recognized globally and by providing institutional support for forensic science research-driven initiatives.

To enhance the critical mass of specialists in forensic genetic investigations, it is essential to gather a sufficient number of highly trained forensic scientists and professionals who can meet the scientific and technical requirements. Additionally, academic efforts should concentrate on maintaining a strong pipeline of skilled individuals, guaranteeing a steady influx of qualified professionals to meet the increasing demands of forensic science.

8.4.3 Formulating international guidelines on forensic ethics

An analysis of international legal systems reveals that forensic DNA regulations vary due to differences in legal frameworks, cultural influences, and governance structures. Institutions must recognize shared sound principles, especially those that demonstrate sensitivity to diverse cultures while retaining scientific credibility and providing ethical protection for individuals, particularly those from vulnerable populations. While differing data privacy laws affect access to personal genetic information across countries, developing standardized procedural guidelines is crucial for maintaining consistency in forensic investigation practices, many of which transcend borders and involve individuals from diverse backgrounds.

Successful forensic policy harmonization has been evident at the regional level. For example, the Prüm Convention of the European Union promotes cross-border sharing of forensic data, ensuring that DNA databases in member states adhere to uniform privacy and security standards (Toom et al., 2019). Similarly, the Interpol DNA Database supports international forensic collaboration, balancing effective law enforcement with ethical oversight (https://www.interpol.int/). In Southeast Asia, forensic networks have sought regional conformity, with groups like the Asian Forensic Sciences Network (AFSN) encouraging the exchange of knowledge and best practices among diverse legal systems (Yap & Cheng, 2024).

Professional forensic organizations, such as the International Society for Forensic Genetics (ISFG), the American Academy of Forensic Sciences (AAFS), the Australian and New Zealand Forensic Science Society (ANZFSS), and AFSN, play a crucial role in developing global forensic guidelines that transcend cultural and legal differences. These organizations contribute to the establishment of scientific standards, ethical policies, and forensic education, ensuring that the application of forensic DNA maintains a balance between public safety and individual rights.

Improving support for these guidelines requires strategic partnerships and government advocacy, which will help keep forensic policies transparent, accountable, and aligned with human rights protections. As forensic technologies advance, unified governance frameworks will be essential for maintaining public trust, ensuring ethical integrity, and promoting the responsible development of forensic science.

Areas covered by international guidelines can be determined through consensus meetings and workshops hosted by organizations recognized for their objectivity, fairness, and scientific integrity. The Forensic Genetics Policy Initiative (2017) recommended nine general areas of best practices that emerged from seven years of discussions about civil society concerns regarding the global surge of forensic DNA databases and related technology (Forensic Genetics Policy Initiative, 2017). These general areas include:

1. Clarification of whose data is included in a forensic DNA database, defining parameters for collection with consent from volunteers or without consent from persons of interest and convicted individuals, and stipulations for the destruction of DNA and related data.

2. Safeguards for the DNA collection process, including data collection, informing data owners, and reducing potential "racial" or ethnic bias, backed by a human rights framework.
3. Safeguards for DNA analysis of specimens gathered from crime scenes, laboratory quality assurance, lab oversight, DNA profiling standards, and establishing elimination databases (specimens that could have contaminated the crime scene);
4. Safeguards for the storage and use of DNA and related data include restrictions on DNA profiles to non-coding DNA, separation of criminal and non-criminal databases (e.g., missing persons), a transparent governance system, established access criteria and restrictions, and criteria-based limitations on the use of both stored data and familial searching.
5. Safeguards for using DNA evidence in court, outlining the criteria and conditions for access and actual use.
6. Safeguards for the international sharing of DNA evidence highlight the lack of a universal consensus on how this can be facilitated, while advocating for discretion guided by principles of autonomy, necessity, and proportionality.
7. Legal foundations provided by clear laws and regulations that include suitable penalties, enhancing public understanding and confidence that such databases respect human rights.
8. Enforce strict oversight of police access to genetic databases established for non-criminal purposes, including requirements for court authorization or a warrant, along with specific criteria for approving such access requests.
9. Evidence-based (or research-based) foundation for resource allocation in establishing forensic DNA databases aligned with national objectives.

8.4.4 Educating stakeholders and the public

In addition to building infrastructure and developing molecular, ethical, and legal expertise, engaging the public in fostering communication and encouraging individuals to ask meaningful questions and participate in finding answers is essential. The greatest barrier to overcome is the highly technical nature of forensics itself. The goal is to democratize access to information by enabling an ordinary person to understand forensic DNA technology well enough to make informed decisions.

Often, the public is concerned about the trustworthiness of forensic science and the vulnerability of genetic information to misuse. In this context, educational institutions and various social media platforms play an important role in promoting truthful and accurate information about real-life cases that the public can relate to. People watch television, spend time on Facebook, TikTok, or YouTube, which can be leveraged to communicate vetted information that shapes a meaningful and genuine understanding of forensic DNA technology. Admittedly, it will take time for such a strategy to take effect, but with enough time and advocates, these initiatives can encourage a culture of service for the public good. Gradually, increasing public knowledge will ultimately serve as protection against its abuse.

8.5 Conclusion

With advances in forensic DNA technologies, it is essential to uphold their significant role in human identification with strict ethical, normative, and legal safeguards. Major concerns regarding informed consent, data ownership, privacy protections, and the secondary use of genetic information highlight the importance of governance frameworks that balance community needs with individual liberties. Establishing robust forensic genetic databases and defining clear protocols for sample retention and investigation methods, especially concerning familial DNA searches and forensic genetic genealogy testing, is crucial for maintaining ethical standards.

When forensic resources are limited, the costly nature of DNA analysis necessitates careful prioritization and responsible use. This is particularly significant in developing countries that face numerous demands on their national budgets and resources, such as health and social welfare. Ethical governance should consider technological advancements and the broader implications of forensic DNA policies, ensuring that justice systems function are founded on principles of transparency, accountability, and the protection of human rights.

Beyond individual policies, partnerships and collaborations based on a genuine belief in the value of forensic DNA methodologies for protecting individuals and communities promote a balanced approach that combines ethical foresight with investigative efficiency. As forensic DNA continues to shape how nations tackle crime and its adverse effects on our communities, a commitment to objectivity, accuracy, fairness, and equality remains crucial for guiding its responsible use. By acknowledging the need to balance personal autonomy with the common good, stakeholders can ensure that DNA analysis upholds justice while safeguarding the fundamental human rights of all individuals, regardless of culture, religion, economic status, or social background.

Acknowledgments

The authors thank Erlyn B. Jimenez for creating **Figure 8.1** using CorelDRAW®, Ces Archae D. Buenavista for technical support, and Jose M. Jose, LLB, for engaging discussions. MCADU presented a portion of this work at the 12th World Conference of Science Journalists in Medellín, Colombia, with support from the Philippine National Academy of Science and Technology through the Scientific Career System.

Disclaimer

This chapter has undergone AI-assisted refinement to enhance coherence and clarity. While AI (Chat GPT) was utilized to check grammar and improve readability, the substantive content, analysis, and conclusions remain the authors' original work. All ethical considerations, arguments, and perspectives reflect the authors' research and insights.

References

Abdulla, M. A., Ahmed, I., Assawamakin, A., Bhak, J., Brahmachari, S. K., Calacal, G. C., Chaurasia, A. K., Chen, C.-H., Chen, J., Chen, Y.-T., Chu, J., Cutiongco-de la Paz, E. M., De Ungria, M. C. A., Delfin, F. C., Edo, J., Fuchareon, S., Ghang, H., Gojobori, T., Han, J., … Zilfalil, B. A. (2009). Mapping human genetic diversity in Asia. *Science, 326*, 1541–1545.

Al-Snan, N. R. (2024). Chapter 1 - Transition of capillary electrophoresis to next generation sequencing for forensic DNA analysis: Need of the hour. In: H. R. Dash, K. M. Elkins, & N. R. Al-Snan (Eds.), *Next Generation Sequencing (NGS) Technology in DNA Analysis* (pp. 3–20). Academic Press. https://doi.org/10.1016/B978-0-323-99144-5.00023-8

Baker, H., & Rattray-Te Mana, H. (2024). Intersections of forensics and indigenous peoples for future bicultural research. *Australian Journal of Forensic Sciences*, 1–17. https://doi.org/10.1080/00450618.2024.2442373

Bernot, A. (2020). Building of the world's largest DNA database: The China case. In P. Shrivastava, H. R. Dash, J. A. Lorente, & J. Imam (Eds.) *Forensic DNA Typing: Principles, Applications and Advancements* (pp. 639–658). Springer https://doi.org/10.1007/978-981-15-6655-4_33

Berry, H. (2024). Applications of NGS in analysis of autosomal STRs. In: H. R. Dash, K. M. Elkins, & N. R. Al-Snan (Eds.), *Next Generation Sequencing (NGS) Technology in DNA Analysis* (pp. 175–184). Academic Press. https://doi.org/10.1016/B978-0-323-99144-5.00012-3

Boonderm, N., Sangpueng, S., Nettakul, A., & Waiyawuth, W. (2020). Genetic polymorphism study on 21 autosomal STR loci of populations from six geographical regions in the Kingdom of Thailand. *Forensic Science International: Genetics, 47,* 102297. https://doi.org/10.1016/j.fsigen.2020.102297

Brown, T. R. (2019). Why we fear genetic informants: Using genetic genealogy to catch serial killers. *Columbia Science and Technology Law Review, 21*(1), 114–181.

Budowle, B., & Sajantila, A. (2023). Revisiting informed consent in forensic genomics in light of current technologies and the times. *International Journal of Legal Medicine, 137*(2), 551–565. https://doi.org/10.1007/s00414-023-02947-w

Budowle, B., & van Daal, A. (2008). Forensically relevant SNP classes. *Biotechniques, 44*(5), 603–608, 610. https://doi.org/10.2144/000112806

Butler, J. M. (2014). *Advanced Topics in Forensic DNA Typing: Interpretation*. Elsevier.

Butler, J. M., Coble, M. D., & Vallone, P. M. (2007). STRs vs. SNPs: Thoughts on the future of forensic DNA testing. *Forensic Science, Medicine, and Pathology, 3*(3), 200–205. https://doi.org/10.1007/s12024-007-0018-1

Caulfield, T., McGuire, A. L., Cho, M., Buchanan, J. A., Burgess, M. M., Danilczyk, U., Diaz, C. M., Fryer-Edwards, K., Green, S. K., Hodosh, M. A., Juengst, E. T., Kaye, J., Kedes, L., Knoppers, B. M., Lemmens, T., Meslin, E. M., Murphy, J., Nussbaum, R. L., Otlowski, M.,... Timmons, M. (2008). Research ethics recommendations for whole-genome research: Consensus statement. *PLoS Biology, 6*(3), e73. https://doi.org/10.1371/journal.pbio.0060073

Cheong, P. Y., Liew, P. V. O., Ibrahim, H., & Budowle, B. (2008). A genetic database for DNA-based forensic analysis in Brunei Darussalam. *Brunei Darussalam Journal of Health, 3,* 1–6.

Chien, A., Edgar, D. B., & Trela, J. M. (1976). Deoxyribonucleic acid polymerase from the extreme thermophile Thermus aquaticus. *Journal of Bacteriology, 127*(3), 1550–1557. https://doi.org/10.1128/jb.127.3.1550-1557.1976

Cho, M. K., & Sankar, P. (2004). Forensic genetics and ethical, legal and social implications beyond the clinic. *Nature Genetics, 36*(11 Suppl), S8–12. https://doi.org/10.1038/ng1594

Cooper, S. L. (2016). Challenges to fingerprint identification evidence: Why the courts need a new approach to finality. *Mitchell Hamline Law Review, 42*(2), 756–790.

Corte-Real, F. (2004). Forensic DNA databases. *Forensic Science International, 146*(Suppl), S143–S144. https://doi.org/10.1016/j.forsciint.2004.09.043

Daly, M., Rioux, J., Schaffner, S., Hudson, T., & Lander, E. (2001). High-resolution haplotype structure in the human genome. *Nature Genetics, 29,* 229–232.

De Knijff, P. (2019). From next generation sequencing to now generation sequencing in forensics. *Forensic Science International: Genetics, 38,* 175–180. https://doi.org/10.1016/j.fsigen.2018.10.017

De La Chica, A., Birkett, J., Akwei, C., Lamont, D., & Dawnay, N. (2025). Improving the forensic genetic workflow for countries with small geographical areas: What are the options and how cost effective are they? *Forensic Science International: Genetics, 74,* 103171. https://doi.org/10.1016/j.fsigen.2024.103171

De Ungria, M. C. A., & Jimenez, E. B. (2022). Ethical governance of forensic DNA databases in Southeast Asia. In: H. R. Dash, P. Shrivastava, & J. A. Lorente (Eds.), *Handbook of DNA Profiling* (pp. 1–39). Springer Singapore. https://doi.org/10.1007/978-981-15-9364-2_9-1

De Ungria, M. C. A., & Jose, J. M. (2018). Forensic science in challenging environments: The Philippine experience. In I. A. C. Caguioa, P. E. L. Balisong, J. S. B. Pangilinan, V. C. S. Lo, & M. C. E. Tecson (Eds.) *Tending Life* (pp. 6–17). Anti-Death Penalty Task Force.

De Ungria, M. C. A., Sagum, M. S., Calacal, G. C., Delfin, F. C., Tabbada, K. A., Dalet, M. R. M., Te, T. O., Diokno, J. I., Diokno, M. S. I., & Asplen, C. A. (2008). Forensic DNA evidence and the death penalty in the Philippines. *Forensic Science International: Genetics, 2*(4), 329–332. https://doi.org/10.1016/j.fsigen.2008.04.004

Edwards, A., Civitello, A., Hammond, H., & Caskey, C. T. (1991). DNA typing and genetic mapping with trimeric and tetrameric tandem repeats. *American Journal of Human Genetics, 49,* 746–756.

Fatumo, S., Chikowore, T., Choudhury, A., Ayub, M., Martin, A. R., & Kuchenbaecker, K. (2022). A roadmap to increase diversity in genomic studies. *Nature Medicine, 28*(2), 243–250. https://doi.org/10.1038/s41591-021-01672-4

Federal Bureau of Investigations. (1994). *PCR-Based Typing Protocols: Laboratory Manual.*

Forensic Genetics Policy Initiative. (2017). *Establishing best practices for forensic DNA databases.*
Gettings, K. B., Aponte, R. A., Vallone, P. M., & Butler, J. M. (2015). STR allele sequence variation: Current knowledge and future issues. *Forensic Science International: Genetics, 18,* 118–130. https://doi.org/http://dx.doi.org/10.1016/j.fsigen.2015.06.005
Goldstein, J. (2013). New York examines over 800 rape cases for possible mishandling of evidence. *The New York Times.*
Gomes, C., & Arroyo-Pardo, E. (2021). Usefulness of the X-chromosome on forensic science. In: H. R. Dash, P. Shrivastava, & J. A. Lorente (Eds.), *Handbook of DNA Profiling* (pp. 1–24). Springer Singapore. https://doi.org/10.1007/978-981-15-9364-2_17-1
Goodwin, W. (2015). DNA profiling: The first 30 years. *Science and Justice, 55*(6), 375–376. https://doi.org/10.1016/j.scijus.2015.11.001
Greely, H. T., Riordan, D. P., Garrison, N. A., & Mountain, J. L. (2006). Family ties: The use of DNA offender databases to catch offenders' kin. *Journal of Law, Medicine & Ethics, 34*(2), 248–262.
Greytak, E. M., Moore, C., & Armentrout, S. L. (2019). Genetic genealogy for cold case and active investigations. *Forensic Science International, 299,* 103–113. https://doi.org/10.1016/j.forsciint.2019.03.039
Gusmão, L., Antão-Sousa, S., Faustino, M., Abovich, M. A., Aguirre, D., Alghafri, R., Alves, C., Amorim, A., Arévalo, C., Baldassarri, L., Barletta-Carrillo, C., Berardi, G., Bobillo, C., Borjas, L., Braganholi, D. F., Brehm, A., Builes, J. J., Cainé, L., Carvalho, E. F.,... Pinto, N. (2025). X-chromosomal STRs: Metapopulations and mutation rates. *Forensic Science International: Genetics, 76,* 103232.
Hares, D. R. (2012). Expanding the CODIS core loci in the United States. *Forensic Science International: Genetics, 6*(1), e52–e54. https://doi.org/10.1016/j.fsigen.2011.04.012
Hashom, M., Khan, H., Lalung, J., Nelson, B., Chambers, G., & Edinur, H. (2020). Autosomal STR profiling and data banking in Malaysia: Current status and future prospects. *Genes, 11*(10). https://doi.org/10.3390/genes11101112
Hill, S. Y. (2017). Data sharing: Guard the privacy of donors. *Nature, 548*(7667), 281. https://doi.org/10.1038/548281a
Holland, M. M., & Parsons, T. J. (1999). Mitochondrial DNA sequence analysis - validation and use for forensic casework. *Forensic Science Review, 11,* 21–50.
Huong, N. T. L., Bao, V. G., & Truong, M. Q. (2024). Genetic characteristics of a 22 a-STR loci in the Vietnamese population. In *Hội nghị Khoa học Toàn quốc về Công nghệ Sinh học 2024* [National Scientific Conference on Biotechnology 2024] (pp. 19–25). Forensic Medicine Center of Ho Chi Minh City.
Iphofen, R. (Ed.) (2020). *Handbook of Research Ethics and Scientific Integrity.* Springer
Jeffreys, A. J., Wilson, V., & Thein, S. L. (1985). Individual-specific 'fingerprints' of human DNA. *Nature, 316*(6023), 76–79.
Jobling, M. A. (2013). Curiosity in the genes: The DNA fingerprinting story. *Investigative Genetics, 4*(1), 1–3. https://doi.org/10.1186/2041-2223-4-20
Jobling, M. A., & Gill, P. (2004). Encoded evidence: DNA in forensic analysis. *Nature Reviews Genetics, 5*(10), 739–751. https://doi.org/10.1038/nrg1455
Kayser, M. (2015). Forensic DNA phenotyping: Predicting human appearance from crime scene material for investigative purposes. *Forensic Science International: Genetics, 18,* 33–48. https://doi.org/10.1016/j.fsigen.2015.02.003
Kayser, M. (2017). Forensic use of Y-chromosome DNA: A general overview. *Human Genetics, 136*(5), 621–635. https://doi.org/10.1007/s00439-017-1776-9
Kennett, D. (2019). Using genetic genealogy databases in missing persons cases and to develop suspect leads in violent crimes. *Forensic Science International, 301,* 107–117. https://doi.org/10.1016/j.forsciint.2019.05.016
Kidd, J. R., Friedlaender, F. R., Speed, W. C., Pakstis, A. J., Vega, F. M., & Kidd, K. K. (2011). Analyses of a set of 128 ancestry informative single-nucleotide polymorphisms in a global set of 119 population samples. *Invest Genet, 2*(1), 1. https://doi.org/10.1186/2041-2223-2-1
Krimsky, S., & Simoncelli, T. (2011). *Genetic Justice: DNA Data Banks, Criminal Investigations, and Civil Liberties.* Columbia University Press.
Laporte, G. M. (2018). *Wrongful convictions and DNA exonerations: Understanding the role of forensic science.* National Institute of Justice. https://nij.ojp.gov/topics/articles/wrongful-convictions-and-dna-exonerations-understanding-role-forensic-science
Levenson, M. (2020, August 21, 2020). Golden state killer sentenced to life in prison without parole. *The New York Times.*
Lunshof, J. E., Chadwick, R., Vorhaus, D. B., & Church, G. M. (2008). From genetic privacy to open consent. *Nature Reviews Genetics, 9*(5), 406–411. https://doi.org/10.1038/nrg2360

Machado, H., & Granja, R. (2020). DNA databases and big data. In: H. Machado, & R. Granja (Eds.), *Forensic Genetics in the Governance of Crime* (pp. 57–70). Springer Singapore. https://doi.org/10.1007/978-981-15-2429-5_5

Machado, H., & Silva, S. (2015). Public participation in genetic databases: Crossing the boundaries between biobanks and forensic DNA databases through the principle of solidarity. *Journal of Medical Ethics*, *41*(10), 820–824. https://doi.org/10.1136/medethics-2014-102126

Mullis, K. B., & Faloona, F. A. (1987). Specific synthesis of DNA in vitro via a polymerase-catalyzed chain reaction. *Methods in Enzymology*, *155*, 335–350. https://doi.org/10.1016/0076-6879(87)55023-6

Murphy, E. (2013). Familial searches of DNA databases. In M. Viñado (Ed.), *Bases de datos policiales de ADN: ¿son una herramienta realmente eficaz en la lucha contra la criminalidad grave nacional y transfronteriza?* (pp. 221–240). Aranzadi.

Murphy, E. E., & Tong, J. H. (2020). The racial composition of forensic DNA databases. *California Law Review*, *108*(6), 1847–1904.

Myall, J. (2020). *An evaluation of the accuracy, use and relevancy of CODIS in criminal investigations in the US* (p. 4). Bournemouth University.

National Research Council. (2009). *Strengthening Forensic Science in the United States: A Path Forward*. The National Academies Press. http://www.nap.edu/catalog/12589.html

National Research Council. (2014). *The Growth of Incarceration in the United States: Exploring Causes and Consequences*. The National Academies Press.

Nelson, A., & Robinson, J. H. (2020). The social life of DTC genetics: The case of 23andMe. In D. L. Kleinman & K. Moore (Eds.) *Routledge Handbook of Science, Technology and Society* (pp. 108–123). Routledge. https://doi.org/10.4324/9780203101827-16

Nothnagel, M., Szibor, R., Vollrath, O., Augustin, C., Edelmann, J., Geppert, M., Alves, C., Gusmao, L., Vennemann, M., Hou, Y., Immel, U. D., Inturri, S., Luo, H., Lutz-Bonengel, S., Robino, C., Roewer, L., Rolf, B., Sanft, J., Shin, K. J., … Hering, S. (2012). Collaborative genetic mapping of 12 forensic short tandem repeat (STR) loci on the human X chromosome. *Forensic Science International: Genetics*, *6*(6), 778–784. https://doi.org/10.1016/j.fsigen.2012.02.015

Nuffield Council on Bioethics. (2007). The forensic use of bioinformation ethical issues. https://www.nuffieldbioethics.org/publications/the-forensic-use-of-bioinformation-ethical-issues

Nurk, S., Koren, S., Rhie, A., Rautiainen, M., Bzikadze, A. V., Mikheenko, A., Vollger, M. R., Altemose, N., Uralsky, L., Gershman, A., Aganezov, S., Hoyt, S. J., Diekhans, M., Logsdon, G. A., Alonge, M., Antonarakis, S. E., Borchers, M., Bouffard, G. G., Brooks, S. Y., … Phillippy, A. M. (2022). The complete sequence of a human genome. *Science*, *376*(6588), 44–53. https://doi.org/10.1126/science.abj6987

Pakstis, A. J., Speed, W. C., Kidd, J. R., & Kidd, K. K. (2007). Candidate SNPs for a universal individual identification panel. *Human Genetics*, *121*(3–4), 305–317. https://doi.org/10.1007/s00439-007-0342-2

Phillips, C. (2015). Forensic genetic analysis of bio-geographical ancestry. *Forensic Science International: Genetics*, *18*, 49–65. https://doi.org/http://dx.doi.org/10.1016/j.fsigen.2015.05.012

Phillips, C., McNevin, D., Kidd, K. K., Lagace, R., Wootton, S., de la Puente, M., Freire-Aradas, A., Mosquera-Miguel, A., Eduardoff, M., Gross, T., Dagostino, L., Power, D., Olson, S., Hashiyada, M., Oz, C., Parson, W., Schneider, P. M., Lareu, M. V., & Daniel, R. (2019). MAPlex - A massively parallel sequencing ancestry analysis multiplex for Asia-pacific populations. *Forensic Science International: Genetics*, *42*, 213–226. https://doi.org/10.1016/j.fsigen.2019.06.022

Phillips, C., Parson, W., Lundsberg, B., Santos, C., Freire-Aradas, A., Torres, M., Eduardoff, M., Borsting, C., Johansen, P., Fondevila, M., Morling, N., Schneider, P. M., Carracedo, A., & Lareau, M; EUROFORGEN-NoE Consortium. (2014). Building a forensic ancestry panel from the ground up: The EUROFORGEN Global AIM-SNP set. *Forensic Science International: Genetics*, *11*, 13–25. https://doi.org/10.1016/j.fsigen.2014.02.012

Piquado, T., Matthies, C. F., Strang, L., Anderson, J. M.; RAND Social and Economic Well-Being (Program). Justice Policy., Rand Corporation., & National Institute of Justice (U.S.). (2019). *Forensic familial and moderate stringency DNA searches: Policies and practices in the United States, England, and Wales*. https://www.rand.org/pubs/research_reports/RR3209.html

Pisani, E., Aaby, P., Breugelmans, J. G., Carr, D., Groves, T., Helinski, M., Kamuya, D., Kern, S., Littler, K., Marsh, V., Mboup, S., Merson, L., Sankoh, O., Serafini, M., Schneider, M., Schoenenberger, V., & Guerin, P. J. (2016). Beyond open data: Realising the health benefits of sharing data. *BMJ*, *355*, i5295. https://doi.org/10.1136/bmj.i5295

Purps, J., Siegert, S., Willuweit, S., Nagy, M., Alves, C., Salazar, R., Angustia, S. M. T., Santos, L. H., Anslinger, K., Bayer, B., Ayub, Q., Wei, W., Xue, Y., Tyler-Smith, C., Bafalluy, M. B., MartÃnez-Jarreta, B., Egyed, B., Balitzki, B., Tschumi, S., … Roewer, L. (2014). A global analysis of Y-chromosomal haplotype diversity for 23 STR loci. *Forensic Science International: Genetics*, *12*, 12–23. https://doi.org/http://dx.doi.org/10.1016/j.fsigen.2014.04.008

Ralf, A., Zandstra, D., van Wersch, B., Köksal, Z., Larmuseau, M. H. D., Rosa, A., Jobling, M. A., D'Amato, M. E., Courts, C., Gysi, M., Haas, C., Flores, R., Neis, M., Wetton, J. H., Kiesler, K., Ameur, A., Azonbakin, S., Bôžiková, A., Choma, A., ... Kayser, M. (2025). UYSD: A novel data repository accessible via public website for worldwide population frequencies of Y-SNP haplogroups. *European Journal of Human Genetics, 33*(7), 1476–5438. https://doi.org/10.1038/s41431-025-01854-5

Ralf, A., Zieger, M., & Kayser, M. (2024). Considerations on expanding criminal offender DNA databases with Y-STR profiles. *Journal of Law and the Biosciences, 11*(2), https://doi.org/10.1093/jlb/lsae017.

Rashid, M. N. A., Mahat, N. A., Khan, H. O., Wahab, R. A., Maarof, H., Ismail, D., Alwi, A. R., & SyedHassan, S. R. (2020). Population data of 21 autosomal STR loci in Malaysian populations for human identification. *International Journal of Legal Medicine, 134*(5), 1675–1678. https://doi.org/10.1007/s00414-020-02279-z

Reilly, P. (2001). Legal and public policy issues in DNA forensics. *Nature Reviews Genetics, 2*(4), 313–317.

Rodriguez, J. J. R. B., Cuales, J. M. D., Herrera, M. J. B., Zubiri, L. A. M., Muallil, R. N., Ishmael, A. I., Salvador, J.-A. R., Villamor, L. P., Jose, J. M., & De Ungria, M. C. A. (2022). Ethical challenges in genetic research among Philippine Indigenous Peoples: Insights from fieldwork in Zamboanga and the Sulu Archipelago. *Frontiers in Genetics, 13*. https://doi.org/10.3389/fgene.2022.901515

Rodriguez, J. J. R. B., Salvador, J. M., Calacal, G. C., Laude, R. P., & De Ungria, M. C. A. (2015). Allele frequencies of 23 autosomal short tandem repeat loci in the Philippine population. *Legal Medicine, 17*(4), 295–297. https://doi.org/http://dx.doi.org/10.1016/j.legalmed.2015.02.005

Roewer, L. (2019). Y-chromosome short tandem repeats in forensics—Sexing, profiling, and matching male DNA. *Wiley Interdisciplinary Reviews: Forensic Science, 1*(4), e1336. https://doi.org/10.1002/wfs2.1336

Salvador, J. M., Rodriguez, J. J. B., Carandang, L. C. D. L., Agmata, A. B., Hondrado, M. L. D. H., Delfin, F. C., & De Ungria, M. C. A. (2019). Filipino DNA variation at 36 Y-chromosomal short tandem repeat (STR) marker units. *Philippine Journal of Science, 148*(S1), 43–52.

Santos, F., Machado, H., & Silva, S. (2013). Forensic DNA databases in European countries: Is size linked to performance? *Life Sciences, Society and Policy, 9*(12), 1–13.

Schiermeier, Q. (2021). Crime-solving DNA databases faces ethical scrutiny. *Nature, 594*, 320–322.

Sessa, F., Chisari, M., Esposito, M., Karaboue, M. A. A., Salerno, M., & Cocimano, G. (2024). Ethical, legal and social implications (ELSI) regarding forensic genetic investigations (FGIs). *Journal of Academic Ethics*. https://doi.org/10.1007/s10805-024-09582-z

Shringarpure, S. S., & Bustamante, C. D. (2015). Privacy risks from genomic data-sharing beacons. *American Journal of Human Genetics, 97*(5), 631–646. https://doi.org/10.1016/j.ajhg.2015.09.010

Shrivastava, P. (2019). Performance comparison of two most recent Y-STR multiplex systems for real forensic casework analysis. *Forensic Science International: Genetics Supplement Series, 7*(1), 262–263. https://doi.org/10.1016/j.fsigss.2019.09.100

Shriver, M. D., Parra, E. J., Dios, S., Bonilla, C., Norton, H., Jovel, C., Pfaff, C., Jones, C., Massac, A., Cameron, N., Baron, A., Jackson, T., Argyropoulos, G., Jin, L., Hoggart, C. J., McKeigue, P. M., & Kittles, R. A. (2003). Skin pigmentation, biogeographical ancestry and admixture mapping. *Human Genetics, 112*(4), 387–399. https://doi.org/10.1007/s00439-002-0896-y

Silvia, A. L., Shugarts, N., & Smith, J. (2017). A preliminary assessment of the ForenSeq FGx system: Next generation sequencing of an STR and SNP multiplex. *International Journal of Legal Medicine, 131*(1), 73–86. https://doi.org/10.1007/s00414-016-1457-6

Skene, L. (2014). The current approach of the courts. *Journal of Medical Ethics, 40*(1), 10–13. https://doi.org/10.1136/medethics-2012-100994

Subramanian, G., Adams, M. D., Venter, J. C., & Broder, S. (2001). Implications of the human genome for understanding human biology and medicine. *JAMA, 286*(18), 2296–2307.

Tillmar, A., Sturk-Andreaggi, K., Daniels-Higginbotham, J., Thomas, J. T., & Marshall, C. (2021). The FORCE panel: An all-in-one SNP marker set for confirming investigative genetic genealogy leads and for general forensic applications. *Genes, 12*(12), 1968.

Toom, V., Granja, R., & Ludwig, A. (2019). The prüm decisions as an aspirational regime: Reviewing a decade of cross-border exchange and comparison of forensic DNA data. *Forensic Science International: Genetics, 41*, 50–57. https://doi.org/10.1016/j.fsigen.2019.03.023

Tuazon, O. M. (2021). Universal forensic DNA databases: Acceptable or illegal under the European court of human rights regime? *Journal of Law and the Biosciences, 8*(1), lsab022. https://doi.org/10.1093/jlb/lsab022

Tuazon, O. M., Wickenheiser, R. A., Ansell, R., Guerrini, C. J., Zwenne, G.-J., & Custers, B. (2024). Law enforcement use of genetic genealogy databases in criminal investigations: Nomenclature, definition and scope. *Forensic Science International: Synergy, 8*, 100460. https://doi.org/10.1016/j.fsisyn.2024.100460

UN General Assembly. (2007). *United Nations declaration on the rights of indigenous peoples* (UN Wash, Issue).
Ursin, L. (2008). Personal autonomy and informed consent. *Medicine, Health Care, and Philosophy*, *12*, 17–24. https://doi.org/10.1007/s11019-008-9144-0
Venables, S. J., Daniel, R., Sarre, S. D., Soedarsono, N., Sudoyo, H., Suryadi, H., Van Oorschot, R. A. H., Walsh, S. J., Widodo, P. T., & McNevin, D. (2016). Allele frequency data for 15 autosomal STR loci in eight Indonesian subpopulations. *Forensic Science International: Genetics*, *20*, 45–52. https://doi.org/10.1016/j.fsigen.2015.09.014
Verogen. (2020). *How next generation sequencing resolved a difficult case, leading to the first criminal conviction of its kind.*
Wall, S. (2019). Prevention of antibiotic resistance – an epidemiological scoping review to identify research categories and knowledge gaps. *Global Health Action*, *12*(sup1), 1756191. https://doi.org/10.1080/16549716.2020.1756191
Willuweit, S., & Roewer, L. (2015). The new y chromosome haplotype reference database. *Forensic Science International: Genetics*, *15*, 43–48. https://doi.org/10.1016/j.fsigen.2014.11.024
Yap, A., & Cheng, N. (2024). The journey of the Asian forensic sciences network (AFSN) through fifteen years. *Forensic Science International*, *359*, 112042. https://doi.org/10.1016/j.forsciint.2024.112042
Yong, R. Y. Y., Aw, L. T., & Yap, E. P. H. (2004). Allele frequencies of 15 STR loci of three main ethnic populations in Singapore using an in-house marker panel. *Forensic Science International*, *141*(2–3), 175–183. https://doi.org/10.1016/j.forsciint.2004.01.004

9

DNA Barcoding and Its Potential Applications in Forensic Investigations

Amandeep Kaur Bhambara and Mukesh Kumar Thakar

9.1 Introduction to DNA barcoding

9.1.1 Definition and scope

DNA barcoding is a technique used to identify and classify species of an unknown biological sample based on species-specific differences in short standardized regions of DNA. This technique generates a unique genetic "barcode" for each species or individual that will aid in accurate identification and comparison. In forensic investigations, DNA barcoding has been used for a long time in solving criminal cases by identifying victims, and verifying the origins of biological samples recovered from the scene of crime. (Hebert et al., 2003). It uses state-of-the-art biotechnology to identify plant and animal species in a rapid, accurate, and cost-effective manner, as its underlying principle lies in the notion that each species has a unique genetic signature that is encoded in specific regions of its genome (Cowan et al., 2006).

Ideally, a DNA barcode allows definite species identification due to considerably sufficient sequence variation between the species. This has become an indispensable tool for the identification of the species due to the existence of a 'barcoding gap'. This gap exists because levels of divergence among individuals of the same species are significantly much lower than those of the closely-related species.

9.1.2 Historical background

The remarkable journey that DNA barcoding has taken to evolve has fundamentally transformed the understanding about the concept of biodiversity (**Figure 9.1**).

In 1977, Fred Sanger and colleagues discovered 'molecular systematics', a branch of molecular biology which deals with studying the evolutionary relationships among individuals and species. With this advent, various researchers started exploring the use of molecular markers, such as DNA and proteins, to determine the evolutionary links between different species (Blaxter, 2004). This led to the discovery of methods for DNA-based identification, such as protein electrophoresis and DNA hybridization.

During the 1980s–1990s, continual developments in DNA sequencing methods were observed. The field of molecular biology experienced a significant transformation with the advent of Sanger sequencing. The sequencing of genes and regions of interest took place during this

DOI: 10.4324/9781003474296-9

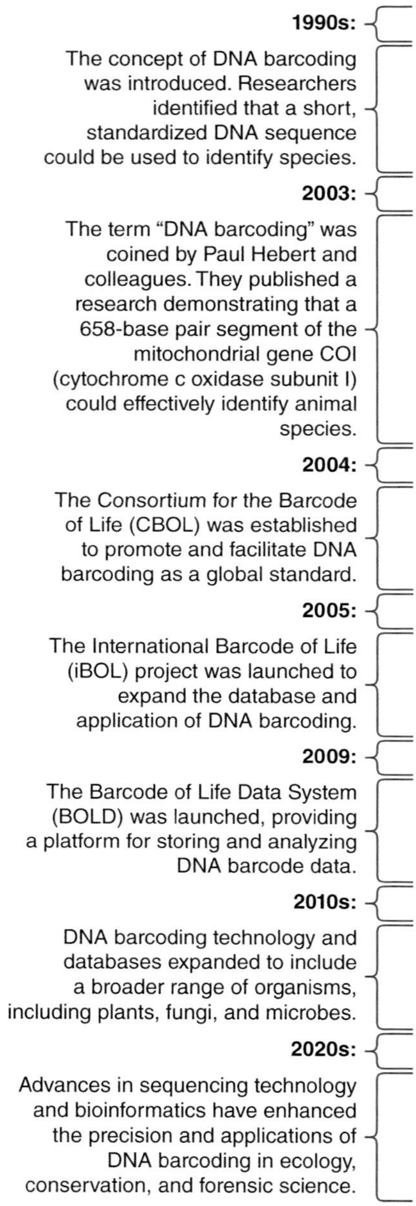

Figure 9.1 *Brief history of DNA barcoding.*

period, which laid the foundation for more discoveries/innovations in the field of DNA barcoding (Ferri et al., 2009a, 2009b).

The technique introduced by Carl Woese and colleagues was applied to arthropods in the 1980s (Hebert et al., 2003). Following this, the concept of DNA barcoding was introduced in 2003 by Dr. Paul Hebert (**Figure 9.2**) (now a Canadian Biologist, Founder and Director of the Centre of Biodiversity at the University of Guelph in Ontario, Canada) with the establishment of the DNA Barcoding Initiative (Hebert et al., 2003). The goal of this project, led by Paul

Figure 9.2 *Dr. Paul Hebert: father of DNA barcoding.*

Hebert and his colleagues, was to provide a universally accepted method for the purpose of species identification by using short DNA sequences. Their work on the Cytochrome c oxidase subunit I (COI) gene as a universal barcode marker was extensively acknowledged and accepted worldwide.

The term 'DNA barcode' was coined in 2003 by Dr. Paul Hebert. According to Hebert, DNA barcodes are unique standardized sequences of DNA with 400–800 base pairs, that help in the identification or classification of organisms, followed by PCR amplification, sequencing and eventually comparison with the existing reference database from different species. At species level, DNA barcoding can help to identify novel and unknown species with more accuracy.

By the year 2004, various researchers and scientists started exploring a variety of DNA barcoding markers for various specific taxonomic groups (Hebert et al., 2004). For instance, plants (rbcL, matK), fungi (ITS) and animals (COI) were identified as standard barcoding regions.

To enrich the database, the International Barcode of Life (iBOL) initiative was launched as a global standard in 2005, succeeding the establishment of the Consortium for the Barcode of Life (CBOL) in 2004. These initiatives emerged as global networks that nurtured the collaboration and data sharing amongst various researchers. Following this, a comprehensive platform was created in 2009 named the Barcode of Life Data System (BOLD), for storing and analyzing DNA barcode data. In BOLD, not only DNA sequences, but the related projects can be managed, files can be traced, specimens in the herbarium can be scanned and pictures stored can be maintained. Plants, fungi, and microorganisms are just a few of the varied range of creatures to which DNA barcoding technology was applied during the 2010s. Presently, DNA barcoding has become increasingly precise and useful in domains including forensic research, ecology, and conservation, thanks to developments in bioinformatics and sequencing technology.

9.2 Fundamentals of DNA barcoding

9.2.1 Concept of DNA barcoding

DNA barcoding involves analysis of a short, standardized segment of DNA to generate a unique genetic identifier, commonly known as 'barcode'. DNA barcodes are generated by a process that begins with the collection of a biological sample from the target organism, followed by DNA extraction from the collected sample to isolate the genetic material. Then, a specific region of the DNA, known as the barcode region, is amplified using Polymerase Chain Reaction (PCR) to produce a considerable number of copies of DNA for nucleotide/genetic sequencing. The resulting sequence is then compared with reference databases, such as the BOLD, to identify and recognize the species. Moreover, if the sequenced species is newly discovered (novel or not previously catalogued), it needs to be subsequently added to the library of databases to expand the repository of DNA barcodes. This standardized method enables precise species identification and also contributes to the global biodiversity database.

In addition to this, a gene region in the DNA barcode must satisfy the following criteria:

- It must exhibit species-level genetic variations and divergence,
- It must possess conserved/flanking adjoining sites to generate universal PCR primers,
- It must have a short sequence length for precise DNA extraction and amplification.

Consequently, a short DNA sequence of 600 bp in the mitochondrial gene for cytochrome *c* oxidase subunit 1 (COI) satisfies the abovementioned requisites and hence has been accepted as a practical, standardized species-level barcode for animals (Hebert et al., 2003).

9.2.2 DNA barcoding techniques

The techniques employed for generating DNA barcodes have been extensively used to identify species with the help of reliable methods standardized on the basis of the genetic information they contain, and for classifying the diverse forms of species. In general, this method involves extracting DNA from a sample, amplifying a specific, short segment of the genome known for its variability among species – such as mitochondrial gene, the COI for animals or various chloroplast genes for plants, matK and rbCL and thereby sequencing the segment to determine its exact nucleotide sequence (**Figure 9.3**). This technique has become a universal, objective, and rapid tool for species identification that significantly enhances the ability to catalogue biodiversity, monitor ecosystems and address challenges in wildlife conservation and forensic investigations.

9.2.2.1 Standardized DNA barcode regions

The commonly used barcode regions are as follows:

- **COI (Cytochrome c oxidase subunit I):**
  The COI barcode, or cytochrome c oxidase subunit I barcode, is a widely used genetic marker in DNA barcoding that helps in species identification. It is basically a mitochondrial gene that encodes a component of the electron transport chain, essential for cellular respiration. Typically, the COI barcode region is a 658-base pair segment near the 5′-end of the mitochondrial gene. The COI barcode has been used in identifying animals, including insects, fish, and mammals. It has also been instrumental in various applications such as monitoring biodiversity, detecting illegal wildlife trade, and

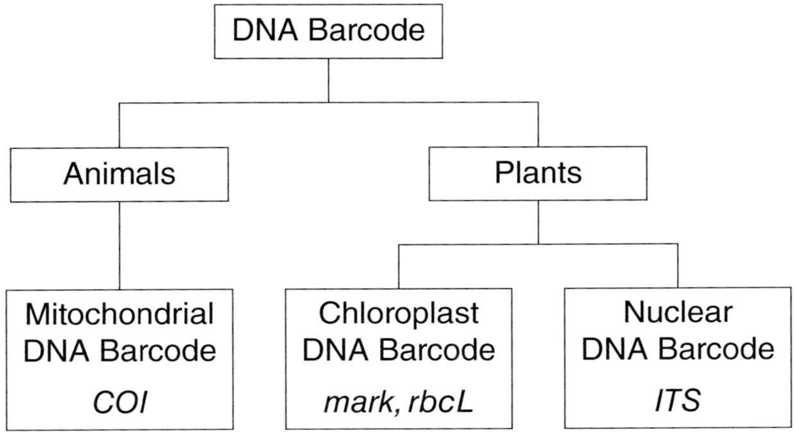

Figure 9.3 *Standardized DNA barcodes in animals and plants.*

conducting ecological research. Its widespread adoption and effectiveness have established the COI barcode as a cornerstone in the field of molecular taxonomy. Though the COI barcode is more reliable for use in animals, but it cannot work for plants, due to its slower mutation rate.

- **rbcL (Ribulose-1,5-bisphosphate carboxylase/oxygenase large subunit)**:
  The rbcL barcode is a primary key genetic marker used to identify the plant species and compare only at the family and genus level, and not at the species level. It was the first gene sequenced from a chloroplast gene that corresponds to a segment of the gene encoding the large subunit of ribulose-1,5-bisphosphate carboxylase/oxygenase (RuBisCO), an enzyme essential for photosynthesis. This gene, located in the chloroplast genome, is highly conserved among plant species and exhibits enough variability to distinguish between different plant taxa. Typically, the rbcL barcode region contains approximately 600–800 base pairs of the gene. The rbcL barcode is a robust region and can be easily amplified, making it a reliable tool for understanding plant systematics, biodiversity assessments and ecological studies.
- **matK (Maturase K):**
  The matK barcode, or maturase K gene barcode, is another essential genetic marker used in the identification of plant species and taxonomy. It has been used at the family, the genus and even at the species level for the identification purposes. It is also a chloroplast gene that corresponds to a segment of the matK gene, which encodes a maturase K, an enzyme involved in the processing of chloroplast pre-mRNAs, in splicing of type II introns. This gene is also known for its variability among different plant species, making it a valuable tool to distinguish between closely related plants. Typically, the matK barcode region has about 800–1,000 base pairs of the gene. The matK barcode is valued for its high discriminatory power and ability to resolve species that are challenging to differentiate using other markers, especially in higher complex plant families.
- **ITS (Internal Transcribed Spacer):**
  In plant molecular taxonomy, nuclear ribosomal DNA (rDNA) genes have been widely used. The rDNA cistrons consist of the 18S, 26S and 5.8S coding regions, encoding the rRNA core of ribosome. These genes are separated by ITS. The ITS region is particularly useful in understanding fungi and plant taxonomy, and to differentiate species

that are morphologically similar or difficult to distinguish. The ITS barcode refers to the Internal Transcribed Spacer regions, which are segments of the ribosomal RNA (rRNA) gene cluster found in the nuclear genome of eukaryotes. Specifically, the ITS regions comprise ITS1 and ITS2, which are located between the rRNA coding regions. These regions are transcribed into precursor rRNA but are then removed during the processing of rRNA.

The ITS barcode is extensively used to identify plant species and fungi, making it a powerful phylogenetic marker. It exhibits a high level of variation among species, which allows it to be used for precise differentiation between closely related taxa. The ITS region consists of approximately 600–800 base pairs, depending on the organism. It also finds its applications in ecological and conservation studies to monitor biodiversity and identify species in environmental samples. Its ability to provide detailed species-level identification has made the ITS barcode a critical tool in molecular taxonomy and biodiversity research. There are certain limitations to the use of ITS as a universal DNA barcode, including gene paralogues and secondary structural problems. Despite of these drawbacks, ITS is superior to chloroplast DNA barcodes.

These DNA barcode regions are fundamental to the effectiveness of DNA barcoding as a tool for species identification and understanding taxonomy. Their importance lies in their ability to provide universal, reliable and precise genetic markers that further help in the accurate differentiation of species (Cowan et al., 2006).

9.2.2.2 Advanced sequencing technologies

Advanced sequencing technologies have significantly improved and enhanced DNA barcoding by increasing the speed, accuracy and extensiveness of genetic analysis. These technologies include:

- **Next-Generation Sequencing (NGS)**: NGS platforms allow high-throughput sequencing (HTS) that is able to process millions of DNA fragments simultaneously. With the development of this technology, the use of chloroplast (cp) genome has been introduced for phylogenetic analysis. The cp genome, if used as a marker, could help avoid problems with gene deletions and difficulties in sequence retrieval. This is because the cp genome offers more diversity, that enables discrimination between closely related plants and hence considered super barcode for plant species. Due to this capability, comprehensive barcode analysis can be performed across a wide range of species that will provide high-resolution data for more precise species identification (Johnson et al., 2014). NGS is particularly useful for large-scale biodiversity surveys and complex ecological studies. For example, Illumina and PacBio are few to list, which are used in DNA Barcode analysis.
- **Metabarcoding**: This method is used in combination with NGS, which helps in the simultaneous identification of multiple species within a single environmental sample by simultaneously amplifying and sequencing multiple DNA barcode regions from mixed DNA samples. It is widely applied in studies related to environmental DNA (eDNA) to assess biodiversity in various ecosystems (Kress, 2017). This technique provides a versatile and precise approach for food analysis, detection of adulterants, in ecological and forensic investigations.

These advanced sequencing technologies can enhance the resolution, efficiency, and scope of DNA barcoding, facilitating much precise and comprehensive species identification.

9.2.2.3 *Reference databases*

Reference databases are repository resources in which the newly discovered genetic or, more specifically, nucleotide sequences are submitted and stored in the form of digital libraries. This stored information can be assessed via different worldwide platforms and repositories available online. Some of the widely known databases include GenBank, BOLD and more, which are used to assess the existing genetic sequences for identification and comparison. These platforms are not only used to identify and compare, but also augmented with the addition of more newly discovered sequences. Few reference databases used for DNA Barcoding are detailed as follows:

- **Barcode of Life Data Systems (BOLD), Canada:**
  BOLD is a comprehensive database widely used in the process of DNA barcoding. It is operated by the University of Guelph in Ontario, Canada (BOLD – The Barcode of Life Data Systems). It provides different tools for species identification, analysis and comparison. BOLD allows global sharing and comparison of barcode data, thereby serving as a universal starter for species identification and comparison with specialized databases.
- **GenBank, USA**:
  International Nucleotide Sequence Database Collaboration (INSDC) involves three major databases – GenBank, USA, European Molecular Biology Laboratory – European Bioinformatics Institute, Europe (EMBL-EBI), and DNA Data Bank of Japan (DDBJ) – that collectively provide access to a vast array of nucleotide sequence data, including DNA barcodes. This platform is maintained by the National Centre for Biotechnology Information (NCBI), a well-known digital library. GenBank is a publicly accessible database that stores a large collection of nucleotide sequences from different species/organisms. The barcode sequences stored can be used for genetic research and species identification in future studies (Kress, 2017).
- **Medicinal Materials DNA Barcode Database (MMDBD), People's Republic of China:**
  This DNA barcoding database has been developed only for medicinal plants, which contains all plastid DNA and nuclear ITS barcode information. It can be accessed via https://137.189.42.34/mherbsdb/index.php, which consists of different DNA sequences and references to medicinal records of pharmacopoeia of the People's Republic of China, American Herbal pharmacopoeia and other references of medicinal plants.

  These databases help in storing, accessing and sharing genetic sequence data, facilitating precise species identification and comparative analysis. By integrating extensive collections of DNA barcode sequences, these platforms provide a wide range of scientific applications, from conservation efforts to ecological monitoring and evolutionary studies. Their continued development and maintenance can help enhance the understanding of biodiversity and improving the effectiveness of DNA barcoding as a tool for species identification and classification in forensic investigation.

9.3 Applications in forensic investigations

The use of DNA barcoding in forensic investigations has enabled precise species identification through standardized DNA sequences and proved to be of great value in providing valuable insights for criminal investigations. The potential of DNA barcoding in forensic palynology provides some advantages over traditional methods, such as improved taxonomic resolution and faster identification processes, emphasizing recent advancements in HTS. This technique is found to be highly versatile in identifying forensically significant insects and monitoring illegal wildlife trade.

In forensic investigations, this method is used to identify individuals by comparing DNA barcodes from evidence samples to those from suspects or databases and finally by matching DNA barcodes from biological samples to known species. For instance, in wildlife crimes, DNA barcoding is widely used to identify the species of illegal wildlife products, such as ivory or bushmeat, seized from poaching or trafficking.

9.3.1 Crime scene investigation

In crime scene investigations, the most reliable method to solve the case is through DNA profiling, which will generate a unique DNA profile of an individual from specific regions of genome. Apart from DNA profiling, DNA barcoding is also employed in crime scene investigations to gather evidence from eDNA through metabarcoding, which involves analysis of multiple species. By comparing DNA barcodes from evidence with those from suspects or databases, forensic scientists can establish connections or exclusions. Hence, this method has been instrumental in solving numerous criminal cases, particularly those involving degraded or mixed samples (Mitra et al., 2018).

9.3.2 Wildlife forensics

The origin of illegal wildlife trade is often hard to trace. In wildlife forensics, DNA barcoding becomes an invaluable tool in combating illegal poaching and trafficking of wildlife products. By identifying species from samples, forensic scientists can help enforce conservation laws and track the origins of illegal goods. This application is found to be highly instrumental in protecting endangered species and addressing illegal wildlife trade at the national and international levels (Mitra et al., 2018; Nithaniyal et al., 2021).

Hence, DNA barcoding has proven to be a transformative tool in forensic investigations by solving wildlife crimes, such as poaching and illegal trade, by accurately identifying the species of seized animal/plant products. Additionally, DNA barcoding also helps in cases involving mislabelling and fraud by verifying the authenticity of samples. The technique is able to provide clear and unambiguous genetic profiles that enhance the credibility and efficacy of forensic evidence, thereby supporting both legal proceedings and conservation efforts.

9.4 Ethical and legal considerations

The generation of genetic information by the process of DNA barcoding raises significant privacy issues. Its application in the forensic context raises several ethical and legal considerations that are crucial to address to ensure responsible use of the technology. Ethically, the collection and use of samples for DNA barcoding must follow the rights of individuals and communities, particularly in cases involving human DNA or samples collected from protected or culturally significant species. Informed consent is a fundamental principle. Additionally, there are concerns about privacy and the potential misuse of genetic information.

Legally, the use of DNA barcoding must comply with regulations that govern the collection, handling and storage of biological samples, including adherence to national and international laws related to biodiversity conservation, such as the Convention on Biological Diversity (CBD) and the Nagoya Protocol. They regulate access to genetic resources and ensure fair and equitable sharing of benefits derived from their use (Nagoya Protocol, 2010; United Nations, 1992). Furthermore, the chain of custody for samples and the integrity of the barcoding process must be effectively maintained to avoid contamination/errors that could further compromise legal outcomes (Staats et al., 2016).

Therefore, ethical and legal considerations are highly significant in the precise implementation of DNA barcoding, encouraging confidence in the technology and ensuring its benefits, while minimizing potential risks to the integrity of evidence(s).

9.5 Challenges and limitations

DNA barcoding, being a powerful tool, especially for species identification and biodiversity monitoring, faces several challenges and limitations that impact its effectiveness. One major challenge is the incomplete and uneven representation of barcode reference databases. Many species, particularly those that are less studied or from underrepresented regions, lack sufficient barcode data. This limits the ability to identify unknown samples accurately, thereby restricting the information from being used globally. For example, databases like BOLD and GenBank have extensive collections, but they still lack many taxa, particularly from biodiversity-rich but poorly studied regions like tropical rainforests (Suriya et al., 2020).

Another significant limitation lies in the variability in barcode efficacy across different taxa. For instance, the COI gene is highly effective for almost all animals, while the other barcode regions, such as the ribulose-1,5-bisphosphate carboxylase/oxygenase (rbcL) gene, may not be as effective for all plant species, particularly those with complex or divergent plastid genomes (Kress et al., 2005). Similarly, for fungi, the ITS regions can exhibit high variability, posing difficulty at the species-level identification (Solano et al., 2020).

Additionally, in DNA barcoding, the issues of contamination and degradation of genetic material are very common. In forensic settings, samples can be compromised by environmental factors, leading to degraded/mixed DNA that is again challenging to amplify and sequence accurately, which will further impact the reliability of species identification (DeSalle et al., 2005).

These challenges highlight the need for continuing improvements in reference databases, refinement of barcode markers, and strict protocols for addressing sample contamination and degradation. Therefore, the accuracy and applicability of DNA barcoding can be enhanced only if the limitations could be looked after and possible solution could be implemented.

9.6 Future directions

The future of DNA barcoding promises advancements in species identification and biodiversity research through several key developments. One major advancement is the expansion and enhancement of reference databases. As the current databases, such as BOLD and GenBank, lack many species, future efforts will be more focussed on expanding these repositories with a more comprehensive representation of global biodiversity. This expansion requires collaborative international initiatives to collect and sequence barcodes for diminished species, ensuring that the databases are more inclusive and accurate for a wider range of organisms.

Another important advancement includes integration of DNA barcoding with other emerging technologies such as eDNA and HTS. eDNA allows for the detection of species from environmental samples like soil or water, which is a non-invasive method for monitoring biodiversity (Shadrin, 2021). HTS allows analysis of large volumes of DNA data rapidly, and technologies can greatly enhance the efficiency and scope of biodiversity surveys, allowing researchers to identify species from more complex and mixed samples (Suriya et al., 2020).

Further, advances in computational tools and bioinformatics help improve the accuracy and precision of DNA barcoding analyses. The development of more sophisticated algorithms for sequence alignment, error correction, and species identification can increase the reliability of barcode data. Additionally, machine learning and artificial intelligence can play increasingly important roles in automating, making it easier to handle the large amounts of data generated (Shadrin, 2021).

Additionally, the application of DNA barcoding to new fields such as personalized medicine and agricultural biotechnology can help in identifying genetic markers for disease susceptibility and treatment response, and of course, in the identification and management of plant and animal species, simultaneously (Robertson et al., 2020).

Overall, the continued development of DNA barcoding technology and its collaboration with other scientific advancements is expected to expand its applications and improve its effectiveness, providing valuable tools for conservation, ecology, and forensic investigations.

9.7 Real-world applications

DNA barcoding is a versatile technique and has shown significant impact in forensic and environmental studies through various case studies and real-world applications across multiple fields. One notable application is in wildlife conservation and anti-poaching efforts. For instance, DNA barcoding helps in combating illegal wildlife trade by identifying the species of animals and their products, such as ivory and bushmeat, which are often mislabelled or concealed. Another example is the use of DNA barcoding to trace the origin of ivory seizures belonging to specific elephant populations, helping law enforcement agencies and conservation organizations to target their efforts more effectively (Staats et al., 2016).

Similarly, in agriculture-related studies, DNA barcoding is used for identifying plant species and verifying the authenticity of crop varieties, which is crucial for both food safety and agricultural trade. For instance, the identification of mislabelled plant species in herbal medicines, DNA barcoding can successfully detect discrepancies between the labelled and actual plant species, thereby preventing potential health risks associated with fraudulent products (Taberlet et al., 2012). This highlights the importance of DNA barcoding in ensuring the integrity of agricultural and medicinal products.

Furthermore, DNA barcoding has made significant achievements in the field of forensic science. For example, in forensic entomology, DNA barcoding is employed to identify insect species found at crime scenes, which can provide valuable information about the time of death and other aspects of the investigation.

9.8 Conclusion

In conclusion, DNA barcoding has emerged as a pivotal tool in forensic investigations, providing a powerful method for species identification and evidence analysis across various contexts. Its precision and reliability in generating genetic profiles have significantly advanced forensic science, particularly in cases involving wildlife crime, such as poaching and illegal trade. The ability to accurately identify species from complex biological samples – ranging from animal/plants and their products to environmental DNA – has strengthened the efforts of enforcement agencies and supported the conservation initiatives.

The integration of DNA barcoding with advanced sequencing technologies and computational tools continues to enhance its effectiveness, enabling more detailed and comprehensive forensic analyses. As reference databases expand and improve, the scope of DNA barcoding applications will grow further, aiding forensic investigations and broadening its impact on legal and conservation issues. Additionally, the continuous refinement of DNA barcoding protocols and methodologies promises to address current limitations, such as database incompleteness and barcode efficacy across diverse taxa.

Looking ahead, DNA barcoding holds substantial potential for addressing emerging challenges in forensic science, including the need for precise identification in cases involving degraded or mixed samples or products. Its application in environmental monitoring, agricultural safety and personalized medicine underscores its versatility and value beyond traditional forensic contexts. As research and technology progress, DNA barcoding will remain a crucial component of forensic investigations and ensuring justice by contributing to a more accurate and effective approach to resolving complex cases and protecting biodiversity. It is expected to advance our understanding of biological evidence in a rapidly changing world.

Bibliography

Abdi, G., Singh, S., Selvakumar, S., Dhar, S. K., Mudgal, G., Swaminathan, P., & Tarighat, M. A. (2024). DNA barcoding and its applications. In: *Advances in Genomics: Methods and Applications* (pp. 91–117). Springer Nature Singapore. https://doi.org/10.1007/978-981-97-3169-5_5

Adams, R. I., Miletto, M., Taylor, J. W., & Bruns, T. D. (2013). Disentangling eukaryotic communities with massively parallel tag sequencing: A case study of fungi in indoor environments. *The ISME Journal*, *7*(2), 274–284. https://doi.org/10.1371/journal.pone.0078866

Anderson, R. M., & May, R. M. (1991). *Infectious Diseases of Humans: Dynamics and Control*. Oxford University Press. https://doi.org/10.1017/S0950268800059896

Antil, S., Abraham, J. S., Sripoorna, S., Maurya, S., Dagar, J., Makhija, S., & Toteja, R. (2023). DNA barcoding, an effective tool for species identification: A review. *Molecular Biology Reports*, *50*(1), 761–775. https://doi.org/10.1007/s11033-022-08015-7

Anuja, N., Kothakota, N. J., Bashar, S., & Vataparthi, P. (2024). Different DNA barcoding techniques in forensic botany: A review. *Biosciences Biotechnology Research Asia*, *21*(3), 935–945. http://dx.doi.org/10.13005/bbra/3275

Bell, K. L., Burgess, K. S., Okamoto, K. C., Aranda, R., & Brosi, B. J. (2016). Review and future prospects for DNA barcoding methods in forensic palynology. *Forensic Science International: Genetics*, *21*, 110–116. https://doi.org/10.1016/j.fsigen.2015.12.010

Benson, D. A., Cavanaugh, M., Clark, K., Karsch-Mizrachi, I., Ostell, J., Pruitt, K. D., & Sayers, E. W. (2018). GenBank. *Nucleic Acids Research*, *46*(D1), D41–D47. https://doi.org/10.1093/nar/gkx1094

Bickford, D., Lohman, D. J., Sodhi, N. S., Ng, P. K., Meier, R., Winker, K., Ingram, K. K., & Das, I. (2007). Cryptic species as a window on diversity and conservation. *Nature*, *448*(7157), 1007–1013. https://doi.org/10.1016/j.tree.2006.11.004

Blaxter, M., Mann, J., Chapman, T., Thomas, F., Whitton, C., Floyd, R., & Abebe, E. (2005). Defining operational taxonomic units using DNA barcode data. *Philosophical Transactions of the Royal Society B: Biological Sciences, 360*(1462), 1935–1943. https://doi.org/10.1098/rstb.2005.1725

Blaxter, M. L. (2004). The promise of a DNA taxonomy. *Philosophical Transactions of the Royal Society B: Biological Sciences, 359*(1444), 669–679. https://doi.org/10.1098/rstb.2003.1447

Chac, L. D., & Thinh, B. B. (2023). Species identification through DNA barcoding and its applications: A review. *Biology Bulletin, 50*(6), 1143–1156. https://doi.org/10.1134/S106235902360229X

Corradini, B., Gianfreda, D., Ferri, G., Ferrari, F., Borciani, I., Santunione, A. L., & Cecchi, R. (2024). Forensic species identification: Practical guide for animal and plant DNA analysis. *International Journal of Legal Medicine, 138*(6), 2271–2280. https://doi.org/10.1007/s00414-024-03284-2

Cowan, R. S., Chase, M. W., Kress, W. J., & Savolainen, V. (2006). 300,000 species to identify: Problems, progress, and prospects in DNA barcoding of land plants. *Taxon, 55*(3), 611–616. https://doi.org/10.2307/25065638

Dawnay, N., Ogden, R., McEwing, R., Carvalho, G. R., & Thorpe, R. S. (2007). Validation of the barcoding gene COI for use in forensic genetic species identification. *Forensic Science International, 173*(1), 1–6. https://doi.org/10.1016/j.forsciint.2006.09.013

de Carvalho, C. B. V. (2014). DNA barcoding in forensic vertebrate species identification. *Brazilian Journal of Forensic Sciences, Medical Law and Bioethics, 4*(1), 12–23. https://doi.org/10.17063/bjfs4(1)y201412

DeSalle, R., Egan, M. G., & Siddall, M. (2005). The unholy trinity: Taxonomy, species concept, and barcoding. *Philosophical Transactions of the Royal Society B: Biological Sciences, 360*(1462), 1905–1916. https://doi.org/10.1098/rstb.2005.1722

DeSalle, R., & Goldstein, P. (2019). Review and interpretation of trends in DNA barcoding. *Frontiers in Ecology and Evolution, 7*, 302. https://doi.org/10.3389/fevo.2019.00302

Edgar, R. C. (2004). MUSCLE: Multiple sequence alignment with high accuracy and high throughput. *Nucleic Acids Research, 32*(5), 1792–1797. https://doi.org/10.1093/nar/gkh340

Fang, T., Liao, S., Chen, X., Zhao, Y., Zhu, Q., Cao, Y., & Zhang, J. (2019). Forensic drowning site inference employing mixed pyrosequencing profile of DNA barcode gene (rbcL). *International Journal of Legal Medicine, 133*, 1351–1360. https://doi.org/10.1007/s00414-019-02075-4

Ferri, G., Alù, M., Corradini, B., & Beduschi, G. (2009a). Forensic botany: Species identification of botanical trace evidence using a multigene barcoding approach. *International Journal of Legal Medicine, 123*, 395–401. https://doi.org/10.1007/s00414-009-0356-5

Ferri, G., Alu, M., Corradini, B., Licata, M., & Beduschi, G. (2009b). Species identification through DNA "barcodes. *Genetic Testing and Molecular Biomarkers, 13*(3), 421–426. https://doi.org/10.1089/gtmb.2008.0144

Ferri, G., Corradini, B., Ferrari, F., Santunione, A. L., Palazzoli, F., & Alu, M. (2015). Forensic botany II, DNA barcode for land plants: Which markers after the international agreement? *Forensic Science International: Genetics, 15*, 131–136. https://doi.org/10.1016/j.fsigen.2014.10.005

Fišer Pečnikar, Ž., & Buzan, E. V. (2014). 20 years since the introduction of DNA barcoding: From theory to application. *Journal of Applied Genetics, 55*, 43–52. https://doi.org/10.1007/s13353-013-0180-y

Frézal, L., & Leblois, R. (2008). Four years of DNA barcoding: Current advances and prospects. *Infection, Genetics and Evolution, 8*(5), 727–736. https://doi.org/10.1016/j.meegid.2008.05.005

Ge, J., Sun, H., Li, H., Liu, C., Yan, J., & Budowle, B. (2014). Future directions of forensic DNA databases. *Croatian Medical Journal, 55*(2), 163. https://doi.org/10.3325/cmj.2014.55.163

Hajibabaei, M., Shokralla, S., Zhou, X., Singer, G. A., & Baird, D. J. (2011). Environmental barcoding: A next-generation sequencing approach for biodiversity analysis. *Molecular Ecology, 20*(23), 4342–4350. https://doi.org/10.1371/journal.pone.0017497

Hajibabaei, M., Singer, G. A. C., & Hebert, P. D. N. (2007). A minimalist barcode can identify a specimen whose DNA is degraded. *Molecular Ecology Notes, 7*(4), 807–814. https://doi.org/10.1111/j.1471-8286.2006.01470.x

Hallerman, E. M. (2021). Applications and limitations of DNA barcoding in environmental biology. *Journal of Environmental Biology, 42*(1), 1–13. http://doi.org/10.22438/jeb/42/1/MRN-1710

Hebert, P. D. N., Cywinska, A., Ball, S. L., & deWaard, J. R. (2003). Biological identification through DNA barcodes. *Proceedings of the Royal Society B: Biological Sciences, 270*(1512), 313–321. https://doi.org/10.1098/rspb.2002.2218

Hebert, P. D. N., & Gregory, T. R. (2005). The promise of DNA barcoding for taxonomy. *Systematic Biology, 54*(5), 852–859. https://doi.org/10.1080/10635150500354886

Hebert, P. D. N., Ratnasingham, S., Zakharov, E. V., Telfer, A. C., Levesque-Beaudin, V., Milton, M. A., Pedersen, S., Jannetta, P., & DeWaard, J. R. (2016). Counting animal species with DNA barcodes: Canadian insects.

Philosophical Transactions of the Royal Society B: Biological Sciences, 371(1702), 20150333. https://doi.org/10.1098/rstb.2015.0333
Ibrahim, M., Liang, T. C., Scott, K., Chakrabarty, K., & Karri, R. (2020). Molecular barcoding as a defense against benchtop biochemical attacks on DNA fingerprinting and information forensics. *IEEE Transactions on Information Forensics and Security, 15*, 3595–3609. https://doi,org/10.1109/TIFS.2020.2994742
Ishak, S., Dormontt, E., & Young, J. M. (2021). Microbiomes in forensic botany: A review. *Forensic Science, Medicine and Pathology, 17*, 297–307. https://doi.org/10.1007/s12024-021-00362-4
Ishfaq, M., & Saleem, M. (2016). Application of molecular markers in forensic botany. *Bulletin of Biological and Allied Sciences Research, 2016*(1), 6–6. https://doi.org/10.54112/bbasr.v2016i1.6
Iyengar, A. (2014). Forensic DNA analysis for animal protection and biodiversity conservation: A review. *Journal for Nature Conservation, 22*(3), 195–205. https://doi.org/10.1016/j.jnc.2013.12.001
Jian, W. U., & Hong-Yan, G. A. O. (2020). DNA barcode technology and its application prospects in forensic medicine. *Journal of Forensic Medicine, 36*(4), 559. https://doi.org/10.12116/j.issn.1004-5619.2020.04.023
Johnson, R. N., Wilson-Wilde, L., & Linacre, A. (2014). Current and future directions of DNA in wildlife forensic science. *Forensic Science International: Genetics, 10*, 1–11. https://doi.org/10.1016/j.fsigen.2013.12.007
Karsch-Mizrachi, I., Takagi, T., & Cochrane, G. (2018). The International Nucleotide Sequence Database Collaboration (INSDC): The European nucleotide archive, the DNA data bank of Japan, and GenBank. *Nucleic Acids Research, 46*(D1), D48–D51. https://doi.org/10.1093/nar/gkx1097
Kress, W. J. (2017). Plant DNA barcodes: Applications today and in the future. *Journal of Systematics and Evolution, 55*(4), 291–307. https://doi.org/10.1111/jse.12254
Kress, W. J., & Erickson, D. L. (2007). A two-locus global DNA barcode for land plants: The coding rbcL gene complements the non-coding matK gene. *PLoS One, 2*(6), e508. https://doi.org/10.1371/journal.pone.0000508
Kress, W. J., Wurdack, K. J., Zimmer, E. A., Weigt, L. A., & Janzen, D. H. (2005). Use of DNA barcodes to identify flowering plants. *Proceedings of the National Academy of Sciences, 102*(23), 8369–8374. https://doi.org/10.1073/pnas.0503123102
Kumar, S., & Stecher, G. (2018). MEGA X: Molecular evolutionary genetics analysis across computing platforms. *Molecular Biology and Evolution, 35*(6), 1547–1549. https://doi.org/10.1093/molbev/msy096
Linacre, A., Gusmão, L., Hecht, W., Hellmann, A. P., Mayr, W. R., Parson, W., & Morling, N. (2011). ISFG: Recommendations regarding the use of non-human (animal) DNA in forensic genetic investigations. *Forensic Science International: Genetics, 5*(5), 501–505. https://doi.org/10.1016/j.fsigen.2010.10.017
Mardis, E. R. (2008). Next-generation DNA sequencing methods. *Annual Review of Genomics and Human Genetics, 9*, 387–402. https://doi.org/10.1146/annurev.genom.9.081307.164359
Mitra, I., Roy, S., & Haque, I. (2018). Application of molecular markers in wildlife DNA forensic investigations. *Journal of Forensic Science and Medicine, 4*(3), 156–160.
Moritz, C., & Cicero, C. (2004). DNA barcoding: Promise and pitfalls. *PLoS Biology, 2*(10), e354. https://doi.org/10.1371/journal.pbio.0020354
Muhammad Tahir, H., & Akhtar, S. (2016). Services of DNA barcoding in different fields. *Mitochondrial DNA Part A, 27*(6), 4463–4474. https://doi.org/10.3109/19401736.2015.1089572
Murray, M. G., & Thompson, W. F. (1980). Rapid isolation of high molecular weight plant DNA. *Nucleic Acids Research, 8*(19), 4321–4325. https://doi.org/10.1093/nar/8.19.4321
Nagoya Protocol. (2010). Nagoya protocol on access to genetic resources and the fair and equitable sharing of benefits arising from their utilization. [Online] Available at: https://www.cbd.int/abs/
Nithaniyal, S., Majumder, S., Umapathy, S., & Parani, M. (2021). Forensic application of DNA barcoding in the identification of commonly occurring poisonous plants. *Journal of Forensic and Legal Medicine, 78*, 102126. https://doi.org/10.1016/j.jflm.2021.102126
Pereira, F., Carneiro, J., & Van Asch, B. (2010). A guide for mitochondrial DNA analysis in non-human forensic investigations. *The Open Forensic Science Journal, 3*(1). http://dx.doi.org/10.2174/1874402801003010033
Quinlan, A. R., & Hall, I. M. (2010). BEDTools: A flexible suite of utilities for comparing genomic features. *Bioinformatics, 26*(6), 841–842. https://doi.org/10.1093/bioinformatics/btq033
Rhoads, A., & Au, K. F. (2015). PacBio sequencing and its applications. *Genomics, Proteomics & Bioinformatics, 13*(5), 278–289. https://doi.org/10.1016/j.gpb.2015.08.002
Robertson, J. M., Damaso, N., & Meiklejohn, K. A. (2020). DNA-based analysis of plant material in forensic investigations. In: *Handbook of DNA Profiling* (pp. 1–32). https://doi.org/10.1007/978-981-15-9364-2_59-1
Shadrin, D. M. (2021). DNA barcoding: Applications. *Russian Journal of Genetics, 57*, 489–497. https://doi.org/10.1134/S102279542104013X

Sharma, S., Jossan, J. K., & Singh, R. (2020). DNA barcoding in forensic mycology: Concepts, limitations, and future prospects. In: *Forensic DNA Typing: Principles, Applications and Advancements* (pp. 431–448). https://doi.org/10.1007/978-981-15-6655-4_22

Sheth, B. P., & Thaker, V. S. (2017). DNA barcoding and traditional taxonomy: an integrated approach for biodiversity conservation. *Genome, 60*(7), 618–628. https://doi.org/10.1139/gen-2015-0167

Solano, J., Anabalón, L., Figueroa, A., & Gangitano, D. (2020). ITS barcoding using high resolution melting analysis of Cannabis sativa drug seizures in Chile: A forensic application. *Forensic Science International, 316*, 110550. https://doi.org/10.1016/j.forsciint.2020.110550

Staats, M., Arulandhu, A. J., Gravendeel, B., Holst-Jensen, A., Scholtens, I., Peelen, T., & Kok, E. (2016). Advances in DNA metabarcoding for food and wildlife forensic species identification. *Analytical and Bioanalytical Chemistry, 408*, 4615–4630. https://doi.org/10.1007/s00216-016-9595-8

Suriya, J., Krishnan, M., Bharathiraja, S., Sekar, V., & Sachithanandam, V. (2020). Implications and utility of DNA barcoding. In: *DNA Barcoding and Molecular Phylogeny* (pp. 3–22). https://doi.org/10.1007/978-3-030-50075-7_1

Taberlet, P., Coissac, E., & Hajibabaei, M. (2012). Environmental DNA. *Molecular Ecology, 21*(8), 1789–1793. https://doi.org/10.1111/j.1365-294X.2012.05542.x

Tamura, K., & Nei, M. (1993). Estimation of the number of nucleotide substitutions in the control region of mitochondrial DNA in humans and chimpanzees. *Molecular Biology and Evolution, 10*(3), 512–526. https://doi.org/10.1093/oxfordjournals.molbev.a040023

United Nations. (1992). *Convention on biological diversity.* Treaty Collection. https://www.cbd.int/convention/text/

Verzeletti, A., Cortellini, V., Cisana, S., Pretto, S., Cerri, N., & De Ferrari, F. (2015). Genetic approach for species identification: Forensic application of DNA barcoding. *Forensic Science International: Genetics Supplement Series, 5*, e576–e578. https://doi.org/10.1016/j.fsigss.2015.09.228

Zaya, D. N., & Ashley, M. V. (2012). Plant genetics for forensic applications. In N. Sucher, J. Hennell, & M. Carles (Eds) *Plant DNA Fingerprinting and Barcoding: Methods and Protocols* (pp. 35–52). Humana Press. https://doi.org/10.1007/978-1-61779-609-8_4

10

Microbial Forensics

Neeti Kapoor, Girraj Sharma, Chhote Raja Patle, and Ashish Badiye

10.1 Introduction

The emergence of biological threats made microbial forensics develop into an essential multidisciplinary discipline where the study of microorganisms in context of justice is done and it enables experts to trace, analyze and interpretate forensically significant evidence which are collected and preserved from different crime scene, such as bioterrorism, biological crime, hoaxes or accidental release of microorganism or toxins (Schmedes & Budowle, 2015; Shrivastava et al., 2023). Moreover, the integration of microbiology, genomics, forensic science, and epidemiology with law enforcement principles facilitates microbial forensics, enabling the identification, characterization, and source attribution of biological agents to trace their origin and hold perpetrators accountable (Budowle et al., 2005; Khan et al., 2020). In addition, it also describes various aspects of scientific investigation of crime scene, including sample handling, safety, transportation, which can play a crucial role to trace its origin and support investigations related to bioterrorism, biological attacks, criminal acts involving microbes, disease outbreaks, pathogen transmission or unintended releases of biological agents or toxins (Budowle et al., 2003; Budowle, Johnson, et al., 2005; Oliveira & Amorim, 2018; Schmedes & Budowle, 2015)

Microbial forensics developed its origin in the early 2000s because of the 2001 anthrax attacks that struck the United States (Dalton, 2008). Multiple deaths occurred among political figures and media personnel after *Bacillus anthracis* spores were delivered via mailed letters, which caused enormous public dread (Dalton, 2008). The important investigation generated through microbial forensic implementation became one of the first large-scale applications that revealed the immediate requirement for dedicated forensic technologies to handle such evidence (Budowle et al., 2014).

The basic aim of microbial forensics involves determining two main points: Which microorganism plays a role and what category it belongs to – natural versus genetically engineered (Mir et al., 2022). The microorganism originates from nature, or does it result from genetic engineering? The organism shows how closely it relates to established strain records (Mir et al., 2022). Moreover, does the origin possess fixable positional information regarding its geographical location or laboratory of origin or its creator? Microbial forensic investigations heavily depend on genomic sequencing and bioinformatics for resolving their important inquiries. DNA and RNA sequencing technologies enable scientists to decode the entire genetic structure of microorganisms, which results in a distinct microorganism-specific identification mark that researchers can crosscheck versus established information databases. The examination of genes between different organisms provides investigators with data about pathogen origins through mutation tracking and evolutionary and molecular signature analysis (Oliveira et al., 2024; Zhang et al., 2023).

DOI: 10.4324/9781003474296-10

The evaluation process of microorganisms for forensic investigation requires complex specialized methods. The DNA analysis process starts by properly collecting and preserving alongside documentation of biological evidence obtained from crime sites, along with victim bodies and environmental elements (Next Generation Sequencing (NGS) Technology in DNA Analysis, 2024).

10.2 Scope and applications

Microbial forensics is one of the versatile and distinct areas (sub-areas) of forensic science that deals with the identification, examination and interpretation of the evidence of microbial life and its genetic profile. This discipline is an integration of microbiology, genomics and forensic methodologies involved in studying biological threats, bioterrorism, biological warfare and even natural outbreaks of disease. Its main aim is the tracing of the origin of bacteria, viruses, fungi and toxins. This branch becomes critical in cases of engineered pathogen dispersal, such as anthrax, or for determining whether a disease was accidentally, or one could have occurred intentionally.

10.2.1 Bioterrorism and biocrime investigations

Microbial forensics serves a crucial function in bioterrorism investigations, along with biocrimes, through scientific procedures to study microbial elements. The field serves to uncover attack origins with their methodology while identifying criminal perpetrators who use biological agents within attacks or crimes. The application of microbial forensics remains crucial for solving cases of bioterrorism and biocrimes through scientific microbial evidence analysis:

- **Identification and characterization of biological agents:** Microbial forensics utilizes techniques like molecular analysis and sequencing to identify the specific microorganisms or toxins used in bioterrorism or biocrime. This could help to differentiate between naturally spread infections and intentionally released one (Mir et al., 2022).
- **Determining the source of the agent:** the comparison of microbial genetic material against reference databases can help investigators to track the origin of the agent, which can identify whether it is laboratory-originated or modified (Oliveira et al., 2024).
- **Attributing to the crime or attack:** Microbial forensics can help attribute a crime or attack to a specific individual or organization by comparing microbial evidence with known samples or potential suspect material (Schmedes & Budowle, 2015).
- **Investigation of biocrime:** The investigation of biocrime can be easily performed by using microbial forensics, which could establish the intentional biological attacks against selected individual in sexual assault cases and intentional poisoning (Oliveira et al., 2020).
- **Preventing future bioterrorism or biocrime**: It can form strategies or defence plan through the identification of attack approach and develop response measures that can stop similar incidents (Mir et al., 2022).

10.2.2 Outbreak tracing

Microbial forensic applications for outbreak tracking enable epidemiologists and public health officials to track infectious disease transmission (Castro & De Ungria, 2022; Khan et al., 2020). Moreover, the genetic investigation of pathogens enables specialists to determine outbreak sources and follow disease transmission routes across food contamination incidents as well as hospital-acquired infections and worldwide developing risks (Quainoo et al., 2017). Moreover, Genetic analysis of microbial genomes from separate patients alongside environmental samples and infected products reveals the infection origins so investigators can draw population and geographic spread maps (Kumari & Rai, 2020; Relman, 2011). In addition, the main objective in outbreak tracking is to find patient zero, who started the infection, since their identification reveals essential information regarding pathogen entry and path of initial transmission (Wilmes et al., 2009). A quick and correct outbreak tracing process protects ongoing infections and provides information that enables public health prevention methods and improves population surveillance systems and emergency response capability (Ibrahim, 2020).

10.2.3 Environmental and agricultural monitoring

Monitoring of environmental and agricultural systems is essential in microbial forensics for identifying irregular microbial patterns and detecting biological threats. It enables the observation of shifts in micro-biomes within soil, water, and air, which may signal intentional biological contamination or attacks. In the agricultural sector, such surveillance can reveal purposeful infection of crops, animals, or food sources (Pirttilä et al., 2021). Information gathered from these observations assists in pinpointing the origin and dissemination of harmful microorganisms. This strategic method strengthens early detection capabilities, supports forensic analysis and boosts biosecurity, allowing for swift action against microbial risks impacting ecosystems, agriculture and public health (Nikolić et al., 2024).

10.3 Key microbial agents

The microbial agents used for biothreats consist of microorganisms that can be easily weaponized deliberately to inflict disease epidemics while spreading panic and creating public disturbances (Venkatesh & Memish, 2003). Moreover, the dangerous effect of biological weapons possesses strong infection capacity and endures in various conditions to inflict fatal or serious sickness (Anderson & Bokor, 2012). For example, *Bacillus anthracis* stands as a vital biothreat agent because it develops resilient spores that persist in outdoor conditions indefinitely while remaining airborne (Binkley et al., 2002). In addition, as a deadly pathogen, *Yersinia pestis* has triggered numerous disastrous epidemics throughout history and exists as a possible infectious agent for biological weapons because it easily spreads between people (Ansari et al., 2020). In continuation, *Clostridium botulinum* is also used frequently in bio terrorism because of its lethal toxins and *Francisella tularensis* because it causes tularaemia with minimal amounts of infection needed to lead to serious illness (Moran et al., 2008). Smallpox viruses, including *Variola major*, and Ebola and Marburg viruses, which are viral haemorrhagic fever agents, pose severe security concerns because they cause high death rates and trigger rapid panic spreading (Cleri et al., 2006). The growing detection of Nipah virus and multidrug-resistant *Mycobacterium tuberculosis* occurs to track their potential as new biothreat agents because of genetic engineering advancements (Jefferson et al., 2014; Kalha et al., 2014). Rapid response to biological threats requires an understanding of significant microorganisms that enables preparedness and quick detection capabilities (Das & Kataria, 2010). The list of key microbial agents is mentioned below.

10.4 Tools and techniques in microbial forensics

10.4.1 DNA sequencing

DNA sequencing refers to the process of determining the exact order of bases within a DNA fragment (Mitchelson, 2005). One of the most widely employed techniques for this purpose is the Sanger sequencing method (Jamuar et al., 2016). This approach utilizes DNA polymerase I along with a polymerase chain reaction (PCR) reaction to repeatedly generate new DNA strands, using the original DNA segment as a template (Jamuar et al., 2016). The synthesis of the strand halts whenever a dideoxynucleotide is incorporated in place of a regular nucleotide. Sanger sequencing can be carried out either manually or with the aid of an automated sequencing machine (Jamuar et al., 2016). Genome sequencing has now become easier and less expensive way of treating and preventing diseases in humans. But it has a very wide range of important applications beyond this (Drmanac et al., 2010). Over time, tremendous advancements have been seen in this technique. The following are some applications of DNA sequencing techniques-

- In medical research, DNA sequencing plays a crucial role in identifying genes linked to hereditary or acquired diseases (Marian, 2011).
- In the field of forensic science, PCR is used to help identify criminals by analysing evidence such as hair, nails, skin cells or blood samples collected from crime scenes (Cavanaugh & Bathrick, 2018).
- DNA sequencing is a very useful test to establish the paternity of a child. Additionally, information gained from DNA sequencing is required to design and plan manipulation of the gene (Giani et al., 2020).
- It is utilized in the creation of restriction endonuclease maps. DNA sequencing also helps identify tandem repeats or inverted repeats, which could lead to hairpin structure formation (Shastri et al., 2018).
- Sequence data can be analyzed to determine the presence of open reading frames (ORFs) that may code for polypeptides (Galperin & Frishman, 1999).
- DNA sequences can be matched with polypeptide sequences from databases or compared with sequences from other organisms to conduct phylogenetic studies (Asim et al., 2025).
- DNA sequencing contributes to building molecular evolution maps (Shendure & Aiden, 2012).

10.4.2 Polymerase chain reaction (PCR)

The PCR is a well-known technique in microbial forensics that is used in detecting, identifying and profiling microbes involved in criminal activities or bioterrorism. It covers such a wide scope that it can be used to analyze tiny or degraded biological traces that could help forensic experts trace the origin of pathogens (Stephenson, 2003). PCR has exceptional sensitivity and specificity that can actually amplify specific DNA or RNA fragments in complex or contaminated samples (McDonald et al., 2024). In microbial forensic investigations, PCR can be used as an indispensable tool for solving crimes. The most prominent applications are as follows:

- **Pathogen detection and identification:** PCR allows forensic scientists to detect and identify microbes (e.g., *Bacillus anthracis* in anthrax attacks) even if they are present in very low quantities. Both DNA and RNA viruses (e.g., Ebola, smallpox) can be detected through conventional PCR or reverse transcription PCR (RT-PCR) (Kim et al., 2023).

- **Strain typing and genetic fingerprinting:** Variations within microbial strains can be analyzed using PCR-based methods (e.g., VNTR analysis, MLVA). This helps determine whether two microbial samples are genetically identical (suggesting a common source) or different (Van Stelten & Nightingale, 2014).
- **Evidence preservation:** PCR requires very little starting material, making it ideal for forensic samples that might be **degraded** or **limited**. It's also fast, meaning results can be obtained quickly for time-sensitive investigations (Bhoyar et al., 2024).
- **Bioterrorism and biocrime investigations:** PCR can detect weaponized biological agents and link them to laboratories, geographic locations or individuals based on unique genetic markers (Mir et al., 2020).
- **Monitoring biothreats:** PCR-based biosensors are employed in fields such as environments to detect and monitor for the presence of biological threats in real time (Fdez-Sanromán et al., 2025).
- **Environmental sample testing:** PCR is able to detect microbes in soil, air, water or on surfaces and can be used to reconstruct the crime scene for biological contamination (Zhang et al., 2021).

10.4.3 Metagenomics

In microbial forensics, PCR serves a vital function by amplifying and analyzing minute amounts of DNA from microbial evidence. This method is crucial for identifying and distinguishing pathogens, helping to trace infection sources, follow transmission routes, and recognize distinct strains or subtypes of microbes (Castro & De Ungria, 2022).

10.4.3.1 Individual identification

- **Microbial signatures**: Each person has a distinct microbiome that can serve as a "biological fingerprint." For instance, the unique composition of salivary microbiota can be utilized to distinguish between individuals (Leake et al., 2016)
- **Trace evidence:** Microbial communities left behind on surfaces or objects handled by a person can be examined to assist in suspect identification, exclusion in criminal investigations, or to determine the origin of a particular sample (Neckovic et al., 2020).
- **Forensic focused research:** Emerging studies indicate that metagenomic profiling of human pubic hair may offer a valuable approach for tracing sexual contact, especially in cases where conventional forensic techniques fall short (Tridico et al., 2014).

10.4.3.2 Post-mortem interval (PMI) estimation

- **Microbial succession:** After death, the microbial communities on a body undergo predictable changes, and tracking these shifts can help estimate the time elapsed since death (Deel et al., 2020).
- **Machine learning models:** Scientists are creating predictive models that combine microbial succession patterns with machine learning techniques to improve PMI estimates, especially for cases involving buried remains (Li et al., 2023)
- **Skeletal remains:** Microbial profiles found on bones can also offer valuable clues about the PMI (Cartozzo et al., 2021).

10.4.3.3 Identification of geolocation

- **Environmental DNA (eDNA):** By examining eDNA (genetic material released into surroundings), metagenomic analysis can help determine the origin of a sample by studying the local microbial communities (Sahu et al., 2023; Thomsen & Willerslev, 2015).
- **Microbial diversity:** Variations in microbial populations across different geographic areas can assist in identifying the location linked to a sample or individual (Clarke et al., 2022).

10.4.3.4 Tissue or fluid identification

- **Substance abuse testing:** Metagenomic approaches may provide fresh insights into how the human microbiome interacts with substance use and abuse (Meckel & Kiraly, 2019).

10.4.4 Bioinformatics

Bioinformatics is essential in microbial forensics, enabling the swift identification and examination of microbial samples to support investigations into outbreaks, bioterrorism and public health emergencies. By uncovering valuable information from complex biological data, it allows forensic specialists to decipher genetic patterns and trace the spread of infectious agents (Ray, 2022; Soni, 2024). These are the most common applications discussed here:

- **Rapid identification:** Bioinformatics tools, such as sequence databases, microbial species, strains and isolates, can be rapidly identified with high precision (Alawi et al., 2019).
- **Sequence analysis:** Bioinformatics supports the examination of microbial genome sequences, helping to detect virulence factors, antibiotic resistance genes and other important functional elements (Jenks, 1998).
- **Data management and analysis:** Bioinformatics plays a vital role in organizing, managing and interpreting the vast datasets produced during microbial forensic investigations (Rather et al., 2023).
- **Source tracing:** Bioinformatics tools enable the tracking of infectious agents' origins and transmission pathways, which is critical for effective outbreak containment and prevention efforts (Bianconi et al., 2023).
- **Forensic investigations:** Microbial-related incidents such as disease outbreaks, bioterrorism and biocrimes are investigated more effectively with the support of bioinformatics techniques (Mir et al., 2022).
- **Public health protection:** By delivering rapid and precise microbial identification and analysis, bioinformatics contributes to public health safety by enabling swift response and containment actions (Zhao et al., 2024).
- **Microbiome analysis:** Bioinformatics facilitates the examination of microbial community structure and diversity, which can aid in personal identification, estimating time since death and determining geographical origin (Singh & Okpeku, 2024).

10.4.5 Isotope and chemical analysis

Isotope and chemical analyses are essential tools in microbial forensics, offering distinct "fingerprints" that help track the origin and transmission pathways of biological agents. These methods assist in distinguishing between microbial strains and can even identify the geographical location from which a sample originated (Barry et al., 2022; Chen & Bontempo, 2025; Saud et al., 2022)

10.4.5.1 Source attribution

- **Isotope analysis**: Isotope ratios, particularly stable isotopes of carbon (^{13}C), nitrogen (^{15}N), oxygen (^{18}O) and hydrogen (^{2}H), can vary depending on the growth medium and water source. By analyzing these ratios in microbial samples, investigators can potentially determine the location where the organism was cultured or produced (Kreuzer-Martin & Jarman, 2007; Saud et al., 2022).

10.4.5.2 Identifying transmission routes

- **Stable isotope ratios**: Stable isotope ratios can also reveal information about how a biological agent was propagated and disseminated. For example, the isotope ratios of water used in a culturing process can be tracked to understand the potential route of exposure (Kreuzer-Martin & Jarman, 2007).
- **Chemical markers:** The presence of specific chemicals or metabolites in a sample can also provide clues about the production process or the route of exposure (Vorkamp et al., 2021).

10.4.5.3 Distinguishing laboratory strains from wild isolates

- **Stable isotope ratios and chemical analysis:** These techniques can help differentiate between laboratory-adapted bacteria and their wild counterparts, which can be crucial in determining the source of a biological agent (Kreuzer-Martin & Jarman, 2007).

10.4.5.4 Investigating environmental impacts

- **Isotope analysis:** Isotope analysis can also help researchers understand the environmental impacts of microbial contamination, such as identifying the source of pollution or tracking the spread of a pathogen (Daskin et al., 2008).

10.4.5.5 Human identification

- **Multi-isotope profiles:** Isotope analysis is also used in forensic anthropology to determine a decedent's region of origin, dietary habits, and travel history (Bartelink & Chesson, 2019).

10.5 Chain of custody and legal considerations

Chain of custody refers to the trail of documentation of handling of physical evidence from the crime scene to court room. Because it is very crucial to maintain the integrity of evidence while collecting, preserving, labelling, storing, transporting, delivering, analyzing, retention, disposal reporting evidence in the courtroom, otherwise it will lose its evidential value. Therefore, the National Institute of Justice framed and defined step-by-step process to reduce human error while handling microbial evidence at scene of crime in terms of the safety, recognition, collection, preservation, and transportation of microbial exhibits (Shrivastava et al., 2023). The main postulates of the process are briefly discussed below:

- Secure, recognize, and define the target area of crime scene.
- The proceeding of crime scene depends on organized, systematic methodical, and collective effort of team who need to understand essential components of crime scene management, such as information, manpower, technology, and logistics management, which will help to maintain the integrity of the crime scene and microbial exhibit and focus on immediate collection and dispatching of transient and conditional evidence.
- Crime scene documentation plays a pivotal role to find out the potential source, location, way of preservation, labelling, and transportation of the microbial evidence and maintains the chain of custody of every action taken for handling of microbial evidence from crime scene to courtroom.
- To prepare and use the list of more appropriate methods and tools for the collection of microbial evidence.
- Use suitable methods for the preservation of microbial evidence (Shrivastava et al., 2023; Smith, 2011).

Chain of custody encompasses the following four key steps, as given in **Figure 10.1**:

1. Crime scene to police station
2. Police station to forensic science laboratory (FSL)
3. FSL to police station
4. Police station to court room

However, legal consideration of chain of custody is significant in handling microbial exhibit because it ensures the legal admissibility and integrity of evidence in the courtroom. Each step of handling of microbial exhibit must be documented in detail, like who collected evidence from where and when, preserved, labelled and forwarded to FSL. However, any inconsistency in the chain of custody can pose a question on the authenticity or inadmissibility in the courtroom. In addition, chain of custody may contain the following legal considerations, as shown in **Figure 10.2**.

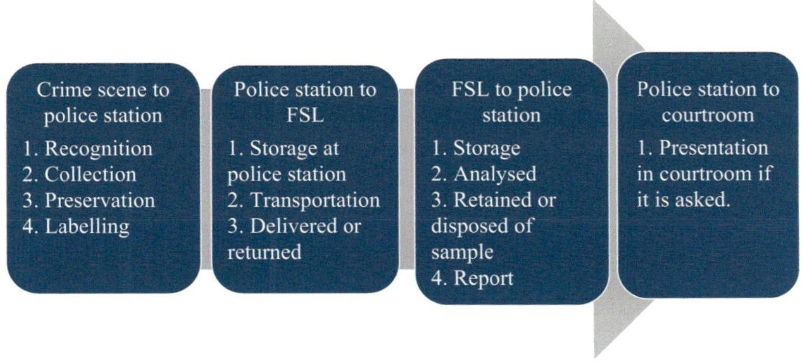

***Figure 10.1** Four key steps of chain of custody.*

Figure 10.2 *Main legal considerations of chain of custody.*

10.6 Challenges in microbial forensics

Microbial forensics investigates various types of microbial exhibits for the court of law. However, it also faces several significant challenges, such as how to differentiate natural from deliberate outbreaks, microbial diversity and mutation rates, limited databases of microbial genomes for comparison, ethical concerns regarding surveillance, data privacy, and potential misuse, etc. Some of them are discussed below.

10.6.1 Differentiating natural from deliberate outbreaks

It is one of the core challenges in microbial forensics to distinguish between natural and deliberate outbreaks because both look similar at the initial stage (Morse & Khan, 2005). However, it could be differentiated at some boundaries by expanding robust forensic and epidemiological methods (Morse & Khan, 2005). In addition, there are a few factors which may play a key role in the identification, such as geographic spread, strain specificity, unusual resistance pattern, target specificity, use of pathogen, manner of introduction, etc. (Budowle et al., 2005; Oliveira et al., 2020).

10.6.2 Microbial diversity and mutation rates

Pathogens like bacteria and viruses show great diversity in their genetic material, and they also exhibit immense mutation rate (Arenas et al., 2018; Brüssow et al., 2004; Korsah et al., 2023). These variations pose significant challenges in the identification and individualization of the strain tracking and phylogenetic profile of the pathogen to the forensic investigator (Singh & Okpeku, 2024). In addition, modifications like mutation and recombination in the genetic material of the pathogen create new variants with distinct genetic profiles, making challenge in the tracking of strain and phylogenetic profile of the pathogen to link to the original strain (Arenas et al., 2018; Markov et al., 2023; Pronyk et al., 2023). These challenges require advanced lab set-up, tools, and techniques for the interpretation of similarities of the pathogen.

10.6.3 Limited databases of microbial genomes

The forensic investigation and analysis of sample is heavily based on the law of comparison, which may be defined as "like can be compared only with like," and generated data can be stored in digital form (Rawtani & Hussain, 2020). In addition, the database could be used for comparison of unknown microbial sample with reference genome. However, there are so many challenges few of which are mentioned below:

- There are many pathogens in the environment that are not included in the existing databases. Therefore, in this situation, it is very difficult to identify strain of pathogen (Medicine & Council, 2002; Mir et al., 2022).
- Poorly characterized regional strain of pathogen can affect forensic identification of pathogen by limiting the ability to pinpoint the origin. This can affect the accuracy of investigations, making it difficult to trace the source of a disease outbreak or connect it to specific individuals or locations (Chancellor et al., 2006).
- Due to a lack of standard data formats and annotations, challenges also arise in data exchange and comparison between two agencies and labs for the identification and individualization of pathogen in forensic settings (Ahuja & Zaheer, 2025; Mir et al., 2023).

10.6.4 Ethical concerns: Surveillance, privacy, and misuse

The microbial data applied in forensic investigation raises several significant ethical and legal concerns, specifically associated with surveillance, privacy, consent, and potential misuse of the information by terrorist or hostile organizations, enemy, etc. (Oliveira & Amorim, 2018). As microbial forensics developing and growing, it is important to frame clear policy, guidelines and regulations to ensure responsible and ethical practices (Budowle et al., 2005; Schmedes & Budowle, 2015). In addition, some of them are discussed below:

- **Surveillance of microorganisms**: Surveillance of microorganisms, their populations and activity are called microbial surveillance. It involves data collection, analysis, interpretation, surveillance system, resistance patterns of microbial populations in any system (Lim et al., 2021). By understanding and responding to potential public health threats, implementing effective infection control measures and advising on effective antimicrobial use, the aim would be to inform appropriate stakeholders during the development of national and regional policy making and provide monitoring across multiple professional sectors (Galvin et al., 2012; Laxmi et al., 2021; World Health Organization, 2022).
- **Microbial forensic data:** From an ethical and security point of view, one of the most important issues mirroring the dual use problem of genomic data is the risk posed by microbial forensics. Scientific information or technologies that have both beneficial and harmful uses are called dual use (Lu et al., 2021; Miller & Selgelid, 2007).

10.6.4.1 Beneficial uses of microbial genomic data

- **Public health surveillance:** Genome sequencing has been valuable for outbreak investigations, where it has been used to track contagious pathogens (coinfections, outbreaks, or pandemics) to identify sources of infection and to calibrate human exposures during the study of transmission pathways and selection of mitigation strategies, and in monitoring the emergence of antibacterial resistance (Gilchrist et al., 2015; Quainoo et al., 2017).

- **Environmental monitoring:** Microbial forensics supports environmental monitoring and detecting endpoint contamination or bioremediation efforts, as well as ecosystem health assessments (Ghannam & Techtmann, 2021).
- **Medical research:** Microbial genetics opens avenues of research into probiotics, personalized medicine, novel therapies, etc. (Cunningham et al., 2021).

10.6.4.2 Malicious uses of microbial genomic data

- **Bioweapon development:** Genomic blueprints obtained through sequencing provide developers of bioweapons the capability to create pathogens that demonstrate more destructive potential and resistance to medical interventions, as well as improved survival in environmental conditions (Van Aken & Hammond, 2003).
- **Synthetic biology risks:** Synthetic biology can be used to reconstruct pathogens from genomic data, even though these diseases have been wiped out from the natural environment, such as smallpox (Duchêne et al., 2020).

10.6.4.3 Informed consent and data ownership in human-associated microbial data collection

The ethical concerns of data ownership and informed consent emerge when scientists analyze human-derived microbial information within microbiome research, particularly for studies involving human-related microbial samples such as gut, skin or oral microbiomes (Ma, 2021; Ponce De Leon-Derecho & Dable-Tupas, 2025).

10.7 Case studies

10.7.1 2001 anthrax attacks (USA)

In 2001, *Bacillus anthracis* attacks became a historic milestone for both microbiological analysis and public health investigation. After the 9/11 attack on America, the postal service delivered letters with *Bacillus anthracis* spores as renditions of terror to both political figures and media organizations. The intentional biological act resulted in five fatalities, together with infections among seventeen people, which produced massive panic throughout the United States (Schmedes & Budowle, 2015).

10.7.2 Use of *Bacillus anthracis* in mailed letters

During their attacks, the perpetrators created *Bacillus anthracis* spores in concentrated and purified forms, so they easily became airborne when the letter recipients opened their packages. The biological agents were packed inside envelopes, which were specifically sent to influential figures to prove that simple processes enable agents to reach public hands (McSweegan, 2004).

10.7.3 Forensic genomics and linking to a research lab

The investigations of the attacks heavily relied on microbial forensic analysis. Genetic analysis of *Bacillus anthracis* spores through genomic sequencing and comparative DNA techniques was performed on the discovered letters. Through extensive use of forensic genetics found that the microorganism was the same as the specimens kept at the U.S. Army Medical Research Institute of Infectious Diseases (USAMRIID) located at Fort Detrick in Maryland.

10.7.4 SARS-CoV-2 origin investigation

Researchers who studied SARS-CoV-2's origin had to use forensic methods that parallel microbial forensic techniques, even though the investigation was not strictly focused on identifying a criminal suspect. The understanding of SARS-CoV-2 emergence and global transmission required both molecular epidemiology and phylogenetic analytical methods specifically. In addition, the role of molecular epidemiology and phylogenetics may be discussed below:

- Molecular epidemiology: It uses genetic information as the basis for pathogen spread tracking, which aids them in creating infection route maps and locating contaminated origins (Riley & Blanton, 2018).
- Phylogenetics: In phylogenetics analysis of time-based genetic mutations in a virus through phylogenetic methods enables researchers to reconstruct its evolutionary pattern to identify its original sources and its human adaptation process (Gorbalenya & Lauber, 2017; Vello et al., 2024).

For SARS-CoV-2, these methods were critical for:

- **Identifying initial clusters**: Scientists have recognized the initial infection concentrations which occurred close to Huanan Seafood Market in Wuhan, China (Worobey et al., 2022).
- **Comparing viral genomes:** Scientists collected virus samples from various patients across the globe, allowing them to observe how the virus underwent population diversity while spreading (Miller et al., 2020).
- **Estimating timelines**: The virus started spreading among humans at least two weeks before authorities acknowledged its existence in the population.

10.7.4.1 Tracing the virus's origins

Through intensive genomic comparison:

- The results linked SARS-CoV-2 with bat coronaviruses, which suggest the virus originated naturally from animals flowing into human populations (Alwine et al., 2023).
- Various experts continue to debate how the spillover origination happened between animals and humans since it took place either in wildlife markets or through an intermediate host or perhaps a laboratory incident (Alwine et al., 2023).

Such investigations confirm that forensic-like methodologies need to be employed to determine the origins of newly discovered pathogens, whether these are deliberate or not.

10.8 Future directions

10.8.1 Future directions in microbial forensics

The field of microbial forensics is an emerging field, but several promising ways are being developed to conduct investigations faster, more accurately, and more globally coordinated. Some of them are mentioned below:

- **Integration of artificial intelligence (AI) and Machine Learning (ML)**
  AI and ML integration is evolving the microbial forensics field through faster, more accurate pathogen identification (Gao & Liu, 2024; Mohseni & Ghorbani, 2024). Moreover, due to the complex nature of microbial genomic data, traditional analytical methods are not suitable for analysis (Gasperskaja & Kicinskis, 2017). However, using AI and ML algorithms to rapidly screen huge sets of genomic sequences, spotting weird patterns that indicate possible signs of a new pathogen or deliberate genetic change (Reddy, 2023). In addition, these sophisticated computational models can predict the origins of pathogens as well as the evolutionary paths and help in getting a critical insight into how it evolved or spread over time. AI allows you to automate the pattern recognition process through an immense amount of data and does it faster and more thoroughly than manual or even conventional computational approaches (Reddy, 2023). This capability results in accelerating the rate needed to identify and classify pathogens and increases the accuracy of source attribution during outbreak, biocrime and bioterrorism events (Gao & Liu, 2024). As the most important factor of things happening in a world which is so connected to biology, it is to reduce that time or response time, and AI-driven microbial forensics can help in dramatically reducing response time and aid in the outbreak containment, and it will strengthen global biosecurity efforts (Nodari et al., 2024).
- **Portable sequencing devices for field forensics**
  The Oxford Nanopore MinION, along with similar portable sequencing tools, developed real-time microbial analysis capabilities through on-site testing that avoids full laboratory dependencies (Tyler et al., 2018; Wang et al., 2021). The small and adaptable sequencing devices perform quick DNA and RNA tests from any challenging location, which includes distant sites as well as limited-resource and dangerous environments (Tyler et al., 2018). The immediate availability of genetic data speeds up sample processing needs that become essential to make swift decisions when dealing with biological emergencies, including bioterrorism attacks or emerging infectious disease outbreaks. The prompt detection and analysis of genetic data accomplished through field-based sequencing eliminates sample transit time to centralized laboratories, thus improving quick reactions from scientific teams and public health authorities in their threat containment efforts and source tracking and control measure implementation (Bianconi et al., 2023). The real-time capabilities from this technology enable investigators to perform superior attributions of biological evidence sources, leading to better criminal investigations and public health response abilities.
- **Expansion of global microbial databases**
  The main technological barrier in microbial forensics consists of insufficient reference data that stops fast and accurate identification of unidentified microbial samples (Gilchrist et al., 2015). Researchers are developing new initiatives that focus on developing extensive global microbial databases to resolve this challenge. The research plans seek to develop databases that include diverse microbial strains from both

controlled laboratory research strains and clinical samples alongside natural environmental and engineered biological organisms (Contreras-Salgado et al., 2024). The addition of detailed metadata elements that include geographic positions and host species data, together with environmental factors serve to enhance database effectiveness for investigative purposes. The scientific community focuses on building open-access data platforms that ensure biological research collaboration through safe data exchanges while avoiding unauthorized access to confidential biological information. A strong microbial database that welcomes participation from across the globe will fundamentally boost forensic identification speed, leading to improved disease outbreak supervision and precise biocrime and emerging infectious diseases attribution (Aplakidou et al., 2024; Caliskan et al., 2023; Cruz et al., 2023).

- **Development of standardized protocols and international collaborations**
  Microbial forensics development depends on standardized protocols together with international collaborations because this will produce reliable investigations which maintain their credibility across all legal systems and regions (Tripathi et al., 2024). In addition, to achieve integrity throughout forensic investigations requires standardized protocols in critical forensic processes involving sample collection, handling and storage, since any evidence degradation presents substantial challenges. Through standardized sequencing methods and bioinformatics protocols, investigators achieve uniform processing of genomic data, which leads to better correctness in their evaluations across different investigations. The legal admission of forensic findings requires strict documentation of microbial evidence, starting from its origin to its endpoint through a system known as chain of custody. The development of microbial forensics depends heavily on international partnerships, which perform an essential role in capability enhancement. International bodies, including INTERPOL and the World Health Organization (WHO), and national public health agencies, help one another exchange informational resources to handle biological emergency cases of natural or intentional origin. The establishment of global collaborations between scientific institutions strengthens both investigation capability and builds credibility and transparency during biological crisis emergencies.

10.9 Conclusion

Microbial forensics bridges the gap between microbiology and criminal justice, offering powerful tools to identify, trace, and attribute microbial agents in both natural and intentional outbreaks. As technology evolves, so too does our ability to respond quickly and accurately to biological threats.

References

Ahuja, S., & Zaheer, S. (2025). Advancements in pathology: Digital transformation, precision medicine, and beyond. *Journal of Pathology Informatics*, *16*, 100408. https://doi.org/10.1016/j.jpi.2024.100408

Alawi, M., Burkhardt, L., Indenbirken, D., Reumann, K., Christopeit, M., Kröger, N., Lütgehetmann, M., Aepfelbacher, M., Fischer, N., & Grundhoff, A. (2019). DAMIAN: An open source bioinformatics tool for fast, systematic and cohort based analysis of microorganisms in diagnostic samples. *Scientific Reports*, *9*(1), 16841. https://doi.org/10.1038/s41598-019-52881-4

Alwine, J. C., Casadevall, A., Enquist, L. W., Goodrum, F. D., & Imperiale, M. J. (2023). A critical analysis of the evidence for the SARS-CoV-2 origin hypotheses. *mSphere*, *8*(2), e00119–e00123. https://doi.org/10.1128/msphere.00119-23

Anderson, P. D., & Bokor, G. (2012). Bioterrorism: Pathogens as weapons. *Journal of Pharmacy Practice, 25*(5), 521–529. https://doi.org/10.1177/0897190012456366

Ansari, I., Grier, G., & Byers, M. (2020). Deliberate release: Plague – A review. *Journal of Biosafety and Biosecurity, 2*(1), 10–22. https://doi.org/10.1016/j.jobb.2020.02.001

Aplakidou, E., Vergoulidis, N., Chasapi, M., Venetsianou, N. K., Kokoli, M., Panagiotopoulou, E., Iliopoulos, I., Karatzas, E., Pafilis, E., Georgakopoulos-Soares, I., Kyrpides, N. C., Pavlopoulos, G. A., & Baltoumas, F. A. (2024). Visualizing metagenomic and metatranscriptomic data: A comprehensive review. *Computational and Structural Biotechnology Journal, 23*, 2011–2033. https://doi.org/10.1016/j.csbj.2024.04.060

Arenas, M., Araujo, N. M., Branco, C., Castelhano, N., Castro-Nallar, E., & Pérez-Losada, M. (2018). Mutation and recombination in pathogen evolution: Relevance, methods and controversies. *Infection, Genetics and Evolution, 63*, 295–306. https://doi.org/10.1016/j.meegid.2017.09.029

Asim, M. N., Ibrahim, M. A., Zaib, A., & Dengel, A. (2025). DNA sequence analysis landscape: A comprehensive review of DNA sequence analysis task types, databases, datasets, word embedding methods, and language models. *Frontiers in Medicine, 12*, 1503229. https://doi.org/10.3389/fmed.2025.1503229

Barry, A., Thomson, S., Dimayuga, I., Chaudhuri, A., & Do, T. (2022). Isotope ratio method: State-of-the-art of forensic applications to CBRNE materials. *Canadian Society of Forensic Science Journal, 55*(3), 115–141. https://doi.org/10.1080/00085030.2022.2054109

Bartelink, E. J., & Chesson, L. A. (2019). Recent applications of isotope analysis to forensic anthropology. *Forensic Sciences Research, 4*(1), 29–44. https://doi.org/10.1080/20961790.2018.1549527

Bhoyar, L., Mehar, P., & Chavali, K. (2024). An overview of DNA degradation and its implications in forensic caseworks. *Egyptian Journal of Forensic Sciences, 14*(1), 15. https://doi.org/10.1186/s41935-024-00389-y

Bianconi, I., Aschbacher, R., & Pagani, E. (2023). Current uses and future perspectives of genomic technologies in clinical microbiology. *Antibiotics, 12*(11), 1580. https://doi.org/10.3390/antibiotics12111580

Binkley, C. E., Cinti, S., Simeone, D. M., & Colletti, L. M. (2002). Bacillus anthracis as an agent of bioterrorism: A review emphasizing surgical treatment. *Annals of Surgery, 236*(1), 9–16. https://doi.org/10.1097/00000658-200207000-00004

Brüssow, H., Canchaya, C., & Hardt, W.-D. (2004). Phages and the evolution of bacterial pathogens: From genomic rearrangements to lysogenic conversion. *Microbiology and Molecular Biology Reviews, 68*(3), 560–602. https://doi.org/10.1128/MMBR.68.3.560-602.2004

Budowle, B., Connell, N. D., Bielecka-Oder, A., Colwell, R. R., Corbett, C. R., Fletcher, J., Forsman, M., Kadavy, D. R., Markotic, A., Morse, S. A., Murch, R. S., Sajantila, A., Schmedes, S. E., Ternus, K. L., Turner, S. D., & Minot, S. (2014). Validation of high throughput sequencing and microbial forensics applications. *Investigative Genetics, 5*(1), 9. https://doi.org/10.1186/2041-2223-5-9

Budowle, B., Johnson, M. D., Fraser, C. M., Leighton, T. J., Murch, R. S., & Chakraborty, R. (2005). Genetic analysis and attribution of microbial forensics evidence. *Critical Reviews in Microbiology, 31*(4), 233–254. https://doi.org/10.1080/10408410500304082

Budowle, B., Murch, R., & Chakraborty, R. (2005). Microbial forensics: The next forensic challenge. *International Journal of Legal Medicine, 119*(6), 317–330. https://doi.org/10.1007/s00414-005-0535-y

Budowle, B., Schutzer, S. E., Einseln, A., Kelley, L. C., Walsh, A. C., Smith, J. A. L., Marrone, B. L., Robertson, J., & Campos, J. (2003). Building microbial forensics as a response to bioterrorism. *Science, 301*(5641), 1852–1853. https://doi.org/10.1126/science.1090083

Caliskan, A., Dangwal, S., & Dandekar, T. (2023). Metadata integrity in bioinformatics: Bridging the gap between data and knowledge. *Computational and Structural Biotechnology Journal, 21*, 4895–4913. https://doi.org/10.1016/j.csbj.2023.10.006

Cartozzo, C., Simmons, T., Swall, J., & Singh, B. (2021). Postmortem submersion interval (PMSI) estimation from the microbiome of Sus scrofa bone in a freshwater river. *Forensic Science International, 318*, 110480. https://doi.org/10.1016/j.forsciint.2020.110480

Castro, A. E., & De Ungria, M. C. A. (2022). Methods used in microbial forensics and epidemiological investigations for stronger health systems. *Forensic Sciences Research, 7*(4), 650–661. https://doi.org/10.1080/20961790.2021.2023272

Cavanaugh, S. E., & Bathrick, A. S. (2018). Direct PCR amplification of forensic touch and other challenging DNA samples: A review. *Forensic Science International: Genetics, 32*, 40–49. https://doi.org/10.1016/j.fsigen.2017.10.005

Chancellor, T. C. B., Holt, J., Villareal, S., Tiongco, E. R., & Venn, J. (2006). Spread of plant virus disease to new plantings: A case study of rice tungro disease. In: *Advances in Virus Research* (Vol. 66, pp. 1–29). Elsevier. https://doi.org/10.1016/S0065-3527(06)66001-6

Chen, L., & Bontempo, L. (2025). Application of isotope ratio mass spectrometry (IRMS) in the geographical determination of selected herbs: A review. *TrAC Trends in Analytical Chemistry, 183*, 118107. https://doi.org/10.1016/j.trac.2024.118107

Clarke, T., Brinkac, L., Greco, C., Alleyne, A. T., Carrasco, P., Inostroza, C., Tau, T., Wisitrasameewong, W., Torralba, M. G., Nelson, K., & Singh, H. (2022). Sampling from four geographically divergent young female populations demonstrates forensic geolocation potential in microbiomes. *Scientific Reports, 12*(1), 18547. https://doi.org/10.1038/s41598-022-21779-z

Cleri, D. J., Ricketti, A. J., Porwancher, R. B., Ramos-Bonner, L. S., & Vernaleo, J. R. (2006). Viral hemorrhagic fevers: Current status of endemic disease and strategies for control. *Infectious Disease Clinics of North America, 20*(2), 359–393. https://doi.org/10.1016/j.idc.2006.02.001

Contreras-Salgado, E. A., Sánchez-Morán, A. G., Rodríguez-Preciado, S. Y., Sifuentes-Franco, S., Rodríguez-Rodríguez, R., Macías-Barragán, J., & Díaz-Zaragoza, M. (2024). Multifaceted applications of synthetic microbial communities: Advances in biomedicine, bioremediation, and industry. *Microbiology Research, 15*(3), 1709–1727. https://doi.org/10.3390/microbiolres15030113

Cruz, G. L. T., Winck, G. R., D'Andrea, P. S., Krempser, E., Vidal, M. M., & Andreazzi, C. S. (2023). Integrating databases for spatial analysis of parasite-host associations and the novel Brazilian dataset. *Scientific Data, 10*(1), 757. https://doi.org/10.1038/s41597-023-02636-8

Cunningham, M., Azcarate-Peril, M. A., Barnard, A., Benoit, V., Grimaldi, R., Guyonnet, D., Holscher, H. D., Hunter, K., Manurung, S., Obis, D., Petrova, M. I., Steinert, R. E., Swanson, K. S., Van Sinderen, D., Vulevic, J., & Gibson, G. R. (2021). Shaping the future of probiotics and prebiotics. *Trends in Microbiology, 29*(8), 667–685. https://doi.org/10.1016/j.tim.2021.01.003

Dalton, R. (2008). Technological advances behind the anthrax investigation. *Nature*. https://doi.org/10.1038/news.2008.1011

Das, S., & Kataria, V. (2010). Bioterrorism: A public health perspective. *Medical Journal Armed Forces India, 66*(3), 255–260. https://doi.org/10.1016/S0377-1237(10)80051-6

Daskin, J. H., Calci, K. R., Burkhardt, W., & Carmichael, R. H. (2008). Use of N stable isotope and microbial analyses to define wastewater influence in Mobile Bay, AL. *Marine Pollution Bulletin, 56*(5), 860–868. https://doi.org/10.1016/j.marpolbul.2008.02.002

Deel, H., Bucheli, S., Belk, A., Ogden, S., Lynne, A., Carter, D. O., Knight, R., & Metcalf, J. L. (2020). Using microbiome tools for estimating the postmortem interval. In: *Microbial Forensics* (pp. 171–191). Elsevier. https://doi.org/10.1016/B978-0-12-815379-6.00012-X

Drmanac, R., Sparks, A. B., Callow, M. J., Halpern, A. L., Burns, N. L., Kermani, B. G., Carnevali, P., Nazarenko, I., Nilsen, G. B., Yeung, G., Dahl, F., Fernandez, A., Staker, B., Pant, K. P., Baccash, J., Borcherding, A. P., Brownley, A., Cedeno, R., Chen, L., … Reid, C. A. A. (2010). Human genome sequencing using unchained base reads on self-assembling DNA nanoarrays. *Science, 327*(5961), 78–81. https://doi.org/10.1126/science.1181498

Duchêne, S., Ho, S. Y. W., Carmichael, A. G., Holmes, E. C., & Poinar, H. (2020). The Recovery, interpretation and use of ancient pathogen genomes. *Current Biology, 30*(19), R1215–R1231. https://doi.org/10.1016/j.cub.2020.08.081

Fdez-Sanromán, A., Bernárdez-Rodas, N., Rosales, E., Pazos, M., González-Romero, E., & Sanromán, M. (2025). Biosensor technologies for water quality: Detection of emerging contaminants and pathogens. *Biosensors, 15*(3), 189. https://doi.org/10.3390/bios15030189

Galperin, M. Y., & Frishman, D. (1999). Towards automated prediction of protein function from microbial genomic sequences. In: *Methods in Microbiology* (Vol. 28, pp. 245–263). Elsevier. https://doi.org/10.1016/S0580-9517(08)70208-3

Galvin, S., Dolan, A., Cahill, O., Daniels, S., & Humphreys, H. (2012). Microbial monitoring of the hospital environment: Why and how? *Journal of Hospital Infection, 82*(3), 143–151. https://doi.org/10.1016/j.jhin.2012.06.015

Gao, Y., & Liu, M. (2024). Application of machine learning based genome sequence analysis in pathogen identification. *Frontiers in Microbiology, 15*, 1474078. https://doi.org/10.3389/fmicb.2024.1474078

Gasperskaja, E., & Kučinskas, V. (2017). The most common technologies and tools for functional genome analysis. *Acta Medica Lituanica, 24*(1), 1–11. https://doi.org/10.6001/actamedica.v24i1.3457

Ghannam, R. B., & Techtmann, S. M. (2021). Machine learning applications in microbial ecology, human microbiome studies, and environmental monitoring. *Computational and Structural Biotechnology Journal, 19*, 1092–1107. https://doi.org/10.1016/j.csbj.2021.01.028

Giani, A. M., Gallo, G. R., Gianfranceschi, L., & Formenti, G. (2020). Long walk to genomics: History and current approaches to genome sequencing and assembly. *Computational and Structural Biotechnology Journal, 18*, 9–19. https://doi.org/10.1016/j.csbj.2019.11.002

Gilchrist, C. A., Turner, S. D., Riley, M. F., Petri, W. A., & Hewlett, E. L. (2015). Whole-genome sequencing in outbreak analysis. *Clinical Microbiology Reviews, 28*(3), 541–563. https://doi.org/10.1128/CMR.00075-13

Gorbalenya, A., & Lauber, C. (2017). Phylogeny of viruses. In: *Encyclopedia of Virology*. Elsevier. https://doi.org/10.1016/B978-0-12-801238-3.95723-4

Ibrahim, N. K. (2020). Epidemiologic surveillance for controlling Covid-19 pandemic: Types, challenges and implications. *Journal of Infection and Public Health, 13*(11), 1630–1638. https://doi.org/10.1016/j.jiph.2020.07.019

Jamuar, S. S., D'Gama, A. M., & Walsh, C. A. (2016). Somatic mosaicism and neurological diseases. In: *Genomics, Circuits, and Pathways in Clinical Neuropsychiatry* (pp. 179–199). Elsevier. https://doi.org/10.1016/B978-0-12-800105-9.00012-3

Jefferson, C., Lentzos, F., & Marris, C. (2014). Synthetic biology and biosecurity: Challenging the â€œMythsâ€. *Frontiers in Public Health, 2*. https://doi.org/10.3389/fpubh.2014.00115

Jenks, P. J. (1998). Science, medicine, and the future: Microbial genome sequencing—beyond the double helix. *BMJ, 317*(7172), 1568–1571. https://doi.org/10.1136/bmj.317.7172.1568

Kalha, C. S., Singh, P. P., Kang, S. S., Hunjan, M. S., Gupta, V., & Sharma, R. (2014). Entomopathogenic viruses and bacteria for insect-pest control. In: *Integrated Pest Management* (pp. 225–244). Elsevier. https://doi.org/10.1016/B978-0-12-398529-3.00013-0

Khan, A. S., Amara, P. S., & Morse, S. A. (2020). Forensic public health. In: *Microbial Forensics* (pp. 105–122). Elsevier. https://doi.org/10.1016/B978-0-12-815379-6.00008-8

Kim, T.-Y., Zhu, X., Kim, S.-M., Lim, J.-A., Woo, M.-A., Lim, M.-C., & Luo, K. (2023). A review of nucleic acid-based detection methods for foodborne viruses: Sample pretreatment and detection techniques. *Food Research International, 174*, 113502. https://doi.org/10.1016/j.foodres.2023.113502

Korsah, M. A., Acquah, C., & Danquah, M. K. (2023). Transmission and intervention dynamics of SARS-CoV-2. In: *Viral Infections and Antiviral Therapies* (pp. 69–83). Elsevier. https://doi.org/10.1016/B978-0-323-91814-5.00009-X

Kreuzer-Martin, H. W., & Jarman, K. H. (2007). Stable Isotope Ratios and Forensic Analysis of Microorganisms. *Applied and Environmental Microbiology, 73*(12), 3896–3908. https://doi.org/10.1128/AEM.02906-06

Kumari, N., & Rai, L. C. (2020). Cyanobacterial diversity: Molecular insights under multifarious environmental conditions. In: *Advances in Cyanobacterial Biology* (pp. 17–33). Elsevier. https://doi.org/10.1016/B978-0-12-819311-2.00002-4

Laxmi, R. V., Ramya, A., Vanaja, S., & Vijayalakshmi, P. (2021). Microbiological surveillance of hospital environment in Chevella, India. *Journal of Pure and Applied Microbiology, 15*(3), 1449–1454. https://doi.org/10.22207/JPAM.15.3.38

Leake, S. L., Pagni, M., Falquet, L., Taroni, F., & Greub, G. (2016). The salivary microbiome for differentiating individuals: Proof of principle. *Microbes and Infection, 18*(6), 399–405. https://doi.org/10.1016/j.micinf.2016.03.011

Li, N., Liang, X., Zhou, S., Dang, L., Li, J., An, G., Ren, K., Jin, Q., Liang, X., Cao, J., Du, Q., Wang, Y., & Sun, J. (2023). Exploring postmortem succession of rat intestinal microbiome for PMI based on machine learning algorithms and potential use for humans. *Forensic Science International: Genetics, 66*, 102904. https://doi.org/10.1016/j.fsigen.2023.102904

Lim, C., Ashley, E. A., Hamers, R. L., Turner, P., Kesteman, T., Akech, S., Corso, A., Mayxay, M., Okeke, I. N., Limmathurotsakul, D., & Van Doorn, H. R. (2021). Surveillance strategies using routine microbiology for antimicrobial resistance in low- and middle-income countries. *Clinical Microbiology and Infection, 27*(10), 1391–1399. https://doi.org/10.1016/j.cmi.2021.05.037

Lu, D., Zhang, Y., Zhang, L., Wang, H., Weng, W., Li, L., & Cai, H. (2021). Methods of privacy-preserving genomic sequencing data alignments. *Briefings in Bioinformatics, 22*(6), bbab151. https://doi.org/10.1093/bib/bbab151

Ma, Y. (2021). Ethical issues of human microbiome research. In: *Encyclopedia of Life Sciences* (1st ed., pp. 1–9). Wiley. https://doi.org/10.1002/9780470015902.a0029126

Marian, A. J. (2011). Medical DNA sequencing. *Current Opinion in Cardiology, 26*(3), 175–180. https://doi.org/10.1097/HCO.0b013e3283459857

Markov, P. V., Ghafari, M., Beer, M., Lythgoe, K., Simmonds, P., Stilianakis, N. I., & Katzourakis, A. (2023). The evolution of SARS-CoV-2. *Nature Reviews Microbiology, 21*(6), 361–379. https://doi.org/10.1038/s41579-023-00878-2

McDonald, C., Taylor, D., & Linacre, A. (2024). PCR in forensic science: A critical review. *Genes, 15*(4), 438. https://doi.org/10.3390/genes15040438

McSweegan, E. (2004). Anthrax and the etiology of the English sweating sickness. *Medical Hypotheses, 62*(1), 155–157. https://doi.org/10.1016/S0306-9877(03)00303-7

Meckel, K. R., & Kiraly, D. D. (2019). A potential role for the gut microbiome in substance use disorders. *Psychopharmacology*, *236*(5), 1513–1530. https://doi.org/10.1007/s00213-019-05232-0

Medicine, I., & Council, N. R. (2002). *Countering Bioterrorism: The Role of Science and Technology*. The National Academies Press. https://doi.org/10.17226/10536

Miller, D., Martin, M. A., Harel, N., Tirosh, O., Kustin, T., Meir, M., Sorek, N., Gefen-Halevi, S., Amit, S., Vorontsov, O., Shaag, A., Wolf, D., Peretz, A., Shemer-Avni, Y., Roif-Kaminsky, D., Kopelman, N. M., Huppert, A., Koelle, K., & Stern, A. (2020). Full genome viral sequences inform patterns of SARS-CoV-2 spread into and within Israel. *Nature Communications*, *11*(1), 5518. https://doi.org/10.1038/s41467-020-19248-0

Miller, S., & Selgelid, M. J. (2007). Ethical and philosophical consideration of the dual-use dilemma in the biological sciences. *Science and Engineering Ethics*, *13*(4), 523–580. https://doi.org/10.1007/s11948-007-9043-4

Mir, T. U. G., Rani, P., & Nawim, A. (2020). Bioterrorism; Dual use of technology and role of law enforcement to combat biological attack. *International Journal of Current Microbiology and Applied Sciences*, *9*(9), 1408–1418. https://doi.org/10.20546/ijcmas.2020.909.178

Mir, T. U. G., Wani, A. K., Akhtar, N., Sena, S., & Singh, J. (2022). Microbial forensics: A potential tool for investigation and response to bioterrorism. *Health Sciences Review*, *5*, 100068. https://doi.org/10.1016/j.hsr.2022.100068

Mir, T. U. G., Wani, A. K., Akhtar, N., Katoch, V., Shukla, S., Kadam, U. S., & Hong, J. C. (2023). Advancing biological investigations using portable sensors for detection of sensitive samples. *Heliyon*, *9*(12), e22679. https://doi.org/10.1016/j.heliyon.2023.e22679

Mitchelson, K. R. (2005). DNA sequencing. In: *Encyclopedia of Analytical Science* (pp. 286–293). Elsevier. https://doi.org/10.1016/B0-12-369397-7/00683-X

Mohseni, P., & Ghorbani, A. (2024). Exploring the synergy of artificial intelligence in microbiology: Advancements, challenges, and future prospects. *Computational and Structural Biotechnology Reports*, *1*, 100005. https://doi.org/10.1016/j.csbr.2024.100005

Moran, G. J., Talan, D. A., & Abrahamian, F. M. (2008). Biological terrorism. *Infectious Disease Clinics of North America*, *22*(1), 145–187. https://doi.org/10.1016/j.idc.2007.12.003

Morse, S. A., & Khan, A. S. (2005). Epidemiologic investigation for public health, biodefense, and forensic microbiology. In: *Microbial Forensics* (pp. 157–171). Elsevier. https://doi.org/10.1016/B978-012088483-4/50011-1

Neckovic, A., Van Oorschot, R. A. H., Szkuta, B., & Durdle, A. (2020). Investigation of direct and indirect transfer of microbiomes between individuals. *Forensic Science International: Genetics*, *45*, 102212. https://doi.org/10.1016/j.fsigen.2019.102212

Nikolić, N., Massaro, S., Tagliapietra, F., Squartini, A., Schiavon, S., & Masin, R. (2024). Method to evaluate the microbial degradation activity in silage, cow rumen with in vitro test, and in manure and slurry. *MethodsX*, *12*, 102550. https://doi.org/10.1016/j.mex.2024.102550

Nodari, R., Arghittu, M., Bailo, P., Cattaneo, C., Creti, R., D'Aleo, F., Saegeman, V., Franceschetti, L., Novati, S., Fernández-Rodríguez, A., Verzeletti, A., Farina, C., & Bandi, C. (2024). Forensic microbiology: When, where and how. *Microorganisms*, *12*(5), 988. https://doi.org/10.3390/microorganisms12050988

Oliveira, M., & Amorim, A. (2018). Microbial forensics: New breakthroughs and future prospects. *Applied Microbiology and Biotechnology*, *102*(24), 10377–10391. https://doi.org/10.1007/s00253-018-9414-6

Oliveira, M., Marszałek, K., Kowalski, M., Frolova, A., Łabaj, P. P., Branicki, W., Madureira-Carvalho, Á, Da Silva, D. D., & Dinis-Oliveira, R. J. (2024). Sequencing technologies in forensic microbiology: Current trends and advancements. *Forensic Sciences*, *4*(4), 523–545. https://doi.org/10.3390/forensicsci4040035

Oliveira, M., Mason-Buck, G., Ballard, D., Branicki, W., & Amorim, A. (2020). Biowarfare, bioterrorism and biocrime: A historical overview on microbial harmful applications. *Forensic Science International*, *314*, 110366. https://doi.org/10.1016/j.forsciint.2020.110366

Pirttilä, A. M., Mohammad Parast Tabas, H., Baruah, N., & Koskimäki, J. J. (2021). Biofertilizers and biocontrol agents for agriculture: How to identify and develop new potent microbial strains and traits. *Microorganisms*, *9*(4), 817. https://doi.org/10.3390/microorganisms9040817

Ponce De Leon-Derecho, C. M., & Dable-Tupas, G. (2025). Ethical considerations in microbiome research. In: *Human Microbiome Drug Targets* (pp. 179–188). Elsevier. https://doi.org/10.1016/B978-0-443-15435-5.00013-X

Pronyk, P. M., De Alwis, R., Rockett, R., Basile, K., Boucher, Y. F., Pang, V., Sessions, O., Getchell, M., Golubchik, T., Lam, C., Lin, R., Mak, T.-M., Marais, B., Twee-Hee Ong, R., Clapham, H. E., Wang, L., Cahyorini, Y., Polotan, F. G. M., Rukminiati, Y., … Sintchenko, V. (2023). Advancing pathogen genomics in resource-limited settings. *Cell Genomics*, *3*(12), 100443. https://doi.org/10.1016/j.xgen.2023.100443

Quainoo, S., Coolen, J. P. M., Van Hijum, S. A. F. T., Huynen, M. A., Melchers, W. J. G., Van Schaik, W., & Wertheim, H. F. L. (2017). Whole-genome sequencing of bacterial pathogens: The future of nosocomial outbreak analysis. *Clinical Microbiology Reviews*, *30*(4), 1015–1063. https://doi.org/10.1128/CMR.00016-17

Rather, M. A., Agarwal, D., Bhat, T. A., Khan, I. A., Zafar, I., Kumar, S., Amin, A., Sundaray, J. K., & Qadri, T. (2023). Bioinformatics approaches and big data analytics opportunities in improving fisheries and aquaculture. *International Journal of Biological Macromolecules*, *233*, 123549. https://doi.org/10.1016/j.ijbiomac.2023.123549

Rawtani, D., & Hussain, C. M. (Eds.). (2020). *Technology in Forensic Science: Sampling, Analysis, Data and Regulations*. Wiley-VCH. https://doi.org/10.1002/9783527827688

Ray, P. K. (2022). Bioinformatics as a modern tool in forensic science for data understanding & investigation in research. *Journal of Forensic Science and Research*, *6*(1), 083–087. https://doi.org/10.29328/journal.jfsr.1001040

Reddy, S. S. P. (2023). Artificial intelligence in the genomics era: A blessing or a curse? *Journal of Regenerative Biology and Medicine*. https://doi.org/10.37191/Mapsci-2582-385X-5(3)-134

Relman, D. A. (2011). Microbial genomics and infectious diseases. *New England Journal of Medicine*, *365*(4), 347–357. https://doi.org/10.1056/NEJMra1003071

Riley, L. W., & Blanton, R. E. (2018). Advances in molecular epidemiology of infectious diseases: Definitions, approaches, and scope of the field. *Microbiology Spectrum*, *6*(6). https://doi.org/10.1128/microbiolspec.AME-0001-2018

Sahu, A., Kumar, N., Pal Singh, C., & Singh, M. (2023). Environmental DNA (eDNA): Powerful technique for biodiversity conservation. *Journal for Nature Conservation*, *71*, 126325. https://doi.org/10.1016/j.jnc.2022.126325

Saud, S., Fahad, S., & Hassan, S. (2022). Developments in the investigation of nitrogen and oxygen stable isotopes in atmospheric nitrate. *Sustainable Chemistry for Climate Action*, *1*, 100003. https://doi.org/10.1016/j.scca.2022.100003

Schmedes, S., & Budowle, B. (2015). Microbial forensics. *Reference Module In Biomedical Sciences*. 10.1016/B978-0-12-801238-3.02483-1

Shastri, N., Tsai, Y.-C., Hile, S., Jordan, D., Powell, B., Chen, J., Maloney, D., Dose, M., Lo, Y., Anastassiadis, T., Rivera, O., Kim, T., Shah, S., Borole, P., Asija, K., Wang, X., Smith, K. D., Finn, D., Schug, J., … Brown, E. J. (2018). Genome-wide identification of structure-forming repeats as principal sites of fork collapse upon ATR inhibition. *Molecular Cell*, *72*(2), 222–238.e11. https://doi.org/10.1016/j.molcel.2018.08.047

Shendure, J., & Aiden, E. L. (2012). The expanding scope of DNA sequencing. *Nature Biotechnology*, *30*(11), 1084–1094. https://doi.org/10.1038/nbt.2421

Shrivastava, P., Lorente, J. A., Srivastava, A., Badiye, A., & Kapoor, N. (Eds.). (2023). *Textbook of Forensic Science* (1st ed.). Springer Nature Singapore. https://doi.org/10.1007/978-981-99-1377-0

Singh, A. A., & Okpeku, M. (2024). Emerging methods of human microbiome analysis and its forensic applications: Review. *Forensic Science International: Reports*, *9*, 100355. https://doi.org/10.1016/j.fsir.2024.100355

Smith, J. A. L. (2011). Collection and preservation of microbial forensic samples. In: *Microbial Forensics* (pp. 379–392). Elsevier. https://doi.org/10.1016/B978-0-12-382006-8.00022-0

Soni, S. (2024). Microbial forensics and bioterrorism: Review of challenges, investigation, and future directions. *International Journal for Research in Applied Science and Engineering Technology*, *12*(11), 297–303. https://doi.org/10.22214/ijraset.2024.65035

Stephenson, F. H. (2003). Forensic science. In: *Calculations for Molecular Biology and Biotechnology* (pp. 278–296). Elsevier. https://doi.org/10.1016/B978-012665751-7/50053-6

Thomsen, P. F., & Willerslev, E. (2015). Environmental DNA – An emerging tool in conservation for monitoring past and present biodiversity. *Biological Conservation*, *183*, 4–18. https://doi.org/10.1016/j.biocon.2014.11.019

Tridico, S. R., Murray, D. C., Addison, J., Kirkbride, K. P., & Bunce, M. (2014). Metagenomic analyses of bacteria on human hairs: A qualitative assessment for applications in forensic science. *Investigative Genetics*, *5*(1), 16. https://doi.org/10.1186/s13323-014-0016-5

Tripathi, P., Render, R., Nidhi, S., & Tripathi, V. (2024). Microbial genomics: A potential toolkit for forensic investigations. *Forensic Science, Medicine and Pathology*, *21*(1), 417–429. https://doi.org/10.1007/s12024-024-00830-7

Tyler, A. D., Mataseje, L., Urfano, C. J., Schmidt, L., Antonation, K. S., Mulvey, M. R., & Corbett, C. R. (2018). Evaluation of Oxford Nanopore's MinION sequencing device for microbial whole genome sequencing applications. *Scientific Reports*, *8*(1), 10931. https://doi.org/10.1038/s41598-018-29334-5

Van Aken, J., & Hammond, E. (2003). Genetic engineering and biological weapons: New technologies, desires and threats from biological research. *EMBO Reports*, *4*(S1). https://doi.org/10.1038/sj.embor.embor860

Van Stelten, A., & Nightingale, K. K. (2014). Microbiological analysis | DNA methods. In: *Encyclopedia of Meat Sciences* (pp. 294–300). Elsevier. https://doi.org/10.1016/B978-0-12-384731-7.00066-0

Vello, F., Filippini, F., & Righetto, I. (2024). Bioinformatics goes viral: I. Databases, phylogenetics and phylodynamics tools for boosting virus research. *Viruses*, *16*(9), 1425. https://doi.org/10.3390/v16091425

Venkatesh, S., & Memish, Z. A. (2003). Bioterrorism—A new challenge for public health. *International Journal of Antimicrobial Agents*, *21*(2), 200–206. https://doi.org/10.1016/S0924-8579(02)00366-7

Vorkamp, K., Castaño, A., Antignac, J.-P., Boada, L. D., Cequier, E., Covaci, A., Esteban López, M., Haug, L. S., Kasper-Sonnenberg, M., Koch, H. M., Pérez Luzardo, O., Osīte, A., Rambaud, L., Pinorini, M.-T., Sabbioni, G., & Thomsen, C. (2021). Biomarkers, matrices and analytical methods targeting human exposure to chemicals selected for a European human biomonitoring initiative. *Environment International*, *146*, 106082. https://doi.org/10.1016/j.envint.2020.106082

Wang, Y., Zhao, Y., Bollas, A., Wang, Y., & Au, K. F. (2021). Nanopore sequencing technology, bioinformatics and applications. *Nature Biotechnology*, *39*(11), 1348–1365. https://doi.org/10.1038/s41587-021-01108-x

Wilmes, P., Simmons, S. L., Denef, V. J., & Banfield, J. F. (2009). The dynamic genetic repertoire of microbial communities. *FEMS Microbiology Reviews*, *33*(1), 109–132. https://doi.org/10.1111/j.1574-6976.2008.00144.x

World Health Organization. (2022). *Global antimicrobial resistance and use surveillance system (GLASS) report 2022* (1st ed).

Worobey, M., Levy, J. I., Malpica Serrano, L., Crits-Christoph, A., Pekar, J. E., Goldstein, S. A., Rasmussen, A. L., Kraemer, M. U. G., Newman, C., Koopmans, M. P. G., Suchard, M. A., Wertheim, J. O., Lemey, P., Robertson, D. L., Garry, R. F., Holmes, E. C., Rambaut, A., & Andersen, K. G. (2022). The Huanan Seafood Wholesale Market in Wuhan was the early epicenter of the COVID-19 pandemic. *Science*, *377*(6609), 951–959. https://doi.org/10.1126/science.abp8715

Zhang, J., Liu, W., Simayijiang, H., Hu, P., & Yan, J. (2023). Application of microbiome in forensics. *Genomics, Proteomics & Bioinformatics*, *21*(1), 97–107. https://doi.org/10.1016/j.gpb.2022.07.007

Zhang, S., Li, X., Wu, J., Coin, L., O'Brien, J., Hai, F., & Jiang, G. (2021). Molecular methods for pathogenic bacteria detection and recent advances in wastewater analysis. *Water*, *13*(24), 3551. https://doi.org/10.3390/w13243551

Zhao, A. P., Li, S., Cao, Z., Hu, P. J.-H., Wang, J., Xiang, Y., Xie, D., & Lu, X. (2024). AI for science: Predicting infectious diseases. *Journal of Safety Science and Resilience*, *5*(2), 130–146. https://doi.org/10.1016/j.jnlssr.2024.02.002

11

The Rising Path of Forensic Archaeology

An Indian Perspective

Gyaneshwer Chaubey, Rahul Mishra, Pankaj Shrivastava, and Chandana Basu Mallick

11.1 Introduction

Archaeology offers crucial techniques for handling burial remains, as well as evaluating and documenting a burial site including its surrounding environment and cultures (Coningham and Young, 2015; Funari et al., 2013; Gamble, 2015; Richards et al., 2003; Shouse, 2001). While forensic anthropology emerged as a field in the early to mid-twentieth century (Dirkmaat et al., 2008; Snow, 1982; Ubelaker, 2018; Yaşar Işcan, 1988), forensic archaeology is a more recent development (Dirkmaat et al., 2008; Groen et al., 2015; Passalacqua et al., 2021). Over time, it has become evident that trained archaeologists can greatly enhance the understanding of crime scenes and that forensic work can contribute substantially to archaeological knowledge. Globally, forensic archaeology is gaining recognition as a distinct field (Passalacqua et al., 2021).

The field of archaeology has traditionally been dominated by those with backgrounds in physical anthropology, creating persistent debate over its fit within forensic anthropology and how it connects to the broader discipline. However, the valuable role that archaeology can play in recovering remains from mass casualties and investigating intricate criminal scenes is gaining greater recognition (de Boer et al., 2018; Hanson and Fenn, 2024). The practice of forensic archaeology exists where multiple evidence-based approaches converge, frequently creating disparate and contradictory demands on how archaeological methods should be applied. While this specialty attracts significant media attention and interest among students, forensic archaeology continues to assume a limited role within anthropology on the whole. Despite this, practitioners in forensic archaeology frequently challenge conventional archaeological thinking, often producing conceptual innovations that arise from their research. This can provide a fresh approach to help revitalize anthropology for this modern era (Crossland, 2013).

Archaeological theories and methodologies concentrate on reconstructing human behaviour from incomplete material evidence, while forensics focuses on investigating crimes. Over the past few decades, the integration of these fields has become increasingly recognized as beneficial in forensic investigations. However, the application of forensic archaeology in India is still relatively limited (Aggrawal, 2015).

This chapter aims to first focus on the cutting-edge technique of biogeographical ancestry and further explore the role of forensic archaeology in India, particularly concerning outdoor

DOI: 10.4324/9781003474296-11

crime scenes where human remains are discovered (Sehrawat et al., 2020). By examining current practices, challenges, and methodologies, this work seeks to emphasize the importance of using archaeological techniques in forensic contexts to enhance the understanding of past events. The goal is to improve the recovery and analysis of human remains while also contributing to broader forensic science practices within the country. An interdisciplinary approach that combines expertise from both archaeology and anthropology can aid in the accurate interpretation of evidence at crime scenes. With continued recognition and application, forensic archaeology has the potential to significantly advance the resolution of criminal cases and provide closure for affected families.

11.2 Biogeographical ancestry in forensic archaeology of India

Ancestry research has gained significant traction, particularly regarding understanding the genetic makeup of populations (Paschou et al., 2010). Ancestry involves the genetic material inherited from various ancestral groups, shaping the genetic profile of an individual. Researchers use Ancestry Informative Markers (AIMs) to assess and interpret this genetic inheritance, providing insights into the geographic or ethnic origins of individuals based on DNA evidence (Enoch et al., 2006). This is especially valuable when personal lineage is unknown or when individuals choose not to disclose their ancestral backgrounds (Phillips et al., 2016).

Various markers can be utilized, including Short Tandem Repeats (STRs) and Single Nucleotide Polymorphisms (SNPs), with a preference for biallelic SNPs. These markers are favoured for their numerous advantages, including their presence throughout the genome, ease of genotyping, and balanced distribution (Cheung et al., 2019; Enoch et al., 2006). AIMs are crucial across multiple domains, particularly in admixture research and identifying an individual's biogeographical ancestry (I-BGA) (Halder et al., 2008).

Research studies reveal that analyzing genetic admixture often requires a significantly higher number of biallelic markers, sometimes four to ten times more than multi-allelic markers, to achieve comparable effectiveness and precision (**Figure 11.1**). This challenge can be addressed through analytical methods like principal component analysis to reduce dimensionality in genetic data (Morin and McCarthy, 2007). Notably, Lewis et al. illustrated that the required number of genetic markers for accurate ancestry interpretation could be dramatically reduced to about 1.5% of the initial SNP count, with an impressive accuracy of 92% in ancestry identification (Lewis et al., 2011).

Admixture panels that leverage AIMs are essential for encapsulating significant genetic data spread across the genome. The design of these panels is influenced by the size of linkage disequilibrium (LD) blocks that arise following admixture events. Over generations, the LD diminishes, leading to smaller ancestral LD blocks and necessitating a denser marker distribution to discern changes in ancestry across chromosomes due to meiotic crossover (Winkler et al., 2010).

To construct admixture mapping panels effectively, it is crucial to include a diverse array of genetic markers that adequately inform on ancestral origins associated with specific alleles at particular loci. These markers should be uniformly spread across the genome and operate independently to ensure effective ancestral distinction (Winkler et al., 2010).

Given the vast and diverse landscape of India, characterized by rich sociocultural traditions, the exploration of biogeographical ancestry markers is a complex endeavour (Reich et al.,

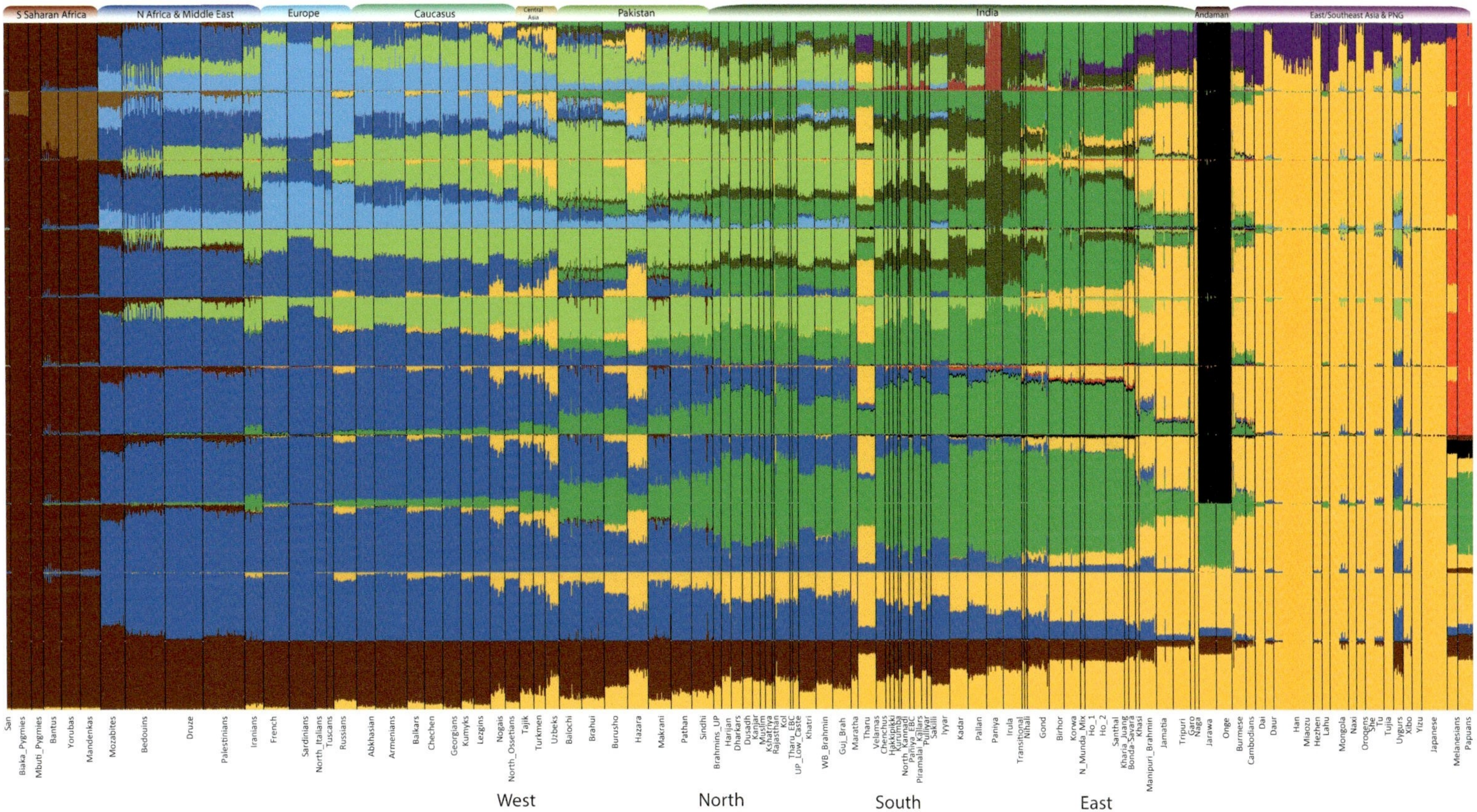

Figure 11.1 *Unsupervised ADMIXTURE analysis at K2-11 reveals various ancestral components among global populations. The distribution of these components is primarily influenced by geographical separation.*

2009). This undertaking demands an extensive repository of genomic sequences, not only from various regions but also from the myriad of distinct ethnic groups that call India home (The Department of Biotechnology (@DBTIndia), n.d.). The significant effective population size, bolstered by numerous founding lineages, further amplifies the need for a comprehensive and nuanced genetic analysis to uncover the intricate patterns of ancestry that reside within this vibrant subcontinent (Nakatsuka et al., 2017).

The growing availability of region-specific genome datasets in India is expected to enable the creation of highly targeted genetic markers for identification. By leveraging population genetics, forensic experts will gain unprecedented accuracy in tracing an individual's geographic roots—a breakthrough with significant implications for law enforcement. Analyzing biogeographical ancestry from DNA recovered at crime scenes can provide critical investigative leads, helping authorities focus on likely suspects when traditional DNA matches fail. Beyond criminal investigations, this approach also holds promise for identifying victims of disasters and missing persons cases, offering a powerful tool for forensic resolution in challenging scenarios.

11.3 Forensic archaeology of a murder that happened 400 years ago

Nestled between Eastern Europe and Western Asia, the stunning Caucasus Mountains encompass the country of Georgia (Aydogmusoglu, 2010). In 1614, the Kingdom of Kakheti, situated in the mountainous eastern region of Georgia, faced an incursion by Shah Abbas I, the emperor of Persia (modern-day Iran) (Klein and Vlachopoulou, 2023). Shah Abbas I is regarded as a pivotal figure of the Safavid dynasty, one of the most significant rulers in Iranian history (Aydogmusoglu, 2010). With a depleted military, King Teimuraz, the ruler of Kakheti, concluded that the only feasible method to thwart the invasion was to offer his mother, Queen Ketevan, along with her two sons, Prince Alexander and Prince Levan, as hostages to the Safavid court (Suny, 1994). Queen Ketevan, who acted as regent of Kakheti, hoped that their sacrifice would prevent a major catastrophe, but her efforts ultimately proved in vain. Tragically, both Prince Alexander and Prince Levan endured brutal torture, leading to their deaths, while Queen Ketevan was imprisoned for a decade (Rai et al., 2014). She was eventually given the choice to convert to Islam and join Shah Abbas I's harem, a proposition she resolutely declined.

According to historical accounts, Queen Ketevan was executed by strangulation in 1624 after refusing to renounce her faith. Before her death, two Augustinian monks from India journeyed to Shiraz. They were to establish a mission in Persia, but, upon meeting the queen during her imprisonment, they were deeply moved by her conviction. As a result, they became her confidants, choosing to serve her. After Queen Ketevan's martyrdom, the monks retrieved her remains. In 1627, they carried a fragment—her right humerus—to the Church of St. Augustine in Old Goa, where it was enshrined in a stone sarcophagus in the Chapter Chapel.

Recognized as a saint by the Georgian Orthodox Church, Queen Ketevan's sacrifice has been commemorated in Western writings (Flannery, 2008). In 1973, historian Roberto Gulbenkian identified the precise resting place of her relics in Goa using Portuguese archives. After the USSR dissolved in 1991, Georgian delegations partnered with India's Archaeological Survey (ASI) to locate her remains (ლორთქიფანიძევაჟა, 2021). By that point—and because the church had deteriorated, with many artefacts of not had vanished or been sold off—the search proved quite challenging.

Undeterred despite setbacks, ASI's superintendent archaeologist Nizamuddin Taher led a meticulous excavation, guided by historical texts. The team focused on finding the stone coffin (Klein and Vlachopoulou, 2023; Rai et al., 2014). While clearing debris and excavating a fallen wall, they uncovered a long arm bone lodged in the sediment. The relic sarcophagus fashioned from stone had shattered when the wall collapsed; however, its lid survived intact, indicating its likely original placement near the second window. Nearby, two additional skeletal remains were found, along with two undamaged stone coffins and their lids (Rai et al., 2014).

With sufficient archaeological findings in hand, a new question emerged: which of these three bones belonged to the Saint queen? To determine this, the remains were transported to CSIR-CCMB in Hyderabad. Scientist Dr. K. Thangaraj and his colleagues, including myself, analyzed ancient DNA to ascertain the sex of the three samples. This sex determination utilized amelogenin markers designed for forensic purposes, which yield two peaks (corresponding to X and Y chromosomes) in males and one peak (an X chromosome) in females (Dutta et al., 2017). Among the three samples, two exhibited two peaks, indicating they belonged to males, while the remaining sample displayed a single peak, confirming it as that of a female.

Following this, the mitochondrial DNA (mtDNA) sequences of the three samples were analyzed. Based on the mitochondrial mutations, the female sample displayed a mutation pattern of 16182C-16183C-16186T-16189C-16207d-16249C-16327T, which corresponds to the West Eurasian haplogroup U1b (Rai et al., 2014). The sequence was later compared with 22000 mtDNA samples from India, revealing that this particular mutation stretch was nearly absent in all instances. This strongly indicated that the female sample under study was not of Indian origin. In contrast, the other two samples matched individuals from India, with one belonging to haplogroup R6a (16129A-16266T-161274A-16309G), while the other was assigned to U2a (16051G-16206C-16230G). Both of these haplogroups are found in significant numbers in western regions of India. Further analysis revealed that the same mutation was identified in 2 out of 20 samples tested from Georgia. Given that the queen had no direct relatives for comparison, the complete absence of haplogroup U1b in the Indian subcontinent, alongside its presence in moderate to high frequencies in Georgia and its neighbouring areas, confirmed that the sample belonged to Queen Ketevan. On 10 July 2021, nearly four centuries after the death of Saint Queen. India's External Affairs Minister, Mr S. Jaishankar, officially took the remains of the Saint Queen, which had been discovered in 2005 at the historic and dilapidated St. Augustine Church complex in Old Goa. The remains were then handed over to the Georgian government, and thus, the happy ending of a brutal murder that occurred 400 years ago was resolved, providing the Saint Queen with a final verdict where the role of science was paramount.

11.4 The Ajnala massacre, a mass murder that took place 165 years ago

The study of mtDNA and stable isotopes in a set of 164-year-old human skeletal remains has yielded significant insights into the ancestry and biogeographical origin of sepoys who participated in the Sepoy Mutiny of 1857 (Greenfield, 2024; Kayastha, 2024; Sehrawat et al., 2022). These remains were discovered by historian Surinder Kochcher in an abandoned well in Punjab, India, and the research combined genetic and isotopic analyses to trace the origins of these unidentified soldiers (Condos, 2023; Sehrawat et al., 2020).

The Sepoy Mutiny marked a crucial uprising against British colonial rule, and while many soldiers lost their lives, their identities remained largely unknown for over a century

(Condos, 2023). Recently, efforts to uncover the genetic identities and origins of these individuals have gained momentum with advances in forensic science, particularly in the extraction and analysis of ancient DNA. mtDNA is particularly valuable in forensic investigations due to its maternal inheritance pattern and its higher recovery rate from ancient remains compared to nuclear DNA. Being present in multiple copies per cell, mtDNA also allows for greater detection chances in degraded or limited samples. This characteristic makes mtDNA a powerful tool in both forensic anthropology and the exploration of historical population movements and demographics (Brandstatter et al., 2007).

The skeletal remains of 50 individuals underwent DNA extraction and mtDNA haplotyping to determine genetic lineages. Various techniques were applied to ensure the integrity of the samples. The results from mtDNA analysis were cross-validated using stable oxygen isotope analysis (δ18O values) conducted on 85 additional archaeological skeletal remains.

Isotope analysis is instrumental in determining geographical origins, as oxygen isotopes in bones reflect local environmental conditions and can indicate regions of birth or long-term habitat (Sehrawat et al., 2024). This dual approach of using genetic and isotopic data enhanced the reliability of the findings.

The mtDNA haplogroup analysis identified at least 12 haplogroups among the pool of analyzed samples. Haplogroup M39 and R32 were the most common lineages observed among the exhibits. The haplogroup distribution suggested a closer link with northern India. It indicated a distinctive genetic connection between the sepoys and populations located east of Punjab, suggesting that these individuals likely originated from regions corresponding to the Gangetic plains of northern India (Sehrawat et al., 2022). This finding challenged previous assumptions of local ancestry, pushing back against the notion that they were primarily of Punjabi descent.

The isotopic data corroborated the mitochondrial analysis, reinforcing the hypothesis of a common geographic ancestry. The δ18O values aligned with those of the Gangetic plains, offering further evidence that these individuals had roots in the eastern part of India. This supports historical migration patterns and demographic shifts during the era.

mtDNA has emerged as an essential tool in contemporary forensic anthropology. The survival of mtDNA from human remains, particularly in challenging conditions where nuclear DNA is often lost or degraded, underscores its significance. Past research has highlighted the various conditions affecting DNA recovery, with mtDNA offering a more robust alternative when dealing with ancient biological samples (Endicott et al., 2006; Maca-Meyer et al., 2004). This study also reinforces the advantage of mtDNA in degraded samples.

The methodologies applied have vital implications for forensic anthropology and bioarchaeology. The understanding that different biological tissues, such as bones and teeth, possess distinct recovery characteristics opens new avenues for research into ancient populations and their societal structures. Teeth, in particular, were shown to be of higher quality for DNA extraction, making them ideal for studies involving skeletal remains recovered from various contexts, including crime scenes, archaeological sites, and disaster scenarios (Adler et al., 2011; Damgaard et al., 2015).

The preference for minimally destructive sampling methods, such as direct drilling of teeth, over pulverizing can significantly impact the quantity and quality of DNA extracted. This consideration is essential when designing forensic protocols for extracting DNA from fragile remains, which are often subjected to numerous environmental challenges (Adler et al., 2011).

The in-depth analysis of mtDNA and stable isotopes from the skeletal remains of sepoys offers essential insights into their historical identities and origins. Utilizing advanced scientific methodologies, this research not only reconstructs the life histories of individuals who played a significant role in a pivotal moment in Indian history but also enhances our broader understanding within the fields of genetics, anthropology, and archaeology.

This study exemplifies the power of integrating genetic, isotopic, and forensic technologies to unravel complex historical narratives. Future research may include more extensive sampling and investigations to find the direct relatives of these martyrs. It will further enrich our understanding of shared ancestry and the impacts of socio-political upheavals throughout history. The intersection of forensic science and historical inquiry represents a promising frontier for exploring the lives of those who greatly influenced our history.

Overall, the findings have implications not only for historical scholarship but also for contemporary discussions about identity, ancestry, and the atrocities of colonialism in India. The integration of genetic and isotopic data can be instrumental in recognizing ancestral connections in populations today, bridging past narratives with present identities.

Thus, the future of forensic archaeology in India holds great promise as the field continues to develop and gain recognition. With advancements in technology and interdisciplinary collaboration, forensic archaeology is set to enhance the recovery and analysis of human remains, leading to a deeper understanding of historical contexts and unresolved crimes. The integration of cutting-edge tools such as DNA analysis, geographical information systems (GIS), and 3D modelling will enable forensic archaeologists to investigate outdoor crime scenes more effectively.

Moreover, as the Indian society becomes increasingly aware of the significance of forensic science in legal matters, the demand for trained professionals in this field is likely to grow. Educational institutions may start offering specialized programs in forensic archaeology, fostering a new wave of experts who can address the unique challenges posed by India's diverse cultural landscape.

Additionally, focusing on human rights violations and social justice, forensic archaeology can play a crucial role in bringing closure to families affected by historical injustices. By bridging the gap between anthropology, archaeology, and forensic science, the future of forensic archaeology in India promises not only advancements in solving crimes but also a greater understanding of the past, contributing to societal healing and justice.

References

Adler, C. J., Haak, W., Donlon, D., Cooper, A., and Consortium, G. (2011). Survival and recovery of DNA from ancient teeth and bones. *J. Archaeol. Sci.* 38, 956–964.

Aggrawal, A. (2015). "Forensic archaeology: An Indian perspective," in *Forensic Archaeology*, eds. W. J. M. Groen, N. Márquez-Grant, and R. C. Janaway (Wiley), 287–291. doi: 10.1002/9781118745977.ch34

Aydogmusoglu, C. (2010). Sah Abbas (1587–1629) Zamaninda Isfahan. *Türk Dünya. Araştırmaları*. Available at: https://openurl.ebsco.com/EPDB%3Agcd%3A13%3A32201933/detailv2?sid=ebsco%3Aplink%3Aresult-item&id=ebsco%3Agcd%3A52796826&bquery=Aydogmusoglu%2C%20C.%20(2010).%20Sah%20Abbas%20(1587%E2%80%931629)%20Zamaninda%20Isfahan.%20T%C3%BCrk%20D%C3%BCnya.%20Ara%C5%9Ft%C4%B1rmalar%C4%B1.&page=1&link_origin=none&crl=f (Accessed March 4, 2025).

Brandstatter, A., Niederstatter, H., Pavlic, M., Grubwieser, P., and Parson, W. (2007). Generating population data for the EMPOP database - An overview of the mtDNA sequencing and data evaluation processes considering 273 Austrian control region sequences as example. *Forensic Sci. Int.* 166, 164–175.

Cheung, E. Y., Phillips, C., Eduardoff, M., Lareu, M. V., and McNevin, D. (2019). Performance of ancestry-informative SNP and microhaplotype markers. *Forensic Sci. Int. Genet.* 43, 102141.

Condos, M. (2023). "The Ajnala massacre of 1857 and the politics of colonial violence and commemoration in contemporary India," in *Colonial Violence and Monuments in Global History*, (Routledge), 104–121. Available at: https://www.taylorfrancis.com/chapters/oa-edit/10.4324/9781003397458-7/ajnala-massacre-1857-politics-colonial-violence-commemoration-contemporary-india-mark-condos (Accessed February 21, 2025).

Coningham, R., and Young, R. (2015). *The Archaeology of South Asia: From the Indus to Asoka, c. 6500 BCE–200 CE.* Cambridge: Cambridge University Press. doi: 10.1017/CBO9781139020633

Crossland, Z. (2013). Evidential regimes of forensic archaeology. *Annu. Rev. Anthropol.* 42, 121–137. doi: 10.1146/annurev-anthro-092412-155513

Damgaard, P. B., Margaryan, A., Schroeder, H., Orlando, L., Willerslev, E., and Allentoft, M. E. (2015). Improving access to endogenous DNA in ancient bones and teeth. *Sci. Rep.* 5, 11184.

de Boer, H. H., Blau, S., Delabarde, T., and Hackman, L. (2018). The role of forensic anthropology in disaster victim identification (DVI): Recent developments and future prospects. *Forensic Sci. Res.* 4, 303–315. doi: 10.1080/20961790.2018.1480460

Dirkmaat, D. C., Cabo, L. L., Ousley, S. D., and Symes, S. A. (2008). New perspectives in forensic anthropology. *Am. J. Phys. Anthropol.* 137, 33–52. doi: 10.1002/ajpa.20948

Dutta, P., Bhosale, S., Singh, R., Gubrellay, P., Patil, J., Sehdev, B., et al. (2017). Amelogenin gene-the pioneer in gender determination from forensic dental samples. *J. Clin. Diagn. Res.* 11, ZC56.

Endicott, P., Metspalu, M., Stringer, C., Macaulay, V., Cooper, A., and Sanchez, J. J. (2006). Multiplexed SNP typing of ancient DNA clarifies the origin of Andaman mtDNA haplogroups amongst south Asian tribal populations. *PLoS One* 1, e81.

Enoch, M.-A., Shen, P.-H., Xu, K., Hodgkinson, C., and Goldman, D. (2006). Using ancestry-informative markers to define populations and detect population stratification. *J. Psychopharmacol.* 20, 19–26.

Flannery, J. (2008). The Martyrdom of Queen Ketevan of Georgia: An episode in the relations between the church of Rome and the Georgian Church. Available at: https://scholarhub.balamand.edu.lb/handle/uob/6320 (Accessed March 4, 2025).

Funari, P. P. A., Jones, S., and Hall, M. (2013). "Introduction: Archaeology in history," in *Historical Archaeology,* (Routledge), 1–20. Available at: https://www.taylorfrancis.com/chapters/edit/10.4324/9780203208816-1/introduction-pedro-paulo-funari-si%C3%A2n-jones-martin-hall (Accessed February 21, 2025).

Gamble, C. (2015). *Archaeology: The Basics.* Routledge. Available at: https://www.taylorfrancis.com/books/mono/10.4324/9781315728384/archaeology-basics-clive-gamble (Accessed February 21, 2025).

Greenfield, S. (2024). "The rising: Mangal Pandey and India's First War of Independence," In (Ed.)S. Machura *Law and War in Popular Culture,* (Nomos Verlagsgesellschaft mbH & Co. KG), 213–234. doi: 10.5771/9783748918295-213

Groen, W. M., Marquez-Grant, N., and Janaway, R. (2015). *Forensic Archaeology: A Global Perspective.* John Wiley & Sons..

Halder, I., Shriver, M., Thomas, M., Fernandez, J. R., and Frudakis, T. (2008). A panel of ancestry informative markers for estimating individual biogeographical ancestry and admixture from four continents: Utility and applications. *Hum. Mutat.* 29, 648–658.

Hanson, I., and Fenn, J. (2024). A review of the contributions of forensic archaeology and anthropology to the process of disaster victim identification. *J. Forensic Sci.* 69, 1637–1657. doi: 10.1111/1556-4029.15553

Kayastha, B. (2024). The Sepoy Mutiny (1857): A landmark event in the modern Indian history. *Res. Found. Int. India* 33, 1–12.

Klein, D., and Vlachopoulou, A. eds. (2023). *Transottoman Biographies, 16th–20th c.,* 1st Edn. Göttingen: V&R Unipress. doi: 10.14220/9783737011662

Lewis, J., Abas, Z., Dadousis, C., Lykidis, D., Paschou, P., and Drineas, P. (2011). Tracing cattle breeds with principal components analysis ancestry informative SNPs. *PLoS One* 6, e18007. doi: 10.1371/journal.pone.0018007

Maca-Meyer, N., Arnay, M., Rando, J. C., Flores, C., Gonzalez, A., Cabrera, V. M., et al. (2004). Ancient mtDNA analysis and the origin of the Guanches. *Eur. J. Hum. Genet.* 12, 155–162.

Morin, P. A., and McCarthy, M. (2007). Highly accurate SNP genotyping from historical and low-quality samples. *Mol. Ecol. Notes* 7, 937–946. doi: 10.1111/j.1471-8286.2007.01804.x

Nakatsuka, N., Moorjani, P., Rai, N., Sarkar, B., Tandon, A., Patterson, N., et al. (2017). The promise of discovering population-specific disease-associated genes in South Asia. *Nat. Genet.* 49, 1403.

Paschou, P., Lewis, J., Javed, A., and Drineas, P. (2010). Ancestry informative markers for fine-scale individual assignment to worldwide populations. *J. Med. Genet.* 47, 835–847.

Passalacqua, N. V., Pilloud, M. A., and Congram, D. (2021). Forensic anthropology as a discipline. *Biology* 10, 691.

Phillips, C., Santos, C., Fondevila, M., Carracedo, Á, and Lareu, M. V. (2016). "Inference of ancestry in forensic analysis I: Autosomal ancestry-informative marker sets," In W. Goodwin (Ed.) *Forensic DNA Typing Protocols* (Vol. 1420, 233–253). Humana Press. doi: 10.1007/978-1-4939-3597-0_18

Rai, N., Taher, N., Singh, M., Chaubey, G., Jha, A. N., Singh, L., et al. (2014). Relic excavated in western India is probably of Georgian Queen Ketevan. *Mitochondrion* 14, 1–6.

Reich, D., Thangaraj, K., Patterson, N., Price, A. L., and Singh, L. (2009). Reconstructing Indian population history. *Nature* 461, 489–494.

Richards, M. P., Schulting, R. J., and Hedges, R. E. M. (2003). Archaeology: Sharp shift in diet at onset of neolithic. *Nature* 425, 366.

Sehrawat, J. S., Agrawal, S., Kenney, A. P., Grimes, V., and Rai, N. (2024). Use of strontium isotope ratios in potential geolocation of Ajnala skeletal remains: A forensic archeological study. *Int. J. Legal Med.* 138, 615–626.

Sehrawat, J. S., Agrawal, S., Sankhyan, D., Singh, M., Kumar, S., Prakash, S., et al. (2022). Pinpointing the geographic origin of 165-year-old human skeletal remains found in Punjab, India: Evidence from mitochondrial DNA and stable isotope analysis. *Front. Genet.* 13, 813934.

Sehrawat, J. S., Rai, N., and Meier-Augenstein, W. (2020). Identification of Ajnala skeletal remains using multiple forensic anthropological methods and techniques: A bioarchaeological report. *J. Archaeol. Sci. Rep.* 32, 102434.

Shouse, B. (2001). Archaeology. Spreading the word, scattering the seeds. *Science* 294, 988–989.

Snow, C. C. (1982). Forensic anthropology. *Annu. Rev. Anthropol.* 97–131.

Suny, R. G. (1994). *The Making of the Georgian Nation*. Indiana University Press.

The Department of Biotechnology (@DBTIndia) (n.d.). Available at: https://dbtindia.gov.in/ (Accessed February 21, 2025).

Ubelaker, D. H. (2018). A history of forensic anthropology. *Am. J. Phys. Anthropol.* 165, 915–923. 10.1002/ajpa.23306

Winkler, C. A., Nelson, G. W., and Smith, M. W. (2010). Admixture mapping comes of age. *Annu. Rev. Genomics Hum. Genet.* 11, 65–89. doi: 10.1146/annurev-genom-082509-141523

Yaşar Işcan, M. (1988). Rise of forensic anthropology. *Am. J. Phys. Anthropol.* 31, 203–229. 10.1002/ajpa.1330310510

ლორთქიფანიძევაჟა (2021). About searching holy relics of Ketevan the Martyr. მაცნე. Available at: http://www.macne.org.ge/index.php/macne/article/view/17 (Accessed March 4, 2025).

12

Wildlife Forensics

From the Science of Conservation to the Courtroom

Ajit Kumar, Dimple Bhatia, and Rajinder Singh

12.1 Introduction

The Indian subcontinent is a mega-diverse nation with approximately 10.78% of the world's floral and 7.31% of the world's faunal population (Thatheyus, 2006). Most of these species have dwelled in certain regions considered biodiversity hotspots, including the Western Ghats, Himalayas, Indo-Burma region, and Sunda land region (Kumar and Verma, 2017). Due to the abundance of biodiversity, India is vulnerable to several wildlife-related crimes, including the unauthorized taking, trading, killing, possessing, and transporting of flora and fauna, and derivative items (Negi, 2013). Illegal wildlife trade (including illegal fishing and timber trade) is the second largest illicit trade in the world, with a turnover of approximately 53 billion USD. It poses a significant threat to global conservation efforts, and its magnitude has steadily increased over time (Andersson et al., 2021). This trade has already caused the extinction of several species and pushed many others to the brink of permanent disappearance, further exacerbating ecological imbalance and threatening a country's socioeconomic status (Agnew et al., 2009; Bhatia et al., 2024; EUROPOL OCTA, 2011).

Wildlife forensics is a field that combines different scientific disciplines, such as genetics, morphology, and toxicology, to investigate wildlife crimes and enforce laws related to wildlife conservation. The journey of wildlife forensics can be traced from its origins in conservation sciences to its current role as a critical tool in conserving endangered species and prosecuting wildlife crimes (Haines et al., 2021). Wildlife forensics initially focused on identifying species and individuals and determining the geographic origin of wildlife products. Furthermore, it plays a crucial role in ensuring justice for wildlife by providing scientific evidence in legal proceedings that strengthen cases against wildlife criminals and ensure their accountability for their actions.

Wild animals are illegally traded as live pets and poached for their body parts such as skin, claws, hairs, teeth, tusks, bones, horns, antlers, and derivative products. Plants, however, are illegally traded for flowers, timber, and for medicinal, aromatic, or cosmetic preparations. According to the World Wildlife Crime Report (2020), a taxonomy-wise analysis of 1,80,000 seizures made across 149 countries between 1999 and 2018 illustrated, that mammals are most extensively traded species, comprising 23% of the seizures, followed by reptiles (21.3%), plants (14.6%), birds (8.5%), molluscs (7.9%), boney fish (4.7%), and other (5.7%). The TRAFFIC report highlighted that from 2011 to 2020, there were 141 wildlife seizure incidents involving 146 wildlife species recorded from 18 Indian airports. With a share of 46%, reptiles accounted for the majority of these wildlife seizures, followed by mammals (18%), timber (13%), aquatic species (10%), birds (8%), insects (4%), and amphibians (1%) (Badola and Gautam, 2022).

DOI: 10.4324/9781003474296-12

National Crime Record Bureau (NCRB) report states that in the year 2022, 535 new cases were registered under the Wildlife (Protection) Act, 1972, and a 65.3% conviction rate was observed with a pendency rate of 93.6% (including pending cases from the previous year and reopened cases). This low conviction and high pendency rate are due to the lack of forensic evidence and their inaccurate collection and handling at the crime scene, resulting in evidence degradation and improper identification of species. Moreover, inappropriate charge framing, judiciary's lagging approach to these crimes, and lacunas in laws protecting wildlife such as inadequate penalties and punishment for wildlife offenses also contribute to the rise of wildlife crimes in India. In addition, limited forensic laboratories and specialists are working exclusively on wildlife crimes as these offenses aren't considered serious in our nation (Gouda et al., 2020; Karmakar, 2013; Soni, 2020).

This chapter will detail the methods and steps that must be followed throughout investigations, as well as the numerous tools and strategies for analyzing such evidence and submitting the report in a court of law. This will assist security officers, crime scene investigators, law enforcement agencies, and forensic examiners in investigating the case and analyzing the evidence promptly and thoroughly, increasing the conviction rate and reducing case pendency to combat wildlife crimes.

12.2 History of wildlife forensics

The field of wildlife forensics has a relatively short but impactful history. Approximately 110 years ago, wildlife forensics emerged in response to the overexploitation of wildlife for commercial purposes. This period marked the beginning of national and international laws aimed at combating various forms of wildlife crime, such as international wildlife trafficking, poaching of animals like elephants and rhinos for their ivory and horns, and harvesting native species (Neme, 2009).

One of the earliest documented cases of wildlife forensic analysis occurred in 1984 when DNA analysis was used to identify the species of ivory seized in a smuggling case (Huffman and Wallace, 2012). It began a new era in wildlife forensics, where scientific techniques were used to investigate and prosecute wildlife crimes. In the following decades, technological advancements, such as polymerase chain reaction (PCR) and next-generation sequencing (NGS), revolutionized wildlife forensics. These techniques allowed for more accurate species identification and individual profiling, even from small or degraded samples. These days, population genetics, relatedness analysis, and geographic origin determination are all included in the field of wildlife forensics, which was once limited to species identification. These advancements have been crucial in understanding the genetic diversity of populations, identifying poaching hotspots, and tracing the origins of seized wildlife products. Today, wildlife forensics plays a critical role in conservation efforts worldwide. Recently, there has been a growing emphasis on capacity building and training in wildlife forensics, particularly in developing countries where wildlife crimes are prevalent. International collaborations and partnerships have also been established to enhance the exchange of knowledge and expertise in this field.

Looking ahead, the future of wildlife forensics lies in the continued development of new techniques and methodologies, such as environmental DNA (eDNA) analysis and stable isotope analysis. These advancements will further improve our ability to combat wildlife crimes and protect endangered species for generations.

12.3 India's commitment to wildlife conservation: Legislation and international agreements

India has long been committed to wildlife conservation, with various agencies and legislation supporting wildlife law enforcement. The Wildlife (Protection) Act is one such legislation that the Indian Parliament passed on September 9, 1972, and was last amended in 2023 to strengthen laws and policies over time. It offers varying levels of protection to 1,800 plant and animal species under six schedules (I-VI). In addition to national legislation, India ratified the CITES (Convention on International Trade in Endangered Species of Wild Fauna and Flora) in 1976. This international agreement aims.to protect 35,000 flora and fauna species from international trade by listing them in three appendices (I-III) (Sinha, 2010). Through these concerted efforts, India continues to play a significant role in global wildlife conservation.

12.4 Crime scene investigation and evidence collection

Wildlife crimes are distinct from other types of conventional crimes because, in conventional crimes, the police file a First Information Report (FIR) and subsequently gather physical evidence from the crime scene. In contrast, when it comes to wildlife crimes, suspected wildlife products are seized/collected first from the individual/crime scene, and only then a case is filed. In India, the first report to initiate an investigation is filed in a jurisdictional court by an authorized person, such as an officer from the revenue, customs, state, or forest department. This report is prepared in accordance with section 50(4) of the Wildlife (Protection) Act, 1972, and may be termed a Wildlife Offence Report (WLOR). WLOR contains the following information (i) the date, time, and location of the seizure; (ii) the name, age, and address of the accused or suspect, if known; (iii) the details of the seizure, such as the species name, the type and quantity of wildlife products, and the Schedule in which the species is included; and (iv) the investigating officer's details (Negi, 2013). Different methods for crime scene investigation and evidence gathering may be used while dealing with poached wild animals and seized wildlife products or live animals.

12.4.1 In the case of poached wild animals

Poaching can be described as the illegal killing, capturing, and removal of fish or plants from protected areas where such activities are prohibited. Criminals have used a variety of methods, including electrocution, wire traps, leg traps, poisoned arrows, and firearm firing, to poach wild animals. Wildlife crimes in the field are usually spotted by forest guards and forest rangers, so they are considered as first responding officers, as they first encounter the crime scene. The first responding officers figure out whether any human or animal involved in the crime requires medical or veterinary treatment. If such action is required, it should be done in such a way that ensures minimal to no alteration of the crime scene and contamination of the physical evidence. Following that, the first responding officer seals the crime scene using crime scene tape, establishing a boundary that restricts the movement of any unauthorized person or vehicle. Securing the scene of crime is one of the key measures since it protects the crime scene and physical evidence from being manipulated and destroyed. The first responding officer relinquishes all control of the crime scene to an appointed investigating officer (IO). The IO primarily surveys the whole crime scene to establish boundaries and provide tools and strategies for searching for evidence. A thorough record of the crime scene is kept through handwritten notes, sketches, photos, and videos. Three sorts of photos should be taken at the

Table 12.1 Sample Collection and Storage Techniques

Sample Types	Collection Procedure	Storage
Tissue	Wear gloves and use a sterile scalpel or scissors to collect a small sample (1 cm^3) of muscle tissue in a container and label it.	Fill half of the container with 70% ethanol and store it in a cool, dry place or freeze it at -20°C for long-term storage.
Blood	Wear gloves, collect blood using a syringe or capillary tube, and transfer the blood to a labelled Vacutainer® tube	Store the tube in a refrigerator at 4°C for short-term storage or freeze it at −20°C for long-term storage
Saliva (Buccal swab)	Wear gloves, use a sterile cotton swab to collect saliva from the inside of the mouth and place the swab in a labelled tube containing a preservation solution (e.g., RNA later)	Store the tube at room temperature or 4°C or according to the manufacturer's instructions
Hair	Wear gloves, use sterile scissors and forceps to cut a small sample of hair close to the skin, or pluck the hair with root	Place the hair in a clean, labelled envelope or container. Store the container in a cool, dry place.
Teeth	Wear gloves and extract a tooth using sterile instruments. Place the tooth in a labelled container with a preservation solution (e.g., 70% alcohol).	Store the container in a cool, dry place or freeze it at −20°C for long-term storage.
Bone	Wear gloves and use a sterile saw or drill to collect a small sample of bone. Place the bone sample in a clean, labelled container.	Store the container in a cool, dry place or freeze it at −20°C for long-term storage
Faeces/Scat	Wear gloves and use a sterile forceps or spatula to collect a small sample of faeces. Place the faeces in a clean, labelled container.	Fill half of the container with 70% ethanol and store it in a cool, dry place or freeze it at −20°C for long-term storage.

crime scene: wide-angle shots that give an overview of the crime scene, middle-range shots that show the relative position of the evidence, and close-up shots that give a comprehensive picture of the evidence. IO and its team members conduct a systematic and comprehensive inspection of the crime scene utilizing the desired search method, which includes the spiral, wheel, grid, strip, and zone methods. Physical evidence that can be found at the crime scene includes footprints, tyre marks, bullets, cartridge shells, cloths/fibres, tools (leg traps, arrows, wires, and spade), empty liquor bottles, and bidi/cigarettes. To facilitate future investigations, each physical evidence should be precisely marked in the sketch and documented with a scale and an evidence label by photography and videography. Physical evidence should be packed separately to prevent contamination and marked with the date and location of collection, and given a unique number for effective record keeping. In the presence of animal carcasses, a necropsy (post-mortem) must be performed to ascertain the probable cause and time of death. Biological samples (body fluids, keratinized tissues, and viscera) must be appropriately gathered, packed, and labelled for subsequent forensic analysis as mentioned in **Table 12.1**. The chain of custody for physical evidence should be properly documented and maintained until it is presented in court (FWG, 2014; Negi, 2013; Sinha, 2010).

12.4.2 In the case of seized wildlife products

When seizures of wild plants, animals, or items derived from them occur, the officer in charge must verify certain details, such as whether the species can be positively identified and, if so, what its name is. Is it protected under the Wildlife (Protection) Act, 1972, and CITES? If so, where is the species listed in the schedule or appendix and its serial number? Verify whether the owner of these wildlife specimens has the required documents or permissions. To ensure the authenticity of the documents or permissions, they must be examined and verified by contacting the granting body. If documents or permissions are found, the details of the products listed in the documents should match the confiscated products. Since some species can be poisonous, caution must be given when handling live animals, and professional veterinary assistance should be obtained if necessary. The individual should be taken under custody, and the confiscated wildlife or products should be forwarded to the appropriate institution or organization for additional examination following the completion of proper documentation, which includes handwritten notes, photos, videos, and audio recordings (FWG, 2014; Huffman and Wallace, 2012; Linacre, 2009; Negi, 2013; Sinha, 2010).

12.5 Laboratory analysis

Wildlife forensics employs a variety of analytical techniques to identify the species concerned with the crime and establish a link between the victim, suspects, and crime through the collected physical evidence. The majority of the time, forensic wildlife examiners are primarily asked to identify the species of seized wildlife specimens, parts, or derivative items. However, they may also be asked to determine the sample's age, geographical location, and individuality (FWG, 2014; Sinha, 2010). Wildlife forensics faces unique challenges, unlike human forensics, due to its broad taxonomic scope and specific legal framework (Huffman and Wallace, 2012). In legal proceedings, species identification is necessary as it establishes if the confiscated wildlife or derived product belongs to a protected species. Species-level identification is a straightforward task when dealing with whole animals. However, it becomes a laborious and challenging task when dealing with body parts or derived products that have been altered or modified to the point where they lack distinguishing characteristics. This calls for a wide range of scientific tools and methods (**Figure 12.1**), such as microscopic, spectroscopic, chromatographic, and molecular methods for species identification (Gouda et al., 2020; Nishant et al., 2017).

12.5.1 Microscopic analysis

Generally, various species of animals, plants, or their parts possess distinct morphological characteristics that may be recognized and studied for their identification. Microscopic techniques are routinely used to compare numerous morphological as well as histological characteristics of the questioned samples with those of reference samples/databases. This technique is economical and time-effective, and for a wide range of samples, microscopic examination is the initial step before applying any other sophisticated techniques. Microscopy can be used to examine a wide variety of wildlife items, such as hair, beaks, claws, skin, bone, ivory, wood,

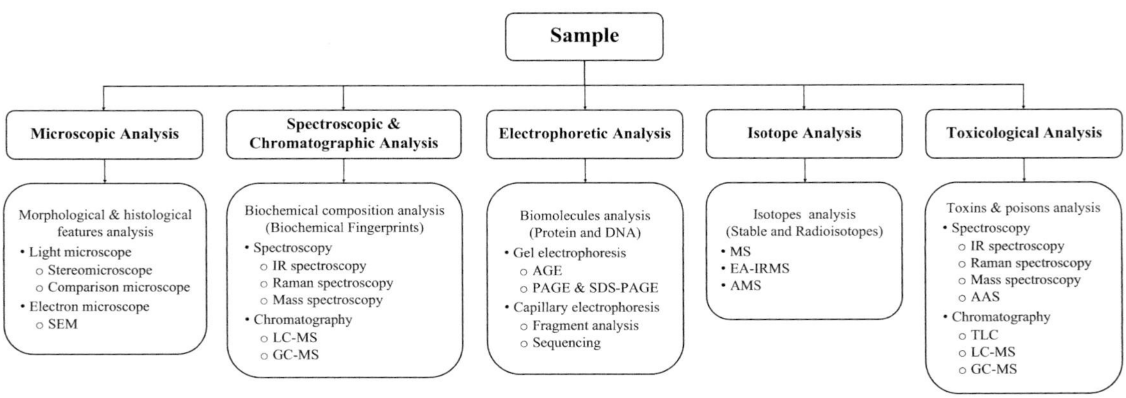

Figure 12.1 *Flow chart for analysis of wildlife-related samples.*

etc. Light microscopes used to examine these wildlife products comprise a stereomicroscope, which offers a low-magnified 3-dimensional image, and a comparison microscope, which allows for a side-by-side comparison of reference and questioned samples. The scanning electron microscope (SEM) can also be used if a highly magnified image with a very high resolution is necessary, although it is not often used in ordinary casework because of its high cost (Linacre, 2009).

There have been several published studies that use microscopic methods to identify the species from wildlife products. Four Indian bear species and ten Indian bovid species can be distinguished from one another using guard hair morphological characteristics, including texture, colour, thickness, cuticle scale pattern, scale margins, scale count index, medullary index, medulla type, and hair cross-section (Sahajpal et al., 2008, 2009). Guard hairs of six endangered Indian wild cats were differentiated by Chandel et al. (2008) based on varying morphological features using light microscopy. Yates et al. (2010) discovered that the tail hairs of African elephants, Asian elephants, and giraffes can be distinguished from one another using three microscopic characteristics, such as pigment distribution and density, and cross-sectional shape. Dehury et al. (2019) used a light microscope and an SEM to compare the medullary index and interscale distance of hair samples from Indian mouse deer, Blackbuck, Hog deer, Spotted deer, Sambar deer, and Four-horned antelope. Ivories from the archaeological sites can be differentiated from bone or antlers by examining histological characteristics utilizing SEM, high-resolution X-ray tomography, and light microscopy. Examination of the Schreger angle by light microscopy can be utilized for the differentiation of ivory from other similar materials (bone and antlers), elephant ivory from other animal ivory (walrus and hippopotamus), and ivory from different species of elephant (Asian and African) (Cartier et al., 2020; Tolksdorf et al., 2015; Turner-Walker and Xu, 2014). In 2020, Dizaj et al. examined the microscopic characteristics (shape, focus, expansion, lateral and posterior filed, number of continuous and discontinuous striae, and orientation and form of radii) of scales belonging to 21 clupeid species and found that these morphological features are consistent with phylogenetic relationships of the 21 clupeid species and hence could be used for genus and species level identification. Some researchers have also utilized scanning electron microscopy (SEM) to study the surface morphology of hair. For instance, Rollins and Hall (1999) analyzed scale patterns in the wool hairs of Ibex, Pashmina, and Tibetan antelope using SEM. Bahuguna and Mukherjee (2000) conducted an SEM study on the cuticular pattern of guard hair of the *P. hodgsonii*. Furthermore, various guides have been published for species identification based on hair characteristics of mammals from different regions, such as Mayer (1952) for mammals of California, Adorjan and Kolenosky (1969) for mammals of Ontario, Brunner and Coman (1974) for mammals of Australia, Moore et al. (1974) for mammals of North America, and Bahuguna et al. (2010) for mammals of India. Morphological examination of the nasal skin using SEM can also be utilized for species identification of some selected animal species (Madkour and Abdelsabour-Khalaf, 2022). Mohammed et al. (2022) utilized light microscopy to identify various species-specific histological characteristics of skin to differentiate a few domestic animals. Microscopic examination of thin cross-sectional studies of wood can be utilized for timber identification. Several studies have been performed for timber identification and discrimination by examining and comparing various microscopic features (shape, size, and arrangement of various cells and tissues) using computer-assisted tools (convolutional neural networks, principal component analysis, and linear discriminant analysis) for fast and reliable identification in an objective manner (Lens et al., 2020; Macchioni, 2011; Rosa da Silva et al., 2017). The microscopic method has some limitations, such as requiring experience in the field and a well-maintained reference database. Additionally, it failed in the case of modified samples where morphological characteristics are compromised.

12.5.2 Spectroscopic and chromatographic analysis

Spectroscopic and chromatographic techniques can be employed to examine the biochemical makeup of various wildlife samples including hair, bone, ivory, antlers, shells, scales, horns, etc. These techniques can provide a comprehensive overview of the sample's entire chemical makeup (biochemical fingerprints) that can be employed for species identification and discrimination (Severcan et al., 2013). IR spectroscopy with distinct sampling accessories has been used in a variety of research published worldwide to distinguish between natural and artificial hair (Manheim et al., 2016), elephant and giraffe tail hair (Espinoza et al., 2008), dried blood stains (Sharma et al., 2023), sea turtle and bovid keratin artifacts (Espinoza et al., 2007), ivories from various animals (Shimoyama et al., 2004), etc. Sharma et al. performed two independent studies to differentiate Royal Bengal tiger and Indian leopard claws (2019a, 2019b), and shahtoosh wool from pashmina/cashmere wool (2022) using ATR-FTIR spectroscopy. Recently, Bhatia et al. (2024) utilized ATR-FTIR spectroscopy to distinguish between hair samples of Royal Bengal tiger, Indian leopard, and Snow leopard. IR spectroscopy can also assist in combating illegal timber logging by recognizing and discriminating different timber species and their geographic provenance as the chemical makeup of trees, even those belonging to the same species, varies depending on where they are located (Dormontt et al., 2015; Lazarescu et al., 2017; Sandak et al., 2011; Sharma et al., 2020). Similarly, Raman spectroscopy has also gained prominence in the field of wildlife forensics and can be used to identify and discriminate species from dried blood (Virkler and Lednev, 2009), ivory specimens from various animals and archaeological sites (Edwards et al., 1997; Edwards et al., 2006a, 2006b), wood specimens (Gerasimov et al., 2016; Yang et al., 1999), mammalian and avian keratinized tissues (Akhtar and Edwards, 1997), mammalian and reptilian keratinized tissues (Edwards et al., 1998), etc. In order to determine the species of origin of the furs on clothing items of the Tyrolean Iceman Oetzi, Hollemeyer et al. (2012) used TOF-MS (time of flight mass spectrometry) to construct a PMF (peptide mass fingerprint) of keratin protein using the MALDI (matrix-assisted laser desorption ionization) technique. In 2013, Paolella et al. used peptide analysis by enzymatically breaking down wool keratin to distinguish between cashmere, yak, and sheep wool fibres using ultra-performance liquid chromatography-mass spectroscopy (UPLC/MS) with electron spray ionization (ESI). Proteomic analysis was performed using LC-MS/MS to determine the taxonomy of several ivory and bone objects that were obtained from the Metropolitan Museum of Art and Mammalogy (American Museum of Natural History). Since proteins tend to persist longer in mineralized tissues than DNA, they are more valuable for taxonomic and phylogenetic research of ancient samples compared to DNA sequencing (Gilbert et al., 2024). In order to distinguish African elephant ivory from other dentine (dugong, sperm whale, and Asian elephant) and bone (koala, cow, and sheep) samples, Ueland et al. (2020) successfully performed volatilome (volatile organic compounds) profiling using 2D gas chromatography (GC × GC) in conjunction with the time-of-flight mass spectrometry (TOF-MS) and chemometric tools. The Australian Government's Department of Environment and Energy provided eight suspected ivory samples, which were subjected to volatilome and DNA profiling analyses. Six suspected samples were successfully recognized as African ivory products by volatilome analysis, while seven samples were identified as African ivory products by DNA analysis. Additionally, this approach was used to identify and discriminate white and black rhinoceros horn samples (Ueland et al., 2016). When combined with artificial intelligence-based methods like chemometrics and machine learning, spectroscopic and chromatographic techniques can rapidly, reliably, and objectively analyze huge intricate datasets without biases in the outcomes.

12.5.3 Molecular analysis

The application of electrophoresis allows for the species identification of suspected wildlife or derived goods through the examination of proteins (including allozyme) and nucleic acids (DNA). These methods can be used to identify and distinguish between species, breeds, and cultivars as well as to authenticate plant and animal foods (Mayer and Fiechter, 2013). Prior to the emergence of DNA-based approaches, protein profiling was frequently utilized for molecular-based species identification (Pereira et al., 2008). Mardini (1984) used agarose gel electrophoresis to study protein-banding patterns in the skeletal muscles of different animals. The study's findings suggested that protein banding patterns can potentially be used to differentiate Cervidae species from each other as well as from other domestic and wild mammals; however, samples from bison, sheep, goats, and muskox could not be distinguished from each other using these patterns. Protein band patterns generated from electrophoretic separation in conjugation with a computer-assisted imaging tool from hair keratin were utilized by Marshall et al. (1985) to identify and distinguish hairs from a variety of species, including merino sheep, man, mouse, guinea pig, kangaroo, rat, elephant, dog, and cat. Day in 1972 employed polyacrylamide disc electrophoresis to study the hair proteins of 23 different animals. He found that while keratin patterns varied throughout species, there were some similarities between closely related species. The use of protein-based approaches is limited by the fact that proteins denature in stressful situations (such as heat treatment) and various proteins express differently in different tissues (Pereira et al., 2008; Saez et al., 2004).

DNA exhibits several advantages over proteins that make it a suitable candidate for species identification as it is found in all biological materials having nucleated cells or non-nucleated cells having plasmid or mitochondria, it is a stable molecule and can be extracted from the biological materials that have been subjected to stress, and DNA carries more information than protein due to codon degeneracy and non-coding regions (Pereira et al., 2008). DNA analysis is a highly effective and accepted tool in wildlife forensics, bolstering prosecution and law enforcement efforts. Consequently, ensuring confidence in the investigative results is paramount (Linacre and Tobe, 2013; Ogden et al., 2009). Traditional morphological methods have sometimes struggled to identify the source of origin of illegally collected wildlife products. To address this challenge, DNA analysis has become increasingly important and effective in monitoring the implementation of national and international agreements regulating wildlife harvest and trade (Haines et al., 2021). The 1980s and 1990s witnessed the emergence of nuclear DNA (nDNA) and mitochondrial DNA (mtDNA) profiling for forensic analysis, which was crucial for the illegal wildlife trade investigations (Taupin, 2004). mtDNA analysis is recommended over nDNA analysis for assessing genetic makeup to identify species because of its large copy number in a cell and improved protection from degradation due to the stiff mitochondrial membrane. This makes it more readily available in biological materials that have been treated, buried, keratinized, or extensively damaged (Bellis et al., 2003; Mitra et al., 2018). mtDNA is a circular DNA molecule that measures 16.5 kb in the majority of animals and more than 200 kb in plants. The majority of plants' bigger genome sizes are made up of non-coding portions that have nothing to do with gene expression. mtDNA contains 37 genes, 13 of which encode proteins required for oxidative phosphorylation, 22 of which encode tRNAs, and two of which encode 16S and 12S rRNA. Animal mtDNA markers include the cytochrome *b* (Cyt *b*) gene, cytochrome c oxidase subunit I (COI) gene, control region or D loop, 12S, and 16S rRNA regions, with the Cyt *b* and COI genes being the most commonly employed. In the case of plants, typically two plastid markers, rbcL (ribulose-1,5-bisphosphate carboxylase/oxygenase), and matK (Maturase K), as well as one nuclear DNA marker, ITS2

Table 12.2 Commonly Used Molecular Techniques in Wildlife DNA Forensics

Molecular Genetics Techniques	Applications in Wildlife Forensics
Mitochondrial DNA Sequencing	Species identification, maternal lineage tracing, geographic origin, phylogenetic relationships, studying population structure and diversity, evolutionary relationships
Y chromosome analysis	Sex determination, paternal testing and lineages, tracing paternal lineages, determining the sex of individuals, kinship analysis, and paternal bio-geographic analysis.
Microsatellite genotyping	Includes both short tandem repeats (STR), single nucleotide polymorphisms (SNP) genotyping for individual identification, population genetics, phylogenetics, studying genetic variations between populations, determining genetic structure relationships

(internal transcribed spacer) are utilized (Cooper and Adams, 2022; Guide for Forensic Timber Identification, 2016; Mitra et al., 2018).

DNA-based techniques involve different steps: DNA extraction, quantification, PCR amplification of the required nuclear or mtDNA marker (**Table 12.2**), sequencing, and finally comparing the amplified marker sequence to the reference database or DNA databanks (NCBI/EMBL/DDBJ). Routinely, DNA-based markers are utilized to identify species of various wildlife products, including scales, hair, claws, ivory, blood, horn, traditional medicines, meat/poultry products, timber, etc. With the help of the COI gene and control region, Jun et al. (2011) were able to link hair samples of suspected Asiatic black bear to *Ursus thibetanus* (Asiatic black bear) and seized Felidae leathers to canine leather using the Cyt *b* gene. Abe et al. (2012) used nDNA (AgGT17 locus) and mtDNA (COI gene) markers to identify the species of 15 confiscated feather specimens. Only 10 samples yielded the COI gene sequence, which was shown to have similarities with the Ara macao sequence using a BOLD search. The distribution pattern of AgGT17 alleles in the confiscated samples matched with that of *Ara macao*, confirming the findings of the COI gene. Lee et al. (2006) amplified 271 bp region of the cyt *b* gene from 4 wool samples (shahtoosh, pashmina, cashmere, and sheep wool) and compared the sequences with GenBank and EMPL data. It was discovered that shahtoosh wool showed 100% similarities with Pantholops hodgsonii, Pashmina and cashmere wool showed 100% similarities with Capra hircus, and sheep wool showed similarities with Ovis aries. A molecular test using DNA sequencing of the Cyt ***b*** gene from guard hair in positive shahtoosh shawls confirmed the hair as belonging to *Pantholops hodgsonii* (Gupta et al., 2022).

On the contrary, STR and SNP markers are frequently utilized for individualization and sex determination of animals (Guidelines on Methods and Procedures for Ivory Sampling and Laboratory Analysis, 2014). When it comes to timber identification, SNP markers are favoured over STR markers because polyploidy levels in plants are simply unknown and DNA yield from timber is typically not optimal for STR typing (Guide for Forensic Timber Identification, 2016). A multiplex genotyping (*tigris* Plex) technique consisting of 22 STR loci and one SRY gene (sex identification) was developed by Zou et al. (2015) to identify tiger subspecies, sex, and even individuals to track the origin of seized wildlife items. White et al. (2012) created a population database for white and red-tailed black cockatoos (*Calyptorhynchus* spp.) using 20 polymorphic STR loci, which they subsequently utilized to solve three different cases. Using scats and hair as a source of DNA, Eiken et al. (2009) identified and used 12 STR loci to build a population database of *Ursus arctos* (brown bears) throughout Northern Europe. Dormontt et al. (2020) genotyped 394 individuals of *Acer macrophyllum* (Bigleaf Maple), an illicitly traded wood species in the United States (US), to create a population database containing 128 SNPs and 3 insertion/deletion (INDEL) markers. Using data from 1081 individuals taken from 27 forest reserves in Peninsular Malaysia, Tnah et al. (2010) developed a 12 STRs-based DNA database of *Neobalanocarpus heimii* (Chengal). The use of the DNA-based methodology in wildlife forensics has grown as a result of substantial advancements in DNA analysis technology, including automated DNA extraction, speedy DNA quantification and amplification, and powerful sequencing methods.

12.5.4 Isotopes analysis

Isotopes are distinct forms of an element that vary in the number of neutrons they possess, which affects the atom's molecular mass. The vast majority of isotopes on Earth are stable, but a few are radioactive and decay over time, with a known half-life. In nature, the ratios of stable isotopes, including sulphur (^{34}S), strontium (^{90}Sr), nitrogen (^{15}N), oxygen (^{18}O), hydrogen (^{2}H), and carbon (^{13}C), vary and are influenced by biological, climatic, and geographical factors. The location, food, mobility, and behaviour of the wild animal are all possible to discover through the analysis of these isotopes. Geographic location identification can also be used to pinpoint hotspot areas with a high frequency of wildlife crimes (Bell, 2011). Vogel et al. (1990) used mass spectroscopy and multivariate analysis to determine the isotopic composition of ^{13}C, ^{15}N, $^{86}Sr/^{87}Sr$, and $^{204}Pb/^{206}Pb/^{207}Pb/^{208}Pb$ to differentiate between African elephant bones and ivory samples that were gathered from several places in southern Africa. Ziegler et al. (2016) developed isotopic fingerprinting data of five stable isotopes (^{13}C, ^{15}N, ^{18}O, ^{2}H, and ^{34}S) from 507 ivory samples (African and Asian elephant range) using continuous flow isotope ratio mass spectroscopy that can be used to predict the geographic origins of unknown ivory samples. Using EA-IRMS (elemental analysis isotopic ratio mass spectrometry), Watkinson et al. (2022) identified timber samples obtained from two forest concessions in Gabon based on isotopic analysis of ^{13}C, ^{15}N, ^{18}O, ^{2}H, and ^{34}S. Radioisotope ^{14}C, also known as radiocarbon, is a naturally occurring isotope that decays with a half-life of 5730 ± 40 years to generate stable isotope ^{14}N. All living things take in carbon during their lifetimes; once it is absorbed, the carbon stays within the body of the organism. The radiocarbon age of wildlife products can be determined by calculating the ^{14}C and ^{12}C atom ratios (for mass-dependent fractionation adjustment, the ^{13}C and ^{12}C ratios should be computed separately) and comparing them with the known standards (Guide for Forensic Timber Identification, 2016). Through the use of accelerator mass spectroscopy (AMS), Dunbar et al. (2016) detail the main processes (sample preparation to sample analysis) used by the Scottish Universities Environmental Research Centre (SUERC) Radiocarbon Dating Laboratory for ^{14}C dating of various samples, including some forensic samples (human remains, rhinoceros horn, ivory, and endangered animals' fur). Quarta et al. (2021) compared the ^{14}C dating data for bone and ivory samples from three laboratories in Italy, Switzerland, and Hungary. Italy laboratory used a 3MV Tandetron™ type accelerator, whereas Switzerland and Hungary laboratories used the Mini Carbon Dating system (MICADAS) AMS technology. This comparative experiment produced highly reproducible results. This information is invaluable for law enforcement, conservation initiatives, and government agencies combating wildlife crimes involving various body parts of scheduled and endangered species. However, there are certain drawbacks associated with this technique, including the possibility of sample contamination from other carbon-containing materials, selection of the sample, labour-intensive sample pre-treatment procedures, and reservoir effects (Wright, 2017).

12.5.5 Toxicological analysis

Forensic toxicology can be helpful in identifying poisonous or toxic substances in case of accidental/deliberate animal poisoning and identification of chemical substances (perfumes, traditional Chinese medicines, drugs) illegally synthesized from animal body parts (Lee et al.,

2016). Depending on the circumstances and the examiner's request, a variety of samples, including blood, tissue, organs, stomach contents, food material, water samples, and confiscated wildlife goods, may be used for toxicological examination. Atomic absorption spectroscopy (AAS), thin layer chromatography (TLC), GC, MS, HPLC, and GC are a few analytical instruments that can be used for toxicological examination. Poisoning poses a severe threat to wildlife species in two ways: first, by using poisoned baits for illegal hunting of animals, and second, by consuming pests that have been poisoned by pesticides. However, the decline in the population of non-targeted species (scavengers and raptorial birds) due to the consumption of poison raised serious concerns among biologists and conservationists.

The majority of wildlife poisoning instances in India are related to pesticides such as carbamate, organochloride, organophosphate, and rodenticides. However, poison-related deaths can go unreported due to the absence of samples for analysis, as carcasses are often discovered in advanced stages of decomposition, and poisoning is not always thought to be a plausible cause of death. A study from Kerala that examined animal fatalities that occurred at the Periyar Tiger Reserve between January 2011 and March 2013 attempted to address the issue of wildlife poisoning. It was noted that 11 mortality instances were reported, of which 3 cases had poisoning identified by TLC and GC-MS, and 1 case had poisoning identified by visual inspection of the stomach contents (Radhakrishnan, 2018). Mateo-Tomas et al. surveyed a forest-covered region in Northwest Spain between 2000 and 2010, observing that 112 poisoning cases were reported, including 228 individuals from 25 different species, the majority of which were predator birds, followed by mammalian carnivores. Wildlife poisoning can also result from the conflict between humans and wildlife, as the unregulated expansion of human settlements inside the wildlife habitat and dependence on the forest for fuel wood and fodder bring them close to the wild animals. Human-wildlife conflict (HWC) involves crop loss, livestock depredation, and injuries, including fatal ones, that cause humans to use poisons such as pesticides and rodenticides to the killing of problematic animals (Gureja et al., 2019; Millins et al., 2014; Ragothaman and Chirukandoth, 2011).

Furthermore, cooperative work by biologists, chemists, and toxicologists over the past 50 years has resulted in notable improvements in understanding the exposure and toxicology of polychlorinated biphenyls (PCBs), polychlorinated di-benzodioxins (PCDDs), and polychlorinated dibenzofurans (PCDFs) in birds (Rattner et al., 2011). This interdisciplinary effort has been particularly successful in identifying the chick oedema factor and investigating various aspects of these compounds' effects on avian species.

12.5.6 Forensic entomology

In wildlife forensics, entomology plays a crucial role in legal investigations involving wildlife by applying insect biology and studying arthropod relatives. This field is particularly important in cases related to poaching, illegal wildlife trade, animal attacks, and other wildlife-related crimes. It contributes significantly to the investigation process, including estimating the time since death, analyzing poisoning and toxicology, studying animal attacks, investigating wildlife trafficking, and examining habitat and migration patterns (Haines et al., 2021). Therefore, proper preservation, collection, and documentation of entomological evidence are essential to ensure its admissibility in court and its usefulness in assisting investigators and legal professionals in solving crimes (Sanford et al., 2019). One-way entomologists contribute to wildlife forensics is by studying the geographic distribution of insects. This enables them to determine the geo-location of wildlife parts or bodies and identify whether remains have been moved. The field of entomology in wildlife forensics is already well-established in criminal and civil courts and can seamlessly integrate with investigations focused on protected wildlife.

In addition, insects can provide evidence of trauma or foul play. For instance, certain species of flies attracted to open wounds can indicate injuries sustained before or after death. To ensure the admissibility of entomological evidence, forensic entomologists must follow established protocols, maintain accurate records, and demonstrate their expertise in the field (Hall, 2019).

12.6 DNA reference database

The accuracy and reliability of results derived from any database depend on the quality of the reference data. Therefore, validating reference datasets is crucial to ensure dependable outcomes (Ogden et al., 2009). The establishment of DNA databases in wildlife forensics has revolutionized the fight against wildlife crime by providing essential tools for species identification, individual profiling, and population monitoring. These databases leverage genetic markers such as mitochondrial DNA sequences and microsatellites, enabling accurate species identification through techniques like DNA barcoding (Hebert et al., 2003; Kumar et al., 2017). Initiatives like the Barcode of Life Data Systems (BOLD) have developed comprehensive repositories of DNA barcodes, which are essential for identifying confiscated wildlife products. Within five years, and upon completion of Phase 1, the International Barcode of Life Consortium (iBOL) successfully generated DNA barcodes from five million specimens, encompassing at least 500,000 species (The International Barcode of Life Consortium, 2024). Moreover, the NCBI (National Centre for Biotechnology Information), EMBL (European Molecular Biology Laboratory), and DDBJ (DNA Data Bank of Japan) databases are central repositories for genetic sequence data, including DNA and protein sequences. These databases are widely used by researchers worldwide for various purposes, including the construction of localized reference data sets in wildlife forensics. This process involves retrieving relevant sequences from the NCBI, EMBL, or DDBJ databases and aligning them to create a comprehensive reference data set. This reference data set can be used for genetic analysis, species identification, and population studies in wildlife forensics, as shown in **Figure 12.2**. This vast amount of data will contribute significantly to developing an exceptional identification system, which can be applied to address real-world challenges such as forensics, conservation efforts, species management, disease control, and ecosystem monitoring. Databases like the Rhinoceros DNA Index System (RhODIS) in India (Ghosh et al., 2021) and South Africa (Harper, 2021) are examples of how individual and population profiling of rhinos is applied.

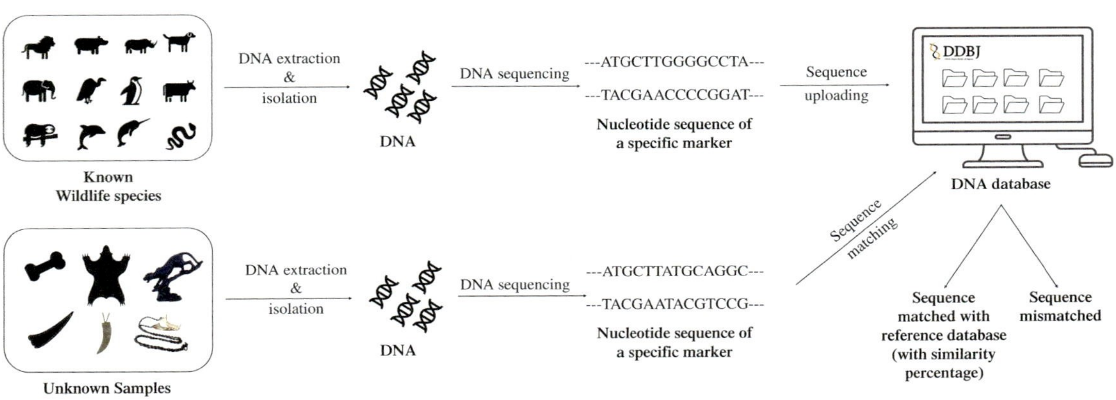

Figure 12.2 *Working of DNA databases.*

RhODIS, a forensic tool, is expected to enhance wildlife crime investigations by providing DNA profiles of individual rhinos, which facilitates tracking and identification of animals involved in poaching incidents.

Additionally, the Convention on Biological Diversity implemented the Nagoya Protocol in 2014 (Adachi et al., 2014), emphasizing the importance of fair and equitable access to genetic resources (Pérez-Espona and CryoArks, 2021). As a result, biobanks are now widely recognized as valuable national assets for research purposes and their contribution to wildlife conservation (Lee et al., 2016). These databases have proven invaluable in legal contexts, providing strong scientific evidence linking seized wildlife products to specific poaching events, thereby supporting prosecutions and strengthening conservation efforts. Integrating these advanced genetic tools in wildlife forensics highlights the importance of interdisciplinary collaboration and continuous development to improve the effectiveness of wildlife protection laws and initiatives.

12.7 Trial of cases

The Wildlife (Protection) Act 1972 states that wildlife is state property and that anyone who commits a crime against it faces penalties. Case trials are held based on complaints filed under section 55 of the Wildlife (Protection) Act 1972, as per the guidelines provided by sections 244 to 248 of the Criminal Procedure Code. To prevent delays in the case proceedings, both the public prosecutor and the investigatory officer (IO) must be present in the courtroom during case trials. In the courts, witnesses, other pieces of evidence, and reports from wildlife forensic examiners must be submitted against the person being charged. The forensic examiner's report includes the identification of the species involved in the crime, as well as its origin, age, cause of death, and potential suspects. The court can also summon forensic experts to testify as expert witnesses. The trial court issues a judgment of conviction or acquittal based on the arguments, facts, and witnesses presented by the public prosecutor and the defence. This judgment should be obtained and reviewed by the concerned deputy/assistant conservator of forest (DCF/ACF) (Negi, 2013).

12.8 Law enforcement

Law enforcement in wildlife forensics involves applying forensic techniques and principles to investigate wildlife-related crimes, such as poaching, illegal trade, and habitat destruction. It plays a crucial role in identifying, apprehending, and prosecuting individuals involved in wildlife crimes. The illicit exploitation of many endangered species is the result of increasing demand for traditional Chinese medicines (TCMs), ornaments and decorative items, skin as a leather product, eggs and meat for local consumption, superstitious beliefs, aphrodisiac use, and retaliatory killings (Koch et al., 2013; Negi, 2013; Sharma et al., 2019a, 2019b). The existing legal framework to regulate and restrict illegal trading is robust; however, there is a pressing need for even stronger and stricter regulations, coupled with effective implementation, to protect biodiversity adequately. The Hailey National Park Act of 1936, which established Hailey National Park (now known as Corbett National Park) in Uttar Pradesh, is considered one of the earliest comprehensive laws to protect wildlife and its habitat. The government has enforced certain action plans in addition to the Acts above to protect and safeguard the rights of these animals and plants. However, stronger implementation forces are needed to address wildlife trade effectively. A brief account of these laws for wildlife protection (Mehra, 2012) is provided below.

12.8.1 Elephant Preservation Act, 1879

The Elephant Preservation Act, 1879 (Act No. 6 of 1979) was an early legislative measure enacted in British India to conserve elephants. The main objective of the Act was to prohibit the indiscriminate killing and capturing of elephants, acknowledging their cultural and environmental significance. The key provisions of the Act included the regulation of hunting, whereby the Act prohibited the hunting, killing, or capturing of wild elephants unless in defence of himself or someone else. This was intended to curb the overexploitation of elephant populations. Additionally, the Act established a licensing system, requiring individuals who wished to hunt or capture elephants to obtain a license. This system aimed to control and monitor the number of elephants being taken from the wild. To deter illegal activities, the Act imposed penalties for violations, including fines (up to five hundred rupees) and imprisonment for killing, injuring, or capturing elephants. This Act was an important step towards wildlife conservation in colonial India, setting a precedent for future wildlife protection laws.

12.8.2 The Wild Birds Protection Act, 1887

The Wild Birds Protection Act, 1887, was one of the early legislative measures aimed at safeguarding avian species in British India. The Act primarily focused on protecting wild birds and prohibiting the sale of any kind of birds during the breeding season to prevent their numbers from declining due to hunting and egg collection. Violations of the Act were subject to fines, serving as a deterrent against illegal activities. This legislation marked a significant step towards the conservation of bird species, recognizing their ecological importance and setting the groundwork for future wildlife protection laws. However, the Act had notable limitations, such as its jurisdiction was restricted to territories within cantonments and municipalities, leaving vast areas unprotected. Additionally, it did not cover a large number of bird species, thereby limiting its effectiveness in curbing the widespread killing of birds. Despite these shortcomings, the Act was an early attempt to address wildlife conservation in British India.

12.8.3 Wildlife (Protection) Act, 1972

The Wildlife (Protection) Act (WPA), 1972, (53 of 1972) marks the first comprehensive legal framework for the safeguarding and management of wildlife and their habitats. It was passed by parliament and received agreement from the President on September 9, 1972. This Act applies to all Indian states, except for Jammu and Kashmir. However, Jammu and Kashmir has its own law known as the Jammu and Kashmir Wildlife (Protection) Act, 1978. The objective of this convention is to ensure that international trade in specimens of wild animals and plants does not threaten their survival. Further, the WPA, 1972 has undergone many amendments: The Wildlife (Protection) Amendment Act of 1982 (Act 23 of 1982), the Wildlife (Protection) Amendment Act of 1986 (Act 28 of 1986), the Wildlife (Protection) Amendment Act of 1991 (Act 44 of 1991), the Wildlife (Protection) Amendment Act of 2003 (16 of 2003), the Wildlife (Protection) Amendment Act, 2006. The latest amendment, the Wildlife (Protection) Amendment Act of 2022, has been in effect since April 1st, 2023. The Wildlife (Protection) Amendment Act of 2022 introduced significant changes to the Act. It revised the schedule status for species, ensuring better protection and management of these species. The amendment also introduced specific provisions for regulating and managing invasive alien species, recognizing their threat to native biodiversity and ecosystems. Additionally, it proposed a new schedule for species listed in the Appendices under the CITES, aligning India's wildlife protection framework with international conservation standards.

In light of the urgent need for conservation, the Act prohibits the hunting of various species of wild animals and birds and regulates the trade of wildlife products. It establishes protected areas, such as wildlife sanctuaries, national parks, and biosphere reserves, to ensure the preservation of biodiversity. Additionally, the Act provides for the creation of the National and State Wildlife Boards, which are responsible for overseeing the implementation of conservation strategies. To discourage illegal activities, the Act imposes strict penalties, including fines and imprisonment. The Wildlife Protection Act, 1972 represents a significant milestone in India's commitment to preserving its valuable wildlife heritage and has played a crucial role in the country's conservation efforts.

12.9 Landmark judgments and case studies

The Indian judiciary has been instrumental in combating illegal wildlife trade within the country. In recent years, numerous high-profile cases involving wildlife crimes have been adjudicated in Indian courts. These cases have not only set significant legal precedents but also heightened public awareness regarding the seriousness of wildlife trafficking. Below is a summary of some key judicial cases related to illegal wildlife trade in India.

12.9.1 M/S Ivory Trader and Manufacturer Association and Other vs. Union Bank of India and Other. AIT 1997 in Civil Writ Petition No. 1016 of 1992

This case was a landmark judgment by the Supreme Court of India that reinforced the country's commitment to wildlife conservation, specifically addressing the trade in ivory. The Ivory Traders and Manufacturers Association challenged the government's ban on the ivory trade, arguing that it adversely affected their livelihood and involved stocks of legally acquired ivory before the ban. They proposed that the government should regulate, rather than ban, the trade of pre-ban stocks. In response, the Union of India emphasized the critical need to protect elephants from poaching, driven by the demand for ivory. They highlighted India's obligations under the CITES to enforce strict measures against the ivory trade. The government argued that any allowance for ivory trade would fuel illegal activities and undermine conservation efforts. The Supreme Court upheld the ban, prioritizing the conservation imperative and the protection of elephants, thereby setting a significant precedent for wildlife protection in India.

12.9.2 Wildlife vs. Ashok Kumar & Others

In the bustling corridors of the Delhi District Court, a significant case unfolded on the morning of April 11, 2018. The defendants, Ashok Kumar, and two others, stood accused of a grave offense: the illegal possession and trade of leopard skin, an animal strictly protected under Schedule I of the Wildlife Protection Act, 1972. The allegations were serious, implicating the accused in activities that threatened the survival of endangered species.

The prosecution presented a compelling case, armed with forensic evidence that included a detailed report identifying the leopard skin. The forensic experts meticulously analyzed the microscopic hair characteristics of the seized skin and compared them with reference samples, conclusively proving the identity of the animal. This scientific evidence formed the cornerstone of the prosecution's argument, establishing a direct link between the accused and the illegal wildlife trade.

The defence, in turn, sought to challenge the prosecution's case, questioning the integrity of the evidence and the procedures followed during the seizure and analysis. They argued that there were gaps in the chain of custody and inconsistencies in the witness testimonies. Despite their efforts, the defence struggled to create a reasonable doubt about the guilt of the accused.

After thorough deliberation and careful examination of all the evidence, the court delivered its judgment. Recognizing the severity of wildlife crimes and the imperative to uphold legal protections for endangered species, the court found all three accused guilty. They were convicted under Sections 49 and 49(B) of the Wildlife Protection Act, 1972, punishable under Section 51.

The judgment underscored the importance of stringent enforcement of wildlife protection laws and served as a stern warning to those involved in wildlife crimes. The case not only highlighted the meticulous work of forensic experts but also reinforced the judiciary's role in conserving India's rich biodiversity. The detailed judgment provided further insights into the court's reasoning, emphasizing the need for vigilant protection of endangered species and the harsh consequences of violating these crucial laws.

12.9.3 State of Madhya Pradesh v. Rajesh Garg & Others, M.Cr.C. No. 26988 of 2019

This case is a landmark legal proceeding addressing wildlife crime in India. Originating from the illegal hunting and trade of a Pangolin in Odisha, the case involved serious charges against Rajesh Garg and other accused parties for poaching and trading wildlife parts. The prosecution presented substantial evidence, including forensic analysis and witness statements, demonstrating the organized nature of the wildlife trafficking network. Despite the defence's attempts to question the evidence and procedures, the Madhya Pradesh High Court found the accused guilty under the Wildlife Protection Act, 1972. The court sentenced them to significant prison terms and imposed hefty fines, emphasizing the severity of their crimes. This judgment set a critical legal precedent, underscoring the judiciary's role in enforcing wildlife laws and protecting endangered species. The case also raised public awareness about the illegal wildlife trade and served as a strong deterrent against future offenses, marking a significant victory for conservation efforts in India.

12.9.4 Wildlife Protection Society of India vs. Ministry of Environment and Forest & Others, W.P. (C) No. 337 of 2005 related to Sariska Tiger Reserve Case

This case is a pivotal legal proceeding in India's wildlife conservation efforts. The Wildlife Protection Society of India (WPSI) filed a public interest litigation against the Ministry of Environment and Forest, challenging the government's inadequate enforcement of wildlife protection laws and the failure to curb the poaching of tigers in the Sariska Tiger Reserve, Rajasthan. The WPSI argued that despite the Wildlife Protection Act of 1972, rampant poaching and trafficking continued, threatening India's biodiversity. The petition sought judicial intervention to compel the government to implement stricter measures and more effective enforcement strategies. The High Court, in response, directed the Ministry to take immediate and concrete actions, including strengthening anti-poaching units, enhancing surveillance in protected areas, and ensuring stricter implementation of existing laws. The court also called for increased coordination between various enforcement agencies and regular monitoring of wildlife trade activities. This case underscored the judiciary's proactive role in addressing environmental issues and reinforced the need for robust governmental action to protect wildlife. It highlighted the importance of judicial oversight in ensuring accountability and effectiveness in wildlife conservation efforts, marking a significant step towards stronger environmental governance in India.

12.10 Future prospects and challenges

Wildlife forensics is on the brink of significant advancements but also faces numerous challenges as it adapts to meet the demands of conservation and law enforcement. To move forward, developing more comprehensive reference databases for morphology and DNA forensics is crucial. These databases will accurately identify species, geographical origin, and individual animals, even from minimal and degraded samples. By harnessing advancements in wildlife forensics and bioinformatics tools, we can better trace the origins of wildlife products, identify poaching hotspots, and uncover trafficking routes. Additionally, integrating environmental DNA (eDNA) analysis will allow for real-time, non-invasive monitoring of biodiversity and detecting illegal activities.

However, these exciting advancements come with their own set of challenges. Standardizing forensic methodologies and protocols ensures consistency and reliability across various laboratories and jurisdictions. Increased funding and resources are needed to expand forensic capabilities and personnel training. Collaboration among scientists, law enforcement, and policymakers is vital to bridge the gap between scientific research and its practical application in the courtroom. Legal frameworks must also evolve to accommodate new forensic techniques and ensure that scientific evidence is admissible and effectively used in prosecutions.

Furthermore, it is crucial to address the ethical implications of forensic research, including handling sensitive data and the potential impact on indigenous communities. International cooperation and sharing of information and expertise will be essential in combating the transnational nature of wildlife crime. Despite the challenges, integrating cutting-edge science with robust legal strategies holds great promise for strengthening wildlife conservation efforts and protecting endangered species worldwide.

References

Abe H, Hayano A, Inoue-Murayama M (2012) Forensic species identification of large macaws using DNA barcodes and microsatellite profiles. *Molecular Biology Reports* 39: 693–699. DOI: 10.1007/s11033-011-0787-1.

Adachi K, Meyer H, Eugui DV, et al. (2014) The convention on biological diversity and the Nagoya protocol: Intellectual property implications. In A. RainerEngels (Ed.) *A Handbook on the Interface Between Global Access and Benefit Sharing Rules and Intellectual Property*. United Nations Conference on Trade and Development, p. 215.

Adorjan AS, Kolenosky GB (1969) A manual for the identification of hairs of selected Ontario mammals. (Research Report [Wildlife] No. 90). Ontario Department of Lands and Forests.

Agnew DJ, Pearce J, Pramod G, et al. (2009) Estimating the worldwide extent of illegal fishing. *PLoS One* 4(2): e4570. DOI: 10.1371/journal.pone.0004570.

Akhtar W, Edwards HGM (1997) Fourier-transform Raman spectroscopy of mammalian and avian keratotic biopolymers. *Spectrochimica Acta Part A: Molecular and Biomolecular Spectroscopy* 53(1): 81–90. DOI: 10.1016/s1386-1425(97)83011-9.

Andersson AA, Tilley HB, Lau W, et al. (2021) CITES and beyond: Illuminating 20 years of global, legal wildlife trade. *Global Ecology and Conservation* 26: e01455. DOI: 10.1016/j.gecco.2021.e01455.

Badola S, Gautam A (2022) High flying: Insight into wildlife trafficking through India's airports, TRAFFIC Report.

Bahuguna A, Mukherjee SK (2000) Use of SEM to recognise Tibetan antelope (Chiru) hair and blending in wool products. *Science & Justice* 40(3): 177–182. DOI: 10.1016/S1355-0306(00)71973-3.

Bahuguna A, Sahajpal V, Goyal SP, et al. (2010) Species Identification from Guard Hair of Selected Indian Mammals: A Reference Guide. Dehradun, India: Wildlife Institute of India.
Bell LS (2011) Forensic science in support of wildlife conservation efforts-morphological and chemical approaches (Global trends). *Forensic Science Review* 23(1): 30–36.
Bellis C, Ashton KJ, Freney L, et al. (2003) A molecular genetic approach for forensic animal species identification. *Forensic Science International* 134(2–3): 99–108. DOI: 10.1016/s0379-0738(03)00128-2.
Best Practice Guide for Forensic Timber Identification (2016) UNODC, UN, New York.
Bhatia D, Sharma CP, Sharma S, et al. (2024) Species discrimination from hair using ATR-FTIR spectroscopy: Application in wildlife forensics. *Science & Justice* 64(3): 314–321. DOI: 10.1016/j.scijus.2024.04.002.
Brunner H, Coman BJ (1974) *The Identification of Mammalian Hair.* Inkata Press.
Cartier LE, Krzemnicki MS, Gysi M, et al. (2020) A case study of ivory species identification using a combination of morphological, gemmological and genetic methods. *Journal of Gemmology* 37(3). DOI: 10.15506/JoG.2020.37.3.282.
Chandel RS, Thakar MK, Goyal SP, et al. (2008) Forensic hair characterization of six endangered felid species of India. *Journal of Forensic Identification* 58(3): 327–341.
Cooper GM, Adams K (2022) *The Cell: A Molecular Approach.* Oxford University Press.
Day TH (1972) Interspecific variation in the hair proteins. *Comparative Biochemistry and Physiology Part B: Comparative Biochemistry* 43(2): 361–367. DOI: 10.1016/0305-0491(72)90295-7.
Dehury S, Mishra UK, Hembram A, et al. (2019) Light and scanning electron microscopy analysis of hair samples of some wild animals for individual identification. *Journal of Entomology and Zoology Studies* 7(6): 900–904.
Dizaj LP, Esmaeili HR, Teimori A, et al. (2020) Comparative microscopic examination of scales in 21 clupeid species from the Caspian Sea and the Indo-Pacific regions. *Micron* 137: 102911. DOI: 10.1016/j.micron.2020.102911.
Dormontt EE, Boner M, Braun B, et al. (2015) Forensic timber identification: It's time to integrate disciplines to combat illegal logging. *Biological Conservation* 191: 790–798. DOI: 10.1016/j.biocon.2015.06.038.
Dormontt EE, Jardine DI, van Dijk KJ, et al. (2020) Forensic validation of a SNP and INDEL panel for individualisation of timber from bigleaf maple (*Acer macrophyllumPursch*). *Forensic Science International: Genetics* 46: 102252. DOI: 10.1016/j.fsigen.2020.102252.
Dunbar E, Cook GT, Naysmith P, et al. (2016) AMS 14C dating at the Scottish Universities Environmental Research Centre (SUERC) radiocarbon dating laboratory-corrigendum. *Radiocarbon* 58(1): 233. DOI: 10.1017/RDC.2015.2.
Edwards HGM, Brody RH, Hassan NFN, et al. (2006a) Identification of archaeological ivories using FT-Raman spectroscopy. *Analytica Chimica Acta* 559(1): 64–72. DOI: 10.1016/j.aca.2005.11.067.
Edwards HGM, Farwell DW, Holder JM, et al. (1997) Fourier-transform Raman spectra of ivory III: Identification of mammalian specimens. *Spectrochimica Acta Part A: Molecular and Biomolecular Spectroscopy* 53(13): 2403–2409. DOI: 10.1016/s1386-1425(97)00180-7.
Edwards HGM, Hassan NN, Arya N (2006b) Evaluation of Raman spectroscopy and application of chemometric methods for the differentiation of contemporary ivory specimens I: Elephant and mammalian species. *Journal of Raman Spectroscopy* 37(1–3): 353–360. DOI: 10.1002/jrs.1458.
Edwards HGM, Hunt DE, Sibley MG (1998) FT-Raman spectroscopic study of keratotic materials: Horn, hoof and tortoiseshell. *Spectrochimica Acta Part A: Molecular and Biomolecular Spectroscopy* 54(5): 745–757. DOI: 10.1016/s1386-1425(98)00013-4.
Eiken HG, Andreassen RJ, Kopatz A, et al. (2009) Population data for 12 STR loci in Northern European brown bear (*Ursus arctos*) and application of DNA profiles for forensic casework. *Forensic Science International: Genetics Supplement Series* 2(1): 273–274. DOI: 10.1016/j.fsigss.2009.07.007.
Espinoza EO, Baker BW, Berry CA (2007) The analysis of sea turtle and bovid keratin artefacts using drift spectroscopy and discriminant analysis. *Archaeometry* 49(4): 685–698. DOI: 10.1111/j.1475-4754.2007.00328.x.
Espinoza EO, Baker BW, Moores TD, et al. (2008) Forensic identification of elephant and giraffe hair artifacts using HATR FTIR spectroscopy and discriminant analysis. *Endangered Species Research* 9(3): 239–246. DOI: 10.3354/esr00125.
EUROPOL OCTA, EU Organised Crime Threat Assessment (2011) The Hague, Netherlands.
Forensic Working Group (2014) Wildlife crime a guide to the use of forensic and specialist techniques in the investigation of wildlife crime. Partnership for Action Against Wildlife Crime.
Gerasimov VA, Gurovich AM, Kostrin DK, et al. (2016) Raman spectroscopy for identification of wood species. *Journal of Physics: Conference Series* 741(1): 012131. DOI: 10.1088/1742-6596/741/1/012131.

Ghosh T, Sharma A, Mondol S (2021) Optimisation and application of a forensic microsatellite panel to combat greater-one horned rhinoceros (*Rhinoceros unicornis)* poaching in India. *Forensic Science International: Genetics* 52: 102472. DOI: 10.1016/j.fsigen.2021.102472.

Gilbert C, Krupicka V, Galluzzi F, et al. (2024) Species identification of ivory and bone museum objects using minimally invasive proteomics. *Science Advances* 10(4): eadi9028. DOI: 10.1126/sciadv.adi9028.

Gouda S, Kerry RG, Das A, et al. (2020) Wildlife forensics: A boon for species identification and conservation implications. *Forensic Science International* 317: 110530. DOI: 10.1016/j.forsciint.2020.110530.

Guidelines on Methods and Procedures for Ivory Sampling and Laboratory Analysis (2014) UNODC, UN, New York.

Gupta SK, Sharma CP, Singh B, Kumar A (2022). Detection of fraudulent in contraband Shahtoosh trade: Unfair deal in illicit wildlife market. *Forensic Science International: Genetics* 59: 102713. DOI: 10.1016/j.fsigen.2022.102713.

Gureja N, Kumar A, Saigal S, et al. (2019) Human-Wildlife Conflict in India, Report prepared for National Biodiversity Strategy and Action Plan (NBSAP), New Delhi.

Haines AM, Webb SL, Wallace JR (2021) Conservation forensics: The intersection of wildlife crime, forensics, and conservation. In: Underkoffler SC, Adams HR (Eds.) *Wildlife Biodiversity Conservation: Multidisciplinary and Forensic Approaches.* 1st ed. Cham: Springer International Publishing, pp. 125–146. DOI: 10.1007/978-3-030-64682-0_6.

Hall RD (2019) The forensic entomologist as expert witness. In: Byrd JH, Tomberlin JK (Eds.) *Forensic Entomology.* 3rd ed. CRC Press, pp. 333–348.

Harper CK (2021) RhODIS®(the rhinoceros DNA index system): The application of simple forensic and genetic tools help conserve African rhinoceros. In: Underkoffler SC, Adams HR (Eds.) *Wildlife Biodiversity Conservation: Multidisciplinary and Forensic Approaches.* 1st ed. Springer International Publishing, pp. 463–485.

Hebert PD, Cywinska A, Ball SL, et al. (2003) Biological identifications through DNA barcodes. Proceedings of the Royal Society of London. *Series B: Biological Sciences* 270(1512): 313–321. DOI: 10.1098/rspb.2002.2218.

Hollemeyer K, Altmeyer W, Heinzle E, et al. (2012) Matrix-assisted laser desorption/ionization time-of-flight mass spectrometry combined with multidimensional scaling, binary hierarchical cluster tree and selected diagnostic masses improves species identification of Neolithic keratin sequences from furs of the Tyrolean Iceman Oetzi. *Rapid Communications in Mass Spectrometry* 26(16): 1735–1745. DOI: 10.1002/rcm.6277.

Huffman JE, Wallace JR (2012) *Wildlife Forensics: Methods and Applications.* Hoboken, New Jersey: John Wiley & Sons.

Jun J, Han SH, Jeong TJ, et al. (2011) Wildlife forensics using mitochondrial DNA sequences: Species identification based on hairs collected in the field and confiscated tanned Felidae leathers. *Genes & Genomics* 33(6): 721–726. DOI: 10.1007/s13258-011-0080-7.

Karmakar AS (2013) The lacunae in wildlife protection laws in India. Law Information Center (eLaw India). http://www.elaw.in/forest/wildlife/lacunaewildlife-Akash.pdf

Koch A, Ziegler T, Boehme W, et al. (2013) Pressing problems: Distribution, threats, and conservation status of the monitor lizards (*Varanidae*: *Varanus* spp.) of Southeast Asia and the Indo-Australian archipelago. *Herpetological Conservation and Biology* 8(3): 1–62.

Kumar A, Ghazi MGU, Singh B, et al. (2017) Conserve primers for sequencing complete ungulate mitochondrial cytochrome c oxidase I (COI) gene from problematic and decomposed biological samples. *Mitochondrial DNA Part B* 2(1): 64–66. DOI: 10.1080/23802359.2016.1247672.

Kumar A, Verma AK (2017) Biodiversity loss and its ecological impact in India. *International Journal on Biological Sciences* 8(2): 156–160.

Lazarescu C, Hart F, Pirouz Z, et al. (2017) Wood species identification by near-infrared spectroscopy. *International Wood Products Journal* 8(1): 32–35. DOI: 10.1080/20426445.2016.1242270.

Lee JCI, Tsai LC, Yang CY, et al. (2006) DNA profiling of shahtoosh. *Electrophoresis* 27(17): 3359–3362. DOI: 10.1002/elps.200600062.

Lee S, Jung PE, Lee Y (2016) Publicly-funded biobanks and networks in East Asia. *Springer Plus* 5: 1–11. DOI: 10.1186/s40064-016-2723-2.

Lens F, Liang C, Guo Y, et al. (2020) Computer-assisted timber identification based on features extracted from microscopic wood sections. *IAWA Journal* 41(4): 660–680. DOI: 10.1163/22941932-bja10029.

Linacre A (Ed) (2009) *Forensic Science in Wildlife Investigations.* CRC press.

Linacre A, Tobe S (2013). Wildlife DNA Analysis: Applications in Forensic Science. Wiley-Blackwell. https://doi.org/10.1002/9781118496411

Macchioni N (2011) Species identification. In Kasal B, Tannert T (Eds) *Situ Assessment of Structural Timber: State of the Art Report of the RILEM Technical Committee 215-AST,* pp. 105–107. Springer

Madkour FA, Abdelsabour-Khalaf M (2022) Scanning electron microscopy of the nasal skin in different animal species as a method for forensic identification. *Microscopy Research and Technique* 85(5): 1643–1653. DOI: 10.1002/jemt.24024.

Manheim J, Doty KC, McLaughlin G, et al. (2016) Forensic hair differentiation using attenuated total reflection Fourier transform infrared (ATR FT-IR) spectroscopy. *Applied Spectroscopy* 70(7): 1109–1117. DOI: 10.1177/0003702816652321.

Mardini A (1984) Species identification of tissues of selected mammals by agarose gel electrophoresis. *Wildlife Society Bulletin* (1973–2006) 12(3): 249–251. DOI: 10.1016/j.meatsci.2007.02.006.

Marshall RC, Gillespie JM, Klement V (1985) Methods and future prospects for forensic identification of hairs by electrophoresis. *Journal of the Forensic Science Society* 25(1): 57–66. DOI: 10.1016/s0015-7368(85)72363-8.

Mayer HK, Fiechter G (2013) Electrophoretic techniques. In M. de la Guardia & A. Gonzálvez (Eds.) *Comprehensive Analytical Chemistry*. Elsevier 60, pp. 251–278. DOI: 10.1016/B978-0-444-59562-1.00010-4.

Mayer WV (1952) The hair of California mammals with keys to the dorsal guard hairs of California mammals. *The American Midland Naturalist* 48(2): 480–512. DOI: 10.2307/2422262.

Mehra S (2012) *Study and Practice of Wild Life Laws in India: Concepts, Acts, Rules and Notifications*. Natraj Publishers.

Millins C, Howie F, Everitt C, et al. (2014) Analysis of suspected wildlife crimes submitted for forensic examinations in Scotland. *Forensic Science, Medicine, and Pathology* 10: 357–362. DOI: 10.1007/s12024-014-9568-1.

Mitra I, Roy S, Haque I (2018) Application of molecular markers in wildlife DNA forensic investigations. *Journal of Forensic Science and Medicine* 4(3): 156–160.

Mohammed ES, Madkour FA, Zayed M, et al. (2022) Comparative histological analysis of the skin for forensic investigation of some animal species. *EXCLI Journal* 21: 1286. DOI: 10.17179/excli2022-5335.

Moore TD, Spence LE, Dugnolle CE (1974) *Identification of the Dorsal Guard Hairs of Some Mammals of Wyoming*. Wyoming Game and Fish Department.

Negi SB (2013) *Wildlife Crime Investigation a Hand Book for Wildlife Crime Investigation Officers*. New Delhi: Wildlife Crime Control Bureau, Ministry of Environment and Forests Government of India.

Neme LA (2009) *Animal Investigators*. New York: Scribner Publishers.

Nishant K, Vrijesh KY, Ajay KR (2017) Wildlife forensic: Current techniques and their limitations. *Journal of Forensic Science & Criminology* 5(4): 402.

Ogden R, Dawnay N, McEwing R (2009) Wildlife DNA forensics—Bridging the gap between conservation genetics and law enforcement. *Endangered Species Research* 9(3): 179–195.

Paolella S, Bencivenni M, Lambertini F, et al. (2013) Identification and quantification of different species in animal fibres by LC/ESI-MS analysis of keratin-derived proteolytic peptides. *Journal of Mass Spectrometry* 48(8): 919–926. DOI: 10.1002/jms.3222.

Pereira F, Carneiro J, Amorim A (2008) Identification of species with DNA-based technology: Current progress and challenges. *Recent Patents on DNA & Gene Sequences* (Discontinued) 2(3): 187–200. DOI: 10.2174/187221508786241738.

Pérez-Espona S, CryoArks C (2021) Conservation-focused biobanks: A valuable resource for wildlife DNA forensics. *Forensic Science International: Animals and Environments* 1: 100017. DOI: 10.1016/j.fsiae.2021.100017.

Quarta G, Molnár M, Hajdas I, et al. (2021) 14C inter comparison exercise on bones and ivory samples: Implications for forensics. *Radiocarbon* 63(2): 533–544. DOI: 10.1017/RDC.2020.142.

Radhakrishnan S (2018) A note on wildlife poisoning cases from Kerala, South India. *European Journal of Wildlife Research* 64(5): 58. DOI: 10.1007/s10344-018-1218-6.

Ragothaman V, Chirukandoth S (2011) Mitigating human-wildlife conflict and retaliatory poisonings in India to preserve biodiversity and maintain sustainable livelihoods. In: Richards N (ed) *Carbofuran and Wildlife Poisoning: Global Perspectives and Forensic Approaches*. 1st ed. John Wiley & Sons, Inc, pp. 99–131.

Rattner BA, Scheuhammer AM, Elliott JE (2011) History of wildlife toxicology and the interpretation of contaminant concentrations in tissues. In W. N. Beyer & J. P. Meador (Eds.) *Environmental Contaminants in Wildlife, Interpreting Tissue Concentrations*. CRC Press, pp. 9–44.

Rollins CK, Hall DM (1999) Using light and scanning electron microscopic methods to differentiate ibex goat and Tibetan antelope fibers. *Textile Research Journal* 69(11): 856–860. DOI: 10.1177/004051759906901109.

Rosa da Silva N, De Ridder M, Baetens JM, et al. (2017) Automated classification of wood transverse cross-section micro-imagery from 77 commercial Central-African timber species. *Annals of Forest Science* 74: 1–14. DOI: 10.1007/s13595-017-0619-0.

Saez R, Sanz Y, Toldrá F (2004) PCR-based fingerprinting techniques for rapid detection of animal species in meat products. *Meat Science* 66(3): 659–665. DOI: 10.1016/S0309-1740(03)00186-4.

Sahajpal V, Goyal SP, Jayapal R, et al. (2008) Hair characteristics of four Indian bear species. *Science & Justice* 48(1): 8–15. DOI: 10.1016/j.scijus.2007.05.001.

Sahajpal V, Goyal SP, Thakar MK, et al. (2009) Microscopic hair characteristics of a few bovid species listed under Schedule-I of Wildlife (Protection) Act 1972 of India. *Forensic Science International* 189(1–3): 34–45. DOI: 10.1016/j.forsciint.2009.04.008.

Sandak A, Sandak J, Negri M (2011) Relationship between near-infrared (NIR) spectra and the geographical provenance of timber. *Wood Science and Technology* 45: 35–48. DOI: 10.1007/s00226-010-0313-y.

Sanford MR, Byrd JH, Tomberlin JK, et al. (2019) Entomological evidence collections methods: American Board of forensic entomology approved protocols. In: Byrd JH, Tomberlin JK (Eds.) *Forensic Entomology*. 3rd ed. CRC Press, pp. 63–85.

Severcan F, Haris PI, Aksoy C, et al. (2013) Progress in vibrational spectroscopy in diagnosis and screening. *Biomedical Spectroscopy and Imaging* 2(1): 73–81. DOI: 10.3233/BSI-130036.

Sharma CP, Kumar A, Sharma V, et al. (2019a) Online selling of wildlife part with spurious name: A serious challenge for wildlife crime enforcement. *International Journal of Legal Medicine* 133: 65–69. DOI: 10.1007/s00414-018-1795-7.

Sharma CP, Sharma S, Rawat GS, et al. (2022) Rapid and non-destructive differentiation of Shahtoosh from Pashmina/Cashmere wool using ATR FT-IR spectroscopy. *Science & Justice* 62(3): 349–357. DOI: 10.1016/j.scijus.2022.04.002.

Sharma CP, Sharma S, Sharma V, et al. (2019b) Rapid and non-destructive identification of claws using ATR-FTIR spectroscopy – A novel approach in wildlife forensics. *Science & Justice* 59(6): 622–629. DOI: 10.1016/j.scijus.2019.08.002.

Sharma CP, Sharma S, Singh R (2023) Species discrimination from blood traces using ATR FT-IR spectroscopy and chemometrics: Application in wildlife forensics. *Forensic Science International: Animals and Environments* 3: 100060. DOI: 10.1016/j.fsiae.2022.100060.

Sharma V, Yadav J, Kumar R, et al. (2020) On the rapid and non-destructive approach for wood identification using ATR-FTIR spectroscopy and chemometric methods. *Vibrational Spectroscopy* 110: 103097. DOI: 10.1016/j.vibspec.2020.103097.

Shimoyama M, Morimoto S, Ozaki Y (2004) Non-destructive analysis of the two subspecies of African elephants, mammoth, hippopotamus, and sperm whale ivories by visible and short-wave near infrared spectroscopy and chemometrics. *Analyst* 129(6): 559–563. DOI: 10.1039/b401003e.

Sinha S (2010) *Handbook on Wildlife Law Enforcement in India*. India: Natraj Publishers.

Soni VK (2020) Wildlife conservation in India: Issues and challenges. *Journal of Interdisciplinary Cycle Research* 12(10): 796–802.

Taupin JM (2004) Forensic hair morphology comparison – A dying art or junk science? *Science & Justice: Journal of the Forensic Science Society* 44(2): 95–100. DOI: 10.1016/S1355-0306(04)71695-0.

Thatheyus AJ (2006) Threats to Indian Biodiversity. In S. J. William (Ed.) *Biodiversity: Life to Our Mother Earth*. Chennai: Loyola College, pp. 217–230.

The International Barcode of Life Consortium (2024). International Barcode of Life project (iBOL). Occurrence dataset https://doi.org/10.15468/inygc6 accessed via GBIF.org on 2024-06-14.

Tnah LH, Lee SL, Ng KKS, et al. (2010) Forensic DNA profiling of tropical timber species in peninsular Malaysia. *Forest Ecology and Management* 259(8): 1436–1446. DOI: 10.1016/j.foreco.2010.01.017.

Tolksdorf JF, Veil S, Kuzu I, et al. (2015) Ivory or bone? A report on practical experience determining material from the Mesolithic site Klein Breese (Northern Germany). *Archaeological and Anthropological Sciences* 7: 351–360. DOI: 10.1007/s12520-014-0197-0.

Turner-Walker G, Xu BY (2014) Identification of animal hard tissues using Fourier transform infrared spectroscopy. In *Natural History Collections*. Preprints ICOM-CC 17th Triennial Conference (2014, Melbourne). Melbourne (pp. 1–11).

Ueland M, Brown A, Bartos C, et al. (2020) Profiling volatilomes: A novel forensic method for identification of confiscated illegal wildlife items. *Separations* 7(1): 5. DOI: 10.3390/separations7010005.

Ueland M, Ewart K, Troobnikoff AN, et al. (2016) A rapid chemical odour profiling method for the identification of rhinoceros horns. *Forensic Science International* 266: e99–e102. DOI: 10.1016/j.forsciint.2016.05.011.

Virkler K, Lednev IK (2009) Blood species identification for forensic purposes using Raman spectroscopy combined with advanced statistical analysis. *Analytical Chemistry* 81(18): 7773–7777. DOI: 10.1021/ac901350a.

Vogel JC, Eglington B, Auret JM (1990) Isotope fingerprints in elephant bone and ivory. *Nature* 346(6286): 747–749. DOI: 10.1038/346747a0.

Watkinson CJ, Rees GO, Gwenael MC, et al. (2022) Stable isotope ratio analysis for the comparison of timber from two forest concessions in Gabon. *Frontiers in Forests and Global Change* 4: 650257. DOI: 10.3389/ffgc.2021.650257.

White NE, Dawson R, Coghlan ML, et al. (2012) Application of STR markers in wildlife forensic casework involving Australian black-cockatoos (*Calyptorhynchus* spp.). *Forensic Science International: Genetics* 6(5): 664–670. DOI: 10.1016/j.fsigen.2011.10.003.

Wildlife (Protection) Act (1972) Government of India.

World Wildlife Crime Report: Trafficking in protected species (2020) UNODC, UN, New York.

Wright DK (2017) Accuracy vs. precision: Understanding potential errors from radiocarbon dating on African landscapes. *African Archaeological Review* 34(3): 303–319. DOI: 10.1007/S10437-017-9257-z.

Yang H, Lewis IR, Griffiths PR (1999) Raman spectrometry and neural networks for the classification of wood types. 2. Kohonen self-organizing maps. *Spectrochimica Acta Part A: Molecular and Biomolecular Spectroscopy* 55(14): 2783–2791. DOI: 10.1016/S1386-1425(99)00088-8.

Yates BC, Espinoza EO, Baker BW (2010) Forensic species identification of elephant (*Elephantidae*) and giraffe (*Giraffidae*) tail hair using light microscopy. *Forensic Science, Medicine, and Pathology* 6(3): 165–171. DOI: 10.1007/s12024-010-9169-6.

Ziegler S, Merker S, Streit B, et al. (2016) Towards understanding isotope variability in elephant ivory to establish isotopic profiling and source-area determination. *Biological Conservation* 197: 154–163. DOI: 10.1016/j.biocon.2016.03.008.

Zou ZT, Uphyrkina OV, Fomenko P, et al. (2015) The development and application of a multiplex short tandem repeat (STR) system for identifying subspecies, individuals and sex in tigers. *Integrative Zoology* 10(4): 376–388. DOI: 10.1111/1749-4877.12136.

13

Forensic DNA Phenotyping

Techniques, Successes, and Future Directions

Charles E. "Chuck" MacLean

13.1 Introduction

Forensic DNA phenotyping (FDP) represents a significant advancement in forensic science, bridging the gap between genetic research and practical criminal investigations. Unlike traditional DNA profiling, which relies on matching a suspect's DNA to evidence found at a crime scene, FDP is predictive; it enables the inference of externally visible characteristics (EVCs) such as eye color, hair color, skin tone, and even facial morphology from a DNA sample. This capability provides forensic investigators with a valuable tool, especially in cases where conventional methods fail to produce leads or identify suspects.

This chapter provides an accessible exploration of FDP, its current state, and its potential future directions (Haddrill, 2021; Phillips, 2023). We begin with an overview of FDP, tracing its historical development and evolution within the broader context of forensic genetics. The chapter then delves into the key techniques employed in FDP, including the use of single nucleotide polymorphisms (SNPs), the HIrisPlex-S system, next-generation sequencing (NGS), and the application of machine learning (ML) algorithms. Following this, we examine the successes and real-world applications of FDP in criminal investigations, highlighting notable case studies and discussing its impact on law enforcement practices. The chapter then addresses the challenges and limitations of FDP, focusing on issues of accuracy and reliability, ethical and legal concerns, and population-specific challenges (Jobling, 2022; Wienroth & McCartney, 2024). Finally, we explore future directions in FDP, including technological advancements, efforts to improve phenotypic predictions, and the push for global standardization and best practices.

Throughout, this chapter aims to provide a balanced view of FDP, acknowledging its potential while critically examining the technical, ethical, and societal challenges it presents.

13.2 Overview of forensic DNA phenotyping (FDP)

FDP represents a significant advancement in forensic science, bridging the gap between genetic research and practical criminal investigations. Unlike traditional DNA profiling, which relies on matching a suspect's DNA to evidence found at a crime scene, FDP is predictive; it enables the inference of EVCs such as eye color, hair color, skin tone, and even facial morphology from a DNA sample. This capability provides forensic investigators with a valuable tool, especially in cases where conventional methods fail to produce leads (Aggarwal, 2020; Bukayev et al., 2024).

DOI: 10.4324/9781003474296-13

The foundation of FDP lies in the study of genetic variation across populations (Branicki et al., 2011; Jeffreys et al. 1985; Liu et al., 2009). Over the past few decades, advancements in genomics and the availability of large-scale genetic data have facilitated a more detailed understanding of how specific genetic markers correlate with physical traits. The journey of FDP is deeply intertwined with advancements in DNA technology and the quest for more precise tools to assist in criminal investigations.

Initially, DNA profiling techniques, such as restriction fragment length polymorphism (RFLP) and polymerase chain reaction-short-tandem repeats (PCR-STR), were primarily used for identification purposes, matching DNA samples from crime scenes with genotypes from known persons in DNA databases. These early methods focused on analyzing non-coding regions of the DNA, which do not directly influence visible traits (MacLean, 2013).

The 1990s marked a significant turning point in forensic genetics with the introduction of PCR-STR, a technique that required smaller samples and provided faster results compared to RFLP (Butler & Willis, 2020). This advancement laid the groundwork for the exploration of DNA's potential in predicting phenotypic traits (Aggarwal, 2020; Van Laan, 2023). The progression from merely identifying individuals to predicting their physical appearance from genetic data marked the birth of FDP.

One of the pivotal moments in the history of FDP was the completion of the Human Genome Project in the early 2000s. This monumental achievement provided a comprehensive map of the human genome, enabling researchers to identify specific genetic markers associated with various phenotypic traits. The project's success fueled interest in using these genetic markers for forensic purposes, leading to the development of tools like the HIrisPlex system, which became instrumental in predicting eye, hair, and skin color from DNA samples (Durkin, 2024; Kadam et al., 2023).

Of course, the relationships between genetic markers and EVCs are often very complex, with many markers working together to create EVCs. Further, environmental influences (nutrition, isolation, trauma, hair dye, contact lenses, and so on) yield sometimes extensive modifications to predicted EVCs based on FDP alone. That said, of course, although identical twins – even those raised apart – are often remarkably similar in EVC and psychological traits, the identical twins' environmental influences are often different enough to affect some EVCs. Even monozygotic (colloquially, identical) twins, for example, do not display identical fingerprints due to both genetic and epigenetic influences (Hwa et al., 2023).

As FDP technology evolved, it became clear that the technique could provide valuable insights in cases where traditional DNA profiling methods fell short (Kayser, 2015; Walsh & Kayser, 2023). For example, in cases with no suspect or where DNA from a crime scene did not match any profiles in existing databases, FDP offered a new avenue for investigation by predicting the appearance of the unknown individual (Kushwaha et al., 2023; MacLean, 2013). And in the medical field, DNA phenotyping is widely used to diagnose, design surgical interventions, and treat physical ailments to which sufferers are genetically predisposed (Hopkins, 2023). Finally, DNA genotyping and phenotyping are beginning to make strides in psychological interventions and treatments for persons with genetic predispositions to psychiatric challenges (Andreassen et al., 2023).

However, the evolution of FDP has not been without challenges. The technique's reliance on SNPs to predict phenotypic traits introduced a layer of complexity, as these predictions are probabilistic rather than definitive. This probabilistic nature of FDP has raised concerns about its accuracy and reliability, particularly in diverse populations where the genetic markers associated with specific traits may vary (Durkin, 2024; Machado & Granja, 2020).

Recent advancements in ML and artificial intelligence (AI) have further enhanced the capabilities of FDP (Sessa et al., 2024; Xavier et al., 2022). These technologies have allowed for more sophisticated analysis of genetic data, improving the accuracy of phenotypic predictions and enabling the exploration of more complex traits (Barash et al., 2024; Sessa et al., 2024). Additionally, the incorporation of epigenetic factors, particularly for age prediction, has expanded the scope of FDP beyond traditional EVCs (Montesanto et al., 2020; Refn et al., 2023).

The ethical and legal implications of FDP have been subjects of intense debate. The ability to predict a person's appearance from their DNA raises questions about privacy, consent, and the potential for misuse. These concerns have led to stringent regulations in some countries, limiting the use of FDP to specific cases or banning it altogether (M'charek, 2020; MacLean, 2013). The global adoption of FDP varies significantly, with some nations embracing the technology while others remain cautious or outright reject its application (Atwood et al., 2021; Granja & Machado, 2023; Zieger, 2022).

In conclusion, FDP has emerged as a powerful tool in criminal investigations, offering new possibilities for generating leads in challenging cases. However, its use must be carefully balanced against ethical considerations and potential limitations. As the field continues to evolve, ongoing research and development aim to improve the accuracy and reliability of FDP across diverse populations, while addressing the ethical and legal challenges associated with its use.

13.3 Techniques in forensic DNA phenotyping

13.3.1 Single nucleotide polymorphisms (SNPs)

SNPs are the cornerstone of FDP (The International SNP Map Working Group, 2001; Kaczmarek & Zysek, 2022; Pośpiech et al., 2022). These genetic markers represent variations at single positions in DNA sequences and are the most common form of genetic diversity among individuals (Kaczmarek & Zysek, 2022). In FDP, SNPs are crucial for predicting EVCs such as eye color (Brancato et al., 2023), hair color, skin tone, and even more complex traits like adult height, facial morphology (Aggarwal, 2020; Alshehhi et al., 2023; Bukayev et al., 2024; Kataria et al., 2023; Liu et al., 2012; Liu et al., 2019), and freckling (Fridman et al., 2022; Kukla-Bartoszek et al., 2019).

The role of SNPs in FDP primarily involves the identification of phenotype-informative SNPs (PISNPs). These SNPs are linked to genes that influence visible traits, allowing forensic scientists to infer physical appearance based on genetic profiles. For example, SNPs in the OCA2 and HERC2 genes are strongly associated with eye color, while SNPs in the MC1R gene are linked to hair color (Branicki et al., 2011; Bukayev et al., 2024; Liu et al., 2009).

One of the primary advantages of SNPs over other genetic markers, such as STRs, is their abundance and distribution throughout the genome (Chakravarty & Pandya, 2022). SNPs occur approximately once every 1,000 base pairs, providing a vast amount of data for analysis. Their bi-allelic nature also simplifies their analysis and interpretation (Bukayev et al., 2024; Kadam et al., 2023).

The technical implementation of SNP analysis in FDP involves several steps, starting with DNA extraction from biological samples (Carratto et al., 2024). SNP genotyping is then performed using techniques such as PCR, SNaPshot™ minisequencing, or massively parallel sequencing (MPS). These methods allow for the simultaneous analysis of multiple SNPs, increasing the efficiency and accuracy of the process (Breslin et al., 2019; Kadam et al., 2023).

Recent advancements in NGS technologies have further enhanced the capabilities of SNP analysis in FDP (Browne & Freeman, 2024; Sapan et al., 2024). NGS allows for the high-throughput sequencing of SNPs, enabling more comprehensive genetic profiles that can be used for both identification and phenotyping purposes (Browne & Freeman, 2024; Carratto et al., 2024).

Despite their advantages, the use of SNPs in FDP faces challenges. The bi-allelic nature of SNPs provides less genetic diversity per marker compared to STRs, necessitating the analysis of a larger number of SNPs to achieve the same level of discrimination. Additionally, the probabilistic nature of SNP-based predictions means that results are expressed as likelihoods rather than certainties, which must be carefully communicated and interpreted in forensic contexts (Durkin, 2024; MacLean, 2013).

A significant concern in ensuring the reliability of FDP is the potential for population bias in prediction models. Most FDP models have been developed using data from populations of European descent, which can limit their accuracy when applied to individuals from other ethnic backgrounds. To address this issue, there is a growing emphasis on developing more inclusive datasets that represent a broader range of populations (Bukayev et al., 2024; Phillips et al., 2022).

13.3.2 The HIrisPlex-S system

The HIrisPlex-S system is a comprehensive tool used in FDP to predict eye, hair, and skin color from genetic material (Walsh & Kayser, 2023; Walsh et al., 2011). Developed as an extension of the earlier IrisPlex system, which focused solely on eye color prediction, HIrisPlex-S incorporates additional SNPs to allow for the simultaneous prediction of hair and skin pigmentation traits.

The HIrisPlex-S system now includes 41 SNPs, with 36 specifically predictive for skin color, sourced from 16 pigmentation genes (Chaitanya et al., 2018; Walsh et al., 2013). The HIrisPlex-S model for skin color has been tested in diverse populations, including admixed groups, with prediction accuracies ranging from 83% to 97%, depending on the skin tone categories used (Bukayev et al., 2024; Chaitanya et al., 2018). Its efficacy has been validated in various populations, making it a valuable tool in forensic investigations (Bukayev et al., 2024; Kadam et al., 2023; Rahat et al., 2023).

HIrisPlex-S operates by analyzing SNPs using various genotyping methods, including SNaPshot™ minisequencing and MPS. These methods allow forensic laboratories to process multiple SNPs simultaneously, which is critical for generating comprehensive phenotypic profiles from often limited and degraded DNA samples found at crime scenes (Breslin et al., 2019; Kadam et al., 2023).

While HIrisPlex-S has demonstrated high accuracy in predicting pigmentation traits, particularly in European populations, its performance can vary in other population groups (Kukla-Bartoszek et al., 2019; Phillips et al., 2022). This highlights the need for continued research and development of more inclusive genetic databases and prediction models (Durkin, 2024; Phillips et al., 2022).

13.3.3 Next-generation sequencing (NGS)

NGS, also known as MPS, has revolutionized FDP by enabling the simultaneous analysis of multiple genetic markers, including SNPs, STRs, and mitochondrial DNA (mtDNA) in a single experiment (Breslin et al., 2019; Butler & Willis, 2020).

NGS platforms like the MiSeq FGx Forensic Genomics System (Illumina) and Ion Torrent (ThermoFisher Scientific) are commonly used in forensic labs for FDP (Carratto et al., 2022). These platforms allow for the detailed analysis of phenotype-informative SNPs (PISNPs), enhancing the accuracy of phenotype predictions (Bukayev et al., 2024; Kadam et al., 2023).

One of the most significant applications of NGS in FDP is its ability to analyze complex DNA mixtures. Traditional STR-based methods often struggle to deconvolute mixtures containing DNA from multiple individuals, particularly when one contributor is present at low levels. NGS, with its high resolution and deep sequencing capabilities, can more effectively separate and analyze these mixtures, leading to more accurate identifications (Bukayev et al., 2024; Butler, 2023; Pośpiech et al., 2020).

NGS has also expanded the scope of biogeographical ancestry (BGA) analysis (Frégeau, 2021; Phillips et al., 2022). By analyzing ancestry-informative SNPs (AISNPs), NGS can provide insights into the geographical origins of an individual, which can be invaluable in narrowing down suspect pools in forensic investigations (Frégeau, 2021; MacLean, 2013).

Despite its numerous advantages, NGS is not without challenges. The massive amounts of data generated by NGS platforms require sophisticated bioinformatics tools and algorithms to ensure accurate interpretation and reporting. Additionally, the lack of standardized nomenclature and reporting practices across forensic laboratories can complicate the comparison of NGS data across jurisdictions (Butler, 2023; Kadam et al., 2023).

13.3.4 Machine learning algorithms in FDP

Machine learning (ML) algorithms are increasingly becoming integral to FDP, enhancing the ability to predict EVCs from genetic data (Barash et al., 2024; Pośpiech et al., 2022). These algorithms can handle vast amounts of data, including SNPs, improving the predictive accuracy of traits even in challenging scenarios such as mixed or degraded DNA samples (Barash et al., 2024; Sessa et al., 2024).

Several types of ML algorithms are employed in FDP, each with its strengths and limitations. Supervised learning is the most common approach, where the algorithm is trained on a labeled dataset that pairs genetic data with known phenotypes. Unsupervised learning algorithms, on the other hand, identify patterns and clusters within the data without predefined categories. Deep learning, a subset of ML, utilizes neural networks with multiple layers to model complex relationships in the data, showing promise in predicting more complex traits such as facial morphology (Bukayev et al., 2024; Kadam et al., 2023).

Recent advancements in ML have allowed for the exploration of more complex traits and the integration of various types of data, including epigenetic information (Llobet et al., 2023; Montesanto et al., 2020). This has expanded the scope of FDP beyond traditional EVCs, enabling more accurate predictions of traits such as age and other age-dependent characteristics (Montesanto et al., 2020; Refn et al., 2023).

While ML algorithms offer significant improvements in FDP, they also present challenges. The potential for bias in the algorithms, particularly if the training data is not representative of the broader population, remains a concern. Additionally, the interpretability of ML models, especially deep learning algorithms, can be challenging, which may complicate their use in legal contexts (Durkin, 2024; Scudder et al., 2019).

In conclusion, the techniques used in FDP continue to evolve, with advancements in SNP analysis, the development of comprehensive systems like HIrisPlex-S, the adoption of NGS technologies, and the integration of ML algorithms. These developments have significantly enhanced the capabilities of FDP, while also presenting new challenges that must be addressed to ensure the reliable and ethical application of these techniques in forensic investigations (Durkin, 2024; Wienroth, 2023).

13.4 Successes and applications of forensic DNA phenotyping

FDP has evolved into a pivotal tool in criminal investigations, especially in cases where traditional methods have failed to yield results (Kayser et al., 2023; Simayijiang & Yan, 2023). By predicting EVCs such as eye color, hair color, skin tone, and BGA, FDP has provided crucial leads in numerous cold cases and complex investigations (Robertson, 2024). Parabon Nanolabs (2022) claims its FDP tools have helped solve over 200 criminal cases.

13.4.1 Case studies and real-world applications

1. *The Golden State Killer Case (2018)*: While this case is most famously solved through forensic genetic genealogy, FDP played a supporting role. The ability to predict ancestry and certain phenotypic traits from DNA helped narrow down the suspect pool, ultimately leading to the identification of Joseph James DeAngelo as the perpetrator. This case demonstrates the potential of combining FDP with other forensic tools to solve decades-old cases (Bukayev et al., 2024; Durkin, 2024; Hernandez, 2023; Olds, 2023).
2. *The French Homicides (2012)*: In this case, DNA found at the crime scene couldn't be matched with any known profiles. Parabon NanoLabs' Snapshot® DNA Phenotyping Service provided a detailed phenotypic profile of the suspect, leading investigators to reconsider familial connections and ultimately identify the murderer within a victim's family (Matheson, 2016; Parabon NanoLabs, 2022).
3. *The Chisholm Cold Case (2013):* In Minnesota, a sexual assault case had stalled due to the lack of identifiable suspects. FDP provided a phenotypic profile that suggested key traits of the suspect, helping law enforcement refocus their investigation and leading to the successful identification and conviction of the perpetrator (Aggarwal, 2020; Bukayev et al., 2024; Dabas et al., 2022).
4. *The Identification of Oregon's Missing Person (1959/2023):* FDP was crucial in identifying the remains of a woman missing since 1959. Traditional identification methods had failed, but phenotypic analysis of her DNA provided clues about her appearance. Coupled with genealogical research, this led to her identification, showcasing FDP's utility in cold cases and missing persons investigations (Alvarez & Gay, 2020; Bukayev et al., 2024; Durkin, 2024).
5. *The Baton Rouge Serial Killer (2003)*: In this case, eyewitnesses initially reported seeing a White male. However, when FDP was applied, it revealed that the suspect was likely of African and Native American ancestry. This shift in focus ultimately led to the identification and arrest of Derrick Todd Lee, demonstrating FDP's ability to provide crucial investigative direction even when it contradicts initial assumptions (Bukayev et al., 2024; MacLean, 2013; MacLean & Lamparello, 2014).

Table 13.1 Selected Criminal Cases Materially Aided by FDP

Case	Year	Location	FDP Contribution	Sources
Rape and Murder	2021	New Jersey USA	FDP generated leads	Parabon Nanolabs (2021)
Homicide	2018	Texas USA	FDP sketch led to confession	Parabon Nanolabs (2021)
Double Homicide	2015	North Carolina USA	FDP profile generated leads	Parabon Nanolabs (2021)
Serial Creeper	2015	Florida USA	FDP profile led to suspect	Aggarwal (2020); Bukayev et al. (2024)
Minstead Rapist	2009	London UK	Ancestry prediction	Bukayev et al. (2024); Durkin 2024
Landowski Murder	2008	Virginia USA	FDP analysis narrowed pool	Bukayev et al. (2024); Durkin (2024)
Serial Murderer	2003	Louisiana USA	FDP biogeography narrowed pool	Aggarwal (2020); Bukayev et al. (2024)

6. *The Faith Hedgepeth Murder Case (2012):* The unsolved murder of Faith Hedgepeth in North Carolina was one of the most high-profile cases where FDP was used. Parabon NanoLabs produced a composite image of the suspect, predicting key physical traits. While the case remained unsolved for some time after this, the profile provided new investigative directions (Bukayev et al., 2024; Durkin, 2024).

Table 13.1 presents just a few of the many other criminal cases materially aided by FDP.

13.4.1.1 Impact on law enforcement

FDP has significantly impacted law enforcement practices by providing detailed phenotypic profiles of unknown suspects, breathing new life into cold cases, and advancing investigations where traditional DNA profiling methods have failed (Koops & Schellekens, 2008). FDP also helps facilitate the process of winnowing the suspect pool and eliminating otherwise suspects who FDP has eliminated by phenotype (Dash et al., 2024). The technology has proven especially effective in identifying unknown human remains, offering a modern solution to long-standing challenges in forensic anthropology and pathology (Alvarez & Gay, 2020; Bukayev et al., 2024; MacLean, 2013; Sulley et al., 2022).

The combination of FDP with investigative genetic genealogy, as seen in the Golden State Killer case, represents a powerful synergy between emerging forensic techniques. This combination has shown that even the most elusive suspects can be identified by leveraging the strengths of both phenotypic predictions and familial DNA searches (Hernandez, 2023; Olds, 2023; Robertson, 2024).

FDP has also demonstrated potential in reducing investigative time and resources (Taylor et al., 2024). By providing more focused leads, law enforcement agencies can streamline their investigations, eliminating individuals who do not match the phenotypic profile generated from DNA evidence. This approach not only saves time but also allows for more efficient allocation of resources to the most promising leads (Bukayev et al., 2024; Butler, 2023).

13.4.1.2 Global adoption and ethical considerations

The adoption of FDP varies significantly across the globe, shaped by varying legal frameworks, cultural contexts, and ethical concerns. Countries like the Netherlands and Slovakia explicitly allow the use of FDP in criminal investigations, while others like Germany and Belgium have largely prohibited its use, citing concerns over privacy and potential misuse (Amelung & Machado, 2021; Butler, 2023; Tozzo et al., 2021; Zieger, 2022).

In the United States, FDP is increasingly being adopted by law enforcement agencies, particularly in cold cases. However, it operates in a largely unregulated space, with no federal laws specifically addressing its application in criminal investigations (Durkin, 2024; Tozzo et al., 2021).

The ethical implications of FDP are a major factor influencing its global adoption. Privacy concerns, the potential for discrimination, and issues of consent are frequently cited ethical issues. Critics argue that FDP could lead to a "genetic surveillance state" where individuals are routinely profiled based on their DNA without their knowledge or consent (Bartram et al., 2023; Bouchagiar, 2021; Durkin, 2024; Tozzo et al., 2021).

The use of FDP to predict BGA is particularly contentious, as it could result in racial profiling or discrimination, especially in diverse populations. This has led to calls for clear guidelines and regulations to govern the use of FDP in criminal investigations (Butler, 2023; M'charek et al., 2020).

13.4.1.3 Challenges and future directions

Despite its successes, FDP faces several challenges. The accuracy and reliability of predictions can vary across different populations due to genetic diversity and the underrepresentation of certain groups in genetic databases (Bukayev et al., 2024; Jobling, 2022; Phillips et al., 2022).

To address these issues, there is a growing emphasis on developing more inclusive datasets that represent a broader range of populations. Researchers are also exploring the use of admixture models, which take into account the genetic contributions from multiple ancestral populations to improve the accuracy of predictions in mixed-ancestry individuals (Butler, 2023; Durkin, 2024; Pośpiech et al., 2022; Schneider et al., 2019).

The integration of ML and AI with FDP holds promise for improving the accuracy of phenotypic predictions and enabling the exploration of more complex traits. These technologies have the potential to revolutionize FDP by automating the interpretation of complex sequencing data and improving the accuracy of phenotype predictions (Barash et al., 2024; Sessa et al., 2024).

In conclusion, while FDP has demonstrated significant potential in aiding criminal investigations, its continued success will depend on addressing technical challenges, ethical concerns, and legal issues. As the technology evolves, ongoing research, development of robust ethical guidelines, and clear legal frameworks will be crucial to ensure that FDP remains a valuable and responsibly used tool in forensic science (Bukayev et al., 2024; Durkin, 2024; Samuel & Prainsack, 2019c; Tozzo et al., 2021; Wienroth & McCartney, 2024).

13.5 Challenges and limitations

13.5.1 Accuracy and reliability

The accuracy and reliability of FDP are critical concerns that directly influence its effectiveness and acceptance in forensic practice (Branicki, 2024; Katsara et al., 2021). Several factors impact these attributes:

1. *Sample Quality and Data Sufficiency:* Forensic samples often differ significantly from high-quality, research-grade samples used to develop FDP models. DNA recovered from crime scenes may be degraded, contain inhibitors, or be present in low quantities, all of which can compromise the accuracy of FDP predictions (Butler, 2023;

Durkin, 2024). The stochastic effects during amplification of low-quantity DNA can introduce significant variability, reducing the reliability of the predictions (Bukayev et al., 2024; MacLean, 2013; Melchionda et al., 2022).

2. *Genetic Markers:* The choice of SNPs plays a crucial role in determining FDP accuracy. Not all SNPs have a strong correlation with the traits they are used to predict (Pośpiech et al., 2022; Sguazzi et al., 2022). Studies have shown that the accuracy of predictions can vary widely depending on the genetic markers used (Durkin, 2024; MacLean, 2013).
3. *Prediction Models:* These models are developed using large datasets of known genotypes and phenotypes. However, their accuracy can be affected by the quality of the input data and the specific characteristics of the population from which the data were derived. Models developed using data from European populations may not perform as well when applied to non-European populations (al Dosari, 2020; Bukayev et al., 2024; Bülbül et al., 2018; Durkin, 2024).
4. *Population Diversity:* A significant challenge in ensuring the reliability of FDP is the potential for population bias in prediction models. Most FDP models have been developed using data from populations of European descent, which can limit their accuracy when applied to individuals from other ethnic backgrounds (Butler, 2023; Durkin, 2024).

Recent studies have emphasized the need for population-specific models to improve predictive accuracy and reliability (al Dosari, 2020; Bülbül et al., 2018; Liu et al., 2019; Phillips et al., 2022; Schneider et al., 2019). Efforts are being made to develop more inclusive datasets and refine prediction models for diverse populations (Bukayev et al., 2024; Phillips et al., 2022).

13.5.2 Ethical and legal concerns

FDP raises significant ethical and legal concerns that must be carefully considered (M'charek, 2008; Samuel & Prainsack, 2019a; Tozzo et al., 2021):

1. *Privacy:* FDP allows for the prediction of physical traits and ancestry, which can be highly personal and sensitive information. Critics argue that FDP could lead to invasions of privacy, as individuals' genetic information could be used to infer characteristics that they may not wish to disclose (Bouchagiar, 2021; Durkin, 2024; Tozzo et al., 2021; White & Lee, 2020).
2. *Discrimination and Racial Profiling:* The ability to predict BGA has raised fears that FDP could be used to target specific racial or ethnic groups, leading to unfair treatment and stigmatization (Bartram et al., 2023; Butler, 2023; van Oorschot & M'charek et al., 2021; Skinner, 2020).
3. *Consent:* In many cases, DNA samples used for FDP are collected without the individual's explicit consent, particularly when samples are obtained from crime scenes or public genealogy databases (Budowle & Sajantila, 2023; Durkin, 2024; Tozzo et al., 2021).
4. *Misuse and Over-Reliance:* The probabilistic nature of FDP means that its predictions are not definitive. There is a risk that law enforcement agencies might over-rely on FDP results, treating them as conclusive evidence rather than as investigative leads (Butler, 2023; Durkin, 2024).

Legal frameworks for FDP vary significantly across jurisdictions. While some countries have specific laws regulating their use, in many others, the legal framework is ambiguous or non-existent (Bukayev et al., 2024; Tozzo et al., 2021). Recent studies have called for more comprehensive regulations that ensure transparency, accountability, and the protection of individuals' privacy and rights (Budowle & Sajantila, 2023; Oosthuizen et al., 2023; Samuel & Prainsack, 2020).

13.5.3 Population-specific issues

The effectiveness and ethical implications of FDP vary significantly depending on the population in question (Jobling, 2022; M'charek et al., 2020):

1. *Genetic Diversity:* Human populations exhibit significant genetic variation, which can influence the prediction of EVCs and BGA. FDP models developed using data from specific populations may have varying predictive power across groups (Bukayev et al., 2024; Butler, 2023; Liu et al., 2019; Phillips et al., 2022).
2. *Underrepresentation in Genetic Databases:* Most large-scale genomic studies and databases have historically focused on populations of European descent, leading to a lack of comprehensive data for many other ethnic groups (Bukayev et al., 2024; Butler, 2023; Durkin, 2024; Tan et al., 2022).
3. *Potential Biases in Technology and Methodology:* The choice of genetic markers and the design of prediction algorithms may inadvertently favor certain populations over others, leading to systematic biases in the results (Bukayev et al., 2024; Butler, 2023; Hopman, 2022; Wienroth, 2020).

Addressing these challenges requires ongoing research and development to improve the accuracy of FDP across diverse populations, including the expansion of genetic databases to include more representative data from a wider range of ethnic groups (Durkin, 2024; Granja & Machado, 2023; Kayser et al., 2023; Tozzo et al., 2021).

In conclusion, while FDP offers significant potential in forensic investigations, its accuracy, ethical implications, and legal standing present ongoing challenges. Addressing these issues is crucial for the responsible and effective use of FDP in forensic science.

13.6 Future directions in forensic DNA phenotyping

The field of FDP continues to evolve rapidly, driven by technological advancements and expanding research (Butler, 2023; Kayser et al., 2023). This section explores emerging trends and potential future developments in FDP, highlighting areas of innovation and addressing ongoing challenges.

13.6.1 Advancements in sequencing technologies

NGS technologies have already transformed FDP, and future developments promise even greater capabilities. Third-generation sequencing technologies, such as nanopore sequencing, are expected to further reduce the time and cost associated with DNA sequencing while increasing the accuracy and depth of analysis (Bukayev et al., 2024; Butler, 2023). These technologies could enable the routine use of FDP in a wider range of forensic cases, including those involving highly degraded or limited DNA samples (Browne & Freeman, 2024; Carratto et al., 2024; Kayser et al., 2023).

Recent advancements in Oxford Nanopore Sequencing, for instance, offer improved speed, accuracy, and the ability to analyze longer DNA sequences, potentially revealing more complex genetic relationships relevant to phenotype prediction (Sapan et al., 2024). This technology's portability also opens up possibilities for on-site DNA analysis in forensic investigations (Sapan et al., 2024).

13.6.2 Expanding trait predictions

While current FDP techniques focus primarily on pigmentation traits (eye, hair, and skin color), research is underway to expand the range of predictable characteristics. Recent studies have shown promise in predicting traits such as facial morphology, height, and even male pattern baldness (Liu et al., 2019; Peng et al., 2019; Şenişik et al., 2023).

Elkins et al. (2023) demonstrated the potential for predicting additional EVCs using commercially available forensic genomics kits. This expansion of predictable traits could significantly enhance the utility of FDP in generating more detailed suspect descriptions (Pandkar et al., 2022; Yuan et al., 2024).

13.6.3 Integration of epigenetics

The incorporation of epigenetic data into FDP is an emerging area that holds great promise for improving prediction accuracy, particularly for age-related traits. Epigenetic markers, such as DNA methylation patterns, can provide information about an individual's biological age and lifestyle factors that influence appearance (Llobet et al., 2023; Refn et al., 2023).

Montesanto et al. (2020) developed a robust epigenetic model for forensic age prediction, demonstrating the potential of this approach. As research in this area progresses, the integration of epigenetic data with genetic information could lead to more comprehensive and accurate phenotypic profiles (Montesanto et al., 2020; Vanneste et al., 2023).

13.6.4 Advanced machine learning and AI applications

The application of sophisticated ML and AI algorithms in FDP is expected to grow significantly. These technologies have the potential to improve the accuracy of phenotypic predictions by identifying complex patterns in genetic data that may not be apparent through traditional statistical methods (Barash et al., 2024; Jiang et al., 2021; Sessa et al., 2024).

Deep learning approaches, in particular, show promise in predicting complex traits like facial morphology. As these algorithms become more sophisticated and are trained on larger, more diverse datasets, they could dramatically enhance the predictive capabilities of FDP (Bukayev et al., 2024; Kadam et al., 2023; Pośpiech et al., 2022).

13.6.5 Population-specific models and diverse databases

Addressing the limitations of current FDP models in diverse populations remains a critical area for future development. Research is focused on creating population-specific models and expanding genetic databases to include a wider range of ancestral groups (Bukayev et al., 2024; Phillips et al., 2022).

The development of tools like Eurasiaplex-2, which focuses on SNPs with high population specificity, represents a step toward more accurate ancestry and phenotype predictions across diverse populations (Phillips et al., 2022). Future advancements in this area will be crucial for improving the global applicability and fairness of FDP (Atwood et al., 2021; Jobling, 2022).

13.6.6 Ethical and legal frameworks

As FDP technologies advance, there is a growing need for comprehensive ethical and legal frameworks to govern their use. Future developments in this area are likely to focus on balancing the investigative potential of FDP with privacy concerns and the risk of genetic discrimination (Durkin, 2024; Oosthuizen et al., 2023; Samuel & Prainsack, 2019b; Tozzo et al., 2021).

The development of standardized guidelines for the use of FDP in criminal investigations, such as those proposed by Samuel and Prainsack (2020), will be crucial. These frameworks will need to address issues such as consent, data protection, and the appropriate contexts for FDP use (Budowle & Sajantila, 2023; Oosthuizen et al., 2023).

13.6.7 Interdisciplinary collaboration

The future of FDP will likely involve increased collaboration between forensic scientists, geneticists, bioinformaticians, ethicists, and legal experts (Granja & Machado, 2023; National Academies of Sciences, Engineering, and Medicine, 2024). This interdisciplinary approach is essential for addressing the complex challenges associated with FDP and ensuring its responsible development and application (Granja & Machado, 2023; Wienroth, 2023).

The future of FDP holds great promise, with advancements in sequencing technologies, expanded trait predictions, and sophisticated AI applications poised to enhance its capabilities significantly. However, these developments must be balanced with careful consideration of ethical implications and the need for inclusive, population-specific approaches. As FDP continues to evolve, ongoing research, robust ethical guidelines, and clear legal frameworks will be essential to ensure its effective and responsible use in forensic investigations (Bukayev et al., 2024; Durkin, 2024; Tozzo et al., 2021; Wienroth & McCartney, 2024).

13.7 Conclusion

FDP has emerged as a powerful and innovative tool in the field of forensic science, offering unprecedented capabilities in predicting EVCs from genetic material. This chapter has explored the multifaceted aspects of FDP, from its technical foundations to its practical applications, ethical implications, and future directions.

The development of FDP techniques, particularly the use of SNPs and systems like HIrisPlex-S, has significantly enhanced our ability to infer physical traits from DNA evidence. The integration of NGS technologies and ML algorithms has further expanded the scope and accuracy of these predictions. Real-world applications, as evidenced by numerous case studies, demonstrate FDP's potential to provide crucial leads in challenging investigations and cold cases.

However, the path forward for FDP is not without challenges. The accuracy and reliability of predictions, especially across diverse populations, remain ongoing concerns. The underrepresentation of certain ethnic groups in genetic databases and the potential for population-specific biases in prediction models highlight the need for more inclusive research and development approaches. Ethical and legal considerations, including privacy concerns, the potential for discrimination, and issues of consent, necessitate careful navigation and the establishment of robust regulatory frameworks.

Looking to the future, several key areas of development promise to shape the evolution of FDP:

1. Advancements in sequencing technologies, such as nanopore sequencing, are expected to improve the speed, accuracy, and accessibility of genetic analysis in forensic contexts (Sapan et al., 2024).
2. Expansion of predictable traits beyond basic pigmentation characteristics to include more complex features like facial morphology and age-related traits (Elkins et al., 2023; Liu et al., 2019).
3. Integration of epigenetic data to enhance prediction accuracy, particularly for age estimation and lifestyle-influenced traits (Llobet et al., 2023; Montesanto et al., 2020).
4. Development of more sophisticated ML and AI applications for complex trait prediction and data interpretation (Barash et al., 2024; Sessa et al., 2024).
5. Creation of more inclusive, population-specific models and diverse genetic databases to improve prediction accuracy across different ethnic groups (Bukayev et al., 2024; Phillips et al., 2022).
6. Establishment of comprehensive ethical and legal frameworks to govern the use of FDP in criminal investigations and ensure the protection of individual rights (Durkin, 2024; Oosthuizen et al., 2023).
7. Fostering interdisciplinary collaboration among forensic scientists, geneticists, ethicists, and legal experts to address the complex challenges associated with FDP (Granja & Machado, 2023; Wienroth, 2023).

As FDP continues to evolve, it holds the potential to become an even more valuable tool in forensic investigations. However, its development and application must be guided by a commitment to scientific rigor, ethical considerations, and the pursuit of justice. By addressing current limitations, embracing emerging technologies, and maintaining a balance between investigative potential and individual rights, FDP can continue to advance the field of forensic science in a responsible and effective manner.

The future of FDP lies in striking a delicate balance between harnessing its powerful investigative capabilities and ensuring its ethical and just application. As we move forward, ongoing research, robust ethical guidelines, and clear legal frameworks will be essential to navigate the complex landscape of genetic privacy, consent, and the boundaries of forensic science (Toom et al., 2016). By doing so, we can harness the full potential of FDP while upholding the principles of fairness, privacy, and justice that are fundamental to our legal systems.

References

Aggarwal, K. (2020). Forensic DNA phenotyping: Significance in criminal investigations. *Academic Journal of Forensic Sciences*, *3*(1), 20–25.

al Dosari, W. R. (2020, September). *Application of identity and ancestry single nucleotide polymorphisms to forensic casework analysis in Qatar*. Dissertation (University of Central Lancashire.

Alshehhi, A., Almarzooqi, A., Alhammadi, K., Werghi, N., Tay, G. K., & Alsafar, H. (2023). Advancement in human face prediction using DNA. *Genes*, *14*, 136. doi:10.3390/genes14010136.

Alvarez, A. G., & Gay, M. A. (2020). DNA rewriting our memory: Recovering missing people through their genetic profile. *Metode Science Studies Journal*, *10*, 119–125.

Amelung, N., & Machado, H. (2021). Governing expectations of forensic innovations in society: The case of FDP in Germany. *New Genetics and Society, 40*(4), 498–519. doi:10.1080/14636778.2020.1868987.

Andreassen, O. A., Hindley, G. F. L., Frei, O., & Smeland, O. B. (2023). New insights from the last decade of research in psychiatric genetics: Discoveries, challenges and clinical implications. *World Psychiatry, 22*(1), 4–24.

Atwood, L., Raymond, J., Sears, A., Bell, M., & Daniel, R. (2021). From identification to intelligence: An assessment of the suitability of forensic DNA phenotyping service providers for use in Australian law enforcement casework. *Frontiers of Genetics, 11*, 568701. doi:10.3389/fgene.2020.568701.

Barash, M., McNevin, D., Federenko, V., & Giverts, P. (2024). Machine learning applications in forensic DNA profiling: A critical review. *Forensic Science International: Genetics, 69*, 102994.

Bartram, I., Plümecke, T., & Schultz, S. (2023). Genetic racial profiling: Extended DNA analyses and entangled processes of discrimination. *Science & Technology Studies, 36*(2), 52–72.

Bouchagiar, G. (2021, January 27–29). *Assessing legality, legitimacy and necessity of phenotyping-based surveillance* [Paper presentation]. 14th International Conference on Computers, Privacy & Data Protection (CPDP), Brussels. https://www.cpdpconferences.org/

Brancato, D., Coniglio, E., Bruno, F., Agostini, V., Saccone, S., & Federico, C. (2023). Forensic DNA phenotyping: Genes and genetic variants for eye color prediction. *Genes, 14*, 1604.

Branicki, W. (2024). DNA testing for investigative purposes: Description of the perpetrator. *Problems of Forensic Sciences, 137*, 17–29.

Branicki, W., Liu, F., van Duijn, K., Draus-Barini, J., Pośpiech, E., Walsh, S., & Kayser, M. (2011). Model-based prediction of human hair color using DNA variants. *Human Genetics, 129*(4), 443–454.

Breslin, K., Wills, B., Ralf, A., Ventayol Garcia, M., Kukla-Bartoszek, M., Pospiech, E., & Kayser, M. (2019). HIrisPlex-S system for eye, hair, and skin color prediction from DNA: Massively parallel sequencing solutions for two common forensically used platforms. *Forensic Science International: Genetics, 43*. doi:10.1016/j.fsigen.2019.102152.

Browne, T. N., & Freeman, M. (2024). Next generation sequencing: Forensic applications and policy considerations. *WIREs Forensic Science*, e1531. doi:10.1002/wfs2.1531.

Budowle, B., & Sajantila, A. (2023). Revising informed consent in forensic genomics in light of current technologies and the times. *International Journal of Legal Medicine, 137*, 551–565.

Bukayev, A., Aidarov, B., Fesenko, D., Saidamarova, V., Ivanovsky, I., Maltseva, E., Naizabayeva, D., Bukayeva, A., Faizov, B., Pylev, V., Darmenov, A., Skiba, Y., Balanovska, E., & Zhabagin, M. (2024). Genotype data for 60 SNP genetic markers associated with eye, hair, skin color, ABO blood group, sex, core Y-chromosome haplogroups in Kazakh population. *BMC Research Notes, 17*, 51. doi:10.1186/s13104-024-06712-z.

Bülbül, O., Zorlu, T., & Filoğlu, G. (2018). Prediction of human eye colour using highly informative phenotype SNPs (PISNPs). *Australian Journal of Forensic Sciences*. doi:10.1080/00450618.2018.1484161.

Butler, J. M. (2023). Recent advances in forensic biology and forensic DNA typing: INTERPOL review 2019–2022. *Forensic Science International: Synergy, 6*, 100311. doi:10.1016/j.fsisyn.2022.100311.

Butler, J. M., & Willis, S. (2020). Interpol review of forensic biology and forensic DNA typing 2016–2019. *Forensic Science International: Synergy, 2*, 352–367.

Carratto, T. M. T., de Oliviera, M. L. G., & Mendes-Junior, C. T. (2024). Forensic DNA phenotyping in the next generation sequencing era. In H. R. Dash, K. M. Elkins, & N. R. Al-Snan (Eds.) *Next Generation Sequencing (NGS) Technology in DNA Analysis* (pp. 311–336). Academic Press.

Carratto, T. M. T., Moraes, V. M. S., Recalde, T. S. F., de Oliviera, M. L. G., & Mendes-Junior, C. T. (2022). Applications of massively parallel sequencing in forensic genetics. *Genetics and Molecular Biology, 45*(supp. 1), e20220077.

Chaitanya, L., Breslin, K., Zuñiga, S., Wirken, L., Pośpiech, E., Kukla-Bartoszek, M., & Walsh, S. (2018). The HIrisPlex-S system for eye, hair and skin colour prediction from DNA: Introduction and forensic developmental validation. *Forensic Science International: Genetics, 35*, 123–135.

Chakravarty, M., & Pandya, P. (2022). Introduction to forensic DNA typing and current trends. In Dash, H. R., Shrivastava, P., & Lorente, J. A. (Eds.), *Handbook of DNA Profiling*. Springer. doi:10.1007/978-981-16-4318-7_1.

Dabas, P., Jain, S., Khajuria, H., & Nayak, B. P. (2022). Forensic DNA phenotyping: Inferring phenotypic traits from crime scene DNA. *Journal of Forensic and Legal Medicine, 92*, 102412.

Dash, H. R., Yadav, T., & Arora, M. (2024). Advancements in forensic DNA analysis in generating investigation leads and elimination of innocents. In Shekhar, B., & Pokhariyal, P. (Eds.), *Forensic Justice: A Global Perspective*. Routledge.

Durkin, A. (2024). Estimating a face: What predicting appearance from DNA reveals about the need to regulate genetic investigations. *Washington University Law Review, 101*, 1241–1310.

Elkins, K. M., Garloff, A. T., & Zeller, C. B. (2023). Additional predictions for forensic DNA phenotyping of externally visible characteristics using the ForenSeq and imagen kits. *Journal of Forensic Sciences, 68*(4), 1449–1458.

Frégeau, C. J. (2021). A multiple predictive tool approach for phenotypic and biogeographical ancestry inferences. *Canadian Society of Forensic Science Journal, 55*(2), 71–99. doi:10.1080/00085030.2021.2016206.

Fridman, C., Ferreira, M. A., Marano, L. A., & Forienza, B. S. (2022). Analysis of genetic polymorphisms associated with the presence of freckles for phenotypic prediction. *Forensic International: Genetics Supplement Series, 8*, 26–28.

Granja, R., & Machado, H. (2023). Forensic DNA phenotyping and its politics of legitimation and contestation: Views of forensic geneticists in Europe. *Social Studies of Science, 53*(6), 850–868.

Haddrill, P. R. (2021). Developments in forensic DNA analysis. *Emerging Topics in Life Sciences, 5*, 381–393.

Hernandez, M. (2023, May 19). Advancements in technology to solve cold cases. *Themis: Research Journal of Justice Studies and Forensic Science, 11*, art. 1.

Hopkins, C. E. (2023). Emerging technologies in personalized medicine. *Molecular Aspects of Medicine, 91*, 101182. doi:10.1016/j.mam.2023.101182.

Hopman, R. (2022). *Facing genetics: An ethnography of race in forensic DNA phenotyping practices.* Dissertation (University of Amsterdam).

Hwa, H. L., Lin, C. Y., Yu, Y. J., Linacre, A., & Lee, J. C. (2023). DNA identification of monozygotic twins. *Forensic Science International: Genetics, 69*, 102998. doi:10.1016/j.fsigen.2023.102998.

Jeffreys, A. J., Wilson, V., & Thein, S. L. (1985). Individual-specific 'fingerprints' of human DNA. *Nature, 316*(6023), 76–79. https://doi.org/10.1038/316076a0

Jiang, T., Li, Y., & Motsinger-Reif, A. A. (2021). Knockoff boosted tree for model-free variable selection. *Bioinformatics, 37*(7), 976–983. https://doi.org/10.1093/bioinformatics/btaa770

Jobling, M. A. (2022). Forensic genetics through the lens of Lewontin: Population structure, ancestry and race. *Philosophical Transactions of the Royal Society B, 377*, 20200422. doi:10.1098/rstb.2020.0422.

Kaczmarek, M., & Zysek, W. (2022). DNA-based predictions of human phenotypic traits for the medico-legal and forensic purposes. Postępy Biochemii, *68*(3), 264–286.

Kadam, V., Bhadane, R., Lakhani, P., & AK, G. (2023). Forensic phenotyping. *Journal of Forensic Sciences & Criminal Investigation, 16*(4). doi:10.19080/JFSCI.2023.16.555943

Kataria, S., Dabas, P., Saraswathy, K. N., Sachdeva, M. P., & Jain, S. (2023). Investigating the morphology and genetics of scalp and facial hair characteristics for phenotype prediction. *Science & Justice, 63*, 135–148.

Katsara, M., Branicki, W., Pośpiech, E., Hysiu, P., Walsh, S., Kayser, M., & Nothnagel, M. (2021). Testing the impact of trait prevalence priors in Bayesian-based genetic prediction modeling of human appearance traits. *Forensic Science Internation: Genetics, 50*, 102412. doi:10.1016/j.fsigen.2020.102412.

Kayser, M. (2015). Forensic DNA phenotyping: Predicting human appearance from crime scene material for investigative purposes. *Forensic Science International: Genetics, 18*, 33–48.

Kayser, M., Branicki, W., Parson, W., & Phillips, C. (2023). Recent advances in forensic DNA phenotyping of appearance, ancestry and age. *Forensic Science International: Genetics, 65*, 102870. doi:10.1016/j.fsigen.2023.102870.

Koops, B.-J., & Schellekens, M. (2008). Forensic DNA phenotyping: Regulatory issues. *Columbia Science and Technology Law Review, 9*, 158–202.

Kukla-Bartoszek, M., Pospiech, E., Wozniak, A., Boron, M., Karlowska-Pik, J., Teisseyre, P., Zubańska, M., Bronikowskag, A., Grzybowskih, T., Płoskii, R., Spólnickac, M., & Branicki, W. (2019). DNA-based predictive models for the presence of freckles. *Forensic Science International: Genetics, 42*, 252–259. doi:10.1016/j.fsigen.2019.07.012.

Kushwaha, P., Ahmed, U., & Rana, M. (2023). Forensic DNA phenotyping: A promising tool to predict human appearance for forensic purposes. *GSC Biological and Pharmaceutical Sciences, 23*(2), 34–41. doi:10.30574/gscbps.2023.23.2.0180.

Liu, F., van der Lijn, F., Schurmann, C., Zhu, G., Chakravarty, M. M., Hysi, P. G., & Kayser, M. (2012). A genome-wide association study identifies five loci influencing facial morphology in Europeans. *PLoS Genetics, 8*(9), e1002932.

Liu, F., van Duijn, K., Vingerling, J. R., Hofman, A., Uitterlinden, A. G., Cecile, A., Janssens, J. W., & Kayser, M. (2009). Eye color and the prediction of complex phenotypes from genotypes. *Current Biology, 19*(5), R192–R193.

Liu, F., Zhong, K., Jing, X., Uitterlinden, A. G., Hendriks, A. E. J., Drop, S. L. S., & Kayser, M. (2019). Update on the predictability of tall stature from DNA markers in Europeans. *Forensic Science International: Genetics, 42*, 8–13. doi: 10.1016/j.fsigen.2019.05.006.

Llobet, M. O., Johansson, A., Gyllensten, U., Allen, M., & Enroth, S. (2023). Forensic prediction of sex, age, height, body mass index, hip-to-waist ratio, smoking status and lipid lowering drugs using epigenetic markers and plasma proteins. *Forensic Science International: Genetics, 65*, 102871.

M'charek, A. (2008). Silent witness, articulate collective: DNA evidence and the inference of visible traits. *Bioethics, 22*(9), 519–528.
M'charek, A. (2020). Tentacular faces: Race and the return of the phenotype in forensic identification. *American Anthropologist, 122*(2), 369–380.
M'charek, A., Toom, V., & Jong, L. (2020). The trouble with race in forensic identification. *Science, Technology, & Human Values, 45*(5), 804–828.
Machado, H., & Granja, R. (2020). *Forensic Genetics in the Governance of Crime*. Palgrave Macmillan.
MacLean, C. E. (2013). Creating a wanted poster from a drop of blood: Using DNA phenotyping to generate an artist's rendering of an offender based only on DNA shed at the crime scene. *Hamline Law Journal, 36*(3), 357–386.
MacLean, C. E., & Lamparello, A. (2014). Forensic DNA phenotyping in criminal investigations and criminal courts: Assessing and mitigating the dilemmas inherent in the science. *Recent Advances in DNA and Gene Sequences, 8*, 104–112.
Matheson, S. (2016). DNA phenotyping: Snapshot of a criminal. *Cell, 166*(5), 1061–1064. doi: 10.1016/j.cell.2016.08.016
Melchionda, F., Silvestrini, B., Robino, C., Bini, C., Fattorini, P., Martinez-Labarga, C., De Angelis, F., Tagliabracci, A., & Turchi, C. (2022). Development and validation of MPS-based system for human appearance prediction in challenging forensic samples. *Genes, 13*, 1688. doi:10.3390/genes13101688.
Montesanto, A., D'Aquila, P., Lagani, V., Pararazzo, E., Geracitano, S., Formentini, L., Giacconi, R., Cardelli, M., Privinciali, M., Bellizzi, D., & Passarino, G. (2020). A new robust epigenetic model for forensic age prediction. *Journal of Forensic Sciences*. doi:10.1111/1556-4029.14460.
National Academies of Sciences, Engineering, and Medicine. (2024). Law enforcement use of probabilistic genotyping, forensic DNA phenotyping, and forensic investigative genetic genealogy technologies: Proceedings of a workshop #8212; in Brief. Washington, DC: The National Academies Press. doi:10.17226/27812.
Olds, R. (2023, September 14). Cold case cracked: Woman's murderer apprehended after nearly 30 years. *Deseret News*.
Oosthuizen, T., Howes, L. M., & White, R. (2023). Forensic DNA analysis and legislative provisions: Balancing rights in a time of scientific advancement. *Alternative Law Journal, 48*(3), 178–184.
Pandkar, M., Moe, T.-S., & Barash, M. (2022). 3d facial biometric verification using a DNA sample for law enforcement applications. *International Joint Conference on Web Intelligence and Intelligent Agent Technology*. doi:10.1109/WI-IAT55865.2022.00114.
Parabon NanoLabs. (2021). *Snapshot posters*. https://snapshot.parabon-nanolabs.com/posters (retrieved August 25, 2024)
Parabon Nanolabs. (2022, January 19). Parabon® Tops 200 solved cases: Buoyed by technological advancements, Parabon's Snapshot® team reports 67 IDs in 2021. Parabon Tops 200 Solved Cases (parabon-nanolabs.com)
Peng, F., Zhu, G., Hysi, P. G., Eller, R. J., Chen, Y., Li, Y., Hamer, M. S., Zeng, C., Hopkins, R. L., Jacobus, C. L., Wallace, P. L., Uitterlinden, A. G., Ikram, M. A., Nijsten, T., Duffy, D. L., Medland, S. E., Spector, T. D., Walsh, S., Martin, N. G., Liu, F., & Kayser, M. (2019). Genome-wide association studies identify multiple genetic loci influencing eyebrow color variation in Europeans. *Journal of Investigative Dermatology, 139*, 1601–1605.
Phillips, C. (2023). *Forensic DNA Phenotyping and the Prediction of Externally Visible Physical Characteristics*. Elsevier.
Phillips, C., de la Puente, M., Ruiz-Ramirez, J., Staniewska, A., Ambroa-Conde, A., Freire-Aradas, A., Mosquera-Miguel, A., Rodriguez, A., & Lareu, M. V. (2022). Eurasiaplex-2: Shifting the focus to SNPs with high population specificity increases the power of forensic ancestry marker sets. *Forensic Science International: Genetics, 61*, 102780.
Pośpiech, E., Kukla-Bartoszek, M., Karłowska-Pik, J., Zieliński, P., Woźniak, A., Boroń, M., Dąbrowski, M., Zubańska, M., Jarosz, A., Grzybowski, T., Płoski, R., Spólnicka, M., & Branicki, W. (2020). Exploring the possibility of predicting human head hair greying from DNA using whole-exome and targeted NGS data. *BMC Genomics, 21*, 538. doi:10.1186/s12864-020-06926-y.
Pośpiech, E., Teisseyre, P., Mielniczuk, J., & Branicki, W. (2022). Predicting physical appearance from DNA data—Towards genomic solutions. *Genes, 13*, 121. doi:10.3390/genes13010121.
Rahat, M. A., Saif, S., Shah, M., Rasool, A., Akbar, F., Ali, S., & Israr, M. (2023). Forensic DNA phenotyping. In R. Scendoni & F. De Micco (Eds.) *Forensic and Legal Medicine – State of the Art, Practical Applications and New Perspectives*. InTechOpen. https://doi.org/10.5772/intechopen.108995.
Refn, M. R., Kampmann, M., Morling, N., Tfelt-Hansen, J., Børsting, C., & Pereira, V. (2023). Prediction of chronological age and its applications in forensic casework: Methods, current practices, and future perspectives. *Forensic Sciences Research, 8*, 85–97.

Robertson, D. (2024). Cold cases and forensic breakthroughs: How science is resurrecting investigations. *Journal of Forensic Medicine*, *9*(1), 340.

Samuel, G., & Prainsack, B. (2019a). Forensic DNA phenotyping in Europe: Views "on the ground" from those who have a professional stake in the technology. *New Genetics and Society*, *38*, 119–141. doi:10.1080/14636778.2018.1549984.

Samuel, G., & Prainsack, B. (2019b). Civil society stakeholder views on forensic DNA phenotyping: Balancing risks and benefits. *Forensic Science International: Genetics*, *43*, 102157.

Samuel, G., & Prainsack, B. (2019c). Societal, ethical, and regulatory dimensions of forensic DNA phenotyping. VISAGE (Visual Attributes Through Genetics). www.visage-h2020.eu/PDF/Deliverable_5.2_for_online_publication_vo1.pdf

Samuel, G., & Prainsack, B. (2020). Report on recommendations to address the ethical and societal challenges of FDP. VISAGE (Visual Attributes Through Genetics).

Sapan, V., Sinsek, S. Z., Filoglu, G., & Bülbül, O. (2024). Forensic DNA phenotyping using Oxford nanopore sequencing system. *Electrophoresis*, *2024*, 1–14.

Schneider, P. M., Prainsack, B., & Kayser, M. (2019). The use of forensic DNA phenotyping in predicting appearance and biogeographic ancestry. Deutsches Arzteblatt *International*, *116*, 873–880. doi:10.3238/arztebl.2019.0873.

Scudder, N., Robertson, J., Kelty, S. F., Walsh, S. J., & McNevin, D. (2019). A law enforcement intelligence framework for use in predictive DNA phenotyping. *Australian Journal of Forensic Sciences*. doi:10.1080/00450618.2019.1569132.

Şenişik, M., Bülbül, Ö, & Filoğlu, G. (2023). Forensic DNA phenotyping: Male pattern baldness: Traditional review. *Turkiye Klinikleri Journal of Forensic Medicine and Forensic Sciences*, *20*(1), 64–74. doi:10.5336/forensic.2023-95361.

Sessa, F., Esposito, M., Cocimano, G., Sablone, S., Karaboue, M. A. A., Chisari, M., Albano, D. G., & Salerno, M. (2024). Artificial intelligence and forensic genetics: Current applications and future perspectives. *Applied Science*, *14*, 2113. doi:10.3390/app14052113.

Sguazzi, G., Varrone, D., Cirioni, C., Andrioletti, V., Linarello, P., Salvaderi, L., Lovisolo, F., Procopio, N., Gentile, F., Cherubini, A., Colloca, D., Marino, A., & Gino, S. (2022). Biological identikit": Development of a SNPs-panel for the analysis of forensic DNA phenotyping and ancestry. *Forensic Science International: Genetics Supplement Series*, *8*, 199–201. doi:10.1016/j.fsigss.2022.10.034.

Simayijiang, H., & Yan, J. (2023). Recent developments in forensic DNA typing. *Journal of Forensic Science & Medicine*, *9*(4), 353–359.

Skinner, D. (2020). Forensic genetics and the prediction of race: What is the problem? *BioSocieties*, *15*, 329–349. doi:10.1057/s41292-018-0141-0.

Sulley, Y. S., Oboakoh, I., & Quansah, L. (2022). The plight of unknown and nameless bodies: A proposal for forensic DNA policy action. *Forensic Genomics*, *2*(3), 67–70.

Tan, W. C. D., Stasi, A., & Dhar, B. K. (2022). Forensic DNA profiling in the southern border provinces of Thailand: Ethical and regulatory issues. *Forensic Science International*, *336*, 111322.

Taylor, M., Mayne, C., Coutts, L., Kinnane, A., Avent, I., Cho, K., Rahtouh, M., & Roffey, P. (2024). Kafka's beautiful eyes: Forensic intelligence utilization of phenotypic information. *Forensic Science International*, *361*, 112120. doi:10.1016/j.forsciint.2024.112120.

The International SNP Map Working Group. (2001). A map of human genome sequence variation containing 1.42 million single nucleotide polymorphisms. *Nature*, *409*(6822), 928–933. https://doi.org/10.1038/35057149

Toom, V., Wienroth, M., M'charek, A., Prainsack, B., Williams, R., Duster, T., & Murphy, E. (2016). Approaching ethical, legal and social issues of emerging forensic DNA phenotyping (FDP) technologies comprehensively: Reply to 'Forensic DNA phenotyping: Predicting human appearance from crime scene material for investigative purposes' by Manfred Kayser. *Forensic Science International: Genetics*, *22*, e1–e4.

Tozzo, P., Politi, C., Delicati, A., Gabbin, A., & Caenazzo, L. (2021). External visible characteristics prediction through SNPs analysis in the forensic setting: A review. *Frontiers in Bioscience-Landmark*, *26*(10), 828–850.

Van Laan, M. (2023). The genetic witness: Forensic DNA phenotyping. *Journal of Emerging Forensic Sciences Research*, *3*(1), 1–15.

van Oorschot, I., & M'charek, A. (2021). Keeping race at bay: Familial DNA research, the 'Turkish community,' and the pragmatics of multiple collectives in investigative practice. *BioSocieties*, *16*(4), 553–573. https://doi.org/10.1057/s41292-021-00246-4

Vanneste, M., Hoskens, H., Goovaerts, S., Matthews, H., Aponte, J. D., Cole, J., Shriver, M., Marazita, M. L., Weinberg, S. M., Walsh, S., Richmond, S., Klein, O. D., Spritz, R. A., Peeters, H., Hallgrímsson, B., &

Claes, P. (2023). Syndrome-informed phenotyping identifies a polygenic background for achondroplasia-like facial variation in the general population [Preprint]. *bioRxiv.* doi: 10.1101/2023.12.07.570544

Walsh, S., & Kayser, M. (2023). Prediction of physical characteristics, such as eye, hair, and skin color, based solely on DNA. In Primorac, D., & Schanfield, M. (Eds.), *Forensic DNA Applications: An Interdisciplinary Perspective* (2nd ed.). CRC Press.

Walsh, S., Liu, F., Ballantyne, K. N., van Oven, M., Lao, O., & Kayser, M. (2011). IrisPlex: A sensitive DNA tool for accurate prediction of blue and brown eye colour in the absence of ancestry information. *Forensic Science International: Genetics*, *5*(3), 170–180.

Walsh, S., Liu, F., Wollstein, A., Kovatsi, L., Ralf, A., Kosiniak-Kamysz, A., & Kayser, M. (2013). The HIrisPlex system for simultaneous prediction of hair and eye colour from DNA. *Forensic Science International: Genetics*, *7*(1), 98–115.

White, T. J., & Lee, S. B. (2020). Forensic genetics, ethics, privacy, and public policy. In H. Erlich, E. Stover, & T. J. White (Eds.), *Silent Witness.* Oxford University Press.

Wienroth, M. (2020). Socio-technical disagreements as ethical fora: Parabon NanoLab's forensic DNA Snapshot™ service at the intersection of discourses around robust science, technology validation, and commerce. *BioSocieties*, *15*(1), 28–45.

Wienroth, M. (2023). Technology in policing, policing in a technological society: Special issue brief. *International Journal of Police Science & Management*, *25*(3), 223–225.

Wienroth, M., & McCartney, C. (2024). Noble cause casuistry in forensic genetics. *WIREs Forensic Science*, *6*(1), e1502. doi:10.1002/wfs2.1502.

Xavier, C., de la Puente, M., Mosquera-Miguel, A., Freire-Aradas, A., Kalamara, V., Ralf, A., Revoir, A., Gross, T. E., Schneider, P. M., Ames, C., Hohoff, C., Phillips, C., Kayser, M., & Parson, W. (2022). Development and inter-laboratory evaluation of the VISAGE enhanced tool for appearance and ancestry inference from DNA. *Forensic Science Internation: Genetics*, *61*, 102779.

Yuan, M., Goovaerts, S., Vanneste, M., Matthews, H., Hoskens, H., Richmond, S., Klein, O. D., Spritz, R. S., Hallgrimsson, B., Walssh, S., Shriver, M. D., Shaffer, J. R., Weinberg, S. M., Peeters, H., & Claes, P. (2024). Mapping genes for human face shape: Exploration of univariate phenotyping strategies. *BioRxiv* [Preprint]. doi:10.1101/2024.06.06.597731.

Zieger, M. (2022). Forensic DNA phenotyping in Europe: How far may it go? *Journal of Law and the Biosciences*, 1–22. doi:10.1093/jlb/lsac024.

14

Streamlining Forensic DNA Analysis with RAPID DNA Typing

Current Developments and Future Directions

D.S. Yadav, Roshan Kumar, and Tanya Gupta

14.1 Introduction

Over the past decade, RAPID DNA technology has become more popular due to the demand for forensic analysis that is quicker and more effective. While RAPID DNA systems can provide profiles in less than two hours, traditional DNA profiling methods might take days or even weeks to produce results (Turingan et al., 2016). This speed is particularly beneficial in urgent situations, such as identifying suspects at crime scenes or during mass casualty events (De Roo et al., 2023). Recent studies have demonstrated the effectiveness of RAPID DNA systems in various forensic contexts. For instance, a study by Hares et al. (2020) highlighted the successful implementation of RAPID DNA technology in law enforcement agencies, where it was used to quickly identify suspects in property crimes and sexual assaults. The ability to produce actionable intelligence in real-time has made RAPID DNA a valuable tool in modern forensic investigations. Additionally, a study by Gettings et al. (2015) evaluated the performance of RAPID DNA systems in a laboratory setting, demonstrating their reliability and accuracy in generating DNA profiles.

14.2 Fundamentals of RAPID DNA typing

14.2.1 Instrumentation

RAPID DNA typing systems are compact, automated instruments that integrate multiple steps of the DNA profiling process. These systems typically include components for sample extraction, amplification, and analysis. For example, the Rapid DNA Analysis System (RDAS) developed by the FBI incorporates a microfluidic platform that allows for the simultaneous processing of multiple samples. Other commercially available RAPID DNA systems, such as the Rapid HIT ID System (Integen X) and the DNA scan Rapid DNA Identification System (Thermo Fisher Scientific), have also been widely adopted by law enforcement agencies (Hares et al., 2020).

14.2.2 Sample preparation

Sample preparation is a critical step in RAPID DNA typing. The technology is designed to handle a variety of biological samples, including blood, saliva, and touch DNA. Automated

DOI: 10.4324/9781003474296-14

extraction methods, such as magnetic bead-based techniques, have been developed to efficiently isolate DNA from these samples while minimizing contamination (Turingan et al., 2016). Studies have shown that RAPID DNA systems can successfully process samples with varying DNA concentrations, making them suitable for a wide range of forensic applications (Gettings et al., 2015).

14.2.3 Data processing techniques

Once the DNA is extracted, it undergoes amplification using Polymerase Chain Reaction (PCR) techniques. RAPID DNA systems utilize optimized PCR protocols that enable the amplification of specific genetic markers, particularly Short Tandem Repeats (STRs). The amplified DNA is then analyzed using capillary electrophoresis or other detection methods to generate a DNA profile (Hares et al., 2020). Recent advancements in microfluidic technology have allowed for the integration of these processes within a single, compact instrument, further streamlining the DNA profiling workflow (De Roo et al., 2023).

14.3 Fundamentals of RAPID DNA typing instrumentation

RAPID DNA typing systems are designed to automate the entire DNA profiling process. These systems typically consist of three main components: sample extraction, amplification, and analysis. The automation of these processes minimizes the risk of contamination and human error, which are critical factors in forensic investigations (Turingan et al., 2016).

1. ***Sample Extraction***: The extraction of DNA from biological samples is a crucial step in RAPID DNA typing. Automated extraction methods, such as magnetic bead-based techniques, have been developed to efficiently isolate DNA while minimizing contamination (De Roo et al., 2023).
2. ***DNA Amplification***: The extracted DNA is amplified using PCR techniques. RAPID DNA systems utilize optimized PCR protocols that enable rapid cycling, allowing for the amplification of specific genetic markers, particularly STRs, in a matter of minutes (Laurin et al., 2023). This enhancement is crucial for generating profiles quickly in time-sensitive situations.
3. ***Data Analysis***: The amplified DNA samples are analyzed using capillary electrophoresis or other detection methods to generate DNA profiles. The resulting profiles are then compared against reference databases for identification purposes. The integration of software algorithms for data interpretation has improved the accuracy and reliability of RAPID DNA typing (Hares et al., 2020).

14.4 Significant improvements in DNA extraction tools for RAPID DNA typing

The advancements in DNA extraction tools for RAPID DNA typing have played a crucial role in enhancing the efficiency and reliability of forensic DNA analysis. Key improvements include:

- ***Automation of Extraction Processes***: Recent advancements in automated DNA extraction systems have streamlined the process of isolating DNA from biological samples. These systems reduce the risk of contamination and human error, which are critical factors in forensic investigations. For example, the use of robotic liquid handling systems has enabled the simultaneous processing of multiple samples, significantly increasing throughput (Sarik et al., 2020). Automated platforms, such as the QIAcube (QIAGEN), allow for standardized protocols that enhance reproducibility and reliability in DNA extraction (Ketsekioulafis et al., 2024).

- ***Magnetic Bead-Based Extraction***: One of the most significant advancements in DNA extraction tools is the use of magnetic bead-based methods. These techniques allow for efficient and rapid isolation of DNA from various biological samples, including blood, saliva, and touch DNA. Magnetic beads facilitate the binding of DNA, enabling easy separation from contaminants and inhibitors (De Roo et al., 2023). Studies have shown that magnetic bead-based extraction can yield higher-quality DNA compared to traditional silica-based methods, making it particularly valuable for RAPID DNA applications (Laurin et al., 2023).
- ***Microfluidic Devices***: The integration of microfluidic technology in DNA extraction tools has revolutionized the sample processing workflow. Microfluidic devices allow for the manipulation of small volumes of fluids, enabling efficient DNA extraction and purification in a compact format. This technology reduces the amount of sample required and enhances the speed of the analysis process (Hares et al., 2020). Microfluidic systems can also be designed to automate multiple steps, from lysis to purification, further streamlining the extraction process (Bowman et al., 2022).
- ***Rapid Lysis Techniques***: The development of enhanced lysis techniques has improved the efficiency of DNA extraction from various biological samples. New chemical lysis buffers and enzymatic methods enable faster cell lysis and DNA release, significantly reducing extraction times (Barbaro et al., 2011). These rapid lysis techniques are particularly beneficial when working with challenging samples, such as those with low DNA yield or degraded DNA. Studies have indicated that optimized lysis conditions can improve the overall yield and quality of extracted DNA, making it more suitable for downstream applications (Menchhoff et al., 2019).
- ***Optimized Protocols for Low-Template DNA***: The development of optimized extraction protocols specifically designed for low-template DNA has enhanced the applicability of RAPID DNA technology in forensic investigations. These protocols utilize specialized reagents and conditions that maximize DNA recovery from limited or degraded samples (Laurin et al., 2023). Research has demonstrated that these optimized methods can improve the success rate of DNA profiling in cases where traditional methods may fail.
- ***Integration of Artificial Intelligence***: The integration of artificial intelligence (AI) into DNA extraction processes is an emerging trend that promises to enhance the efficiency and accuracy of forensic DNA analysis. AI algorithms can optimize extraction protocols by analyzing sample characteristics and recommending the most effective extraction methods (Ketsekioulafis et al., 2024). This technology can also assist in real-time monitoring of extraction processes, ensuring that optimal conditions are maintained throughout the procedure (Barash et al., 2023). The use of AI in DNA extraction tools is expected to improve the overall reliability of forensic DNA profiling.

14.5 Various systems available for RAPID DNA analysis and their limitations

Several firms are in the process of launching and updating their instruments for providing faster DNA analysis. Some of the widely used systems include

- ***RapidHIT***: By Thermo-Fisher Scientific, is known for its ease of use and portability, it can quickly generate forensic DNA profiles and is used in various settings, including crime scenes and booking stations.
- ***ANDE***: Often used in disaster victim identification and other forensic applications.
- ***ParaDNA***: This system is also noted for its performance.

14.5.1 Applied biosystems RapidHIT ID system

The RapidHIT ID System, a fully automated "sample-to-answer" platform for Rapid DNA analysis, is a significant development in forensic science. In sharp contrast to the multi-day turnaround time characteristic of traditional laboratory procedures, it significantly cuts the time-to-result for DNA profiles to about 90 minutes. All the necessary steps to generate a DNA profile are automated by this integrated system in a single, standalone device. The procedure necessitates very little human involvement. A comprehensive STR profile is generated by the instrument's software once it has interpreted the CE data. This profile is prepared for quick and safe upload to law enforcement databases or national DNA databases such as the US CODIS (Combined DNA Index System). Although this approach is perfect for reference samples of superior quality acquired in booking stations; however, more complex or degraded forensic casework samples may still require manual processing and expert interpretation in a traditional laboratory setting.

The U.S. Federal Bureau of Investigation (FBI) has officially approved the use of the Applied Biosystems RapidHIT™ ID System, an important step in forensic science, within the strict parameters for its DNA analysis infrastructure. This authorization is only given to forensic DNA labs that have been authorized by the FBI's exacting Quality Assurance Standards (QAS), guaranteeing that every participating facility already complies with stringent validation, process, and quality control standards. The approval's scope is precisely defined: the system is validated for processing **single-source DNA samples**. This refers to evidentiary material, most commonly referenced buccal swabs from arrestees that contain genetic material from only one individual. Such samples yield clean, unambiguous electropherograms, making them ideal for automated analysis. This contrasts with more complex forensic casework samples, such as DNA mixtures from multiple contributors or low-template/degraded samples, which still demand the nuanced interpretation of a highly trained DNA analyst.

Importantly, the FBI's validation demonstrates that the RapidHIT ID's integrated Expert System Software (ESS) is reliable enough to produce profiles that don't require manual interpretation and evaluation when searching against the National DNA Index System (NDIS), the highest tier of the CODIS database. A trained analyst carefully checks the raw data for artifacts such as dye blobs, allelic dropout, and stutter in a conventional approach. By employing proven algorithms to generate accurate genotyping calls for every STR locus, the ESS automates this crucial step. The foundation of the Rapid DNA initiative is this "human-out-of-the-loop" approval for single-source samples, which allows law enforcement to significantly speed up the identification of connections between arrestees and unsolved crimes nationwide.

The DNA profiling evidence plays a significant role in forensic investigations and in solving crimes. However, generating DNA profiles from exhibits is a time taking and tedious exercise. Hence, there was a need for innovative technology that allows for quick DNA profiling and significantly reduces the time it takes to identify suspects and solve cases. Traditionally, DNA testing could take days, weeks, or even months, potentially allowing criminals to evade justice and put more people at risk. Hence, there was a necessity to deal with such demand, and the *Applied Biosystems RapidHIT ID System* is one of such equipment that can generate forensic

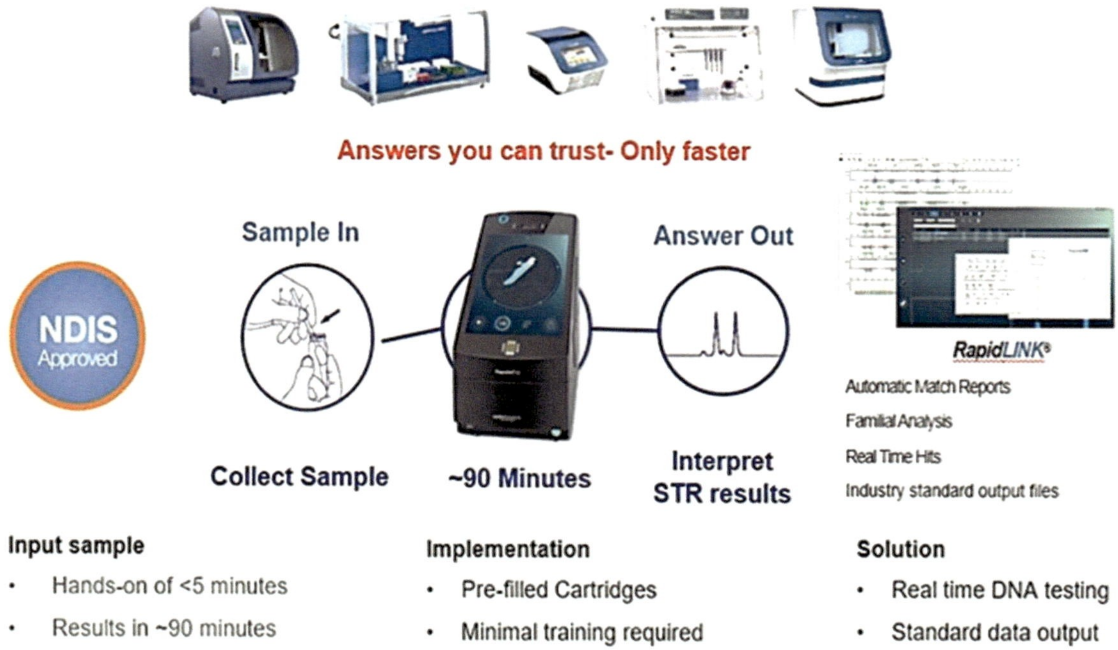

Figure 14.1 *Applied biosystems RapidHIT ID system.*

DNA profiles in as little as 90 minutes, drastically accelerating the investigative process. The device is compact and simple to operate, requiring no specialized technical expertise. This means that law enforcement agencies, even those in remote or resource-limited areas, can quickly access forensic DNA analysis without needing to rely on specialized labs or technicians (**Figure 14.1**).

- **Solve Crime Faster with Rapid DNA**: Fight record-breaking crime rates: With the RapidHIT ID, in as little as 90 minutes, the booking, charging, and investigating agencies are automatically notified if an arrestee's DNA matches DNA from unsolved cases.
- **Generate Timely Investigative Leads**: Low clearance rates are having a lasting impact on public safety.
- **Deter and Prevent Crime**: The RapidHIT ID System strengthens the power of DNA databases to support law enforcement. Registering arrestees saves money: INR5000–INR20,000 saved for each DNA profile included in the database.
- **Hold offenders Accountable**: When lawfully owed arrestee samples are missing from the database, investigations are negatively impacted, and the impact of DNA enrollment on recidivism and criminal justice cost savings is diminished. Studies show recidivism of violent and property crime offenders can be curtailed, with the largest impact on recidivism attributed to offenders enrolled following their first offense. Automated, rapid DNA processing helps ensure all DNA profiles from qualifying arrestees are enrolled while offenders are still early in their criminal careers—where the biggest effects on deterrence and public safety can occur.

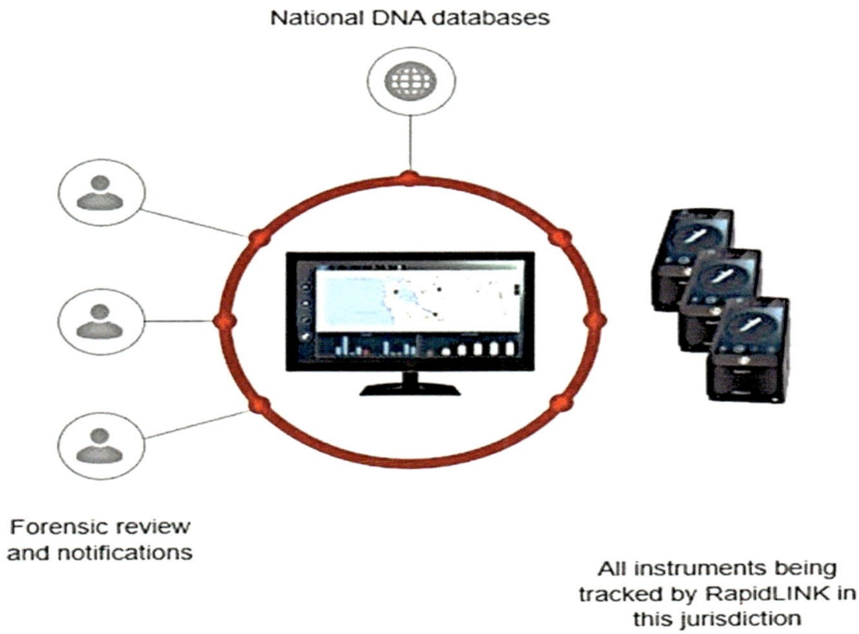

Figure 14.2 *Applications of RapidHIT ID system.*

- **Integrating DNA within Existing Biometric Enrollment Systems**: It ensures DNA is captured for qualifying arrestees early in the investigative process. By using rapid DNA, law enforcement can maximize the strength of these investigative tools and have more direct and immediate impacts on crime-fighting abilities.

14.5.2 The RapidHIT ID system

- The RapidHIT ID system can be used with networked software to allow multiple systems at different locations to be monitored centrally.
- It can be supplied with software allowing administrative users to control the instrument remotely and securely review results.
- Samples processed can be successfully recovered from the processed sample cartridge, enabling retesting with the same system or with other traditional laboratory methods (**Figure 14.2**).

14.5.3 Instrument

The instrument is compact in design, with dimensions of 48 cm in height, 53 cm in length, and 27 cm in width. It is compatible with self-contained, user-friendly ACE and INTEL sample cartridge formats and operates using a primary reagent cartridge that includes all necessary reagents and consumables required for capillary electrophoresis of up to 100 samples.

The system weighs approximately 25.4 kg without the primary cartridge and 28.4 kg with the cartridge installed. It is equipped with a solid-state, long-life laser and offers multiple built-in access control features, including a facial recognition camera, fingerprint reader, and PIN/password authentication.

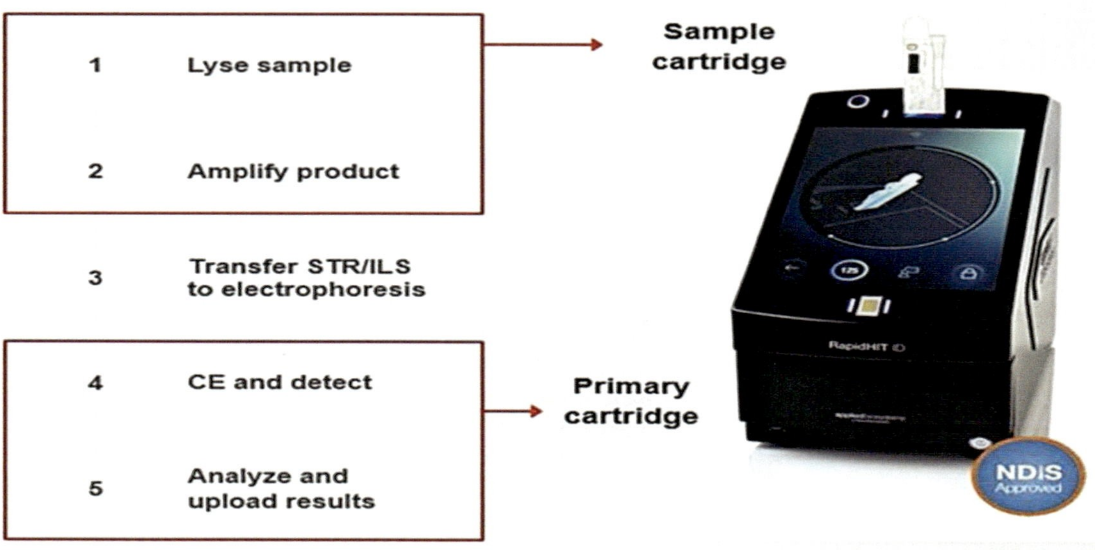

Figure 14.3 *Working with the applied biosystems ACE global filer express and RapidINTEL plus cartridge system.*

Additionally, the instrument features automatic calibration of consumables, reducing user intervention and maintenance costs. It includes an integrated refrigeration system for the primary cartridge gel, allowing storage and usage for up to 6 months on the instrument.

The system supports six-dye detection capability and is capable of generating a complete DNA profile in as little as 90 minutes. It provides high-resolution DNA sizing within the range of 60 to 460 base pairs, with a standard deviation of less than 0.15 bp for a given allele, and achieves single base-pair resolution (**Figure 14.3**).

14.5.3.1 Instrument on-board computer

- The instrument can run in standalone mode, without a computer attached.
- It includes a touchscreen interface that can start the instrument without the need for an attached computer.
- It contains on-board storage capability with memory capacity (100 GB) to store approximately 650 runs.
- Its touch screen interface allows for the creation of user accounts with PIN/password, fingerprint, or facial recognition-protected access.
- Its touchscreen interface primarily uses imagery to instruct users to maximize universal understanding.
- Its touchscreen interface allows for time zone customization.
- - It enables consumables tracking through the Radio-Frequency Identification (RFID) reader on the instrument and RFID chips in the consumables.
- The instrument also allows for manual entry of barcodes of other non-RFID consumables.
- It enables network security through on-instrument encryption, VPN, and Active Directory.

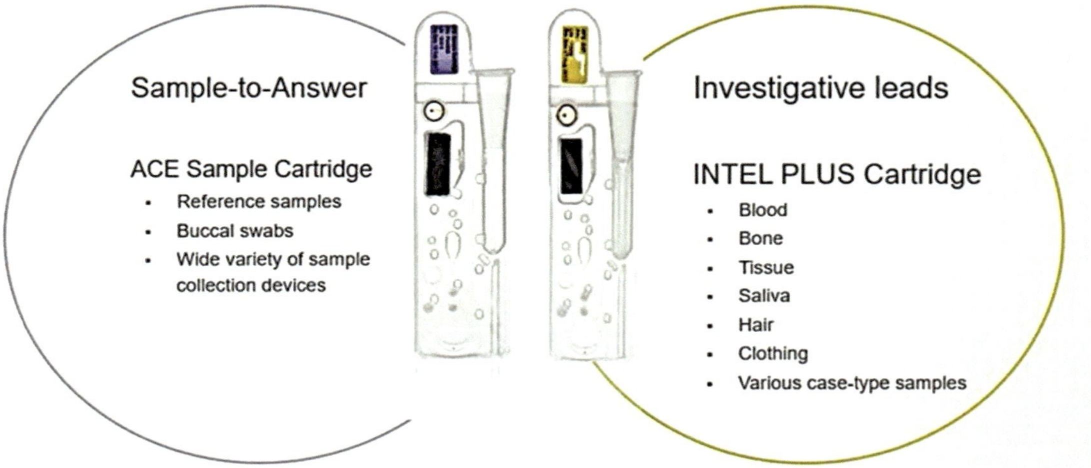

Figure 14.4 *Various sample types analyzed through the respective cartridges.*

14.5.4 Consumables

The Applied Biosystems ACE Global Filer Express and RapidINTEL Plus Sample Cartridges contain all necessary reagents for sample preparation, including Applied Biosystems GlobalFiler Express chemistry. The sample cartridges also have an RFID chip that contains information that includes lot number and expiration date.

- This is a single-use consumable with up to 12 months shelf life.
- The sample cartridges accept standard cotton swabs and do not require specialized swab or sample collection equipment.
- The Applied Biosystems Primary Cartridge includes the capillary array, optimized polymer/gel, and buffers for system installation and 150 sample runs. It has up to 12 months' shelf life from the date of manufacture.
- The Applied Biosystems ACE GlobalFiler Express Positive and Negative control cartridges, when used regularly, give the user confidence that the system is performing to specification.
- The sample cartridges are packaged in colored Mylar pouches to visually differentiate different sample cartridges and to protect sensitive reagents from light.
- The sample cartridges are labeled using colored labels to visually differentiate different sample cartridge types (**Figure 14.4**).

14.5.5 Analysis software

The instrument software is designed with a touchscreen interface for instrument control to regulate the functions of the instrument and perform automatic primary analysis. If the system detects potential issues associated with the data analysis, flags are generated and associated with the run data. Different colored flags denote the following run status:

- Green – profile generated, no flags associated

- Yellow – profile generated with associated flags and may require review in the Applied Biosystems RapidLINK v2.0 (and higher) software.
- Red – no profile generated, proceed with troubleshooting.

The Applied Biosystems RapidLINK v2.0 (and higher) software is an external software that is part of the Applied Biosystems RapidHIT ID System, as it:

- securely manages one or more RapidHIT ID systems from a central location;
- allows the review of DNA profiles in real time prior to uploading to a third-party database;
- enables and disables users in real time;
- collects and allows for the review of metadata;
- allows the system to be Expert System-capable through sample accessioning capable workflows and the creation of RCMF or CMF files;
- contains Match, Familial, Staff Elimination Database, and Kinship software applications;
- is a database that stores and provides the status of all RapidHIT ID system runs; and
- is designed to export data in PDF and CSV format.

The instrument can connect with RapidLINK v2.0 (and higher) software for data upload and instrument monitoring through direct connection (via Ethernet cable) and wired/wireless networks.

14.6 Possible future applications of RAPID DNA technology in forensic DNA analysis

RAPID DNA typing technology has emerged as a transformative tool in forensic science, enabling law enforcement agencies to obtain DNA profiles quickly and efficiently. This chapter explores the current state of RAPID DNA technology, its fundamental principles, applications in criminal investigations, challenges, and potential future advancements.

14.6.1 Future directions

The future of RAPID DNA technology in forensic analysis holds great promise, with ongoing research and development aimed at addressing current limitations and expanding its applications.

1. **Miniaturization and Portability**: Ongoing research aims to develop smaller, more portable RAPID DNA systems that can be deployed in remote locations or at crime scenes. This would enhance the technology's accessibility and utility in various forensic contexts (De Roo et al., 2023). The development of handheld devices for on-site DNA analysis could revolutionize the way forensic investigations are conducted in the field (Turingan et al., 2016).
2. **Increased Multiplexing**: Advances in microfluidic technology may enable the simultaneous analysis of multiple genetic markers, improving the discriminatory power of DNA profiles and enhancing the accuracy of identifications (Hares et al., 2020). Research by Bowman et al. (2022) suggested that increasing the number of STR markers analyzed could provide greater resolution in complex cases, such as those involving mixed samples.

3. **Integration with Other Forensic Disciplines**: Future developments may see RAPID DNA technology integrated with other forensic techniques, such as fingerprint analysis or digital forensics, providing a more comprehensive approach to crime scene investigations (Hares et al., 2020). The combination of multiple forensic methodologies could enhance the overall effectiveness of investigations and lead to more accurate conclusions (Laurin et al., 2023).

RAPID DNA typing represents a significant advancement in forensic DNA analysis, offering rapid, efficient, and accurate profiling capabilities. Key improvements in DNA extraction tools, including automation, magnetic bead-based methods, microfluidic devices, rapid lysis techniques, and optimized protocols for low-template DNA, have enhanced the effectiveness of RAPID DNA systems. As the technology continues to evolve, it will play an increasingly important role in forensic investigations, victim identification, and the administration of justice. Addressing the challenges and limitations associated with RAPID DNA technology will be crucial to ensuring its effective and responsible use in forensic applications. Future research and development efforts should focus on enhancing the technology's capabilities, expanding its applications, and integrating it with other forensic disciplines to maximize its potential impact on the field of forensic science.

14.7 Potential applications of RAPID DNA typing in criminal investigations

RAPID DNA typing offers a range of potential applications in criminal investigations, enhancing the ability to solve crimes and identify individuals involved in criminal activities. The following are some key applications:

- **Immediate Suspect Identification**: One of the most significant applications of RAPID DNA typing is its ability to provide immediate identification of suspects. This capability is particularly useful in cases where a suspect is apprehended at the scene of a crime. Studies have shown that RAPID DNA systems can generate profiles from evidence samples and compare them to known offender databases in real-time, facilitating swift arrests (De Roo et al., 2023; Laurin et al., 2023). For instance, the use of RAPID DNA technology in property crimes has led to successful identifications and subsequent arrests, demonstrating its effectiveness in urgent situations (Gettings et al., 2015).
- **Victim Identification in Mass Casualty Events**: RAPID DNA typing can play a crucial role in identifying victims in mass casualty incidents, such as natural disasters or terrorist attacks. The technology allows for rapid processing of family reference samples and comparison with unidentified remains, significantly speeding up the identification process (Turingan et al., 2016). A study by Hares et al. (2020) highlighted the successful application of RAPID DNA in victim identification during the aftermath of mass casualty events, showcasing its potential to aid in disaster response efforts.
- **Kinship Analysis**: RAPID DNA typing can assist in establishing familial relationships, which is particularly useful in cases of missing persons or paternity disputes. The ability to quickly generate profiles allows investigators to explore potential familial connections (Hares et al., 2020). Research has indicated that RAPID DNA technology can be effectively used in kinship analysis, providing critical information in cases where traditional methods may be insufficient ((Bowman et al., 2022).

- **Cold Case Investigations**: The application of RAPID DNA typing in cold case investigations has the potential to yield new leads and solve previously unsolved crimes. By re-examining evidence from old cases using RAPID DNA technology, law enforcement agencies can generate new profiles and compare them to current databases, potentially identifying suspects who were previously unknown (Laurin et al., 2023). This approach has been successfully implemented in several jurisdictions, leading to breakthroughs in long-standing cases (De Roo et al., 2023).
- **On-Site DNA Profiling**: The portability and efficiency of RAPID DNA systems allow for on-site DNA profiling at crime scenes. This capability enables investigators to obtain DNA profiles from evidence samples in real-time, facilitating immediate decision-making regarding suspects and investigative directions (Gettings et al., 2015). The use of portable RAPID DNA devices can enhance the overall efficiency of crime scene investigations and improve the chances of obtaining actionable intelligence quickly (Turingan et al., 2016).
- **Integration with Other Forensic Techniques**: Future applications of RAPID DNA typing may include its integration with other forensic techniques, such as fingerprint analysis or digital forensics. Combining DNA profiles with other evidence types can provide a more comprehensive understanding of criminal activities and strengthen cases against suspects (Hares et al., 2020). This multidisciplinary approach can lead to more accurate conclusions and improve the overall effectiveness of forensic investigations.

With the development of rapid DNA profiling technology, communities can become safer by solving crimes faster, and suspects can be apprehended earlier, preventing further harm to potential victims and preventing future crimes.

14.8 Challenges and limitations of RAPID DNA technology

Despite its advantages, RAPID DNA technology faces several challenges and limitations that must be addressed to maximize its potential in forensic investigations.

- **Sample Quality**: The efficacy of RAPID DNA typing is highly dependent on the quality and quantity of the DNA sample. Degraded or low-template samples may yield unreliable results, limiting the technology's applicability in certain cases (De Roo et al., 2023). Research by Laurin et al. (2023) indicated that while RAPID DNA systems can process challenging samples, there remains a need for further optimization to enhance performance with degraded DNA.
- **Database Limitations**: The success of RAPID DNA profiling relies on the existence of comprehensive DNA databases for comparison. Variability in database size and quality across jurisdictions can hinder the identification process (Turingan et al., 2016). A study by Hares et al. (2020) noted that the effectiveness of RAPID DNA technology is contingent upon the availability of robust reference databases, which can vary significantly between regions.
- **Cost and Infrastructure**: Implementing RAPID DNA technology requires significant investment in specialized instruments and training for personnel. This can pose a barrier to widespread adoption, particularly in resource-limited settings (Hares et al., 2020). The costs associated with maintaining and operating RAPID DNA systems may also deter some law enforcement agencies from fully integrating this technology into their workflows (Bowman et al., 2022).

References

Barash, M., McNevin, D., Fedorenko, V., & Giverts, P. (2023). Machine learning applications in forensic DNA profiling: A critical review. *Forensic Science International Genetics*, *69*, 102994. https://doi.org/10.1016/j.fsigen.2023.102994

Barbaro, A., Cormaci, P., & Falcone, G. (2011). Validation of BTATM lysis buffer for DNA extraction from challenged forensic samples. *Forensic Science International. Genetics Supplement Series*, *3*(1), e61–e62. https://doi.org/10.1016/j.fsigss.2011.08.030

Bowman, Z., Daniel, R., Gerostamoulos, D., Woodford, N., & Hartman, D. (2022). Rapid DNA from a disaster victim identification perspective: Is it a game changer? *Forensic Science International Genetics*, *58*, 102684. https://doi.org/10.1016/j.fsigen.2022.102684

De Roo, R., Mapes, A., Van Cooten, M., Van Hooff, B., Kneppers, S., Kokshoorn, B., Valkenburg, T., & De Poot, C. (2023). Introducing a rapid DNA analysis procedure for crime scene samples outside of the Laboratory—A field experiment. *Sensors*, *23*(8), 4153. https://doi.org/10.3390/s23084153

Gettings, K. B., Aponte, R. A., Vallone, P. M., & Butler, J. M. (2015). STR allele sequence variation: Current knowledge and future issues. *Forensic Science International Genetics*, *18*, 118–130. https://doi.org/10.1016/j.fsigen.2015.06.005

Hares, D. R., Kneppers, A., Onorato, A. J., & Kahn, S. (2020). Rapid DNA for crime scene use: Enhancements and data needed to consider use on forensic evidence for State and National DNA Databasing – An agreed position statement by ENFSI, SWGDAM and the Rapid DNA Crime Scene Technology Advancement Task Group. *Forensic Science International Genetics*, *48*, 102349. https://doi.org/10.1016/j.fsigen.2020.102349

Ketsekioulafis, I., Filandrianos, G., Katsos, K., Thomas, K., Spiliopoulou, C., Stamou, G., & Sakelliadis, E. I. (2024). Artificial Intelligence in Forensic Sciences: A Systematic review of past and current applications and future perspectives. *Cureus*, *16*(9), e70363. https://doi.org/10.7759/cureus.70363

Laurin, N., Boulianne, H., & Frégeau, C. (2023). Comparative analysis of two Rapid DNA technologies for the processing of blood and saliva-based samples. *Forensic Science International Genetics*, *67*, 102928. https://doi.org/10.1016/j.fsigen.2023.102928

Menchhoff, S. I., Delacruz, M. T., Hytinen, M. E., Cox, J. O., Miller, M. T., & Cruz, T. D. (2019). DNA purification cell lysis and wash step modifications for Low-Template DNA sample processing. *Journal of Forensic Sciences*, *65*(2), 597–600. https://doi.org/10.1111/1556-4029.14203

Sarik, H., Naughton, K., & Miller, K. W. (2020). Validation of a High-Throughput Automated Liquid Handling DNA extraction system to maximize efficiency in forensic casework. *Forensic Genomics*, *1*(1), 16–26. https://doi.org/10.1089/forensic.2020.0001

Turingan, R. S., Vasantgadkar, S., Palombo, L., Hogan, C., Jiang, H., Tan, E., & Selden, R. F. (2016). Rapid DNA analysis for automated processing and interpretation of low DNA content samples. *Investigative Genetics*, *7*(1), 2. https://doi.org/10.1186/s13323-016-0033-7

15

Exploring the Role of Epigenetic Factors in Forensic Science

Chavi Bhardwaj, Jayapriya Panicker, and Ajay Kumar Rana

15.1 Introduction

Epigenetics encompasses the study of how environmental factors influence gene expression through mechanisms such as chromatin remodelling and non-coding RNA regulation. Epigenetic mechanisms regulate gene expression without changing the DNA sequence, playing a crucial role in diverse gene expression and consequential phenotypic characteristics across cell types, tissues, and physical appearance, ranging from unicellular microbes to multicellular organisms, including humans (Moosavi 2016; Wang et al. 2021a). Epigenetics at the molecular level encompasses modifications such as methylation, acetylation, phosphorylation, ribosylation, and ubiquitination as well as gene expression changes through the activity of special RNAs such as long non-coding RNAs and microRNAs (miRNA) (Abdul et al. 2017; Al Aboud et al. 2023; Alegria-Torres et al. 2011; Glynn 2020; Handy et al. 2011; Rocchi et al. 2020; Statello et al. 2020; Yao et al. 2019; Zhang et al. 2020). DNA methylation and miRNA analysis are the most extensively studied and characterized epigenetic markers, which have shown potential for use in identifying biological samples in forensic science (Gerra et al. 2024). DNA methylation in general is catalysed by specific DNA methyltransferases, transferring methyl group from the cofactor S-adenosyl-L-methionine (SAM) to the specific position of DNA's nitrogenous bases. For example, the cytosine base is modified to 5mC and N4-methylcytosine while adenine base is modified to N6-methyladenine (m6A) through their specific methyltransferases (Auclair and Weber 2012; Kumar et al. 2019). Further diversifying the landscape of cytosine modifications are 5-hydroxymethylcytosine (5hmC), 5-formylcytosine (5fC), and 5-carboxycytosine (5caC) (Kitsera et al. 2017). The presence of these modifications, including 5mC, is often tied to cell-specific epigenetic expression patterns or impacted by environmental factors (Szulik et al. 2015).

5mC, the most thoroughly examined methylated base, covers an additional layer of epigenetic information onto the genetic code, serving as a significant marker of apparent phenotypic variances discernible in humans or human tissues (Heyn et al. 2013). Extensive progress in detecting and measuring 5mC has aided in linking epigenetic modifications to a wide range of phenotypic changes, potentially serving as an autonomous aspect of forensic analysis or supporting the findings of existing STR tests (Richards 2006). In humans, around 3% of the DNA is identified as being methylated with 5mC, predominantly occurring at CpG dinucleotides by the activity of three chief DNA methyltransferases, viz. DNMT1, DNMT3a, DNMT3b (Aberg et al. 2012) (**Figure 15.1**). Two mechanisms can eliminate this methylation signature: active DNA demethylation, facilitated by the ten-eleven translocation (TET) family enzymes (Kohli and Zhang 2013), and passive DNA demethylation, which occurs during DNA replication

DOI: 10.4324/9781003474296-15

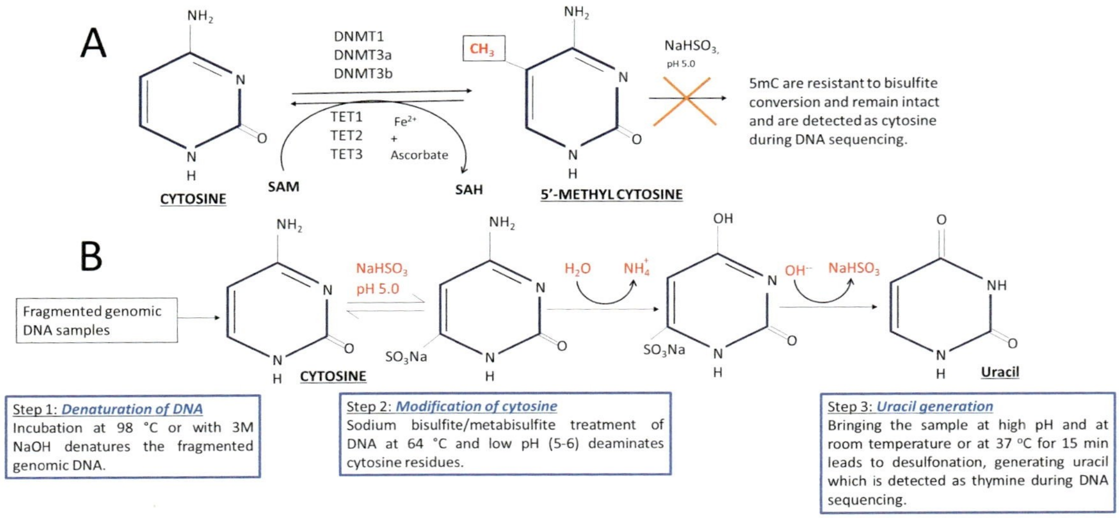

Figure 15.1 *Cytosine DNA methylation and bisulfite conversion:* ***A.*** *5mC DNA methylation in humans catalysed by three distinct DNA methyltransferases: DNMT1, DNMT3a, and DNMT3b using cofactor/methyl donor S-adenosyl-L-methionine (SAM), which is converted to S-adenosyl-L-homocysteine after donating methyl group. The methylation status of cytosine can be reversed by the ten-eleven translocation (TET) enzymes (TET1, TET2, and TET3), and requires ferrous ions and ascorbic acid as cofactors to catalyse the oxidation of 5mC, ultimately leading to its demethylation. Methylated cytosine is resistant to treatment with sodium bisulfite ($NaHSO_3$).* ***B.*** *Bisulfite conversion is carried out on denatured and fragmented DNA. The fragmented DNA is treated at low pH (pH5.0) and high temperature (64°C) with sodium bisulfite/sodium metabisulfite, forming a cytosine-bisulfite adduct. This adduct then undergoes deamination, resulting in the formation of a uracil-bisulfite intermediate. Finally, desulfonation is carried out by raising pH and bringing the sample to room temperature, leading to the conversion of the adduct into uracil. Methylated cytosines resist this conversion due to steric hindrance from the methyl group, allowing for the distinction between methylated and unmethylated cytosines through subsequent DNA sequencing or amplification techniques.*

when there's a lack of DNA methylation maintenance activity during several cycles of DNA replication (Roldán-Arjona and Rariza 2009; Rasmussen and Helin 2016). This process renders DNA methylation marks reversible and dynamic, capable of being adjusted according to cellular needs and environmental cues. Age prediction holds particular relevance in narrowing down the pool of suspects during investigations, made feasible by the fact that DNA methylation patterns undergo changes as individuals age (Bocklandt et al. 2011), and the use of a "DNA methylation clock" to estimate age with high accuracy (Bell et al. 2019). Within human chromosomal DNA, researchers have pinpointed differentially methylated regions (DMRs). These regions reveal variations in methylation patterns not only between individuals but also across different tissues within the same person (Choufani et al. 2011). When these DMRs appear consistently across multiple tissues, they are termed tDMRs, providing crucial molecular insights into epigenetic alterations across diverse biological landscapes. Modern next-generation sequencing (NGS) techniques now facilitate the detailed study of DMRs. Notably, bisulfite conversion of genomic DNA followed by next-generation sequencing (BS-seq) stands out as the preferred and standard method for analyzing the 5mC profile of genomic DNA. 5mC methylation analysis from human mitochondrial DNA (mtDNA) has also been implicated as a candidate for various regulatory roles (Devall et al. 2022; Huang et al. 2022; Liu et al. 2016; Rana 2018b); however, it has been found to be in very low copy number or absent altogether

(Bicci et al. 2021; Guitton et al. 2022), making it a controversial topic of research. Recently, the epigenetic marker m6A in mtDNA has been identified and is now being actively pursued and explored for its potential regulatory roles and applications in the diagnostic field (Hao et al. 2020; Koh et al. 2018).

MicroRNA (miRNA) is another marker involved in epigenesis that is directly controlled by DNA methylation during its biogenesis in mammalian cells (Glaich et al. 2019). The miRNA is synthesized by RNA polymerase II (RNA pol II) in the nucleus (Lee et al. 2004) as pri-miRNA. The pri-miRNA is recognized by DGCR8 and then cleaved by Drosha (class 3 RNase III) into pre-miRNA. The pre-miRNA is then bound by Exportin-5/Ran-GTP complex, which actively translocates pre-miRNA through the nuclear pore complex (NPC) (**Figure 15.2**). In the cytoplasm, the pre-miRNA is cleaved into mature miRNA by class 4 RNase III (Dicer), and subsequently miRNA is incorporated into an RNA-induced silencing complex (RISC), which together interacts with specific mRNA to repress, cleave, or deadenylate the mRNA (Carthew 2006; Murchison and Hannon 2004; Wu et al. 2006). These non-coding single-stranded miRNA range from 18 to 25 nucleotides (average ~22 nucleotides) in length, which are involved in mRNA regulations (Ying et al. 2008). They target mainly on the 3′UTR of the mRNA sequence to inhibit translation (Didiano and Hobert 2008) or through affecting deadenylation of mRNA and its decay (Eulalio et al. 2009). MiRNA analysis has recently emerged as a promising tool

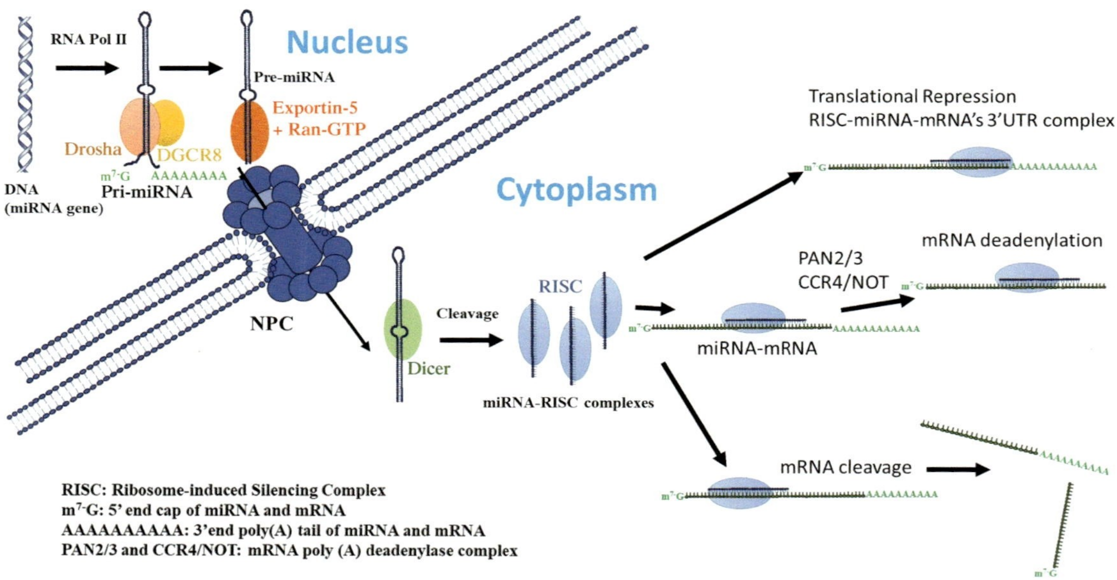

Figure 15.2 *MicroRNA biogenesis and their regulatory pathways for gene regulation: MicroRNA (miRNA) synthesis initiates in the nucleus, where miRNA genes are transcribed by RNA polymerase II to produce primary miRNA (pri-miRNA). The pri-miRNA is recognized by DGCR8, which recruits Drosha to cleave it into precursor miRNA (pre-miRNA). Exportin-5 then recognizes the pre-miRNA and, with the aid of Ran-GTP, translocates it through nuclear pore complex (NPC) from the nucleus to the cytoplasm. In the cytoplasm, the pre-miRNA is cleaved into mature miRNA by the class 4 RNase III enzyme, Dicer, resulting in miRNA forms that are 18 to 25 nucleotides in length. Once processed into mature miRNA, these single-stranded miRNAs are incorporated into an RNA-induced silencing complex (RISC). The miRNA-RISC complex interacts with specific messenger RNA (mRNA) targets, leading to repression, cleavage, or deadenylation of the mRNA. These regulatory actions play crucial roles in gene expression modulation, impacting various cellular processes and functions.*

in forensic science due to its stability and tissue-specific expression patterns in various body fluids and tissues (Peiro-Chova et al. 2013; Seashols-Williams et al. 2016). This approach allows for the identification and discrimination of body fluids like blood, saliva, semen, and vaginal secretions, crucial for crime scene investigations where sample type identification is critical (Fujimoto et al. 2019; Zubakov et al. 2010). MiRNAs also offer potential applications in age estimation and post-mortem interval (PMI) estimation, leveraging their unique expression profiles (Cobos Jimenez et al. 2014; Fang et al. 2020; Na 2020). However, challenges such as variability in miRNA expression due to environmental factors and degradation in forensic samples, coupled with the need for standardized protocols and rigorous validation, underscore the current complexities in integrating miRNA analysis into routine forensic practice. Nonetheless, advancements in technology and increasing understanding of miRNA biology hold promise for enhancing forensic investigations by providing additional layers of specificity and accuracy in sample identification and individual profiling.

Despite challenges like standardization and interpretation, forensic epigenetics shows promise in enhancing forensic investigations by providing a deeper understanding of biological evidence and its implications. In the following sections, we will explore the role of DNA methylation and miRNA, the methods and techniques for their identification, the limitations and challenges they present, and the future prospects of these markers in forensic science to aid in crime investigation.

15.2 Role of DNA methylation and miRNA in crime investigations

DNA methylation and miRNAs play pivotal roles in forensic investigations by providing unique biomarkers for biological evidence analysis (**Figure 15.3**). DNA methylation patterns are instrumental in identifying body fluid or tissue types, discerning MZ twins, authenticating DNA samples, identifying biogeographic ancestry, and can aid in age estimation due to their age-related changes (Auerkari and Ibrahim 2020; Fleckhaus et al. 2023; Li et al. 2013). They also reflect environmental exposures, offering potential insights into an individual's history. On the other hand, miRNAs exhibit tissue-specific expression profiles that enable the identification of body fluids, such as blood or semen, crucial in crime scene analysis (Zubakov et al. 2010). Additionally, miRNA stability post-mortem contributes to estimating the PMI, aiding in forensic timeline reconstruction (Na 2020). Together, these epigenetic markers enhance the precision and depth of forensic investigations, promising advancements in both identification and interpretation of biological evidence in criminal cases.

While practical considerations like sample preservation and advanced technological requirements exist, the application of epigenetic markers offers promising advancements in forensic science and can be used to differentiate different body fluids, including menstrual blood and vaginal secretions, for which no conventional testing method exists (Choung et al. 2021). In coming future, these epigenetic markers bear high potential to become an integral part in forensic genetics for solving personal crimes, as well as understanding the circumstances surrounding them. Here, we outline the critical forensic applications of DNA methylation and miRNA, which can significantly enhance crime investigations.

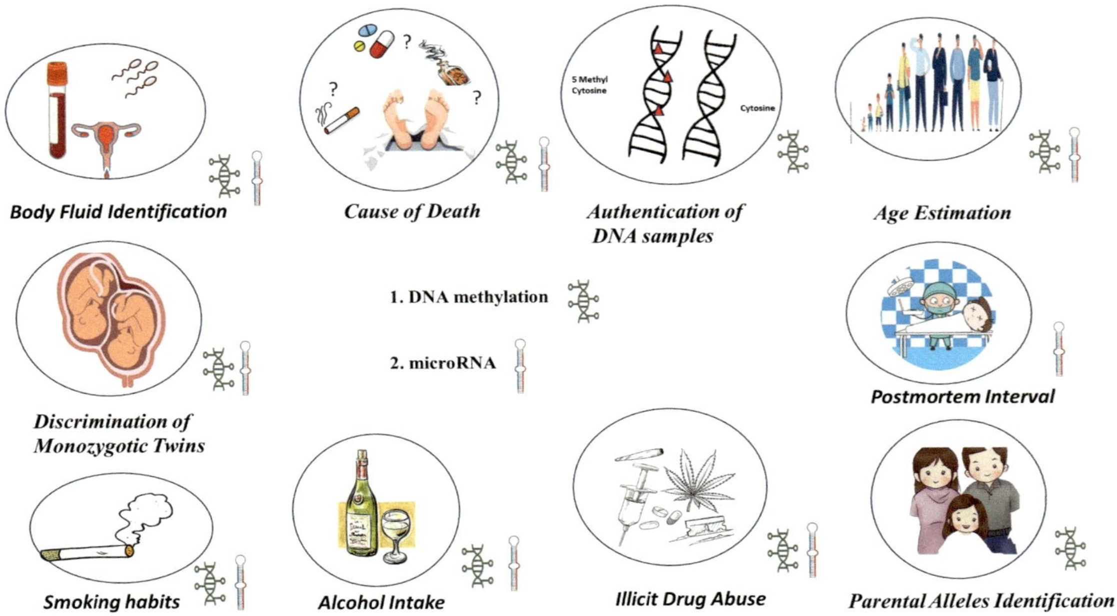

Figure 15.3 *Role of DNA methylation and microRNA in forensic investigations: DNA methylation (symbolized as green structure with projections as methyl group) and microRNAs (symbolized as multi-coloured pre-miRNA structure) play critical roles in forensic science. DNA methylation aids in body fluid identification, monozygotic twin differentiation, sample authentication, determining cause of death, age estimation, postmortem interval, and parental allele identification. miRNAs provide insights into body fluid identification, illicit drug use, alcohol intake, and smoking habits, enhancing the accuracy of forensic investigations.*

15.2.1 Identification of body fluids

DNA methylation patterns and miRNA expression profiles can be highly specific to different types of body fluids, such as blood, saliva, semen, and vaginal secretions. By analysing these epigenetic markers, forensic investigators can accurately identify the origin of biological samples recovered from crime scenes or other forensic contexts. While protein and messenger RNA (mRNA) assays have shown their value due to cell-specific expression (Bamberg et al. 2021; de Beijer et al. 2018; Juusola and Ballantyne 2005), it is important to note their susceptibility to degradation by ubiquitous proteases and ribonucleases, rendering them less stable compared to DNA and miRNA (Boon et al. 2020; Houseley and Tollervey 2009; Peiro-Chova et al. 2013).

First published in 2011, the methylation-based tissue identification report utilized the methylation-sensitive restriction enzyme HhaI (Frumkin et al. 2011). Since then, numerous studies have expanded on this approach, identifying new potential markers. One such study selected eight DNA methylation markers for evaluation: cg06379435 and cg08792630 for blood, cg23521140 and cg17610929 for semen, cg26107890 and cg20691722 for saliva, and cg01774894 and cg14991487 for vaginal secretions (Park et al. 2014). Using pyrosequencing-based DNA methylation analysis, these markers were tested on 80 samples from different body fluids, demonstrating high sensitivity and specificity in distinguishing between blood, saliva, semen, and vaginal secretions.

In another study, instead of examining the methylation status of individual CpG sites, researchers focused on investigating tissue-specific differentially methylated regions (tDMRs) that had been previously reported. DNA methylation profiles were generated through bisulfite sequencing using combined DNA samples from blood, semen, saliva, menstrual blood, and vaginal fluid (Lee et al. 2012). This approach can aid in the reconstruction of crime scenes or sexual assault events. Furthermore, a recent study leveraged bisulfite-modified pyrosequencing to discover PFN3A (Patatin-like phospholipase domain-containing protein 3A), a novel marker capable of distinguishing vaginal epithelial cells from other common forensic body tissue/fluids such as blood, saliva, and semen (Antunes et al. 2016).

In another similar study, five genes (BIK, CYTH4, GAS2L1, MDFI, and OSM) were selected to differentiate venous blood, saliva, semen, menstrual blood, and vaginal fluid using pyrosequencing-based DNA methylation, which also showed high specificity and sensitivity to discriminate the bodily fluids (Fu et al. 2015). Another study was conducted using the Illumina BeadChip Array to examine the stability of methylation markers in small quantities of different body fluids (Forat et al. 2016). This study involved a significant number of samples (~100 samples analysed for each body fluid), and the results obtained from this analysis demonstrated a sensitivity comparable to that of STR analysis, despite the potential interference from both endogenous and exogenous factors. Another forensic technique based on the combination of Combined Bisulfite Restriction Analysis (COBRA) and methylation-specific PCR (MSP) methods has been developed, taking advantage of the differences in methylation patterns found in the DNA of these different body fluids (Lin et al. 2016).

Similarly, miRNA profiling plays a crucial role in body fluid identification due to the tissue-specific expression patterns of these small, non-coding RNA molecules. In an initial study, miRNA markers for blood (miR-144 and miR-185) and for semen (miR-135a and miR-891a) were identified through microarray screening and confirmed through RT-qPCR (Zubakov et al. 2010). Later on, specific miRNA markers for differentiation of blood (miR-126, miR-150, miR-451) and saliva (miR-200c, miR-203, miR-205) were reported (Courts and Madea 2011). While examination of tissue type through miRNA profiling, the miRNA of skin cells can be distinguished using specific markers such as miR-4761-5p for menstrual blood, miR-137 for saliva and peripheral blood, miR-4473 for semen, and miR-585-3p for vaginal secretion (Luo et al. 2015). In a recent study utilizing RT-qPCR, researchers observed that seven effective miRNAs could differentiate between four types of body fluids: miR-144-3p and miR-451a-5p for venous blood, miR-203a-3p and miR-223-3p for saliva, miR-888-5p and miR-891a-5p for semen, and miR-1260b for vaginal secretion. The study also found that while saliva and vaginal secretions are not quantitatively distinguishable, miR-203a-3p and miR-223-3p were more highly expressed in saliva, and miR-1260b was more highly expressed in vaginal secretion compared to the other body fluids. Given the relative levels of these miRNAs in different body fluids, a probabilistic discriminant model was employed to distinguish between the four types of body fluids (Fujimoto et al. 2019). In the future, miRNA analysis along with DNA methylation markers could provide a clearer understanding of body fluid identification.

15.2.2 Age estimation

The imprecise determination of the individual's age from bone and tooth characteristics paves a future potential for epigenetic analysis. This is due to the fact that DNA methylation, as well as miRNA, plays a crucial role in cell differentiation and the aging process (Gentilini et al. 2013; Johnson et al. 2012; Kinser and Pincus 2020; Unnikrishnan et al. 2019; Wang et al. 2024). As cells and tissues undergo differentiation during development, various changes such as

DNA mutations, modifications in organ structure, and shifts in gene expression may occur. In contrast, analysing methylation patterns on the DNA extracted from these tissues can provide a more precise prediction of an individual's age (Bekaert et al. 2015). An individual's DNA methylation is almost zero at the initial stage but changes rapidly from fertilization to implantation (Horvath 2013) and has an inherited effect with age acceleration (Guo et al. 2014). The identification of age-sensitive DNA methylation patterns in human genomic regions enables the accurate prediction of biological age within a margin of 5 years from the chronological age (Yi et al. 2015).

It has been observed that among individuals aged 18 to 70, three gene's promoters regions, viz. EDARADD, TOM1L1, and NPTX2 all varied linearly with increasing age (TOM1L1 is hypermethylated, while EDARADD and NPTX2 are hypomethylated) (Bocklandt et al. 2011). Subsequent research has revealed that, in a variety of human tissues, five genes (NPTX2, TRIM58, GRIA2, KCNQ1DN, and BIRC4BP) continuously lose methyl groups (hypomethylated) as people age (Koch and Wagner 2011), whereas ELOVL2 exhibits an increase in methylation levels with aging (Garagnani et al. 2012). In a series of experiments, the epigenetic markers (cg06304190, cg06979108, and cg12837463) for age prediction using semen samples have been developed through collaborative research (Lee et al. 2015) and has been validated for age prediction using case work samples (Lee et al. 2018). The sperm specific age estimation has been recently improved by employing MethylationEPIC microarray data and incorporating dozens of more age-related CpG markers (AR-CpG) in addition to these three markers, which further improved the age estimation using DNA methylation (Xiao et al. 2023). In a study, a novel age prediction model was built based on a multivariate quantile regression analysis of seven age-related methylation markers, viz. ASPA, PDE4C, ELOVL2, FHL2, CCDC102B, C1orf132, chr16:85395429, and EDARADD, which provides a median absolute age prediction error of ±3.07 years and a 6.3% percentage of prediction error relative to the age (Freire-Aradas et al. 2016).

Similarly, the miRNA shows variations in expression with human age and can be used for age prediction. In most of the cases, the miRNAs are underexpressed with age and in a study using 5221 adult individuals, blood miRNA expression identified 127 miRNAs differentially expressed by age at $P < 3.3 \times 10{-}4$ (Bonferroni-corrected). In another study, it has been observed that miR-146a-5p, miR-146b-5p, and miR-142-5p were lower expressed with increasing age as these miRNAs were associated with the regulation of genes involved in cell cycle regulation, immune system, and DNA damage and repair (Ong et al. 2019). Additionally, researchers identified the top 25 miRNA candidates expressed in peripheral blood cells correlated with chronological age (Huan et al. 2018). Using miRNA profiling from blood cells and machine learning models, researchers predicted human age based on six age-related miRNAs: miR-98-3p, miR-324-3p, miR-32-3p, miR-330-5p, miR-374c-5p, and miR-342-3p (Fang et al. 2020). Both DNA methylation and miRNA are promising tools for age prediction and hold potential for forensic science applications to precisely predict human age.

15.2.3 Authentication of forensic DNA samples

There is no denying the fact that the DNA can be fabricated and planted at the scene of crime by the influential class of people. The authenticity of DNA used as evidence in court is contingent upon the belief that it is derived from authentic biological material found at the crime scene. Nevertheless, through the utilization of established techniques, DNA can be artificially generated, raising the possibility that DNA with any desired genetic profile can be created in a laboratory setting and deliberately placed at crime scenes (Frumkin et al. 2010).

The researcher who introduced the DNA methylation-based tissue identification also pointed out the synthetic DNA fabrication potential in highly sophisticated crime cases, which can be circumvented by the use of DNA methylation identification (Frumkin et al. 2010). The artificial DNA is a laboratory-synthesized entity by mainly commercially available polymerase, which lacks DNA methylation marks caused *in vivo* by the specific DNA methyltransferases at specific sites. In contrast, DNA generated naturally within living organisms contains a combination of unmethylated and methylated regions. Authentication assay has been developed using four non-CODIS loci (NT18, ADD6, MS53, SW14) and a reference CODIS locus (FGA) (Dhanapal et al. 2010). In a mock study to differentiate between artificial and original DNA using one Combined DNA Index System (CODIS) loci (FGAref) and four non-CODIS loci (NT18, ADD6, MS53, SW14), researchers found no amplification of non-CODIS loci, and their methylation status was confirmed by DNA methylation profiling (Dhanapal et al. 2010). To distinguish between *in vitro*-generated artificial DNA and legitimate natural DNA, sodium bisulfite treatment of DNA can be employed, followed by MSP or cloning and Sanger sequencing (Frumkin et al. 2010; Lasken and Egholm 2003).

15.2.4 Discrimination of monozygotic twins

Law enforcement faces a challenge when dealing with monozygotic (MZ) twins since they have the identical genomic sequence and cannot be individually identified through standard forensic STR or any sequence-based analysis. However, despite sharing the same genetic material, twins often exhibit distinct phenotypic variations (Szyf 2011; Wong et al. 2005) due to epigenetic variations accumulating through their aging in different environmental exposures (Reynolds et al. 2020; Talens et al. 2012).

In a study to distinguish MZ twins, DNA extracted from blood samples from 22 pairs of adult MZ twins were examined with a methylation BeadChip including 27,578 CpG sites. The study's findings were highly encouraging, as they revealed notable differences in DNA methylation at 92 CpG sites (Li et al. 2013). However, additional research is required to explore the metastability of CpG methylation and its association with factors such as age, tissue type, and population-specific DNA methylation (Li et al. 2011).

In addition, distinct variations in CpG methylation levels can be observed in male MZ pairs and female MZ pairs, whether it is on autosomal or X-chromosomal DNA (Watanabe et al. 2016). The disparities in the 5mC pattern can be detected through ultra-deep NGS, offering a resolution for paternity and forensic issues related to MZ twins (Auerkari and Ibrahim 2020; Weber-Lehmann et al. 2014). Furthermore, high-resolution melting curve analysis (HRM or HRMA) can be conducted with bisulfite-converted/PCR-amplified DNA from both twins, where the twin with a higher number of methyl groups in their DNA sequence will exhibit a distinct melting temperature (Stewart et al. 2015).

Similarly to DNA methylation, miRNA expression varies among individuals based on lifestyle and behaviours. In a study involving four pairs of MZ twins, 158 miRNAs were detected in each individual using NGS and RT-qPCR, with 14% of these miRNAs differentially expressed within each pair of twins (Fang et al. 2019b). Another study with seven pairs of MZ twins of different ages and sexes identified a total of 545 differentially expressed miRNAs using miRNA microarray analysis on their blood samples (Xiao et al. 2019). Of these, 2, 5, 22, 53, and 132 differentially expressed miRNAs were shared across six, five, four, three, and two pairs of MZ twins, respectively. This variability in miRNA expression underscores its potential alongside DNA methylation as a tool for differentiation of MZ twins.

15.2.5 Identification of the maternal or paternal source of alleles

In cases of paternity testing, the alleged father is typically ruled out if his DNA profile lacks the obligatory STR alleles at the corresponding alleles of the DNA profile of the child when the maternal allele is known. However, in situations where the mother's DNA profile is unavailable, the obligatory alleles cannot be established, leading to a lower exclusion probability. Therefore, identifying the parental origin of alleles in such cases is a challenging task.

Each individual inherits two copies, or alleles, of each autosomal DNA segment or gene—one from their mother and one from their father. Typically, both alleles are active and functional. However, there is a small group of genes where one copy from either parent is turned off due to DNA methylation. This phenomenon is known as genomic imprinting, which is an epigenetic mark based on the parental origin of a gene (Bajrami and Spiroski 2016). The maternal and paternal alleles have different methylation patterns, allowing the parental origin of an allele to be determined through DNA methylation analysis.

In a study, five SNPs (rs220028, rs737380, rs2279533, rs760087, rs2074399) located on five imprinted genes (*SNRPN, SIM2, ZNF215, COL9A3, EVX1*) were targeted for methylation status determination using bisulfite treatment of genomic DNA followed by fragment length discrepant allele-specific PCR (FLDAS-PCR) and pyrosequencing. The methylation status at loci of rs220028, rs760087, and rs2074399 was used in this case to determine the parental origin of alleles owing to significant differences in methylation status of these three alleles (Zhang et al. 2013).

Another study was performed to analyse a set of multiple SNPs located in five imprinted genes, utilizing the bisulfite-sequencing method across different tissue types. The maternal allele was detected from the maternally expressed gene (*H19*), and the paternal alleles from the paternally expressed ones (*HYMAI, MEST, SNRPN, PEG3*) (Nakayashiki et al. 2008). Nevertheless, there were instances where unforeseen variations in DNA methylation levels were observed in select tissues, including nails, hair, and sperm, as well as in samples subjected to elevated temperatures or extended preservation durations (Nakayashiki et al. 2009). Therefore, additional investigation into tissue specificity and the stability of DNA methylation is necessary when considering the application of methylation patterns in forensic contexts.

15.2.6 Cause of death analysis

One of the primary responsibilities of a forensic pathologist is to investigate the cause and manner of death, particularly in cases of sudden, unexpected, or unwitnessed deaths. To understand the general pathophysiological changes involved in the death process, postmortem functional biochemistry and molecular biology can be utilized (Maeda et al. 2011; Zhao et al. 2009). Recent research has shown that abnormal changes in DNA methylation levels are linked to various diseases and cancers. Thus, it will be intriguing to investigate if the DNA methylation status in specific gene promoter regions could be altered in response to acquired predispositions and/or causes of mortality (Vidaki and Kayser 2018).

Studies have found that children exposed to environmental lead (Pb), known to cause neurological disorders, exhibit aberrant increases in CpG methylation of the promoter regions of the *P16* (a tumour suppressor) and *ALAD* (δ-aminolevulinic acid dehydratase) genes (Yohannes et al. 2020). In this investigation, researchers employed the LeadCare II analyser and methylation-specific PCR to detect and measure the levels of lead and the methylation status of the genes in the blood. *P16* promoter methylation is highly prevalent and widespread in individuals who are overexposed to environmental Pb (Kovatsi et al. 2010). Additionally, another study found that certain drugs, such as methamphetamine, cause alterations in circadian rhythmicity by influencing the methylation status of the *Cry1* gene promoter region (Nakatome et al. 2011). Researchers studied the methylation status of promoter regions among individuals and within the same individual across different tissues for nine circadian clock genes (*CLOCK, Cry1, Cry2, Per1, Per2, Per3, Bmal1, Tim,* and *Ck1ε*) using bisulfite conversion followed by methylation-specific PCR. Tissues were collected from various cadavers with known or unknown causes of death (such as arrhythmia, asphyxia, head injury, fire casualty, etc.). These results provide valuable insights into the potential forensic applications of DNA methylation analysis.

The cause of death in autopsied bodies has been investigated using miRNA analysis because of its relatively higher stability compared to other RNA forms (Li et al. 2021). This stability is attributed to its less susceptibility to different types of RNases, such as benzonase, Exonuclease T7, and Exonuclease T, packing in exosomal bodies or tight association with Argonaute-protein complex (Aryani and Denecke 2015; Keller et al. 2011; Riley et al. 2012; Turchinovich et al. 2013). In contrast, other RNA forms are highly susceptible to degradation by human and bacterial RNases, which are abundantly released in the body after death.

RNA is better preserved in the brain, heart, and muscle than in the liver, spleen, and pancreas, likely due to the latter's high RNase content. Specific miRNA markers for certain diseases can be profiled in these tissues to ascertain the cause of death. For instance, in breast cancer, miR-21, miR-27a, and miR-155 (oncomiR) are upregulated in malignant cells compared to normal cells (Wang and Luo 2015), while miR-155 is also upregulated in liver, lung, pancreatic, and prostate cancers (Babar et al. 2011; Cai et al. 2015; Gironella et al. 2007; Zhang et al. 2012). In cases of bacterial infection causing sepsis, specific miRNAs are upregulated; for example, miR-708-5p and miR-325-3p are increased in macrophages with *M. tuberculosis* (Fu et al. 2020; Li and Zhang 2019), and miR-199a expression can be upregulated by lipopolysaccharide (LPS) in alveolar macrophages (Liu et al. 2018b). *Staphylococcus aureus* induces the overexpression of miR-127, which can serve as a potential biomarker for death due to microbial sepsis (Liu et al. 2020). In neurodegenerative diseases, miR-103a-3p, miR-30b-5p, and miR-29a-3p are increased in the blood of L-DOPA-treated Parkinson's disease patients, while miR-34a is increased in Alzheimer's disease. These miRNAs can potentially serve as markers to determine the neurodegenerative cause of death (Bazrgar et al. 2021; Serafin et al. 2015).

15.2.7 Post-mortem interval estimation

Estimating the post-mortem interval (PMI) is a challenging task in forensic science. Typically, entomological studies are conducted to estimate PMI up to 30 days (Amendt et al. 2004); however, these methods can be inaccurate due to varying environmental conditions, individual characteristics, lack of universality, and standardization of the technique (Hu et al. 2023). Several studies have demonstrated that specific miRNAs are differentially expressed during the post-mortem period. For instance, research has shown that certain miRNAs exhibit a consistent pattern of degradation or stability over time. In one study, degenerative changes after death were investigated by examining the expression levels of miRNA-205 and miRNA-21 using RT-qPCR (Ibrahim et al. 2019). The results showed a significant increase in these miRNAs, peaking at 24 hours after death, followed by a marked decline with increasing PMI. Another study analysed 71 patella bones from autopsies with PMIs ranging from 1 day to 2 years. In this study, the miRNAs let-7e and miR-16 were selected as biomarkers to estimate PMI. Using RT-qPCR, the expression levels of let-7e and miR-16 were measured and found to negatively correlate with increasing PMI (Na 2020). Significant differences in the expression of these miRNAs were observed across four distinct groups: Group A (< 1 month), Group B

(1–3 months), Group C (3–6 months), and Group D (> 6 months). These findings suggest that the expression levels of let-7e and miR-16 in bone tissue can be used to estimate PMI.

15.2.8 Illegal drug abuse identification

Epigenetic studies on illicit drug abuse reveal how drug use can cause long-lasting changes in gene expression without altering the DNA sequence (Heinbockel and Csoka 2018). These changes are mediated through mechanisms such as DNA methylation, histone modification, and non-coding RNAs. Alcohol, a commonly consumed substance, can lead to differential DNA methylation both globally and at specific gene loci. For instance, alcohol abuse can cause hypermethylation in the 3′-UTR of the 3′-protein-phosphatase-1G (*PP1mG*) gene (Ruggeri et al. 2015), associated with alcohol use disorder, and hypomethylation in the *leptin* (*LEP*) gene promoter, linked to increased alcohol craving during withdrawal (Hillemacher et al. 2015). In alcohol users, the concentration of plasma miRNAs like miR-193b-3p, miR-122-5p, miR-3937, and miR-4507 has been shown to exceed the Bonferroni-corrected threshold (Karabegovic et al. 2023).

Chronic cocaine use is linked to hypermethylation of the tyrosine hydroxylase (*TH*) gene in the caudate nucleus and hypomethylation of exon 3 of the Iroquois Homeobox A (*IRX2*) gene in neuronal nuclei (Vaillancourt et al. 2021b, a). These epigenetic changes contribute to cocaine dependence through modifications in the 3D chromatin structure in the caudate nucleus. Cocaine also induces miRNA-423-5p, miR-33, and miR-130b, while downregulating four miR-8 family members (miR-200a, miR-200b, miR-200c, and miR-429) (Eipper-Mains et al. 2011).

Heroin use can cause hypermethylation of CpG sites in the promoter region of the *OPRM1* gene, increasing μ-opioid receptor density in brain regions (Nielsen et al. 2009). Methamphetamine abuse is associated with demethylation within the alpha-synuclein (*SNCA*) gene promoter, leading to increased expression and levels of striatal synuclein (Biagioni et al. 2019). Heroin also increases the plasma levels of miRNA-320a and let-7b-5p, which can serve as potential biomarkers for heroin abuse (Liu et al. 2021).

In individuals addicted to cannabis and also using tobacco, the most differentially methylated sites were in the *AHRR* and *F2RL3* genes, consistent with previous studies on tobacco effects (Grieshober et al. 2020; Osborne et al. 2020; Zhang et al. 2014). Additional differentially methylated loci were associated with genes involved in brain function, such as TMEM190, MUC3L, CDC20, and SP9. In smokers and vapers, seven miRNAs (hsa-let-7a-5p, hsa-miR-21-5p, hsa-miR-29b-3p, hsa-let-7f-5p, hsa-miR-143-3p, hsa-miR-30a-5p, and hsa-let-7i-5p) were differentially expressed compared to non-smokers (Singh et al. 2020), serving as potential biomarkers for tobacco abuse.

Each year, thousands of drug-related offenses occur worldwide, necessitating the identification of toxic effects on a molecular level to pinpoint specific drug abuse. Integrating forensic toxicology with epigenetic analysis and profiling of cytochrome P450 monooxygenases promises to be a robust approach for identifying and understanding substance exposure and its impacts on individuals. This comprehensive strategy will enhance the accuracy and scope of forensic investigations, leading to improved legal outcomes and enhanced public safety.

15.2.9 Identifying the lifestyle of the person

An individual's lifestyle encompasses various factors, including nutrition, behaviour, stress, physical activity, work habits, smoking, and alcohol consumption (Alegria-Torres et al. 2011). There is substantial evidence that these lifestyle factors influence epigenetic modifications such as DNA methylation, histone acetylation, phosphorylation, ubiquitination, and miRNA expression (Mahmoud 2022). These epigenetic changes are sensitive to environmental stressors both early in development and later in life (Godfrey et al. 2015). Research has found that individuals with a healthy lifestyle exhibit marked hypomethylation of the *glutamine-fructose-6-phosphate transaminase 2* (*GFPT2*) gene and hypermethylation of the *glutamate rich 1* (*ERICH1*) gene (Klemp et al. 2022).

Despite widespread recognition of the negative effects, a vast majority of the global population still consumes tobacco, alcohol, cigarettes, and drugs. Using forensic epigenetics to trace such individuals can be beneficial in criminal investigations. Tobacco use is known to cause telomere shortening, DNA damage, and epigenetic modifications by impacting DNA methyltransferase expression and DNA methylation patterns (Huang et al. 2013; Satta et al. 2008). In an effort to explore differential DNA methylation linked to smoking, the first epigenome-wide association study (EWAS) in blood identified smoking-related CpG sites in various genes such as *AHRR, ALPP2, GFI1, GPR15,* and *MYO1G* (Breitling et al. 2011). Moreover, the results found in blood were also observed in certain tissues, such as the lungs. Further studies have shown that certain CpG markers can help differentiate between people who have never smoked and former smokers (Vidaki and Kayser 2017).

In a study involving 232 Europeans (90 current smokers, 71 former smokers, and 71 non-smokers), analysis of 13 CpG markers in blood DNA revealed correlations with daily smoking, age, and sex-associated methylation differences (Vidaki et al. 2023). Another study with Chinese cohorts who smoke tobacco found that two CpG loci (cg05575921 and cg01940273) are differentially methylated, aiding in the prediction of individuals' smoking status (Wen et al. 2022). Cannabis smoking similarly increases the mRNA expression of DNA methyltransferases (DNMT1 and DNMT3A) as well as cannabinoid receptor 2 (CB2) in blood lymphocytes (Smith et al. 2022). This modulation of genome DNA methylation in blood lymphocytes warrants further research to distinguish it from the CpG methylation patterns induced by tobacco.

Forensic toxicological tests can detect alcohol metabolites in blood, urine, and hair but cannot reveal a person's typical drinking habits. Chronic alcoholics show significantly elevated levels of global blood methylation and have differentially methylated genes, including the dopamine transporter (DAT), γ-Aminobutyric acid-A receptor delta ($GABA_A$Rdelta), and γ-aminobutyric acid B receptor subunit 1 ($GABA_B$R1) (Liu et al. 2018a; Nieratschker et al. 2014; Yen et al. 2015). In one study involving 24 alcohol-dependent patients and 23 healthy controls, researchers observed differential methylation patterns in the *HECW2* (cg18752527) and *SRPK3* (cg16529483 and cg24496423) genes. This analysis used EWAS profiles of purified CD3+ T-cells with the Illumina 450K array (Bruckmann et al. 2017).

15.3 Methods and techniques of DNA methylation and miRNA analysis

Forensic scientists utilize various molecular techniques to establish sensitive and specific methods for analysing DNA methylation patterns and miRNA variations. Bisulfite treatment is essential, converting unmethylated cytosines to uracils while preserving methylated cytosines, allowing for precise determination of methylation status at individual bases. Methylation-sensitive restriction enzymes selectively cleave DNA based on its methylation status, pinpointing methylated regions for subsequent analysis. Methylated DNA immunoprecipitation (MeDIP) enriches methylated DNA fragments using antibodies specific to 5-methylcytosine, enabling analysis via microarrays or sequencing. High-throughput DNA methylation arrays like the Illumina Infinium BeadChip offer simultaneous profiling of methylation across numerous

CpG sites. Whole genome bisulfite sequencing (WGBS) provides comprehensive coverage of the methylome with single-base resolution across the entire genome, while reduced representation bisulfite sequencing (RRBS) focuses on CpG-rich regions. Quantitative methylation-specific PCR (qMSP) quantifies methylation levels at targeted CpG sites using PCR assays.

MicroRNA analysis from human samples involves miRNA extraction using the Trizol reagent or commercial kits (Zununi Vahed et al. 2016). After extraction, the RNA is subjected to RT-qPCR, which involves reverse transcription followed by amplification using either Taqman probes or SYBR green dyes (Ponchel et al. 2003). High-throughput profiling techniques such as NGS and microarray are then employed to identify and investigate miRNA expression in tissues. Bioinformatic prediction of miRNA targets is conducted using online tools such as miRWalk (Dweep et al. 2014), miRDB (Chen and Wang 2020), and TargetScan (Agarwal et al. 2015), with known miRNA sequences searched through databases like miRBase and mirGate (Andres-Leon et al. 2015; Griffiths-Jones 2010). Here, we discuss such methods in detail, which will collectively enable detailed exploration of DNA methylation patterns and miRNA variations in forensic investigations.

15.3.1 Sequencing-based DNA methylation and miRNA analysis

WGBS allows for the detection of 5mC methylation across the entire genome. This method consists of fragmentation of isolated genomic DNA, end repair, adapter ligation, final repair, and bisulfite conversion. This method, first introduced by Frommer et al., is based on the discovery that cytosine and 5-methylcytosine (5mC) react differently when treated with sodium bisulfite (Frommer et al. 1992). The genomic DNA sample is fragmented into smaller pieces, typically ranging from 200 to 800 bp in size (often achieved through sonication or Covaris S2 system), which undergoes treatment with sodium bisulfite ($NaHSO_3$) at a lower pH (pH 5-6) (Darst et al. 2010), using commercial kit such as EpiTect Fast DNA Bisulfite Kit (Qiagen). Bisulfite treatment converts unmethylated cytosines to uracils by adding a sulfite group to the C6 position of cytosine (**Figure 15.1**), while methylated cytosines remain unchanged. The bisulfite-treated DNA is subjected to PCR amplification using primers specific to the region of interest, which helps in enrichment of the DNA fragments containing the converted cytosines (library preparation). The quality control of the library is usually performed with BioAnalyzer DNA 1000 chips (Agilent) and the Qubit dsDNA High Sensitivity fluoro-metric assay (Invitrogen). The PCR-amplified DNA fragments are sequenced, commonly using Sanger sequencing or NGS techniques such as Illumina HiSeq or MiSeq (Nair et al. 2018; Vidaki et al. 2017). The sequencing data, such as FASTQ files, are then analysed with the Burrows-Wheeler Aligner, Sequence Alignment/Map, and Genome Analysis Toolkit software to determine the methylation status of individual cytosines within the DNA sequence (Legendre et al. 2015; Li and Durbin 2010).

For precise determination of DNA methylation at individual CpG sites in PCR-amplified sequences, bisulfite pyrosequencing stands as a valuable quantitative technique (Tost and Gut 2007). Its operational principle is rooted in a sequencing-by-synthesis cascade: the incorporation of deoxynucleotide triphosphates into bisulfite-treated DNA leads to pyrophosphate (PPi) release, proportional to the nucleotides added. This PPi is then converted to ATP, which in turn fuels a luminescent reaction transforming luciferin into oxyluciferin. The highly quantitative methylation patterns yielded by bisulfite pyrosequencing are instrumental in distinguishing various human body fluids, even from trace DNA evidence (Vidaki et al. 2016).

RRBS is the most frequently used DNA analysis method to analyse CpG methylation by enriching the regions of the genome with CpG content, making it a cost-effective alternative to WGBS. The technique employs MspI restriction enzyme, which cuts DNA at all C^CGG sites i.e. at the CpG dense regions, regardless of their DNA methylation status at the CG site (Meissner et al. 2008). Briefly, the 100ng-1μg of genomic DNA is digested with MspI (usually 40–220 bp size is selected) and purified with a DNA purification kit (Gu et al. 2011). The blunt end products are dA-tailed at 3′ ends with Taq polymerase or Klenow exo--DNA polymerase (Clark et al. 1987) and then ligated to methylated adapters, which are selected and purified through magnetic beads. The purified DNA is treated with bisulfite to convert unmethylated cytosine residues to uracil, leaving the unmethylated residues unchanged. The product is purified again, and then PCR amplified (12–15 cycles) with primers that anneal to the adapters. The amplified products are purified again with magnetic beads, and their quality and concentration are checked with Qubit or Bioanalyzer (Gu et al. 2021). The library is sequenced at NGS platform, such as Illumina, and the data is analysed with bisulfite analysis tool, such as Bismark (Krueger and Andrews 2011). RRBS seems to be a powerful tool in forensic science, which can provide detailed methylation profiles that can be used for a variety of applications, from age estimation to tissue identification and beyond. Its cost-effectiveness and high resolution make it a valuable method for integrating epigenetic analysis into forensic investigations.

MiRNA analysis through sequencing is increasing due to lowered cost in recent years, and analysis of all small RNA genes, in contrast to defined gene analysis through RT-qPCR or microarray. Beginning with RNA extraction using methods like the Trizol reagent or commercial kits, high-quality miRNAs are isolated through length separation on gel electrophoresis or binding to specific proteins. The isolated small RNAs are extended by adapter ligation or by polyadenylation for creating primer binding site for reverse transcriptase. The extended small RNAs are subjected to cDNA library preparation using random primers for sequencing. The sequencing can be carried out by sequencing-by-synthesis technology with DNA polymerase or ligase on platforms like Roche 454, Illumina, Helicos, and PacBio (Pacific Biosciences) platforms, SOLiD (Life Technologies), Complete Genomics, depending on the sample size, accessibility, cost, and nature of miRNA analysis. The resulting data are de-multiplexed, and the resultant fastq files are subjected to quality control, such as discarding of low depth lanes and clipping of adapter sequences using FASTQC or FASTX-toolkit (Lokhande 2023; Potla et al. 2021). The profiling of the demultiplexed and trimmed fastq files is carried out using tools like sRNAbench, miRDeep2, miRExpress, miRNAKey, and differential expression of miRNA is analysed (Bisgin et al. 2018). Recent studies of miRNA using NGS in forensic blood samples have yielded useful information about the storage conditions and the effects of environmental factors, such as oxidants and microbial contamination, on the samples (Fang et al. 2019a; Seashols-Williams et al. 2016).

15.3.2 Methylation-specific PCR

Methylation-specific PCR (MSP) detects 5mC methylation through specially designed PCR primers that anneal to sequences containing methylated cytosines, allowing for selective amplification of methylated DNA fragments following bisulfite treatment (Herman et al. 1996). Additionally, unmethylated-specific primer sets may also be employed as controls or for collecting positive data, depending on the experiment. By using both methylated and unmethylated primer sets, MSP can provide qualitative information about the methylation status of specific genomic regions.

A modified version of MSP called MethyLight MSP combines methylation-specific primer design with quantitative real-time PCR (qPCR) technology (Eads et al. 2000). In this method, methylated-specific primers are designed with a fluorescent reporter molecule, such as a TaqMan probe, which anneals to the region of interest. During qPCR amplification, the fluorescence signal generated is directly proportional to the amount of methylated DNA present in the sample, allowing for quantitative analysis of DNA methylation levels.

Another qPCR technology called EpiDirect® has been developed for the detection of methylation status in real-time without using bisulfite treatment, which employs a special primer called EpiPrimer® (Bendixen et al. 2023). The EpiPrimer® has an anchor, a loop, and a starter sequence, out of which the anchor sequence has a heteroaromatic intercalating DNA analog (oligonucleotide) with pseudo-nucleotide called INA®, which intercalates DNA, exhibits fluorescence, and increases its π-stacking energy, thereby increasing its melting temperature (Tm). The EpiPrimer® binds weakly to the unmethylated CpG site than to the methylated site, causing large differences in Ct values, which helps in evaluation of CpG methylation status of the DNA sequence of interest.

Methylation-sensitive high resolution melting (MS-HRM) is a method used to distinguish MSP-generated DNA fragments containing varying levels of methylation (Wojdacz and Dobrovic 2007). In HRM, post-bisulfite treated DNA and untreated DNA are PCR-amplified and subjected to a melting curve analysis, where the DNA is gradually heated (60°C to 90°C at the rate of 0.2°C/sec), causing the double-stranded DNA to denature. The resulting melting curves can reveal differences in the Tm of methylated and unmethylated DNA fragments, allowing for the quantitative assessment of methylation levels.

15.3.3 The combined bisulfite restriction analysis

Combined Bisulfite Restriction Analysis (COBRA) methylation assay is a technique used to assess the methylation status of specific DNA sequences. Genomic DNA is treated with sodium bisulfite, which converts unmethylated cytosines to uracils while leaving methylated cytosines unchanged. This treatment effectively distinguishes between methylated and unmethylated cytosines in the DNA sequence. Following bisulfite treatment, region-specific PCR primers are used to amplify the target DNA sequence of interest. These primers are designed to anneal specifically to the region of interest, allowing for selective amplification of the target DNA fragment. The PCR-amplified DNA fragments are then purified to remove any excess primers, nucleotides, and enzymes that may interfere with subsequent steps of the assay. The purified DNA fragments undergo restriction enzyme digestion. The choice of restriction enzyme(s) is based on the sequence-specific recognition sites present in the target DNA sequence. For example, to assess the methylation status of CGCG sequences in the DNA region of interest, a methyl-sensitive restriction enzyme such as BstUI can be used. The bisulfite-treated genomic DNA is PCR amplified for the region of interest, where the methylated DNA sequences (^{5m}CG^{5m}CG) remain unchanged except methylation is lost and can be cleaved with BstUI, while unmethylated DNA sequences are converted to TGTG, which has lost the cleaving site for BstUI (Xiong and Laird 1997). Overall, it provides a means to analyse the methylation status of specific DNA sequences by combining bisulfite treatment, PCR amplification, and restriction enzyme digestion. This method offers insights into the methylation patterns of target DNA regions.

15.3.4 The HELP methylation assay

HpaII tiny fragment Enrichment by Ligation-mediated PCR (HELP) assay is a method used to assess the dynamic nature of DNA methylation status of a DNA segment or whole genome in different tissues (Oda and Greally 2009). Genomic DNA undergoes digestion using two isoschizomers, viz. HpaII and MspI. HpaII is methylation-sensitive, specifically cleaving unmethylated 5′-C^CGG-3′ sites and is blocked when internal cytosine is methylated, while MspI is methylation-insensitive, cleaving the same sequence regardless of methylation status (Waalwijk and Flavell 1978). The products of HpaII and MspI digestion are compared to identify methylated and unmethylated regions of the genome (Khulan et al. 2006). HpaII digestion enriches for regions deficient in methylation, as it cleaves unmethylated sites, while MspI digestion serves as a control for DNA accessibility and digestion efficiency. The digested DNA fragments are subjected to ligation-mediated PCR (LM-PCR). This technique selectively amplifies DNA fragments digested by the restriction enzymes, resulting in amplified fragments typically ranging from 200 to 2000 base pairs. The methylation status of each locus is assessed by comparing the patterns generated from digestion with HpaII and MspI. Differential representation between HpaII and MspI digestion indicates differences in DNA methylation status at specific genomic regions. The HELP assay provides insights into the dynamic nature of DNA methylation by selectively amplifying and comparing methylated and unmethylated regions of the genome.

15.3.5 Microarray-based DNA methylation and miRNA analysis

This method is an advanced technique for assessing the methylation status of large genomic DNA, combining restriction enzyme digestion, ligation, PCR amplification, and microarray analysis (Schumacher et al. 2006a, 2006b). Genomic DNA is initially fragmented into smaller pieces using the restriction enzyme MseI, which cuts at 5′-TTAA-3′ sites, leaving CpG islands (CGIs) intact for selective methylation analysis (Yan et al. 2002). These fragmented DNA molecules are then ligated to synthetic linkers, creating priming sites for subsequent PCR amplification. Next, the ligated DNA fragments undergo digestion with CpG-specific restriction enzymes, such as BstUI and/or HpaII, to target methylated CpG dinucleotides specifically. Methylated DNA fragments, protected from digestion, are amplified using linker-dependent PCR to generate sufficient DNA for microarray analysis. The PCR products are labelled with fluorescent dyes (e.g., Cy3 for reference DNA and Cy5 for crime scene DNA in case of forensic samples) to distinguish and visualize DNA methylation patterns between two different samples. The labelled PCR products from both samples are combined and applied to a differential hybridization microarray slide containing human CGI library probes (Schumacher et al. 2006a, 2006b). After rigorous washing to remove unbound DNA, the microarray slide is scanned for fluorescence intensity. The Cy3/Cy5 ratio at each locus indicates the relative methylation status in the reference DNA compared to the crime scene DNA. Differences in methylation patterns between the samples can then be identified and analysed. This method provides a comprehensive approach to assessing DNA methylation patterns in forensic samples, allowing for the identification of differences in methylation status between reference and crime scene DNA.

Microarray-based miRNA analysis of forensic samples begins with miRNA extraction using the Trizol reagent or column-based kits like Qiagen's miRNeasy to ensure high-quality isolation (Sriram et al. 2021). Extracted miRNAs are cDNA synthesized using reverse transcriptase and labelled at their 3' ends with fluorescent dyes such as Cy3 or Cy5 using T4 RNA ligase (Wang et al. 2007) or chemically synthesized special RNA probe called LNA (locked nucleic acid, where 2′ O and the 4′ C elements of ribose are bonded by a methylene bridge, locking it in a C3′-endo conformation like A-form of nucleic acids) can be used for hybridization (Castoldi et al. 2006). These fluorescently labelled miRNAs or LNA probes are hybridized to probes on a glass slide specific to known miRNAs, allowing simultaneous analysis of thousands of miRNAs (Courts and Madea 2011; Zubakov et al. 2010). Hybridization signal data are normalized using the quantile algorithm, followed by hierarchical clustering, ANOVA, and significance analysis using software such as TM4. Differentially expressed miRNAs are quantified and confirmed through RT-qPCR, and their profiles can be used to identify body fluids, estimate the PMI, perform individual identification, determine tissue origin, and infer phenotypic traits.

15.3.6 Illumina Infinium Methylation Assay

The Illumina Infinium Methylation Assay is a sophisticated technique used to analyse DNA methylation patterns across the genome. It employs high-density microarray technology to assess the methylation status of CpG sites with high accuracy and resolution. Infinium platforms can perform both bead array and NGS analyses. In the bead array approach, DNA samples are hybridized to arrays containing probes specific to CpG sites of interest. Alternatively, NGS can be used for sequencing-based methylation analysis. This assay offers extensive coverage of CpG sites. The latest version, Infinium HumanMethylationEPIC version 2.0 (900K EPIC v2), covers over 900,000 CpG probes out of which 77.63% of which are homologous with the first version, and more than 200,000 of which are new CpGs covering open chromatin and enhancer regions identified by projects like ENCODE and FANTOM5 (Davis et al. 2018; Lizio et al. 2015; Noguera-Castells et al. 2023). Even with very limited amounts of genomic DNA—typically around 250 nanograms—it's possible to gather methylation information for these specific CpG sites. Following bisulfite conversion of the genomic DNA, the assay then discerns between methylated and non-methylated CpG positions through either the Infinium I (single-color) or Infinium II (dual-color) chemistry (Dedeurwaerder et al. 2011). Infinium I chemistry relies on two separate site-specific probes: an "M" bead type designed for methylated sites and a "U" bead type for unmethylated sites. A fluorophore is directly attached to the terminating end of each probe at the respective methylated or unmethylated cytosine. Conversely, Infinium II chemistry employs only one probe, which undergoes a single nucleotide extension. During this extension, fluorophores are incorporated to indicate the methylation status. The Illumina GenomeStudio® software generates quantitative values, called β-values, for each CpG site, providing a precise measure of methylation status (Triche et al. 2013). These β-values represent the proportion of DNA methylation at each CpG site, ranging from 0 (completely unmethylated) to 1 (fully methylated). This assay provides high-resolution methylation data, enabling the identification of various tissue types and aiding in the resolution of diverse forensic cases.

15.3.7 Methyl-binding domain sequencing

Methyl-binding domain sequencing (MBD-seq) is an alternative approach for methylome-wise association studies (MWAS) that provides near-complete coverage of the methylome at similar costs as the array-based technologies (Aberg et al. 2012). Genomic DNA is broken into smaller fragments using techniques like sonication or enzymatic digestion. The fragmented DNA undergoes enrichment for methylated sequences. This is achieved by pulling down methylated sequences using a 5mC binding protein like MBD protein or commercial kits such as the MethylMiner Methylated DNA Enrichment Kit (Harris et al. 2010). These methods capture DNA containing methylated CpG sites. Enriched methylated DNA fragments are sequenced using high-throughput techniques like NGS (Lan et al. 2011). This yields millions of short sequence reads corresponding to methylated regions of the genome. The sequenced tags are aligned to the human reference genome to determine their exact positions. By comparing the distribution of these tags to known genomic features like CpG islands, promoters, and gene bodies, the methylation status of CpG-rich regions can be inferred. MBD-seq is a powerful tool for comprehensively analysing DNA methylation patterns across the genome, providing insights into epigenetic regulation in various biological processes and diseases (Fraga et al. 2003).

15.3.8 Methylated DNA immunoprecipitation

MeDIP (methylated DNA immunoprecipitation), also known as DNA immunoprecipitation (DIP), is a method of immunoprecipitation of methylated DNA fragments (CpG and non-CpG sequences throughout the genome) by 5mC or 5hmC monoclonal antibody (Borgel et al. 2012). Initially, the extracted and purified genomic DNA (100 ng to 10 µg, depending on the downstream method and purpose) is fragmented (~100–500 bp) using either sonication (amplitude 20%, pulse on and off 15 sec each for a time duration of 10 min) (Cheung et al. 2012) or enzymatic digestion (e.g., PstI) (Rezaei et al. 2022). The antibody-bound methylated DNA fragments are captured using paramagnetic beads or A/G-coupled agarose beads, followed by elution, separating methylated DNA from non-methylated DNA (Ben Maamar et al. 2021). Purified methylated DNA can be used in different analyses, like MSP, which helps in detecting methylation at specific loci (MeDIP-PCR), methylation-sensitive restriction enzyme analysis, which examines methylation patterns at specific sites, and genome-wide methylation analysis, which utilizes NGS (MeDIP-seq or DIP-seq) or microarrays for profiling of DNA methylation patterns across the genome (Borgel et al. 2012). Alignment software is used to map sequence reads to the reference genome, identifying methylated regions. The majority of mammalian genomes, including human, are having low density CpGs (>90%), and MeDIP is preferable over genome-wide methylation analysis over bisulfite sequencing or MBD-seq, which are biased towards high-density CpG (Ben Maamar et al. 2021).

For methylation analysis of the genomic DNA, fragments containing methylated reads are compared to the human genome using alignment tools like MEDIPS (Chavez et al. 2010), MEDME (Pelizzola et al. 2008), MeQA (http://life.tongji.edu.cn/meqa/) (Huang et al. 2012), Batman (Down et al. 2008), or MeDUSA (Wilson and Beck 2016), while the quality control of the data can be carried out with the R package Repitools (Statham et al. 2010). MeDIP-seq or MeDIP-PCR targets methylated genomic loci using as little as ~1 ng of starting genomic DNA (Zhao et al. 2014), which makes it quite useful for methylation profiling of low template DNA samples.

15.3.9 SMRT sequencing

Single molecule, real-time (SMRT) sequencing, using long-read sequencers developed by Pacific Biosciences, is capable of detecting DNA modifications (Flusberg et al. 2010). It is an advanced DNA sequencing technology that allows for real-time observation of DNA synthesis at the single-molecule level (Ardui et al. 2018). The PacBio Sequel sequencer employs an SMRT cell containing a million wells for this sequencing process. Each well in the SMRT cell holds a complex of a DNA polymerase and a single DNA molecule (Xu and Seki 2020). During SMRT sequencing, the incorporation of fluorescently labelled nucleotides by DNA polymerase is continuously observed in each well. The PacBio Sequel sequencer generates approximately 365,000 reads per SMRT cell, with an average read length of 10–14 kb (Ardui et al. 2018).

SMRT sequencing utilizing zero-mode waveguides (ZMWs) to immobilize DNA polymerase within a nanophotonic confinement of ~70nm x ~100nm (diameter × depth) making a ZMW holes of volume of 20 zeptoliters (20×10^{-21} liters) (Korlach et al. 2008), SMRT sequencing generates ultra-long reads (≥10 kb) in real time, achieves high accuracy (≥95%) through circular consensus sequencing (CCS) (Wenger et al. 2019), and directly detects 5mC methylation (Ni et al. 2023). Thus, immobilized polymerase kinetics altered by the modified base of the native DNA, which is measured by fluorescence pulse signals, time of base sitting for being sequenced (pulse width), and time intervals of sequencing two bases (interpulse duration) (Flusberg et al. 2010).

15.3.10 Nanopore sequencing

Nanopore sequencing is a powerful tool for detecting a wide range of epigenetic modifications, such as 5mC, m6A, etc., directly from DNA as well as RNA, which has been developed by Oxford Nanotechnology (ONT) (Xu and Seki 2020). Recently, ONT unveiled its high-throughput technology on its website, capable of reading DNA/RNA lengths of up to 4Mb in a single run. This technology is available on platforms such as the MinION, the GridION (a compact benchtop device with 5 MinION Flow Cells), and the PromethION (which supports 1 to 48 flow cells as needed) (Wang et al. 2021b).

On a flow cell, the single-stranded DNA is ratcheted through a motor protein channel (phi 29 DNA polymerase) of 3.6 to 6 nm pore size (Ji et al. 2020). The change in current pattern called "squiggle" while passing through the pore is detected, and the identification of the particular nucleotide is done (Bao et al. 2021). Although the single molecule signal is low compared to bulk modification detection, there are several tools like Nanoraw, which can be used to identify various types of DNA modification like 5mC, 6mA and N^4-methylcytosine (4mC) from ONT data (Stoiber et al. 2016). Other tools like Nanopolish, Megalodon, and DeepSignal demonstrate high accuracy in detecting 5mC signals on ONT platforms, but their accuracy for m6A detection is lower compared to PacBio SMRT sequencing. The ONT platform holds significant promise for future applications in detecting diverse DNA modifications using portable handheld devices like the MinION. Recently, researchers have used this sequencing platform to identify age-associated and body fluid-specific CpG methylation signatures from single samples (Yuen et al. 2024). This technology could prove invaluable for human sample or human identification in the field of forensic science.

15.3.11 Single base extension methylation analysis

Researchers can utilize single-base extension (SBE) to determine the precise nucleotide at a specific location along a nucleic acid sequence, as seen with SNPs. Intriguingly, this method also proves valuable for establishing the methylation status of cytosine (Hu 2012). The procedure begins with the design of an oligonucleotide probe that perfectly matches a particular sequence segment. Its 3′ end is then positioned directly next to the target base. An enzyme extends this probe by adding a solitary nucleotide base, and a special terminator immediately halts further extension. This terminator is typically a 2′,3′-dideoxynucleotide triphosphate (ddNTP), which pairs specifically with the target base. The added terminator base can be detected using various methods, such as fluorescence, isotope labelling, or mass spectrometry. The extension products are, however, usually separated based on size using capillary electrophoresis (Boyd and Zon 2008). The labelled terminator bases are detected, and the results are displayed as electropherograms. SBE, commercialized as SNaPshot® Multiplex Kit (Thermofisher), is now multiplexed for up to 10 loci and can be used for methylation analysis in CpG-rich genomic DNA following bisulfite treatment and PCR amplification. This method is useful for detecting specific 5mC methylation status by designing primers corresponding to that specific cytosine marker, which may show variations such as body fluids or tissues, monozygotic twins, or human age (Onofri et al. 2023).

15.4 The challenges and limitations associated with DNA methylation/miRNA analysis in forensic investigations

Epigenetic factors hold great potential for forensic investigations, offering the ability to glean information about an individual's age, tissue type, and environmental exposures from DNA samples. However, several challenges and limitations must be addressed to fully harness this potential.

15.4.1 Complexity of epigenetic mechanisms and variability among individuals

Epigenetic markers, such as DNA methylation and miRNA, are highly complex and influenced by various factors, including age, lifestyle choices, and environmental exposures like diet, smoking, stress, and toxins (Abdul et al. 2017; Alegria-Torres et al. 2011; Francavilla et al. 2021; Klemp et al. 2022; Pointner et al. 2024). The significant inter-individual variability in methylation patterns and miRNA expression is influenced by genetic background, ethnicity, and personal history (Elliott et al. 2022; Flowers et al. 2022). For instance, genetic background can interact with lifestyle choices to create a unique epigenetic profile, while ethnicity-related differences can be further complicated by migration and inter-ethnic marriages. These factors do not act in isolation but rather interact in complex ways. For example, environmental factors such as diet and smoking can have different effects on DNA methylation depending on an individual's age and genetic background (Klemp et al. 2022). Similarly, stress and exposure to toxins may influence miRNA expression differently across various ethnicities due to genetic and environmental interactions (Flowers et al. 2022; Miguel et al. 2018). Epigenetic markers can change over time, complicating the ability to link current epigenetic states to past events accurately (Turner 2009). To use epigenetic markers in forensic applications effectively, all influencing factors must be carefully examined. For example, to assert that a specific CpG site shows a body-fluid-specific DNA methylation pattern, it must be ensured that the pattern is not altered by age, lifestyle choices, or environmental factors. This complexity and variability pose significant challenges in accurately interpreting epigenetic data and developing universal forensic markers that are reliable across diverse populations. Understanding the interactions between these factors is crucial for advancing the use of epigenetic markers in forensic science.

15.4.2 Crime scene samples, DNA degradation, and contamination

Biological samples collected at crime scenes often suffer from degradation or contamination, significantly impacting the quality and integrity of epigenetic data. The degradation of DNA and miRNA can obscure epigenetic patterns, complicating the process of obtaining accurate and reliable results. This issue is particularly significant in DNA methylation analysis, where the original DNA samples must be analysed without amplification, as PCR can erase DNA methylation marks, making the initial amount of extracted DNA critically important (Hernandez and Singleton 2012). Furthermore, the limited quantity of available samples for extracting DNA or miRNA, a frequent challenge in forensic science, poses a significant threat to the analysis. When samples are scarce, they can only be analysed once, making it impossible to repeat or reproduce tests for reconfirmation, especially in cases with only a partial profile of epigenetic markers. The interplay between sample degradation and limited quantity underscores the difficulty in conducting reliable epigenetic analyses, presenting significant challenges in the forensic application of epigenetic markers.

15.4.3 Technical limitations, standardization, validation, and interpretations

The detection and analysis of epigenetic modifications necessitate sophisticated techniques and high-resolution methods. Current technologies, such as bisulfite sequencing, Microarray, RRBS, MeDIP, BeadChip Array, Nanopre, or SMRT sequencing, are costly, time-intensive, and require specialized equipment and expertise (Amarasinghe et al. 2020; Kurdyukov and Bullock 2016). These technical constraints pose challenges to the widespread adoption of epigenetic methods in routine forensic practice. Additionally, there is a notable absence of standardized protocols and validated studies for utilizing epigenetic markers in forensic investigations. Without uniform methods, comparing results across different laboratories and studies becomes problematic. Rigorous validation is crucial to establish the credibility and reproducibility of epigenetic analyses within forensic contexts. Interpreting epigenetic data in forensic settings presents its own set of challenges. Establishing clear correlations between specific epigenetic modifications and forensic traits, such as age or tissue type, is often complex and nuanced. Forensic experts must exercise caution when interpreting epigenetic data, ensuring conclusions are grounded in robust scientific evidence

15.4.4 Legal challenges in the court of law

The use of epigenetic information in forensic investigations introduces significant legal challenges. The admissibility of epigenetic evidence in court demands thorough evaluation, as the scientific community must reach a consensus on its reliability and relevance. DNA methylation analysis is a complex task, and no single method can provide clear and comprehensive data that meets all necessary requirements. Various technical difficulties may arise, and the chosen DNA methylation process often depends on the available tools and expertise. Comprehensive analyses in forensic epigenetics are essential for solving crimes and identifying unknown criminals. These analyses require identifying epigenetic signatures related to lifestyle, tissue specificity, age, and disease, as well as developing advanced methods for extracting high-quality DNA from crime scene evidence. Establishing the validity of forensic epigenetics in court remains a significant challenge due to its highly complex nature.

15.5 Conclusion and future perspective

We have laid here the various roles of DNA methylation and miRNA markers in crime investigation, their analysis using advanced methods, and some limitations of these epigenetic analyses. DNA methylation and miRNA analysis hold significant promise for advancing forensic science through improved age estimation, body fluid/tissue identification, time since death estimation, MZ twins' differentiations, illicit drug abuse identification, and individual's life style identification. This chapter holds potential for investigation of critical cases and samples that are otherwise not possible to solve through conventional STR analysis, and also holds promise for future research and applications (Rana 2018a). Future advancements in sequencing technologies, combined with international collaboration on protocol development, could overcome these hurdles, paving the way for these biomarkers to become integral components of forensic investigations.

Epigenetic applications in forensics are currently limited and relatively new; however, we anticipate swift progress in the field of forensic epigenomics in the near future. Although only a few forensically relevant questions can be answered by epigenetics, we envision a move towards forensic epigenomics to help with further questions and research in epigenetics, which will ultimately lead to the identification of humans through solely epigenetic markers. Various factors will determine the degree to which forensic epigenetics will expand into forensic epigenomics. First, continued scientific advancement in the categorization and comprehension of epigenetic markers influenced by lifestyle and environmental factors. The second step involves determining epigenetic markers and developing and validating statistical models to accurately predict an epigenetic lifestyle. Thirdly, advancements in technology have made it possible to analyse numerous epigenetic markers simultaneously from low-quality or low-quantity DNA (possibly using new technologies like Oxford Nanopore sequencing and SMRT sequencing) and to develop and validate sensitive multiplex analysis assays forensically. In the future, these advanced sequencing technologies hold significant potential for forensic applications. These technologies are capable of detecting a wide range of base modifications, extending beyond the traditional 5mC detection. SMRT sequencing offers real-time analysis of long DNA molecules, allowing for the identification of various modifications, such as 5mC, 5hmC, 4mC, m6A, and other less common epigenetic marks.

Similarly, nanopore sequencing provides real-time, portable sequencing capabilities, which can detect modifications by measuring changes in electrical conductivity as nucleotides pass through a nanopore. Both technologies enable comprehensive epigenetic mapping without the need for bisulfite conversion or other chemical treatments, thus preserving the integrity of the DNA and allowing for more accurate analysis. These advancements can revolutionize forensic science by providing more detailed and precise genetic and epigenetic profiles of individuals and could have a profound impact on forensic science by enabling the routine discrimination of individuals with higher accuracy and detail. The integration of SMRT and nanopore sequencing into forensic workflows could also streamline the analysis process, reduce turnaround times, and increase the robustness of forensic evidence. However, the implementation of these technologies in forensic settings will require overcoming several challenges, including the need for standardized protocols, the development of comprehensive reference databases, and ensuring the legal admissibility of the data generated.

As these sequencing technologies continue to advance and become more accessible, their adoption in forensic science could lead to more precise and reliable individual discrimination, ultimately enhancing the accuracy and effectiveness of forensic investigations. The combined goal of these methods is to direct crime investigations towards the identification of unidentified criminals who cannot be identified through standard forensic DNA profiling.

In the midst of controversy over the existence of CpG/5mC methylation in human mtDNA according to the recent publications (Bicci et al. 2021; Guitton et al. 2022; Hao et al. 2020; Liu et al. 2016), the regulatory role of the identified epigenetic marker, viz. m6A have been implicated (Hao et al. 2020) and may have important repercussions, including the discovery of novel markers for human identification. The mtDNA is well preserved in double-membrane structure and can be isolated even from ancient bones and massively destroyed tissues (Barta et al. 2014; Higgins et al. 2015; Nesheva 2014). The human mtDNA is relatively small, consisting of 16,569 base pairs. Each cell contains a high number of mtDNA copies, ranging approximately from 1,000 to 10,000 depending on the cell type, with each mitochondrion housing around 2–10 copies of mtDNA. Unlike the two copies of autosomal DNA, this abundance increases the likelihood of successfully extracting mitochondrial DNA from forensic samples and, therefore, holds promising prospects for discovering m6A forensic markers in the near future using advanced sequencing technologies.

Lastly, moral and societal debates about the advantages and disadvantages of utilizing such human epigenetic data in forensic work, with potential legal ramifications if needed. When combined with genetic appearance and biogeographic ancestry predictions, as well as epigenetic lifetime age prediction, epigenomic prediction of lifestyle and environmental factors will, if used in forensic practice, improve DNA investigative intelligence. DNA methylation and miRNA analysis hold significant potential to revolutionize forensic science, providing deeper insights into biological samples and enhancing the ability to solve complex cases. By addressing current challenges and leveraging future technological and methodological advancements, these tools can become invaluable assets in forensic investigations, aiding in the pursuit of justice and the protection of public safety.

References

Abdul, Q. A., et al. (2017), 'Epigenetic modifications of gene expression by lifestyle and environment', *Arch Pharm Res*, 40 (11), 1219–37.

Aberg, K. A., McClay, J. L., Nerella, S., Xie, L. Y., Clark, S. L., Hudson, A. D., Bukszar, J., Adkins, D., Consortium, S. S., Hultman, C. M., Sullivan, P. F., Magnusson, P. K., and van den Oord, E. J. (2012), 'MBD-seq as a cost-effective approach for methylome-wide association studies: Demonstration in 1500 case–control samples'. *Epigenomics*, 4(6), 605–621. doi: 10.2217/epi.12.59

Agarwal, V., et al. (2015), 'Predicting effective microRNA target sites in mammalian mRNAs', *Elife*, 4, e05005. doi: 10.7554/eLife.05005.

Al Aboud, N. M., Tupper, C., and Jialal, I. (2023), 'Genetics, epigenetic mechanism'. In *StatPearls*. StatPearls Publishing. https://www.ncbi.nlm.nih.gov/books/NBK532993/

Alegria-Torres, J. A., Baccarelli, A., and Bollati, V. (2011), 'Epigenetics and lifestyle', *Epigenomics*, 3 (3), 267–77.

Amarasinghe, S. L., et al. (2020), 'Opportunities and challenges in long-read sequencing data analysis', *Genome Biol*, 21 (1), 30.

Amendt, J., Krettek, R., and Zehner, R. (2004), 'Forensic entomology', *Naturwissenschaften*, 91 (2), 51–65.

Andres-Leon, E., et al. (2015), 'miRGate: A curated database of human, mouse and rat miRNA-mRNA targets', *Database (Oxford)*, 2015, bav035.

Antunes, J., et al. (2016), 'Forensic discrimination of vaginal epithelia by DNA methylation analysis through pyrosequencing', *Electrophoresis*, 37 (21), 2751–58.

Ardui, S., et al. (2018), 'Single molecule real-time (SMRT) sequencing comes of age: Applications and utilities for medical diagnostics', *Nucleic Acids Res*, 46 (5), 2159–68.

Aryani, A., and Denecke, B. (2015), 'In vitro application of ribonucleases: Comparison of the effects on mRNA and miRNA stability', *BMC Res Notes*, 8, 164.

Auclair, G., and Weber, M. (2012), 'Mechanisms of DNA methylation and demethylation in mammals', *Biochimie*, 94, 2202–11.

Auerkari, C. M., and Ibrahim, E. (2020), 'Age estimation using DNA methylation technique in forensics: A systematic review', *Egypt J Forensic Sci*, *10*(1), 1–15. doi: 10.1186/s41935-020-00214-2.

Babar, I. A., et al. (2011), 'Inhibition of hypoxia-induced miR-155 radiosensitizes hypoxic lung cancer cells', *Cancer Biol Ther*, 12 (10), 908–14.

Bajrami, E., and Spiroski, M. (2016), 'Genomic Imprinting', *Open Access Maced J Med Sci*, 4 (1), 181–4.

Bamberg, M., et al. (2021), 'Development and validation of an mRNA-based multiplex body fluid identification workflow and a rectal mucosa marker pilot study', *Forensic Sci Int Genet*, 54, 102542.

Bao, Y., et al. (2021), 'SquiggleNet: Real-time, direct classification of nanopore signals', *Genome Biol*, 22 (1), 298.

Barta, J. L., et al. (2014), 'Mitochondrial DNA preservation across 3000-year-old northern fur seal ribs is not related to bone density: Implications for forensic investigations', *Forensic Sci Int*, 239, 11–8.

Bazrgar, M., et al. (2021), 'The role of microRNA-34 family in Alzheimer's disease: A potential molecular link between neurodegeneration and metabolic disorders', *Pharmacol Res*, 172, 105805.

Bekaert, B., Kamalandua, A., Zapico, S. C., Van de Voorde, W., and Decorte, R. (2015), 'Improved age determination of blood and teeth samples using a selected set of DNA methylation markers', *Epigenetics,* 10 (10), 922–30. doi: 10.1080/15592294.2015.1080413. Epub 2015 Aug 17.

Bell, C. G., et al. (2019), 'DNA methylation aging clocks: Challenges and recommendations', *Genome Biol*, 20, 249.

Bendixen, K. K., et al. (2023), 'A qPCR technology for direct quantification of methylation in untreated DNA', *Nat Commun*, 14 (1), 5153.

Ben Maamar, M., et al. (2021), 'Genome-wide mapping of DNA methylation 5mC by methylated DNA immunoprecipitation (MeDIP)-sequencing', *Methods Mol Biol*, 2198, 301–10.

Biagioni, F., et al. (2019), 'Methamphetamine persistently increases alpha-synuclein and suppresses gene promoter methylation within striatal neurons', *Brain Res*, 1719, 157–75.

Bicci, I., et al. (2021), 'Single-molecule mitochondrial DNA sequencing shows no evidence of CpG methylation in human cells and tissues', *Nucleic Acids Res*, 49 (22), 12757–68.

Bisgin, H., et al. (2018), 'Evaluation of bioinformatics approaches for next-generation sequencing analysis of microRNAs with a toxicogenomics study design', *Front Genet*, 9, 22.

Bocklandt, S., Lin, W., Sehl, M. E., Sánchez, F. J., Sinsheimer, J. S., Horvath, S., and Vilain, E. (2011), 'Epigenetic predictor of age', *PLoS One*, 6(6), e14821. doi: 10.1371/journal.pone.0014821. Epub 2011 Jun 22.

Boon, L., et al. (2020), 'Protease propeptide structures, mechanisms of activation, and functions', *Crit Rev Biochem Mol Biol*, 55 (2), 111–65.

Borgel, J., Guibert, S., and Weber, M. (2012), 'Methylated DNA immunoprecipitation (MeDIP) from low amounts of cells'. In N. Engel (Ed.) *Genomic Imprinting: Methods and Protocols* (Methods in Molecular Biology, Vol. 925, pp. 149–158). Humana Press. doi: 10.1007/978-1-62703-011-3_9

Boyd, V. L., and Zon, G. (2008), 'Capillary electrophoretic analysis of methylation status in CpG-rich regions by single-base extension of primers modified with N^6-methoxy-2,6-diaminopurine', *Anal Biochem, 380*(1), 13–20. doi: org/10.1016/j.ab.2008.05.017

Breitling, L. P., Yang, R., Korn, B., Burwinkel, B., and Brenner, H. (2011), 'Tobacco-smoking-related differential DNA methylation: 27K discovery and replication', *Am J Hum Genet* 2011 Apr 8;88(4):450–7. doi: 10.1016/j.ajhg.2011.03.003. Epub 2011 Mar 31.

Bruckmann, C., et al. (2017), 'DNA methylation signatures of chronic alcohol dependence in purified CD3(+) T-cells of patients undergoing alcohol treatment', *Sci Rep*, 7 (1), 6605.

Cai, Z. K., et al. (2015), 'microRNA-155 promotes the proliferation of prostate cancer cells by targeting annexin 7', *Mol Med Rep*, 11 (1), 533–8.

Carthew, R. W. (2006), 'Gene regulation by microRNAs', *Curr Opin Genet Dev*, 16 (2), 203–8.

Castoldi, M., et al. (2006), 'A sensitive array for microRNA expression profiling (miChip) based on locked nucleic acids (LNA)', *RNA*, 12 (5), 913–20.

Chavez, L., et al. (2010), 'Computational analysis of genome-wide DNA methylation during the differentiation of human embryonic stem cells along the endodermal lineage', *Genome Res*, 20 (10), 1441–50.

Chen, Y., and Wang, X. (2020), 'miRDB: An online database for prediction of functional microRNA targets', *Nucleic Acids Res*, 48 (D1), D127–31.

Cheung, H. H., et al. (2012), 'Methylation profiling using methylated DNA immunoprecipitation and tiling array hybridization', *Methods Mol Biol*, 825, 115–26.

Choufani, S., Shapiro, J. S., Susiarjo, M., Butcher, D. T., Grafodatskaya, D., Lou, Y., Ferreira, J. C., Pinto, D., Scherer, S. W., Shaffer, L. G., Coullin, P., Caniggia, I., Beyene, J., Slim, R., Bartolomei, M. S., and Weksberg, R. (2011), 'A novel approach identifies new differentially methylated regions (DMRs) associated with imprinted genes', *Genome Res,* 21:465–476.

Choung, C. M., et al. (2021), 'A forensic case study for body fluid identification using DNA methylation analysis', *Leg Med (Tokyo)*, 51, 101872.

Clark, J. M., Joyce, C. M., and Beardsley, G. P. (1987), 'Novel blunt-end addition reactions catalyzed by DNA polymerase I of *Escherichia coli*', *J Mol Biol*, 198 (1), 123–7.

Cobos Jimenez, V., et al. (2014), 'Next-generation sequencing of microRNAs uncovers expression signatures in polarized macrophages', *Physiol Genomics*, 46 (3), 91–103.

Courts, C., and Madea, B. (2011), 'Specific micro-RNA signatures for the detection of saliva and blood in forensic body-fluid identification', *J Forensic Sci*, 56 (6), 1464–70.

Darst, R. P., Pardo, C. E., Ai, L., Brown, K. D., and Kladde, M. P.` (2010), 'Bisulfite sequencing of DNA', *Curr Protoc Mol Biol, 91*(1), 7–9. doi: 10.1002/0471142727.mb0709s91

Davis, C. A., et al. (2018), 'The encyclopedia of DNA elements (ENCODE): Data portal update', *Nucleic Acids Res*, 46 (D1), D794–801.

de Beijer, R. P., et al. (2018), 'Identification and detection of protein markers to differentiate between forensically relevant body fluids', *Forensic Sci Int*, 290, 196–206.

Dedeurwaerder, S., et al. (2011), Evaluation of the Infinium methylation 450K technology', *Epigenomics*, 3 (6), 771–84.

Devall, M., et al. (2022), 'Genome-wide characterization of mitochondrial DNA methylation in human brain', *Front Endocrinol (Lausanne)*, 13, 1059120.

Dhanapal, R., et al. (2010), 'Current trends in forensic genetics', *J Forensic Dent Sci*, 2 (2), 96–8.

Didiano, D., and Hobert, O. (2008), 'Molecular architecture of a miRNA-regulated 3' UTR', *RNA*, 14 (7), 1297–317.

Down, T. A., et al. (2008), 'A Bayesian deconvolution strategy for immunoprecipitation-based DNA methylome analysis', *Nat Biotechnol*, 26 (7), 779–85.

Dweep, H., Gretz, N., and Sticht, C. (2014), 'miRWalk database for miRNA-target interactions', *Methods Mol Biol*, 1182, 289–305.
Eads, C. A., Danenberg, K. D., Kawakami, K., Saltz, L. B., Blake, C., Shibata, D., Danenberg, P. V., and Laird, P. W. (2000), 'MethyLight: A high-throughput assay to measure DNA methylation', *Nucleic Acids Res*, *28*(8). doi: 10.1093/nar/28.8.e32
Eipper-Mains, J. E., et al. (2011), 'microRNA-seq reveals cocaine-regulated expression of striatal microRNAs', *RNA*, 17 (8), 1529–43.
Elliott, H. R., et al. (2022), 'Characterisation of ethnic differences in DNA methylation between UK-resident South Asians and Europeans', *Clin Epigenetics*, 14 (1), 130.
Eulalio, A., et al. (2009), 'Deadenylation is a widespread effect of miRNA regulation', *RNA*, 15 (1), 21–32.
Fang, C., et al. (2019a), 'Massively parallel sequencing of microRNA in bloodstains and evaluation of environmental influences on miRNA candidates using realtime polymerase chain reaction', *Forensic Sci Int Genet*, 38, 32–38.
Fang, C., et al. (2019b), 'MicroRNA profile analysis for discrimination of monozygotic twins using massively parallel sequencing and real-time PCR', *Forensic Sci Int Genet*, 38, 23–31.
Fang, C., et al. (2020), 'Age estimation using bloodstain miRNAs based on massive parallel sequencing and machine learning: A pilot study', *Forensic Sci Int Genet*, 47, 102300.
Fleckhaus, J., et al. (2023), 'Investigation of the impact of biogeographic ancestry on DNA methylation based age predictions comparing a middle East and a Central European population', *Forensic Sci Int Genet*, 67, 102923.
Flowers, E., et al. (2022), The role of racial and ethnic factors in MicroRNA expression and risk for type 2 diabetes, *Front Genet*, 13, 853633.
Flusberg, B. A., Webster, D. R., Lee, J. H., Travers, K. J., Olivares, E. C., Clark, T. A., Korlach, J., and Turner, S. W. (2010), 'Direct detection of DNA methylation during single-molecule, real-time sequencing', *Nat Methods*, 7 (6), 461–5.
Forat, S., Huettel, B., Reinhardt, R., Fimmers, R., Haidl, G., Denschlag, D., and Olek, K. (2016), 'Methylation markers for the identification of body fluids and tissues from forensic trace evidence', *PLoS One*, 11(2), e0147973. doi: 10.1371/journal.pone.0147973. eCollection 2016.
Fraga, M. F., Ballestar, E., Montoya, G., Taysavang, P., Wade, P. A., and Esteller, M. (2003), 'The affinity of different MBD proteins for a specific methylated locus depends on their intrinsic binding properties', *Nucleic Acids Res*, *31*(6), 1765–1774. doi: 10.1093/nar/gkg249
Francavilla, A., et al. (2021), 'Faecal miRNA profiles associated with age, sex, BMI, and lifestyle habits in healthy individuals', *Sci Rep*, 11 (1), 20645.
Freire-Aradas, A., Phillips, C., Mosquera-Miguel, A., Giron-Santamaria, L., Gomez-Tato, A., de Cal, M. C., Alvarez-Dios, J., Ansede-Bermejo, J., Torres-Espanol, M., Schneider, P. M., Pospiech, E., Branicki, W., Carracedo, A., and Lareu, M. V. (2016), 'Development of a methylation marker set for forensic age estimation using analysis of public methylation data and the Agena bioscience EpiTYPER system', *Forensic Sci Int-Genet*, 24. 65–75. CLARE: ELSEVIER IRELAND LTD.
Frommer, M., et al. (1992), 'A genomic sequencing protocol that yields a positive display of 5-methylcytosine residues in individual DNA strands', *Proc Natl Acad Sci U S A*, 89 (5), 1827–31.
Frumkin, D., Wasserstrom, A., Budowle, B., and Davidson, A. (2011), 'DNA methylation-based forensic tissue identification', *Forensic Sci Int Genet*, 5 (5), 517–24. doi: 10.1016/j.fsigen.2010.12.001. Epub 2010 Dec 31.
Frumkin, D., Wasserstrom, A., Davidson, A., and Grafit, A. (2010), 'Authentication of forensic DNA samples', *Forensic Sci Int Genet*, 4 (2), 95–103. doi: 10.1016/j.fsigen.2009.06.009. Epub 2009 Jul 16.
Fu, B., et al. (2020), 'MicroRNA-325-3p facilitates immune escape of mycobacterium tuberculosis through targeting LNX1 via NEK6 accumulation to promote anti-apoptotic STAT3 signaling', *mBio*, 11 (3), e00557–20. doi: 10.1128/mBio.00557-20.
Fu, X. D., et al. (2015), 'Identification of body fluid using tissue-specific DNA methylation markers', *Forensic Sci Int Genet Suppl Ser*, 5, e151–3.
Fujimoto, S., et al. (2019), 'Distinct spectrum of microRNA expression in forensically relevant body fluids and probabilistic discriminant approach', *Sci Rep*, 9 (1), 14332.
Garagnani, P., Bacalini, M. G., Pirazzini, C., Gori, D., Giuliani, C., Mari, D., Di Blasio, A. M., Gentilini, D., Vitale, G., Collino, S., Rezzi, S., Castellani, G., Capri, M., Salvioli, S., and Franceschi, C. (2012), 'Methylation of ELOVL2 gene as a new epigenetic marker of age', *Aging Cell*, 11 (6), 1132–4. doi: 10.1111/acel.12005. Epub 2012 Oct 14.
Gentilini, D., et al. (2013), 'Role of epigenetics in human aging and longevity: Genome-wide DNA methylation profile in centenarians and centenarians' offspring', *Age (Dordr)*, 35 (5), 1961–73.

Gerra, M. C., Dallabona, C., and Cecchi, R. (2024), 'Epigenetic analyses in forensic medicine: Future and challenges', *Int J Legal Med*, 138 (3), 701–19.
Gironella, M., et al. (2007), 'Tumor protein 53-induced nuclear protein 1 expression is repressed by miR-155, and its restoration inhibits pancreatic tumor development', *Proc Natl Acad Sci U S A*, 104 (41), 16170–5.
Glaich, O., et al. (2019), 'DNA methylation directs microRNA biogenesis in mammalian cells', *Nat Commun*, 10 (1), 5657.
Glynn, C. L. (2020), 'Potential applications of microRNA profiling to forensic investigations', *RNA*, 26 (1), 1–9.
Godfrey, K. M., Costello, P. M., and Lillycrop, K. A. (2015), 'The developmental environment, epigenetic biomarkers and long-term health', *J Dev Orig Health Dis*, 6 (5), 399–406.
Grieshober, L., et al. (2020), 'AHRR methylation in heavy smokers: Associations with smoking, lung cancer risk, and lung cancer mortality', *BMC Cancer*, 20 (1), 905.
Griffiths-Jones, S. (2010), 'miRBase: microRNA sequences and annotation', *Curr Protoc Bioinformatics*, *29*(1), 12.9.1–12.9.10. doi: 10.1002/0471250953.bi1209s29.
Gu, H., et al. (2011), 'Preparation of reduced representation bisulfite sequencing libraries for genome-scale DNA methylation profiling', *Nat Protoc*, 6 (4), 468–81.
Gu, H., et al. (2021), 'Smart-RRBS for single-cell methylome and transcriptome analysis', *Nat Protoc*, 16 (8), 4004–30.
Guitton, R., Nido, G. S., and Tzoulis, C. (2022), 'No evidence of extensive non-CpG methylation in mtDNA', *Nucleic Acids Res*, 50 (16), 9190–94.
Guo, H., Zhu, P., Yan, L., Li, R., Hu, B., Lian, Y., Yan, J., Ren, X., Lin, S., Li, J., Jin, X., Shi, X., Liu, P., Wang, X., Wang, W., Wei, Y., Li, X., Guo, F., Wu, X., Fan, X., Yong, J., Wen, L., Xie, S. X., Tang, F., and Qiao, J. (2014), 'The DNA methylation landscape of human early embryos', *Nature*, 511(7511):606–10. doi: 10.1038/nature13544. Epub 2014 Jul 23.
Handy, D. E., Castro, R., and Loscalzo, J. (2011), 'Epigenetic modifications: Basic mechanisms and role in cardiovascular disease', *Circulation*, 123 (19), 2145–56.
Hao, Z., et al. (2020), 'N(6)-deoxyadenosine methylation in mammalian mitochondrial DNA', *Mol Cell*, 78 (3), 382–95 e8.
Harris, R. A., Wang, T., Coarfa, C., Nagarajan, R. P., Hong, C., Downey, S. L., Johnson, B. E., Fouse, S. D., Delaney, A., Zhao, Y., Olshen, A., Ballinger, T., Zhou, X., Forsberg, K. J., Gu, J., Echipare, L., O'Geen, H., Lister, R.,… and Costello, J. F. (2010), 'Comparison of sequencing-based methods to profile DNA methylation and identification of monoallelic epigenetic modifications', *Nat Biotechnol*, *28*(10), 1097–1105. doi: 10.1038/nbt.1682
Heinbockel, T., and Csoka, A. B. (2018), 'Epigenetic effects of drugs of abuse', *Int J Environ Res Public Health*, 15 (10), 2098. doi: 10.3390/ijerph15102098.
Herman, J. G., Graff, J. R., Myohanen, S., Nelkin, B. D., and Baylin, S. B. (1996), 'Methylation-specific PCR: A novel PCR assay for methylation status of CpG islands', *Proc Natl Acad Sci U S A*, *93* (18), 9821–9826. doi: 10.1073/pnas.93.18.9821
Hernandez, D. G., and Singleton, A. B. (2012), 'Using DNA methylation to understand biological consequences of genetic variability', *Neurodegener Dis*, 9 (2), 53–9.
Heyn, H., Moran, S., Hernando-Herraez, I., Sayols, S., Gomez, A., Sandoval, J., Monk, D., Hata, K., Marques-Bonet, T., Wang, L., and Esteller, M. (2013), 'DNA methylation contributes to natural human variation', *Genome Res*, 23, 1363–72.
Higgins, D., et al. (2015), 'Differential nuclear and mitochondrial DNA preservation in post-mortem teeth with implications for forensic and ancient DNA studies', *PLoS One*, 10 (5), e0126935.
Hillemacher, T., et al. (2015), 'DNA methylation of the LEP gene is associated with craving during alcohol withdrawal', *Psychoneuroendocrinology*, 51, 371–7.
Horvath, S. (2013), 'DNA methylation age of human tissues and cell types', *Genome Biol*, 14 (10), R115.
Houseley, J., and Tollervey, D. (2009), 'The many pathways of RNA degradation', *Cell*, 136 (4), 763–76.
Hu, G., et al. (2023), 'A global perspective of forensic entomology case reports from 1935 to 2022', *Int J Legal Med*, 137 (5), 1535–53.
Hu, J. (2012), 'Single base extension reaction-based surface enhanced Raman spectroscopy for DNA methylation assay', *Biosens Bioelectron*, *31*(1), 451–457. doi: 10.1016/j.bios.2011.11.014
Huan, T., Chen, G., Liu, C., Bhattacharya, A., Rong, J., Chen, B. H., Seshadri, S., Tanriverdi, K., Freedman, J. E., Larson, M. G., Murabito, J. M., & Levy, D. (2018), 'Age-associated microRNA expression in human peripheral blood is associated with all-cause mortality and age-related traits', *Aging Cell*, *17*(1). doi: 10.1111/acel.12687
Huang, C. H., et al. (2022), 'Mitochondrial DNA methylation profiling of the human prefrontal cortex and nucleus accumbens: Correlations with aging and drug use', *Clin Epigenetics*, 14 (1), 79.

Huang, J., Okuka, M., Lu, W., Tsibris, J. C. M., McLean, M. P., Keefe, D. L., and Liu, L. (2013), 'Telomere shortening and DNA damage of embryonic stem cells induced by cigarette smoke', *Reprod Toxicol*, 35, 89–95. doi: 10.1016/j.reprotox.2012.07.003. Epub 2012 Jul 21.

Huang, J., Renault, V., Sengenes, J., Touleimat, N., Michel, S., Lathrop, M., and Tost, J. (2012), 'MeQA: A pipeline for MeDIP-seq data quality assessment and analysis', *Bioinformatics*, 28, 587–88.

Ibrahim, S. F., Ali, M. M., Basyouni, H., Rashed, L. A., Amer, E. A. E., & El-Kareem, D. A. (2019), 'Histological and miRNAs postmortem changes in incisional wound', *Egypt J Forensic Sci*, 9 (1). doi: 10.1186/s41935-019-0141-7,

Ji, Z., et al. (2020), 'Insertion of channel of phi29 DNA packaging motor into polymer membrane for high-throughput sensing', *Nanomedicine*, 25, 102170.

Johnson, A. A., et al. (2012), 'The role of DNA methylation in aging, rejuvenation, and age-related disease', *Rejuvenation Res*, 15 (5), 483–94.

Juusola, J., and Ballantyne, J. (2005), 'Multiplex mRNA profiling for the identification of body fluids', *Forensic Sci Int*, 152(1):1–12. doi: 10.1016/j.forsciint.2005.02.020.

Karabegovic, I., et al. (2023), 'Plasma MicroRNA signature of alcohol consumption: The Rotterdam study', *J Nutr*, 152 (12), 2677–88.

Keller, S., et al. (2011), 'Body fluid derived exosomes as a novel template for clinical diagnostics', *J Transl Med*, 9, 86.

Khulan, B., Thompson, R. F., Ye, K., Fazzari, M. J., Suzuki, M., Stasiek, E., Figueroa, M. E., Glass, J. L., Chen, Q., Montagna, C., Hatchwell, E., Selzer, R. R., Richmond, T. A., Green, R. D., Melnick, A., and Greally, J. M. (2006), 'Comparative isoschizomer profiling of cytosine methylation: the HELP assay', *Genome research*, *16* (8), 1046–1055. doi: 10.1101/gr.5273806

Kinser, H. E., and Pincus, Z. (2020), 'MicroRNAs as modulators of longevity and the aging process', *Hum Genet*, 139 (3), 291–308.

Kitsera, N., et al. (2017), 'Functional impacts of 5-hydroxymethylcytosine, 5-formylcytosine, and 5-carboxycytosine at a single hemi-modified CpG dinucleotide in a gene promoter', *Nucleic Acids Res*, 45 (19), 11033–42.

Klemp, I., et al. (2022), 'DNA methylation patterns reflect individual's lifestyle independent of obesity', *Clin Transl Med*, 12 (6), e851.

Koch, C. M., and Wagner, W. (2011), 'Epigenetic-aging-signature to determine age in different tissues', *Aging (Albany NY)*, 3(10):1018–27. doi: 10.18632/aging.100395.

Koh, C. W. Q., et al. (2018), 'Single-nucleotide-resolution sequencing of human N6-methyldeoxyadenosine reveals strand-asymmetric clusters associated with SSBP1 on the mitochondrial genome', *Nucleic Acids Res*, 46 (22), 11659–70.

Kohli, R. M., and Zhang, Y. (2013), 'TET enzymes, TDG and the dynamics of DNA demethylation', *Nature*, 502, 472–9.

Korlach, J., et al. (2008), 'Selective aluminum passivation for targeted immobilization of single DNA polymerase molecules in zero-mode waveguide nanostructures', *Proc Natl Acad Sci U S A*, 105 (4), 1176–81.

Kovatsi, L., Georgiou, E., Ioannou, A., Haitoglou, C., Tzimagiorgis, G., Tsoukali, H., and Kouidou, S. (2010), 'p16 promoter methylation in Pb2+ -exposed individuals', *Clin Toxicol (Phila)*, 48 (2), 124–8. doi: 10.3109/15563650903567091.

Krueger, F., and Andrews, S. R. (2011), Bismark: A flexible aligner and methylation caller for bisulfite-seq applications, *Bioinformatics*, 27 (11), 1571–2.

Kumar, S., Chinnusamy, V., and Mohapatra, T. (2019), 'Epigenetics of modified DNA bases: 5-methylcytosine and beyond', *Front Genet*, 9, 640.

Kurdyukov, S., and Bullock, M. (2016), 'DNA methylation analysis: Choosing the right method', *Biology (Basel)*, 5 (1) doi: 10.3390/biology5010003.

Lan, X., Adams, C., Landers, M., Dudas, M., Krissinger, D., Marnellos, G., Bonneville, R., Xu, M., Wang, J., Huang, T. H., Meredith, G., and Jin, V. X. (2011), 'High resolution detection and analysis of CpG dinucleotides methylation using MBD-Seq technology', *PloS one*, *6*(7), e22226. doi: 10.1371/journal.pone.0022226

Lasken, R. S., and Egholm, M. (2003), 'Whole genome amplification: Abundant supplies of DNA from precious samples or clinical specimens', *Trends Biotechnol*, 21 (12), 531–5. doi: 10.1016/j.tibtech.2003.09.010.

Lee, H. Y., Jung, S.-E., Oh, Y. N., Choi, A., Yang, W. I., and Shin, K.-J. (2015), 'Epigenetic age signatures in the forensically relevant body fluid of semen: A preliminary study', *Forensic Sci Int Genet*, 19, 28–34. doi: 10.1016/j.fsigen.2015.05.014. Epub 2015 May 27.

Lee, H. Y., Park, M. J., Choi, A., An, J. H., Yang, W. I., and Shin, K.-J. (2012), 'Potential forensic application of DNA methylation profiling to body fluid identification', *Int J Legal Med*, 126 (1), 55–62. doi: 10.1007/s00414-011-0569-2. Epub 2011 Apr 6.

Lee, J. W., et al. (2018), 'A validation study of DNA methylation-based age prediction using semen in forensic casework samples', *Leg Med (Tokyo)*, 31, 74–77.

Lee, Y., et al. (2004), 'MicroRNA genes are transcribed by RNA polymerase II', *EMBO J*, 23 (20), 4051–60.

Legendre, C., et al. (2015), 'Whole-genome bisulfite sequencing of cell-free DNA identifies signature associated with metastatic breast cancer', *Clin Epigenetics*, 7 (1), 100.

Li, C., et al. (2013), 'Differences of DNA methylation profiles between monozygotic twins' blood samples', *Mol Biol Rep*, 40 (9), 5275–80.

Li, C., Zhang, S., Que, T., Li, L., and Zhao, S. (2011), 'Identical but not the same: The value of DNA methylation profiling in forensic discrimination within monozygotic twins', *Forensic Sci Int Genet Suppl Ser*, 3, e337–e338, doi: 10.1016/j.fsigss.2011.09.031

Li, H., and Durbin, R. (2010), 'Fast and accurate long-read alignment with burrows-wheeler transform', *Bioinformatics*, 26 (5), 589–95.

Li, W. T., and Zhang, Q. (2019), 'MicroRNA-708-5p regulates mycobacterial vitality and the secretion of inflammatory factors in mycobacterium tuberculosis-infected macrophages by targeting TLR4', *Eur Rev Med Pharmacol Sci*, 23 (18), 8028–38.

Li, Z., et al. (2021), 'mRNA and microRNA stability validation of blood samples under different environmental conditions', *Forensic Sci Int Genet*, 55, 102567.

Lin, Y.-C., Tsai, L.-C., Lee, J. C.-I., Su, C.-W., Tzen, J. T.-C., Linacre, A., and Hsieh, H.-M. (2016), 'Novel identification of biofluids using a multiplex methylation sensitive restriction enzyme-PCR system', *Forensic Sci Int Genet*, 25, 157–65. doi: 10.1016/j.fsigen.2016.08.011. Epub 2016 Sep 3.

Liu, B., et al. (2016), 'CpG methylation patterns of human mitochondrial DNA', *Sci Rep*, 6, 23421.

Liu, C., et al. (2018a), 'A DNA methylation biomarker of alcohol consumption', *Mol Psychiatry*, 23 (2), 422–33.

Liu, H., et al. (2021), 'Increased expression of plasma miRNA-320a and let-7b-5p in heroin-dependent patients and its clinical significance', *Front Psychiatry*, 12, 679206.

Liu, X., et al. (2020), 'MicroRNA-127 promotes anti-microbial host defense through restricting A20-mediated de-ubiquitination of STAT3', *iScience*, 23 (1), 100763.

Liu, Y., et al. (2018b), 'Acute downregulation of miR-199a attenuates sepsis-induced acute lung injury by targeting SIRT1', *Am J Physiol Cell Physiol*, 314 (4), C449–55.

Lizio, M., et al. (2015), 'Gateways to the FANTOM5 promoter level mammalian expression atlas', *Genome Biol*, 16 (1), 22.

Lokhande, H. A. (2023), 'Bioinformatics analysis of miRNA sequencing data', *Methods Mol Biol*, 2595, 225–37.

Luo, X. Y., et al. (2015), 'MicroRNA markers for forensic body fluid identification obtained from miRCURY™ LNA array', *Forensic Sci Int Genet Suppl Ser*, 5, e630–32.

Maeda, H., Ishikawa, T., and Michiue, T. (2011), 'Forensic biochemistry for functional investigation of death: Concept and practical application', *Leg Med (Tokyo)*, 13 (2), 55–67. doi: 10.1016/j.legalmed.2010.12.005.

Mahmoud, A. M. (2022), 'An overview of epigenetics in obesity: The role of lifestyle and therapeutic interventions', *Int J Mol Sci*, 23 (3). doi: 10.3390/ijms23031341

Meissner, A., et al. (2008), 'Genome-scale DNA methylation maps of pluripotent and differentiated cells', *Nature*, 454 (7205), 766–70.

Miguel, V., et al. (2018), 'The role of MicroRNAs in environmental risk factors, noise-induced hearing loss, and mental stress', *Antioxid Redox Signal*, 28 (9), 773–96.

Moosavi, A., & Motevalizadeh Ardekani, A. (2016), 'Role of epigenetics in biology and human diseases', *Iran Biomed J*, *20*(5), 246–258. doi: 10.22045/ibj.2016.01

Murchison, E. P., and Hannon, G. J. (2004), 'miRNAs on the move: miRNA biogenesis and the RNAi machinery', *Curr Opin Cell Biol*, 16 (3), 223–9.

Na, J. Y. (2020), 'Estimation of the post-mortem interval using microRNA in the bones', *J Forensic Leg Med*, 75, 102049.

Nair, S. S., et al. (2018), 'Guidelines for whole genome bisulphite sequencing of intact and FFPET DNA on the Illumina HiSeq X ten', *Epigenetics Chromatin*, 11 (1), 24.

Nakatome, M., Orii, M., Hamajima, M., Hirata, Y., Uemura, M., Hirayama, S., and Isobe, I. (2011), 'Methylation analysis of circadian clock gene promoters in forensic autopsy specimens', *Leg Med*, 13(4):205–9. doi: 10.1016/j.legalmed.2011.03.001. Epub 2011 May 18.

Nakayashiki, N., et al. (2008), 'Studies on differentially methylated parental allele in imprinted genes', *Forensic Sci Int Genet Suppl Ser*, 1 (1), 572–73.

Nakayashiki, N., Takamiya, M., Shimamoto, K., and Aoki, Y. (2009), 'Analysis of the methylation profiles in imprinted genes applicable to parental allele discrimination', *Leg Med (Tokyo)*, 11 (Suppl 1), S471–2. doi: 10.1016/j.legalmed.2009.02.013.

Nesheva, D. (2014), 'Aspects of ancient mitochondrial DNA analysis in different populations for understanding human evolution', *Balkan J Med Genet*, 17 (1), 5–14.

Ni, P., et al. (2023), 'DNA 5-methylcytosine detection and methylation phasing using PacBio circular consensus sequencing', *Nat Commun*, 14 (1), 4054.

Nielsen, D. A., et al. (2009), 'Increased OPRM1 DNA methylation in lymphocytes of methadone-maintained former heroin addicts', *Neuropsychopharmacology*, 34 (4), 867–73.

Nieratschker, V., et al. (2014), 'Epigenetic alteration of the dopamine transporter gene in alcohol-dependent patients is associated with age', *Addict Biol*, 19 (2), 305–11.

Noguera-Castells, A., et al. (2023), 'Validation of the new EPIC DNA methylation microarray (900K EPIC v2) for high-throughput profiling of the human DNA methylome', *Epigenetics*, 18 (1), 2185742.

Oda, M., and Greally, J. M. (2009), 'The HELP assay' In J. M. Greally (Ed.), *Methods in Molecular Biology: Vol. 507. Epigenetics* (pp. 77–87). Humana Press. doi: 10.1007/978-1-59745-522-0_7.

Ong, J., et al. (2019), 'Age-related gene and miRNA expression changes in airways of healthy individuals', *Sci Rep*, 9 (1), 3765.

Onofri, M., et al. (2023), 'Forensic Age Estimation through a DNA Methylation-Based Age Prediction Model in the Italian Population: A Pilot Study', *Int J Mol Sci*, 24 (6), 5381. doi: 10.3390/ijms24065381.

Osborne, A. J., et al. (2020), 'Genome-wide DNA methylation analysis of heavy cannabis exposure in a New Zealand longitudinal cohort', *Transl Psychiatry*, 10 (1), 114.

Park, J. L., et al (2014), 'Identification of body fluid-specific DNA methylation markers for use in forensic science', *Forensic Sci Int Genet*, 13, 147–53.

Peiro-Chova, L., et al. (2013), 'High stability of microRNAs in tissue samples of compromised quality', *Virchows Arch*, 463 (6), 765–74.

Pelizzola, M., et al. (2008), 'MEDME: An experimental and analytical methodology for the estimation of DNA methylation levels based on microarray derived MeDIP-enrichment', *Genome Res*, 18 (10), 1652–9.

Pointner, A., et al. (2024), 'Lifestyle-driven variations in nutrimiromic MicroRNA expression patterns across and beyond genders', *Life (Basel)*, 14 (3), 390. doi: 10.3390/life14030390.

Ponchel, F., et al. (2003), 'Real-time PCR based on SYBR-green i fluorescence: An alternative to the TaqMan assay for a relative quantification of Gene rearrangements, Gene amplifications and micro gene deletions', *BMC Biotechnol*, 3, 18.

Potla, P., Ali, S. A., and Kapoor, M. (2021), 'A bioinformatics approach to microRNA-sequencing analysis', *Osteoarthr Cartil Open*, 3 (1), 100131.

Rana, A. K. (2018a), 'The future of forensic biology', *J Biomed*, 3, 13–18.

Rana, A. K. (2018b), 'Crime investigation through DNA methylation analysis: Methods and applications in forensics', *Egypt J Forensic Sci*, 8 (7). doi: 10.1186/s41935-018-0042-1

Rasmussen, K. D., and Helin, K. (2016), 'Role of TET enzymes in DNA methylation, development, and cancer', *Genes Dev*, 30 (7), 733–50.

Reynolds, C. A., et al. (2020), 'A decade of epigenetic change in aging twins: Genetic and environmental contributions to longitudinal DNA methylation', *Aging Cell*, 19 (8), e13197.

Rezaei, S., et al. (2022), 'GBS-MeDIP: A protocol for parallel identification of genetic and epigenetic variation in the same reduced fraction of genomes across individuals', *STAR Protoc*, 3 (1), 101202.

Richards, E. J. (2006), 'Inherited epigenetic variation–revisiting soft inheritance.', *Nat Rev Genet*, 7, 395–401.

Riley, K. J., Yario, T. A., and Steitz, J. A. (2012), 'Association of Argonaute proteins and microRNAs can occur after cell lysis', *RNA*, 18 (9), 1581–5.

Roldán-Arjona, T., and Rariza, R. R. (2009), 'DNA Demethylation', in Henri Grosjean (ed.), *DNA and RNA Modification Enzymes Structure, Mechanism, Function and Evolution* (1 ed.). CRC Press.

Ruggeri, B., et al. (2015), 'Association of protein phosphatase PPM1G with alcohol use disorder and brain activity during behavioral control in a genome-wide methylation analysis', *Am J Psychiatry*, 172 (6), 543–52.

Satta, R., Maloku, E., Zhubi, A., Pibiri, F., Hajos, M., Costa, E., and Guidotti, A. (2008), 'Nicotine decreases DNA methyltransferase 1 expression and glutamic acid decarboxylase 67 promoter methylation in GABAergic interneurons', *Proc Natl Acad Sci U S A*, 105(42), 16356–61. doi: 10.1073/pnas.0808699105. Epub 2008 Oct 13.

Schumacher, A., et al. (2006a), 'Microarray-based DNA methylation profiling: Technology and applications', *Nucleic Acids Res*, 34 (2), 528–42.

Schumacher, A., Kapranov, P., Kaminsky, Z., Flanagan, J., Assadzadeh, A., Yau, P., Virtanen, C., Winegarden, N., Cheng, J., Gingeras, T., and Petronis, A. (2006b), 'Microarray-based DNA methylation profiling: Technology and applications', *Nucleic Acids Res*, *34*(2), 528–542. doi: 10.1093/nar/gkj461.

Seashols-Williams, S., et al. (2016), 'High-throughput miRNA sequencing and identification of biomarkers for forensically relevant biological fluids', *Electrophoresis*, 37 (21), 2780–88.

Serafin, A., et al. (2015), 'Overexpression of blood microRNAs 103a, 30b, and 29a in L-dopa-treated patients with PD', *Neurology*, 84 (7), 645–53.

Singh, K. P., et al. (2020), 'Exosomal microRNAs are novel circulating biomarkers in cigarette, waterpipe smokers, E-cigarette users and dual smokers', *BMC Med Genomics*, 13 (1), 128.

Smith, R. C., et al. (2022), 'Changes in expression of DNA-methyltransferase and cannabinoid receptor mRNAs in blood lymphocytes after acute cannabis smoking', *Front Psychiatry*, 13, 887700.
Sriram, H., et al. (2021), 'Improved protocol for plasma microRNA extraction and comparison of commercial kits', *Biochem Med (Zagreb)*, 31 (3), 030705.
Statello, L., et al. (2020), 'Gene regulation by long non-coding RNAs and its biological functions', *Nat Rev Mol Cell Biol*, 22 (2), 96–118.
Statham, A. L., et al. (2010), 'Repitools: An R package for the analysis of enrichment-based epigenomic data', *Bioinformatics*, 26 (13), 1662–3.
Stewart, L., Evans, N., Bexon, K. J., van der Meer, D. J., and Williams, G. A. (2015), 'Differentiating between monozygotic twins through DNA methylation-specific high-resolution melt curve analysis', *Anal Biochem,* 476, 36–9. doi: 10.1016/j.ab.2015.02.001. Epub 2015 Feb 10.
Stoiber, M., et al. (2016), 'De novo identification of DNA modifications enabled by genome-guided nanopore signal processing', *bioRxiv*, doi: 10.1101/094672
Szulik, M. W., Pallan, P. S., Nocek, B., Voehler, M., Banerjee, S., Brooks, S., Joachimiak, A., Egli, M., Eichman, B. F., and Stone, M. P. (2015), 'Differential stabilities and sequence-dependent base pair opening dynamics of Watson-Crick base pairs with 5-hydroxymethylcytosine, 5-formylcytosine, or 5-carboxylcytosine', *Biochemistry,* 54, 1294–305.
Szyf, M. (2011), 'DNA methylation, the early-life social environment and behavioral disorders', *J Neurodev Disord*, 3 (3), 238–49.
Talens, R. P., et al. (2012), 'Epigenetic variation during the adult lifespan: Cross-sectional and longitudinal data on monozygotic twin pairs', *Aging Cell*, 11 (4), 694–703.
Tost, J., and Gut, I. G. (2007), 'DNA methylation analysis by pyrosequencing', *Natur Prot*, 2 (9), 2265–2275. doi: 10.1038/nprot.2007.314
Triche, T. J. Jr, Weisenberger, D. J., Van Den Berg, D., Laird, P. W., and Siegmund, K. D. (2013), 'Low-level processing of Illumina Infinium DNA Methylation BeadArrays', *Nucleic Acids Res*, *41*(7), e90.
Turchinovich, A., et al. (2013), 'Circulating miRNAs: Cell-cell communication function? *Front Genet*, 4, 119.
Turner, B. M. (2009), 'Epigenetic responses to environmental change and their evolutionary implications', *Philos Trans R Soc Lond B Biol Sci*, 364 (1534), 3403–18.
Unnikrishnan, A., et al. (2019), 'The role of DNA methylation in epigenetics of aging', *Pharmacol Ther*, 195, 172–85.
Vaillancourt, K., et al. (2021a), 'Methylation of the tyrosine hydroxylase gene is dysregulated by cocaine dependence in the human striatum', *iScience*, 24 (10), 103169.
Vaillancourt, K., et al. (2021b), 'Cocaine-related DNA methylation in caudate neurons alters 3D chromatin structure of the IRXA gene cluster', *Mol Psychiatry*, 26 (7), 3134–51.
Vidaki, A., et al. (2017), 'DNA methylation-based forensic age prediction using artificial neural networks and next generation sequencing', *Forensic Sci Int Genet*, 28, 225–36.
Vidaki, A., et al. (2023), 'Targeted DNA methylation analysis and prediction of smoking habits in blood based on massively parallel sequencing', *Forensic Sci Int Genet*, 65, 102878.
Vidaki, A., Giangasparo, F., and Syndercombe Court, D. (2016), 'Discovery of potential DNA methylation markers for forensic tissue identification using bisulphite pyrosequencing', Electrophoresis, 37, 2767–2779. doi: 10.1002/elps.201600261
Vidaki, A., and Kayser, M. (2017), 'From forensic epigenetics to forensic epigenomics: Broadening DNA investigative intelligence', *Genome Biol,* 18, 238.
Vidaki, A., and Kayser, M. (2018), 'Recent progress, methods and perspectives in forensic epigenetics', *Forensic Sci Int Genet*, 37, 180–95. doi: 10.1016/j.fsigen.2018.08.008.
Waalwijk, C., and Flavell, R. A. (1978), 'MspI, an isoschizomer of hpaII which cleaves both unmethylated and methylated hpaII sites', *Nucleic Acids Res*, 5 (9), 3231–6.
Wang, H., Ach, R. A., and Curry, B. (2007), 'Direct and sensitive miRNA profiling from low-input total RNA', *RNA*, 13 (1), 151–9.
Wang, S., et al. (2024), 'The role of microRNA-23a-3p in the progression of human aging process by targeting FOXO3a', *Mol Biotechnol*, 66 (2), 277–87.
Wang, S. Y., et al. (2021a), 'Role of epigenetics in unicellular to multicellular transition in Dictyostelium', *Genome Biol*, 22 (1), 134.
Wang, W., and Luo, Y. P. (2015), 'MicroRNAs in breast cancer: Oncogene and tumor suppressors with clinical potential', *J Zhejiang Univ Sci B*, 16 (1), 18–31.
Wang, Y., et al. (2021b), 'Nanopore sequencing technology, bioinformatics and applications', *Nat Biotechnol*, 39 (11), 1348–65.
Watanabe, N., Bajgain, P., Sukumaran, S., Ansari, S., Heslop, H. E., Rooney, C. M., Brenner, M. K., Leen, A. M., and Vera, J. F. 1 (2016), 'Fine-tuning the CAR spacer improves T-cell potency', *Oncoimmunology*, *5*(12), e1253656. doi: 10.1080/2162402X.2016.1253656

Weber-Lehmann, J., Schilling, E., Gradl, G., Richter, D. C., Wiehler, J., and Rolf, B. (2014), 'Finding the needle in the haystack: Differentiating "identical" twins in paternity testing and forensics by ultra-deep next generation sequencing', *Forensic Sci Int Genet*, 9, 42–6. doi: 10.1016/j.fsigen.2013.10.015. Epub 2013 Nov 8.

Wen, D., et al. (2022), 'DNA methylation analysis for smoking status prediction in the Chinese population based on the methylation-sensitive single-nucleotide primer extension method', *Forensic Sci Int*, 339, 111412.

Wenger, A. M., et al. (2019), 'Accurate circular consensus long-read sequencing improves variant detection and assembly of a human genome', *Nat Biotechnol*, 37 (10), 1155–62.

Wilson, G. A., and Beck, S. (2016), 'Computational analysis and integration of MeDIP-Seq Methylome data', In J. K. Kulski (Ed.), *Next Generation Sequencing: Advances, Applications and Challenges* (Wellcome Trust-Funded Monographs and Book Chapters; Rijeka (HR)). InTech Open

Wojdacz, T. K., and Dobrovic, A. (2007), 'Methylation-sensitive high resolution melting (MS-HRM): A new approach for sensitive and high-throughput assessment of methylation', *Nucleic Acids Res*, 35 (6), e41.

Wong, A. H. C., Gottesman, I. I., and Petronis, A. (2005), 'Phenotypic differences in genetically identical organisms: The epigenetic perspective', *Hum Mol Genet*, 14 Spec No 1:R11–8.

Wu, L., Fan, J., and Belasco, J. G. (2006), 'MicroRNAs direct rapid deadenylation of mRNA', *Proc Natl Acad Sci U S A*, 103 (11), 4034–9.

Xiao, C., et al. (2019), 'Differences of microRNA expression profiles between monozygotic twins' blood samples', *Forensic Sci Int Genet*, 41, 152–58.

Xiao, C., et al. (2023), 'Improved age estimation from semen using sperm-specific age-related CpG markers', *Forensic Sci Int Genet*, 67, 102941.

Xiong, Z., and Laird, P. W. (1997), 'COBRA: A sensitive and quantitative DNA methylation assay', *Nucleic Acids Res*, *25*(12), 2532–2534. doi: 10.1093/nar/25.12.2532.

Xu, L., and Seki, M. (2020), 'Recent advances in the detection of base modifications using the nanopore sequencer', *J Hum Genet*, 65 (1), 25–33.

Yan, P. S., Chen, C. M., Shi, H., Rahmatpanah, F., Wei, S. H., and Huang, T. H. (2002), 'Applications of CpG island microarrays for high-throughput analysis of DNA methylation', *J Nutr*, *132*(8 Suppl), 2430S–2434S. doi: 10.1093/jn/132.8.2430S

Yao, Q., Chen, Y., and Zhou, X. (2019), 'The roles of microRNAs in epigenetic regulation', *Curr Opin Chem Biol*, 51, 11–17.

Yen, C. H., et al. (2015), 'Reduced dopamine transporter availability and neurocognitive deficits in male patients with alcohol dependence', *PLoS One*, 10 (6), e0131017.

Yi, S. H., Jia, Y. S., Mei, K., Yang, R. Z., and Huang, D. X. (2015), 'Age-related DNA methylation changes for forensic age-prediction', *Int J Legal Med*, 129(2), 237–44. doi: 10.1007/s00414-014-1100-3. Epub 2014 Nov 16.

Ying, S. Y., Chang, D. C., and Lin, S. L. (2008), 'The microRNA (miRNA): Overview of the RNA genes that modulate gene function', *Mol Biotechnol*, 38 (3), 257–68.

Yohannes, Y. B., et al. (2020), 'Blood lead levels and aberrant DNA methylation of the ALAD and p16 gene promoters in children exposed to environmental-lead', *Environ Res*, 188, 109759.

Yuen, Z. W., et al. (2024), 'Profiling age and body fluid DNA methylation markers using nanopore adaptive sampling', *Forensic Sci Int Genet*, 71, 103048.

Zhang, L., Lu, Q., and Chang, C. (2020), 'Epigenetics in health and disease', *Adv Exp Med Biol*, 1253, 3–55.

Zhang, S., Zhao, S., Wang, Z., and Li, C. (2013), 'Investigation of parent-of-origin SNPs in 5 imprinted genes for forensic purpose', *Forensic Sci Int Genet Suppl Ser*, 4, e304–5.

Zhang, Y., et al. (2012), 'Hepatitis C virus-induced up-regulation of microRNA-155 promotes hepatocarcinogenesis by activating Wnt signaling', *Hepatology*, 56 (5), 1631–40.

Zhang, Y., et al. (2014), 'F2RL3 methylation as a biomarker of current and lifetime smoking exposures', *Environ Health Perspect*, 122 (2), 131–7.

Zhao, D., Ishikawa, T., Quan, L., Michiue, T., Zhu, B.-L., and Maeda, H. (2009), 'Postmortem quantitative mRNA analyses of death investigation in forensic pathology: An overview and prospects', *Leg Med (Tokyo)*, 11(Suppl 1), S43–5. doi: 10.1016/j.legalmed.2009.01.066.

Zhao, M. T., et al. (2014), 'Methylated DNA immunoprecipitation and high-throughput sequencing (MeDIP-seq) using low amounts of genomic DNA', *Cell Reprogram*, 16 (3), 175–84.

Zubakov, D., et al. (2010), 'MicroRNA markers for forensic body fluid identification obtained from microarray screening and quantitative RT-PCR confirmation', *Int J Legal Med*, 124 (3), 217–26.

Zununi Vahed, S., et al. (2016), 'A microRNA isolation method from clinical samples', *Bioimpacts*, 6 (1), 25–31.

16

Mass Spectrometry in Forensic DNA Analysis

A New Frontier

Bharti Jain, Rajeev Jain, and Ritesh K. Shukla

16.1 Introduction

DNA (deoxyribonucleic acid) is the molecular blueprint of life, consisting of four fundamental nucleotide bases: adenine (A), guanine (G), thymine (T), and cytosine (C). These bases form a double-stranded helical structure, where each strand is linked by a sugar-phosphate backbone (Tost & Gut, 2006). The specific pairing between purine bases (adenine and guanine) and pyrimidine bases (thymine and cytosine) through hydrogen bonds establishes the foundation of the genetic code. This sequence-driven arrangement of nucleotides forms the basis of all hereditary information, guiding cellular functions and enabling genetic continuity across generations (Gingras et al., 2005).

The completion of the Human Genome Project in 2003 represented a pivotal achievement in molecular biology, providing an almost complete and highly accurate map of the three billion base pairs that make up the human genome. This monumental effort has revolutionized biological and medical research by offering insights into gene function, regulation, and the intricate network of biological pathways that drive cellular processes. It has facilitated an unprecedented focus on exploring genetic variations, such as single nucleotide polymorphisms (SNPs), which play a crucial role in understanding the molecular basis of diseases and in identifying potential targets for therapeutic interventions (Ong & Mann, 2005).

Current advancements in DNA sequencing technologies have enabled the transition from analyzing single genes to examining entire genomes, thus allowing for a comprehensive view of the genetic architecture of organisms. Capillary array DNA sequencing technology, which utilizes laser-induced fluorescence detection, has emerged as the state-of-the-art method for high-throughput sequencing (Jut et al., 1995; Smith et al., 1986). It efficiently addresses the needs of large-scale genomic projects by providing high read lengths and throughput. However, despite its capabilities, this technology faces limitations in accuracy, particularly in detecting genetic variations within GC-rich regions, identifying heterozygous states, and accurately interpreting frameshift mutations caused by insertions or deletions (Edwards et al., 2005).

The inherent limitations of conventional electrophoresis-based sequencing methods, such as the slow migration of samples through gels and the difficulty in resolving complex genetic variations, have prompted the exploration of alternative analytical techniques. In this regard, mass spectrometry (MS) has gained attention as a versatile tool, offering a promising approach

DOI: 10.4324/9781003474296-16

to nucleic acid analysis (Bowling et al., 1991; Yamakawa & Ohara, 1997). Traditional MS initially designed for small molecule analysis were not suitable for DNA due to the fragile nature of its phosphate backbone, which tends to fragment during volatilization. However, the development of "soft" ionization techniques, specifically electrospray ionization (ESI) and matrix-assisted laser desorption/ionization (MALDI), has enabled the direct analysis of nucleic acids without significant degradation (Fenn et al., 1989; Ronaghi et al., 1998; Ronald & Brian, 1989).

Among these, MALDI-time-of-flight (MALDI-TOF) MS has emerged as a highly effective approach for DNA analysis due to its ability to generate predominantly singly charged molecular ions, detect a broad range of mass sizes, and facilitate rapid and automated data collection. This technique has proven advantageous in a variety of applications, from short DNA sequencing and genotyping to the detection of small genetic variations, such as SNPs and indels. Its capacity for high-throughput analysis, combined with the elimination of the need for fluorescent or radioactive labeling, positions MALDI-TOF MS as a powerful tool for genomic studies and clinical diagnostics (Edwards et al., 2005).

MALDI-TOF-MS addresses several challenges posed by conventional sequencing techniques. For instance, it offers clearer differentiation of homopolymeric regions and unambiguous identification of sequence alterations that can otherwise be difficult to detect using pyrosequencing or capillary electrophoresis (CE) methods. Moreover, its high-resolution capabilities and automated sample processing significantly reduce the time required for large-scale genetic analyses, making it highly suitable for high-throughput genomic studies (Nordhoff et al., 2000).

Following the genome sequencing revolution, MS plays a pivotal role beyond traditional DNA sequencing to encompass broader applications in genomics and proteomics, including DNA methylation analysis, RNA expression profiling, and the investigation of protein-DNA interactions. As research continues to delve into the complexities of gene expression and regulation, MALDI-TOF-MS remains at the forefront of technologies driving these discoveries, providing deeper insights into molecular biology and advancing our understanding of genetic and epigenetic mechanisms (Ross et al., 1998).

Lately, MS has become an essential tool for the characterization of biomolecules, including proteins, peptides, and nucleic acids, owing to its ability to provide detailed molecular information and quantify specific analytes within complex mixtures. The flexibility of MS techniques, especially MALDI-TOF, makes them highly suitable for high-throughput DNA analysis, as they eliminate the need for fluorescence or radioactive labeling and bypass the stringent conditions required for hybridization assays (Mirabelli et al., 2013).

In this chapter, we aim to discuss the principles of MS, advancements that led to the rapidly expanding field of MS, methods for the analysis of nucleic acids (DNA/RNA), the dynamic data acquisition approaches, sophisticated data analysis, and provide a brief overview of some of the current and potential forensic applications.

16.2 MS techniques for DNA analysis

16.2.1 ESI

ESI represents a pivotal advancement in the field of molecular biology because it allows the hyphenation of MS with liquid chromatography (LC). This innovation has opened the possibility of direct examination of polar compounds using MS, including DNA nucleosides, nucleotides, and oligonucleotides. In the process of ESI, a liquid sample is forced through a narrow nozzle applied with a high voltage, commonly between 1 and 5 kV; this can be achieved directly at the nozzle or through a liquid-liquid junction. This process creates fine aerosol mist made up of electrospray charged droplets. As these droplets evaporate, their size continues to reduce, eventually forming ions in the gas phase at ambient pressure. Even though the process for the transition of these charged droplets into ions in the gas phase remains highly complex and not fully elucidated, it is thought to be an interplay of ion desorption and droplet fission (Oberacher et al., 2007; Tretyakova et al., 2013).

The transformation of ions from the liquid phase to the gaseous phase is fundamentally competitive, as the charged droplets can only hold a finite amount of ions. As a result, other sample components may compete with or even suppress the ionization of the target analyte, a phenomenon known as ion suppression. This issue is particularly prevalent in trace-level ESI analysis, where the presence of background substances can significantly decrease the detection limits and precision of the assay. To counteract ion suppression in LC-MS protocols designed for DNA modifications, implementing rigorous sample cleanup procedures and employing HR-HPLC separations can be efficient approaches. These techniques help isolate the analyte from interfering substances, thereby enhancing ionization efficiency and improving mass spectrometric detection sensitivity (Buhrman et al., 1996).

Since the generation of positively or negatively charged ions by ESI is dependent on the voltage polarity applied to the ion source, this technique is applicable for analytes having either acidic or basic functional groups. Due to the negative charge of the phosphodiester backbone of DNA, analytical processes are mostly carried out in the negative ion mode where DNA molecules act as polyanions. Analogously, nucleoside monophosphates are generally analyzed with ESI-MS in negative ion mode. However, free DNA nucleobases and nucleosides, respectively, protonate at acidic pH and are consequently typically detected in positive ion mode (Farmer et al., 2005).

16.2.2 MALDI

MALDI has become a crucial technique for the mass spectrometric examination of oligonucleotides and nucleic acids due to its soft ionization capability, generating predominantly singly charged ions. In MALDI, analyte samples are co-crystallized with a UV or IR-absorbing organic matrix, such as 3-HPA (3-hydroxypicolinic acid) for DNA or a mixture of trihydroxyacetophenone derivatives for RNA, to form a solid deposit on the target plate. A pulsed laser irradiates this matrix-analyte mixture, facilitating the desorption and ionization of the analyte molecules while minimizing fragmentation (**Figure 16.1**). This matrix effectively absorbs the laser energy and protects the analyte, ensuring efficient ionization (Griffin & Smith, 2000; Jurinke et al., 2004).

The compatibility of MLADI with TOF-MS is notable, as both techniques are inherently suited to pulsed operation, enabling the analysis of a wide range of ion masses. Initially, MALDI-MS was predominantly utilized for proteins and peptides, with DNA analysis limited to smaller oligonucleotides. However, advancements in matrix selection, particularly the use of 3-HPA, and the integration of delayed extraction significantly improved the resolution and analytical capabilities of the technique (Gut, 2004; Nordhoff et al., 2000).

The challenges associated with larger DNA fragments, such as depurination and the formation of metal ion adducts, posed obstacles for MALDI in sequencing applications. However, the focus gradually shifted toward single nucleotide polymorphism (SNP) genotyping and mutation analysis, where only short DNA sequences are analyzed. To enhance the analytical performance, various purification methods, including reversed-phase techniques, ethanol

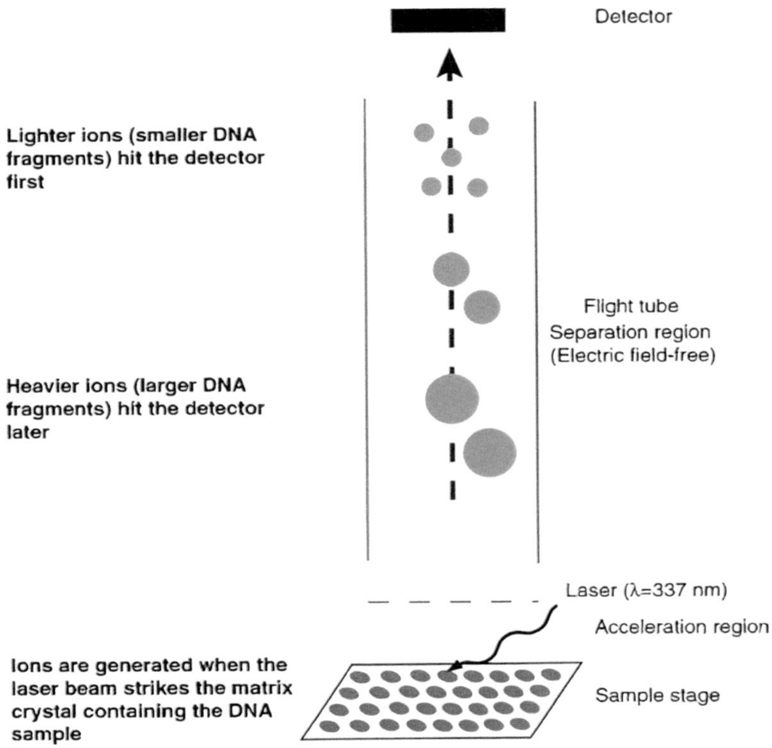

Figure 16.1 *Schematic representation of MALDI-TOF-MS for DNA analysis.*

Reproduced with permission from Edwards et al. (2005). License no. 5898080221838.

precipitation, and ion-exchange bead approaches, were employed to remove interfering substances. This shift has positioned MALDI-MS as a crucial tool in SNP genotyping, leveraging its rapid data acquisition and adaptability to high-throughput analysis while addressing issues like adduct formation and resolution (Edwards et al., 2005; Ross et al., 1998).

16.3 Applications

Traditionally, forensic identity testing has used mitochondrial DNA (mtDNA) by sequencing two hypervariable segments of its control region. To improve efficiency and reduce processing time, SNP assays are being investigated as an alternative. However, these assays generally cover only a limited portion of the target regions, necessitate prior knowledge of SNP locations, and often lack quantitative capabilities. Additionally, the dense clustering of SNPs in mtDNA makes designing effective SNP extension primers or hybridization probes a complex task. To address these issues, Hall et al. (2005) developed an automated ESI-MS technique that can detect multiple clustered SNPs within an amplicon without requiring prior information about specific SNP positions, providing quantitative results. This method determines the base composition of PCR amplicons shorter than 140 nucleotides, identifying the presence of SNPs by comparing base compositions between samples. This approach eliminates the need for additional post-PCR analysis to detect variations within DNA fragments. The assay successfully distinguished nearly 90% of the 2,754 unique mtDNA sequences listed in the public forensic mtDNA database. Furthermore, the precision of MS facilitates the detailed characterization and quantification of heteroplasmic samples or mixtures, supporting the analysis of both mtDNA and other SNP mixtures. The technique's adaptability extends to analyzing genetic variations in the mtDNA coding region and can be automated to efficiently process large sample sets, such as those encountered in mass disaster cases.

In another independent study, Oberacher et al. (2007) used a technique called ion-pair RP-HPLC in combination with ESI-TOF-MS to separately analyze length and nucleotide polymorphisms. The study examined two separate areas within the first (HVS-I) and second (HVS-II) hypervariable segments of the mtDNA control region, which are known for their high frequency of length heteroplasmy in the population. In this study, the above-mentioned mtDNA regions were amplified and examined by ICEMS in a group of 90 mother-offspring pairs (maternally unrelated) from Austria. Results were confirmed by direct sequencing of PCR products. Length heteroplasmic variants that were comprehensively analyzed through ICEMS proved to have better informational yield than those derived from direct sequencing, thereby providing a more enhanced view into these heteroplasmic variations.

Hofstadler et al. (2009) introduced an advanced DNA forensics platform that relies on automated high-throughput ESI-MS. This technique utilizes ESI-MS to precisely measure the mass of DNA markers, allowing for the determination of clear base compositions (Aw, Gx, Cy, Tz), which were subsequently used to generate a profile of DNA for individuals. A series of blinded experiments evaluated the platform's performance for both short tandem repeat (STR) and mtDNA typing. The same system accommodates both types of analysis, offering several advantages over traditional typing methods. This technique is especially useful for examining mtDNA samples with length heteroplasmy in the HV1 and HV2 regions when sequencing is not practical. The platform employs base composition profiles obtained from a 24-primer pair assay, which covers CRS 15924–16428 and 31–576, demonstrating superior resolving capabilities compared to conventional sequencing methods that target 16024–16365 and 73–340, thereby enhancing mixture analysis. Similarly, STR markers can be assessed using this approach, which benefits from high mass accuracy, enabling the detection of SNPs within STR markers. An analysis of samples from diverse population groups indicated that SNP frequencies in several CODIS loci exceed 30%, with SNPs identified in 10 out of the 13 core loci.

Pitterl et al. (2010) conducted an investigation into the molecular masses derived from crude STR amplicons, which were subsequently purified and fractionated utilizing a liquid chromatographic system that is directly integrated with an electrospray ionization mass spectrometer. This scholarly inquiry provided novel population data pertinent to forensically significant STRs within the Khoisan and Yakut demographic groups. These indigenous populations were chosen due to their potential to harbor distinctive STR alleles that are infrequently encountered or entirely absent in contemporary human populations originating from more urbanized locales, particularly the Khoisan, who are recognized as one of the most ancient human cohorts. The examination of STR molecular masses unveiled a more extensive spectrum of allele variability, thereby augmenting their forensic applicability. Consequently, established STR data derived from fragment size analysis can still be effectively deployed in forensic investigations or within intelligence repositories.

In their investigation, Planz et al. (2012) devised an automated STR assay framework predicated on electrospray ionization mass spectrometry (ESI-MS) to augment the discriminatory capacity of specific CODIS core STR loci. This innovative system yielded superior data from both conventional and complex sample types. The information generated by the ESI-MS

STR methodology exhibited complete compatibility with the findings obtained via capillary electrophoresis techniques. In addition, the ESI-MS system was also found to be efficient in uncovering nucleotide-level polymorphisms in STR alleles, thus increasing observable genetic variation per individual locus. The research investigated population-level genetic variation by examining 13 CODIS STR loci in African American, Caucasian, and Hispanic samples, with the focus on the analysis of allele length profiles and sequence-level variations within repeats and surrounding regions.

The division of many nominal-length alleles was prompted by the identification of other polymorphisms in 11 of the 13 loci. Almost 5% of the sample exhibited heterozygosity with alleles of the same length at one or more of five significant CODIS loci, reflecting a considerable rise in overall heterozygosity. Through the enhanced analysis of polymorphism, the seven most polymorphic STR loci reached a level of discrimination equivalent to that of the ten most informative CODIS core loci. Population substructure analysis showed that there was higher expected heterozygosity when alleles were used together without consideration of nominal lengths. Incorporation of SNP-based variation into the allele profile resulted in increased heterozygosity values at D3S1358 (0.0477) and vWA (0.0234) loci, as opposed to significantly lower estimates (0.0145 and 0.01266) obtained from nominal repeat numbers alone. These discrepancies suggested the presence of population-specific allele distributions. Furthermore, a nomenclature framework was instituted to facilitate the databasing, querying, and analytical processes pertaining to these combined data types.

The work by Nordhoff et al. (2000) introduced an improved protocol aimed at the quick sequencing of small DNA pieces (15 to 20 nucleotides) through MALDI-TOF MS. This methodology is predicated upon the Sanger sequencing principle, utilizing double-stranded template DNA. All four sequencing reactions are conducted concurrently within a single vial, thereby enhancing the efficiency of the process. The resultant sequencing products undergo analysis and identification via MALDI-TOF-MS. The nucleotide sequence is subsequently deduced by juxtaposing the observed molecular weight discrepancies with the anticipated values. The protocol was meticulously crafted to be both economically viable and broadly applicable, typically requiring template (300 fmol), primer (10 pmol), and each nucleotide monomer per reaction (200 pmol), obviating the requirement for labeled primers or nucleotide monomers. Furthermore, the technique incorporates swift solid-phase purification, concentration, and sample preparation, facilitating the parallel processing of up to 96 samples within a mere few minutes. The mass spectrometric analysis and the ensuing sequence determination are accomplished within seconds per template, rendering it a highly efficient alternative for high-throughput sequencing.

The research conducted by Ross and others in 1998 introduced a primer extension assay intended for the highly multiplexed genotyping of SNPs within genomic DNA, amplified through the use of multiplex PCR methods. By means of MALDI-TOF-MS, this technique is capable of reliably assessing the weights of short oligonucleotide primers that had been lengthened by a single dideoxynucleotide, skillfully taking advantage of the natural mass variations among DNA bases to support accurate genotyping. In order to secure unequivocal allele identification, a comprehensive examination of the primer composition that was complementary to the target was performed, which included the intentional incorporation of no complementary 5′ bases into the genotyping primers. This technique allowed for the production of unique mass spectra corresponding to interleaved genotyping products. Illustrating the efficacy of the method, the research revealed a 12-fold multiplex SNP analysis employing a model multiplex PCR system, thereby underscoring its potential applicability for high-throughput genotyping endeavors.

Ding and colleagues (2003) put forth a groundbreaking approach for examining gene expression, integrating competitive PCR with MALDI-TOF-MS. This technique necessitated the formulation of a DNA standard incorporating an artificial SNP within the targeted gene. The standard was subsequently introduced into the reverse transcription product prior to the amplification via PCR. Following this step, a base extension reaction was executed at the SNP locus, and the resultant products were quantified utilizing MALDI-TOF-MS. This methodology enabled both relative and absolute quantification of gene expression and exhibited exceptional sensitivity, permitting the identification of as few as five DNA copies. Furthermore, it provided substantial reproducibility and enabled the concurrent quantification of both alleles in heterozygotes, in addition to alternatively spliced genes. The confluence of this technique with the homogeneous mass extension system (Sequenom) culminated in the development of a high-throughput, automated platform for gene expression analysis, capable of accurately quantifying several hundred genes from 20 to 500 distinct samples on a daily basis.

The research by Petkovski et al. (2005) focused on how DNA markers can aid in individual identification in forensic science, particularly emphasizing repeat sequences in nuclear DNA and the significance of hypervariable regions 1 and 2 of mitochondrial DNA. They noted single nucleotide mutations (SNMs) as a hopeful substitute, thanks to their built-in advantages in the assessment of degraded or inferior samples, frequently present in forensic and anthropological work. In order to evaluate the feasibility of SNP analysis in these fields, the study meticulously selected 41 SNPs based on rigorous criteria, including conservation, absence of phenotypic expression, and population frequency. Among the selected SNPs, a total of 36 autosomal SNPs were genotyped across 21 inclusionary and 3 exclusionary paternity cases, whereas the behavior of 5 X-chromosome SNPs was scrutinized within a representative sample of the French population. The methodological structure adopted the multiplex PCR technique to support the amplification process, which was followed by simultaneous detection via primer extension and then further identified by MALDI-TOF MS. The identified autosomal SNPs demonstrated independent inheritance patterns and yielded unequivocal results in paternity assessments. Furthermore, all X-SNPs were determined to be effective markers for both paternity testing and individual identification. The elevated sensitivity, accuracy, and rapidity of PEX and MALDI-TOF MS provided a robust methodology for the forensic and anthropological examination of biallelic markers.

16.4 Conclusion

Forensic scientists have noted the introduction of MS as an adaptable device for the meticulous study of DNA and RNA. Its utilization encompasses polymorphism characterization, sequence analysis, HLA-typing, and investigations pertaining to DNA methylation. MS presents numerous advantages over conventional gel electrophoresis methodologies, such as reduced analytical durations and the capability to identify DNA without the necessity of radioactive or dye-based labeling. Regardless of these perks, MS is still not widely adopted in forensic DNA analysis and has not completely reached its possible impact in this domain. To upgrade its features, anticipated progress must aim at bettering mass resolution to ease the evaluation of long DNA strands, thereby enhancing the accuracy of findings associated with STRs, variable number tandem repeats (VNTRs), and SNPs. Furthermore, the implementation of robotic automation for high-throughput sample processing and the creation of portable devices for field-based DNA analysis would be pivotal in addressing the existing bottlenecks. These advancements will amplify the significance of MS in forensic science, thereby fostering its broader application as a reliable instrument in forensic investigations.

References

Bowling, J. M., Bruner, K. L., Cmarik, J. L., & Tibbetts, C. (1991). Neighboring nucleotide interactions during DNA sequencing gel electrophoresis. *Nucleic Acids Research, 19*(11). https://academic.oup.com/nar/article/19/11/3089/2387780

Buhrman, D. L., Price, P. I., & Rudewicz, P. J. (1996). Quantitation of SR 27417 in human plasma using electrospray liquid chromatography-tandem mass spectrometry: A study of ion suppression. *Journal of the American Society for Mass Spectrometry, 7*(11), 1099–1105. https://doi.org/10.1016/S1044-0305(96)00072-4

Ding, C., & Cantor, C. R. (2003). A high-throughput gene expression analysis technique using competitive PCR and matrix-assisted laser desorption ionization time-of-flight MS. *Proc Nat Acad Sci. 100*(6), 3059–3064. https://doi.org/10.1073/pnas.0630494100

Edwards, J. R., Ruparel, H., & Ju, J. (2005). Mass-spectrometry DNA sequencing. In *Mutation Research - Fundamental and Molecular Mechanisms of Mutagenesis* (Vol. 573, Issues 1–2, pp. 3–12). Elsevier. https://doi.org/10.1016/j.mrfmmm.2004.07.021

Farmer, P. B., Brown, K., Tompkins, E., Emms, V. L., Jones, D. J. L., Singh, R., & Phillips, D. H. (2005). DNA adducts: Mass spectrometry methods and future prospects. *Toxicology and Applied Pharmacology, 207*(2 SUPPL), 293–301. https://doi.org/10.1016/j.taap.2004.12.020

Fenn, J. B., Mann, M., Meng, C. K., Wong, S. F., & Whitehouse, C. M. (1989). Electrospray ionization for mass spectrometry of large biomolecules. *Science, 246*(4926), 64–71.

Gingras, A. C., Aebersold, R., & Raught, B. (2005). Advances in protein complex analysis using mass spectrometry. *Journal of Physiology, 563*(1), 11–21. https://doi.org/10.1113/jphysiol.2004.080440

Griffin, T. J., & Smith, L. M. (2000). Single-nucleotide polymorphism analysis by MALDI-TOF mass spectrometry. *Trends in Biotechnology, 18*(2), 77–84. https://doi.org/10.1016/s0167-7799(99)01401-8.

Gut, I. G. (2004). DNA analysis by MALDI-TOF mass spectrometry. *Human Mutation, 23*(5), 437–441. https://doi.org/10.1002/humu.20023

Hall, T. A., Budowle, B., Jiang, Y., Blyn, L., Eshoo, M., Sannes-Lowery, K. A., Sampath, R., Drader, J. J., Hannis, J. C., Harrell, P., Samant, V., White, N., Ecker, D. J., & Hofstadler, S. A. (2005). Base composition analysis of human mitochondrial DNA using electrospray ionization mass spectrometry: A novel tool for the identification and differentiation of humans. *Analytical Biochemistry, 344*(1), 53–69. https://doi.org/10.1016/j.ab.2005.05.028

Hofstadler, S. A., Hall, T. A., Sannes-Lowery, K. A., Manalili, S., Paulsen, J. E., McCurdy, L. D., Gioeni, L., Penella, T., Eisenberg, A. J., Planz, J. V., & Budowle, B. (2009). Analysis of DNA forensic markers using high throughput mass spectrometry. *Forensic Science International: Genetics Supplement Series, 2*(1), 524–526. https://doi.org/10.1016/j.fsigss.2009.08.140

Jurinke, C., Oeth, P., & Van Den Boom, D. (2004). Molecular biotechnology MALDI-TOF mass spectrometry 147 review 147 MALDI-TOF mass spectrometry a versatile tool for high-performance DNA analysis. *Molecular Biotechnology, 26*(2), 147–164. doi: 10.1385/MB:26:2:147

Jut, J., Ruant, C., Fullert, C. W., Glazer, A. N., & Mathiestl, R. A. (1995). Fluorescence energy transfer dye-labeled primers for DNA sequencing and analysis (Forster energy transfer/oligodeoxynucleotide synthesis/primer labeling/fluorescent tags). *Biophysics, 92*. https://www.pnas.org

Mirabelli, M. F., Chramow, A., Cabral, E. C., & Ifa, D. R. (2013). Analysis of sexual assault evidence by desorption electrospray ionization mass spectrometry. *Journal of Mass Spectrometry, 48*(7), 774–778. https://doi.org/10.1002/jms.3205

Nordhoff, E., Luebbert, C., Thiele, G., Heiser, V., & Lehrach, H. (2000). Rapid determination of short DNA sequences by the use of MALDI-MS. *Nucleic Acids Research, 28*(20), E86. https://doi.org/10.1093/nar/28.20.e86.

Oberacher, H., Niederstätter, H., & Parson, W. (2007). Liquid chromatography-electrospray ionization mass spectrometry for simultaneous detection of mtDNA length and nucleotide polymorphisms. *International Journal of Legal Medicine, 121*(1), 57–67. https://doi.org/10.1007/s00414-006-0117-7

Ong, S. E., & Mann, M. (2005). Mass Spectrometry–Based proteomics turns quantitative. *Nature Chemical Biology, 1*(5), 252–262. https://doi.org/10.1038/nchembio736

Petkovski, E., Keyser-Tracqui, C., Hienne, R., & Ludes, B. (2005). SNPs and MALDI-TOF MS: Tools for DNA typing in forensic paternity testing and anthropology. *Journal of Forensic Sciences, 50*(3).

Pitterl, F., Schmidt, K., Huber, G., Zimmermann, B., Delport, R., Amory, S., Ludes, B., Oberacher, H., & Parson, W. (2010). Increasing the discrimination power of forensic STR testing by employing high-performance mass spectrometry, as illustrated in indigenous South African and Central Asian populations. *International Journal of Legal Medicine, 124*(6), 551–558. https://doi.org/10.1007/s00414-009-0408-x

Planz, J. V., Sannes-Lowery, K. A., Duncan, D. D., Manalili, S., Budowle, B., Chakraborty, R., Hofstadler, S. A., & Hall, T. A. (2012). Automated analysis of sequence polymorphism in STR alleles by PCR and direct electrospray ionization mass spectrometry. *Forensic Science International: Genetics*, *6*(5), 594–606. https://doi.org/10.1016/j.fsigen.2012.02.002
Ronaghi, M., Uhlén, M., & Nyrén, P. (1998). A sequencing method based on real-time pyrophosphate. *Science*, *281*(5375), 363–365. https://doi.org/10.1126/science.281.5375.363
Ronald, B. C., & Brian, C. T. (1989). Matrix-assisted laser-desorption mass spectrometry using 355 nm radiation. *Rapid Communications in Mass Spectrometry*. https://doi.org/10.1002/rcm.1290031208
Ross, P., Hall, L., Smirnov, I., & Haff, L. (1998). High level multiplex genotyping by MALDI-TOF mass spectrometry. *Nature Biotechnology*, *16*. https://doi.org/10.1038/4328
Smith, L. M., Sanders, J. Z., Kaiser, R. J., Hughes, P., Dodd, C., Connell, C. R., Heiner, C., Kent, S. B., & Hood, L. E. (1986). Fluorescence detection in automated DNA sequence analysis. *Nature*, *321*. https://doi.org/10.1038/321674a0
Tost, J., & Gut, I. G. (2006). DNA analysis by mass spectrometry - Past, present and future. *Journal of Mass Spectrometry*, *41*(8), 981–995. https://doi.org/10.1002/jms.1096
Tretyakova, N., Villalta, P. W., & Kotapati, S. (2013). Mass spectrometry of structurally modified DNA. *Chemical Reviews*, *113*(4), 2395–2436. https://doi.org/10.1021/cr300391r
Yamakawa, H., & Ohara, O. (1997). A DNA cycle sequencing reaction that minimizes compressions on automated fluorescent sequencers. In *Nucleic Acids Research* (Vol. 25, Issue 6). Oxford University Press.

17

Role of Proteomics in Criminal Investigations

Satish Kumar, Neelkant Verma, and Aaliya Sachdev

17.1 Introduction

Biomolecular constituents form the basis of all biological processes occurring in living organisms. This chapter discusses proteins, which are among the most versatile and abundant biomolecules. Proteins are major biological macromolecules present in various biological matrices. Twenty-two amino acids embedded in the gene sequence form the basis of the protein structure, which further determines the protein's shape and function.

Proteins have been studied in biochemistry since time immemorial. Proteins are essential in all organisms. Owing to their plethora of diversity, proteins are found in living organisms in the form of enzymes, hormones, immunoglobulins, membrane receptors, collagen, elastin, and keratin. To put it comprehensively, proteins are called "working horses of the cell (Patzelt, 2004)".

The word protein comes from the Greek word PROTEIOS, which means holding first place. Dutch chemist Mueller came up with the idea of using the term protein for nitrogen-containing high-molecular-weight species that are mostly found in flora and fauna (Müller & Fabricius, 1786).

The origin of biochemistry dates back to the 1800s during the era of Parceleus, van Helmont, and Sylvius. Later, scientists were trained to be pharmacists or chemists with the objective of characterising the constituents of plants or living organisms, giving rise to separation chemistry. In the 1900s, the focus shifted to organic and inorganic chemistry with a better understanding of the physiology of chemistry and medicine (Patzelt, 2004).

A certain group of organic compounds baffled scientists in 1900, as their physiology was not well-defined. Here, we refer to lecithin and albuminoid substances such as casein, fibrin, and albumin, which came to be known as proteins, which also provided answers to the mysteries of the 1930s, wherein a substance (named Pepsin by Schwann) affected the digestible protein in gastric juices. The latter half of the 19th century was dedicated to the enzyme theory of life. There are a few references concerning the timeline of protein chemistry research progress that has emerged in protein biochemistry (Patzelt, 2004).

With the advent of science, proteins were the first biomolecules used in criminal investigations of biological evidence. Teichmann, a Polish anatomist, testified to the presence of blood using certain chemicals and haemoglobin (a protein found in blood) in 1853. Related progress was made in 1863 when the enzymatic activity of peroxidase was utilised in the presence of blood. Proteins were then used to detect the presence of various forensically relevant

DOI: 10.4324/9781003474296-17

body fluids, such as saliva and seminal fluid, and to determine species, sex, donor age, and alpha-fetoprotein.

This paved the way for the development of DNA-based individual identification and gave birth to the landmark Human Genome Project, which commenced in October 1990 and reached successful completion in 2003, when the blueprint of the human genome was mapped (Barranco, 2021).

DNA at a crime scene can help to identify the source person from whom the biological evidence matches with great certainty. However, when DNA is contaminated with PCR inhibitors, such as denim, the ability to obtain a clean DNA profile is limited, thus rendering crucial evidence useless. Owing to the perishable nature of DNA, obtaining DNA evidence over time may not be easy. Improperly stored DNA can cause cell death, leading to the loss of important DNA information (Merkley, 2019).

In some cases, the deposition of biological samples is very low, and a partial DNA profile is generated. Thus, a corroborative individual identification technique can lead an investigator to the perpetrator. A developing field of proteomics could help identify the individual, cell, or tissue of biological evidence, and even the environment of exposure in some cases. The success of the Human Genome Project provided a forward push for the development of HUPO, with a vision to launch the HPP in 2010, with the aim of mapping the entire human genome to see how the genome is expressed in phenotype and how proteins help us identify the environment of the cell.

An impartial justice system is the backbone of a thriving civilisation. On that day, people relied on eyewitnesses and evidence as they saw them. Some tricky criminals were used to avoid crime. As society has evolved, people have begun to rely more on evidence than on hearsay or eyewitnesses. Locard's principle of exchange laid the foundation for forensic science, expanding its applications to almost all fields of science and technology. The knowledge obtained from all these fields has further diversified and deepened the knowledge of each branch of modern forensic science. Today, forensic science has branched out into numerous fields, such as forensic Ballistics, Forensic psychology, criminology, forensic engineering, toxicology, questioned documents, fingerprinting, forensic physics, forensic marine biology, forensic biotechnology, forensic serology, forensic odontology, forensic nursing, forensic botany, and forensic anthropology.

A detailed analysis of forensic evidence provides deep insights and leads investigators, which could help in taking the investigation in a different direction. In the past, there have been many cases of wrongful exonerations. The famous Innocence Project is the brainchild of forensic science that has helped release wrongfully convicted individuals. Their innocence was proven based on DNA evidence analysis, which has helped free 250 wrongfully convicted individuals to date, thus building a society that has faith in the justice system (*Our Impact: By the Numbers–Innocence Project*).

17.1.1 Proteomics

Proteomics, the comprehensive study of proteins and their diverse functions, has transformed the field of forensic science by introducing a powerful and precise approach to protein identification and characterisation for a wide range of forensic applications. By employing proteomics, forensic scientists can perform thorough and unique analyses of postmortem bodily fluids, tissues, and organs, thereby significantly enhancing the accuracy and reliability of protein identification and characterisation. This study aimed to explore the extensive application

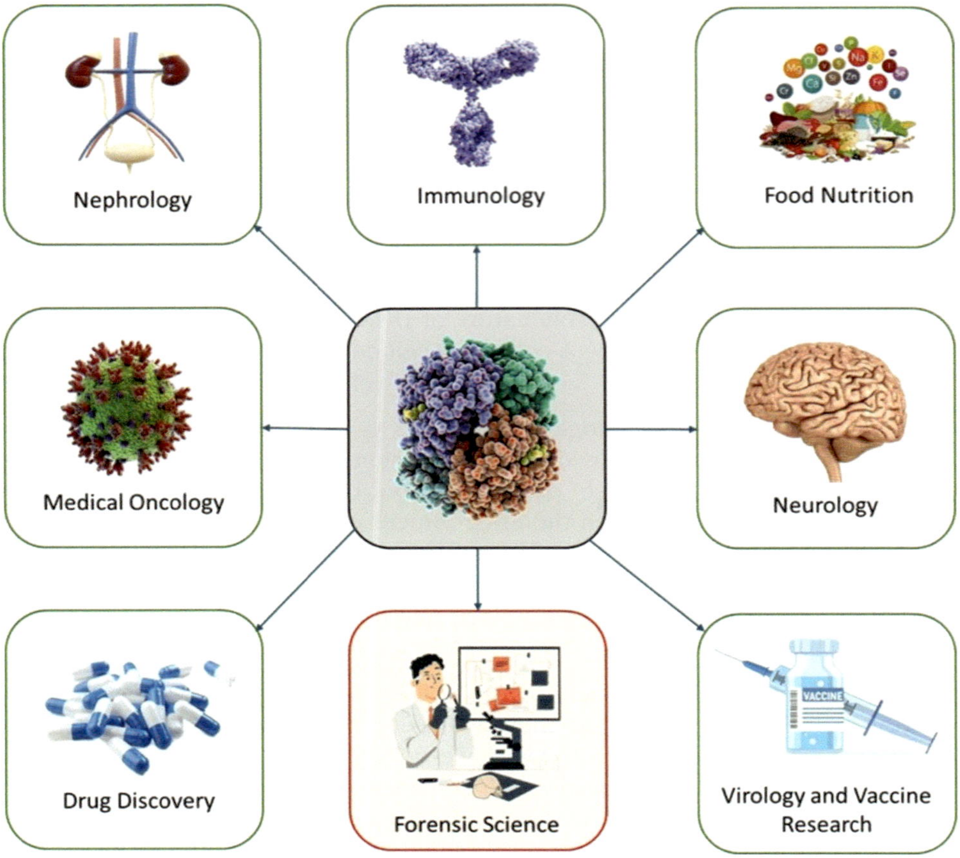

Figure 17.1 *Various applications of proteomics.*

of proteomics in forensic pathology and autopsy analyses. Various applications of proteomics are depicted in **Figure 17.1**.

Cutting-edge advancements in gene technologies and their by-products, such as RNA and proteins, have opened doors to a better understanding of biology and disease. The word "omics" came into popular use in the second half of the 20th century and replaced the word "ome". The word "ome" refers to a dataset regarding a particular individual organism, whereas omics encompasses all datasets regarding that particular domain (Müller & Fabricius, 1786).

Examples of omics technologies include genomics, proteomics, epigenomics, lipidomics, transcriptomics, glycomics, metabolomics, nutrigenomics, pharmacogenomics, toxicogenomics, psychogenomics, connectomics, foodomics, and various possible branches of omics (Institute of Medicine, 2012).

The term proteomics, coined in 1997, was a precursor to the word proteome, used by Marc Wilkins in 1994 to denote the "protein complement of a genome" (Aslam et al., 2017). Commencement of protein studies took place in 1975 when O'Farrell mapped proteins from *E. coli* and Klose and Scheele did the same for mouse and guinea pig, respectively, by using the same technique, i.e., 2-D gel electrophoresis. At that time, no method for identifying proteins was known.

Another milestone in proteomics is protein sequencing using Edman's degradation method, in which the N-terminus of an amino acid is labelled and cleaved off, and the process is repeated for the entire protein to obtain the amino acid sequence. This is similar to the results obtained for other growth technologies (Gobena et al., 2024).

All barriers and limitations were overcome by utilising mass spectrometry to obtain protein data, thus laying the foundation for modern mass-spectrometry-based proteomics. Patterson and Aebersold, notable systems biologists and proteomics experts, described the aim of proteomics as providing a detailed description of the structure, function, and control of biological systems in both health and disease (Patterson & Aebersold, 2003). Mattias Mann revolutionised mass spectrometry-based proteomics by introducing electron spray ionisation and integrated computational algorithm with protein profile analysis. He stated that the foundation of proteomics—and its significance alongside genomics—lies in the concept of proteoforms, which originate from the same gene but differ in amino acid sequences and/or post-translational modifications (Pandey & Mann, 2000).

Pandey et al. combined various omics technologies, including proteomics and computational biology. Their work described proteomics as a large-scale study of proteins, usually using biochemical methods to study them. This allows the identification, quantification, and characterisation of proteins to understand their functional interactions and roles in various cellular processes (Pandey & Mann, 2000). Proteomics, as defined by Gygi, is the systematic analysis of proteins expressed in cells or tissues. This involves studying the entire complement system of proteins in a biological sample. Mass spectrometry is an essential analytical tool in proteomics for the identification and characterisation of proteins (Peng & Gygi, 2001).

17.1.2 Tools and techniques used in proteomics

17.1.2.1 SDS-PAGE

Sodium dodecyl sulphate – polyacrylamide gel electrophoresis technique was developed by Ulrich K. Laemmli in the year 1970 primarily for the separation and analysis of proteins based upon their molecular weight (Laemmli, 1970). This is sometimes referred to as one-dimensional (1D) electrophoresis, a common technique used for protein characterisation. However, the resolving power of 1D SDS-PAGE is limited when analysing complex protein mixtures. A denaturant, such as sodium dodecyl sulphate, is used for protein denaturation to achieve separation. In lieu of this, SDS also imparts a negative charge to the proteins, allowing them to successfully migrate towards the cathode under the influence of the electric charge. The separated proteins were digested and characterised to determine their protein sequences and post-translational modifications.

17.1.2.2 2D-PAGE

Two-dimensional (2-D) PAGE combines isoelectric focusing (IEF) with SDS-PAGE to examine protein mixtures with high precision and accuracy. IEF first differentiates proteins based on their charge (isoelectric point, pI). In the second dimension, SDS-PAGE was used to separate the proteins according to their size.

Large-format gels (20 cm × 20 cm) are commonly used for optimal separation, although minigels can also be effective for this purpose. IEF was performed using an acrylamide gel on a plastic strip (18 cm × 3 mm) in large-format gels. The gel incorporated ampholytes to establish a pH gradient and included 8M urea along with a non-ionic detergent to denature proteins and keep them soluble. Proteins in their denatured state were sorted according to their isoelectric points. The IEF strip is soaked in a buffer containing SDS, which binds to denatured proteins. Subsequently, the samples are subjected to 10% SDS-PAGE, where electrophoresis is used to separate the proteins by size. The entire process included rehydrating the IEF gels overnight, running the first-dimension IEF for 6–8 hours, equilibrating with SDS buffer for 15 min, and running the SDS-PAGE for 5 h.

A typical 2-D gel can resolve between 1000 and 3000 proteins from a cell or tissue extract, with some reports separating up to 10000 proteins (Wilson et al., 2018). This method plays a crucial role in proteomics, allowing for the detailed analysis of complex protein mixtures.

17.1.2.3 Affinity chromatography

Affinity chromatography uses highly specific biological interactions to separate and purify complex mixtures in a single step. It has been established as a purifying enzyme; however, it is now used to purify nucleotides, nucleic acids, immunoglobulins, membrane receptors, whole cells, and cellular fragments. This process involves isolating the material by binding it to a particular ligand that is bound to an insoluble matrix. Typical materials used include crosslinked dextran, agarose, polyacrylamide, polymethacrylate, polystyrene, cellulose, porous glass, and silica. The chemical properties of the ligand are dictated by the biological specificity of the compound being purified. In affinity chromatography, a buffer is employed to provide the cofactors essential for interactions between the ligand and macromolecule. The purified compound was extracted from the ligand using either specific or non-specific elution methods. Non-specific elution can be accomplished by changing the pH or ionic strength of the buffer.

Immunoaffinity chromatography uses antibodies as immobilised ligands to isolate and purify proteins, including viral membrane proteins. This technique involves linking monoclonal antibodies to agarose matrices using cyanogen bromide. Proteins were successfully bound to a neutral buffer solution with a moderate salt concentration. To release the bound proteins, harsh conditions are often necessary to break strong ionic or hydrophobic interactions, which can result in protein denaturation and loss of function. Elution methods may involve the use of high salt concentrations, detergents, urea, guanidine hydrochloride, thiocyanate, perchlorate, trifluoroacetate, or reduction of the pH to prevent denaturation. Metal chelate chromatography is a specialised affinity chromatography that employs an immobilised metal ion or transition metal ion to selectively bind proteins by interacting with imidazole groups in histidine residues, thiol groups in cysteine residues, and indole groups in tryptophan residues. Immobilisation is achieved by the formation of stable coordinate bonds. Dye–ligand chromatography utilises triazine dyes, which contain conjugated rings and ionic groups, to bind proteins, thereby enhancing their attachment to materials such as Sepharose 4 B. The most commonly used dye in this process is Cibacron Blue F3G-A (Wilson et al., 2018).

17.1.2.4 Ion exchange chromatography

High-performance liquid chromatography (HPLC) is a type of chromatography in which an ion exchanger binds oppositely charged stationary phases between an analyte and the stationary phase. With their excellent resolving power and capacity, these substances are ideal for separating and purifying proteins, peptides, nucleic acids, polynucleotides, and other charged molecules. Ion exchangers are divided into two categories: cationic and anionic. Cation exchangers contain negatively charged groups that attract positively charged cations, whereas anion exchangers contain positively charged groups that attract negatively charged anions.

The HPLC matrices comprised polystyrene, cellulose, and agarose. Sulfonate and quaternary ammonium function as strong exchangers because they remain fully ionised across all typical working pH levels, whereas carboxylate and diethylammonium are considered weak exchangers because their ionisation occurs only within a limited pH range. Bonded-phase ion exchangers, designed for HPLC, offer a diverse range of ionic groups. They are particularly beneficial for protein separation.

The selection of an ion exchanger depends on the stability of the analytes being tested, their molecular weight, and specific separation needs. When dealing with biological analytes such as proteins, it is crucial that the chosen exchanger functions within a limited pH range.

To ensure that the analytes adhere to the exchanger, the pH and ionic strength of the initial buffer must be accurately set. Gradient elution, which is more prevalent than isocratic elution, can be achieved through continuous or stepwise variations in pH and ionic strength (Wilson et al., 2018).

17.1.2.5 MALDI-TOF MS

Matrix-assisted laser desorption ionisation (MALDI) is a technique that generates gas-phase protonated ions by exciting sample molecules using the energy of a laser provided by a UV light-absorbing matrix. The matrix is a conjugated organic material, usually a weak organic acid, mixed with the sample. The sample is combined with an excess of the matrix and dried onto the target plate, where they co-crystallise. Laser light pulses of a few nanoseconds generate fast excitation and vaporisation of the crystalline matrix, and the subsequent ejection of the matrix and analyte ions into the gas phase. This induces the emission of the matrix and analyte ions, which are analysed in a TOF mass analyser. MALDI is a soft ionisation technique that does not result in substantial fragmentation.

To obtain optimal sensitivity in MALDI–TOF, the samples were diluted to a certain concentration range. Peptides and proteins offer optimum spectra at approximately 0.1 to 10 pmol mm^{-3}, but glycoproteins may yield superior results at concentrations up to 10 pmol mm^{-3}. Oligonucleotides offer better spectra at approximately 10–100 pmol mm^{-3} than proteins.

Novel hybrid instruments, such as MALDI-Q-TOF, have been designed to combine the best features of both types of ion sources with the best features of all types of analysers to improve tandem MS capabilities and increase sensitivity.

17.2 Forensic proteomics

Proteomics has emerged as a valuable tool in forensic science. Studies have indicated that proteomics is extensively used to examine body fluids and tissues, estimate the time since death, and identify individuals. Recent analyses have focused on human samples, such as blood, hair, bone, and fingernails. Forensic proteomics has a wide array of applications in criminal investigations (**Figure 17.2**). However, challenges such as unidentified samples and contaminants can hinder the application of forensic proteomics (Zapico & Adserias-Garriga, 2022). Additionally, specific procedures are required for sample preparation, limiting the widespread use of proteomic analysis in forensic science. The strength of proteomics lies in its capacity to offer a more detailed analysis of a sample than genomic methods.

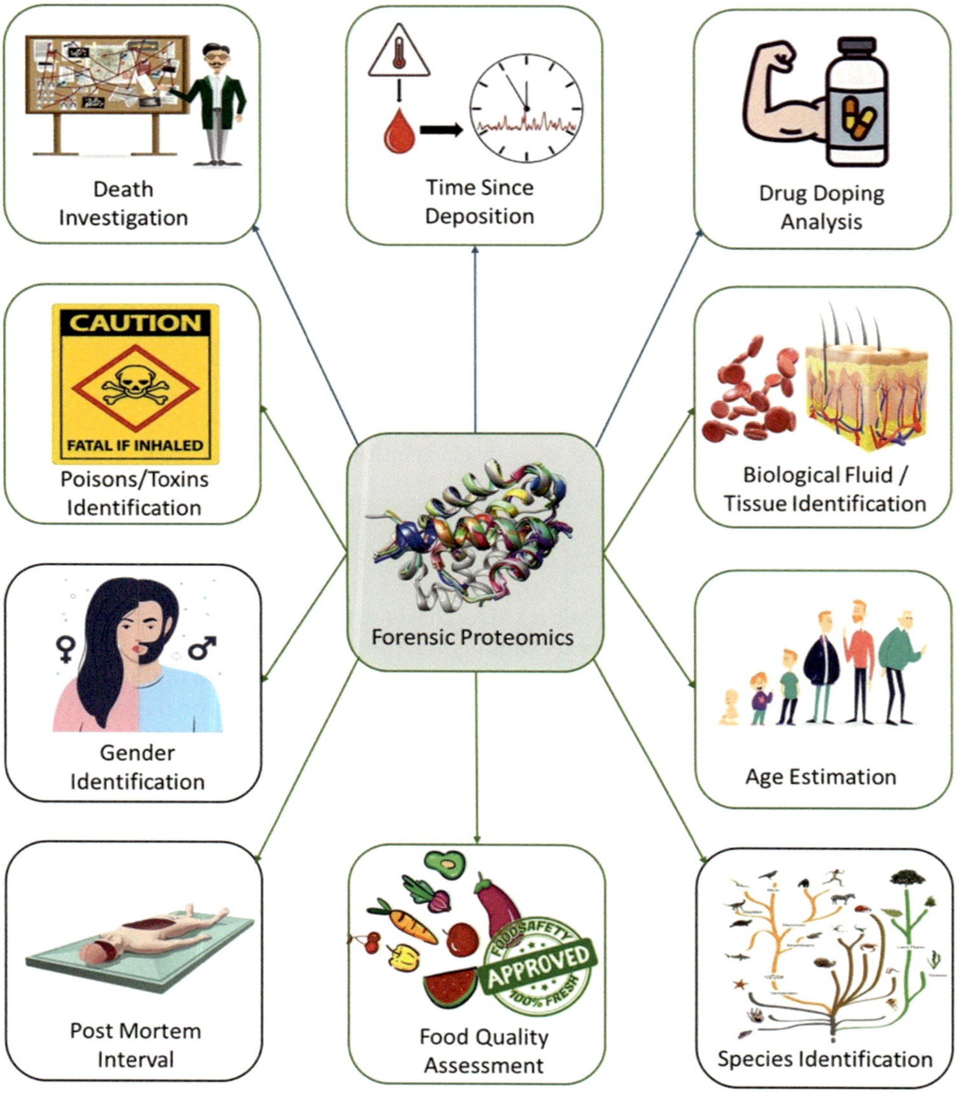

Figure 17.2 *Applications of forensic proteomics in criminal investigations.*

Proteomics is particularly useful in the absence of nucleic acids or their degradation (Zapico & Adserias-Garriga, 2022). Therefore, proteomics provides a viable alternative forensic method when DNA is unavailable, unclear, or absent (Bonicelli et al., 2022). Additionally, the literature highlights other potential applications have been highlighted in the literature. For instance, proteomic analyses of tissue residues from bullets can be used to determine the organs affected by a specific bullet (Zapico & Adserias-Garriga, 2022). The use of proteomics to determine the cause of death is highly beneficial. For example, in drowning cases, apolipoprotein A1 (ApoA1) and alpha-1 antitrypsin levels were measured to construct ROC curves. ApoA1 levels were higher in drowning victims, whereas alpha-1 antitrypsin levels were lower in drowning victims. This indicates that analysing protein biomarkers in blood samples could help identify drowning as a cause of death (Hernández-Romero et al., 2020).

17.2.1 Proteins at scene of crime

Forensic science, being an interdisciplinary science, has joined hands with the evolving field of proteomics. However, this conjunction began when Leon Lattes detected blood in 1901 using the absorption-elution method for ABO blood grouping of dried blood stains.

Evidence for the application of forensic proteomics is as follows:

1. Detection of body fluids like peripheral blood, menstrual blood, semen, saliva, breast milk, vomit, faeces, vaginal secretions
2. Identification of species from hair, horn, bone, and origin of other organ tissues
3. Protein-based toxins or plant toxins, such as ricin and abrin.
4. Proteins expressed by bacteria
5. Human individualisation
6. Presence of protein-based doping or toxic substances.
7. Palaentological or archeological evidence

Traditionally used methods based on proteins include crystal-forming reactions, enzyme-based assays, immunological assays, and microscopic visualisation. Crystal-forming assays, such as the Teichmann and Takayama tests, utilise haeme to confirm the presence of blood. However, their applicability is limited owing to their limited sensitivity and destructive nature, as the samples used cannot be reused for further analysis.

Enzyme-based assays, such as peroxidase activity for blood detection, amylase activity for saliva detection, and acid phosphatase activity for semen detection, suffer from cross-reactivity with nonspecific materials. Horseradish, garlic, and onion cross-react with each other and exhibit peroxidase-like activities. Similarly, β-amylase from plants and bacteria exhibits cross-reactivity in salivary amylase tests. Vaginal fluid, spermicides, feminine hygiene products, and certain plant materials react positively to acid phosphatase tests, limiting their sensitivity.

Immunological tests have also been abundantly used, wherein antibodies are bound to antigens from body fluids using methods such as radial diffusion and Ouchterlony double diffusion. Test kits have been developed to detect specific protein markers in the blood. However, they have shown non-specific binding and a high-dose hook effect, which could yield false negative results (McKiernan et al., 2019).

Another drawback associated with serological assays is that they use samples, leaving very few samples available for DNA identification. Therefore, it is important to develop a multiplexed proteomics-based method to address this issue.

17.2.2 Forensic protein markers of body fluids

Protein-based biomarkers were identified in the six body fluids. These biomarkers aid in the highly specific detection of proteins in body fluids. Currently, five proteome maps have been identified for peripheral blood (Schneider et al., 2022), menstrual blood (Yang et al., 2012), semen, urine, and saliva (Grocholska et al., 2023). These can be detected using a QTOF and nano-electrospray ion trap mass spectrometer. An inclusion list for the targeted ions was developed using 53 peptide targets derived from 23 potential protein biomarkers. This list was crafted to selectively identify fragment candidate biomarkers for the targeted confirmation of the six types of human body fluids (**Table 17.1**). The inclusion list was meticulously designed to establish that the peptide targets were free from known post-translational modifications,

Table 17.1 Protein Biomarkers of Various Body Fluids and Analysis Techniques

Body Fluid	Accession Number	Protein Marker	Technique Used	References
Seminal Fluid	P04279	Semenogelin-1	Immunochromatographic assay	Sato et al. (2004)
	Q02383	Semenogelin-2	Immunochromatographic lateral flow assay strip tests and RT PCR	Lundwall et al. (2002)
	P15309	Prostatic acid phosphatase	Antibody reactions in gel diffusion	Shulman et al. (1964)
	P07288	Prostate-specific antigen	Immunoassay	Healy et al. (2007)
	Q6W4X9	Mucin-6	Histological detection	Walsh and Jass (2000)
	B2R597	Beta-microseminoprotein	SDS PAGE and in silico characterisation	Vickram et al. (2018)
	P09466	Glycodelin	Immunological	Lundwall et al. (2002)
	P54107	Cysteine-rich secretory protein I	2D Electrophoresis and mass spectrometer	Hernández-Silva et al. (2020)
	P6116	Epididymal secretory protein EI	MS/MS analyses	Hu et al. (2017)
	P12273	Prolactin-inducible protein	Quadrupole time-of-flight mass spectrometry	Legg et al. (2017)
Saliva	P04745	Alpha-amylase I	2-D gel electrophoresis, MS	Hirtz et al. (2005)
	P02814	Submaxillary gland androgen-regulated protein 3B	LC-MS	Hartenbach et al. (2020)
	Q6P5S2	Protein LEGI homolog	Mass spectrometry	Mirza et al. (2022)
	P09228	Cystatin-SA	Mass spectrometry	Manconi et al. (2018)
	P28325	Cystatin-D	HPLC-ESI-MS	Serrao (2020)
	P28325	Cystatin-SN	Mass spectrometry	Liori (2018)
	P15515	Histatin-I	High performance liquid chromatography-mass spectrometry	Fanali et al. (2008)
	P02808	Statherin	RP-HPLC ESI-IT MS	Inzitari et al. (2006)
	QI6651	Prostasin	DNA sequencing	Zhu et al. (2008)
	Q8FAX7	Mucin-7	Immunohistochemistry	NguyenHoang et al. (2016)
Peripheral Blood	P68871	Haemoglobin subunit beta	Mass spectrometry	Lithanatudom and Smith (2016)
	P69905	Haemoglobin subunit alpha	Mass spectrometry	Lithanatudom and Smith (2016)
	P02730	Band 3 anion transport protein	DNA sequencing	Bruce et al. (2005)
	P02549	Spectrin alpha chain, erythrocytic 1	MALDI-TOF) MS	Demiralp et al. (2012)
	P01009	Alpha-I-antitrypsin	Immunoelectrophoretic and immunodiffusion techniques	Benkmann et al. (2008)
	P10909	Clusterin	Immunohistochemistry	Yu and Tan (2012)
	P01024	Complement C3	Radioimmunoassay	Whaley (1980)
	P02790	Hemopexin	Radioimmunoassay	Winter et al. (2021)
Vaginal Fluid	Q9UBG3	Cornulin	Immunohistochemistry	Karumuri et al. (2022)
	Q9Y6R7	IgGFc-binding protein	Quadrupole time-of-flight mass spectrometry	Legg et al. (2017)
	P07476	Involucrin	Multidimensional liquid chromatography, tandem mass spectrometry, enzyme immunoassay	Hitti et al. (2010)

(Continued)

Table 17.1 *(Continued)* Protein Biomarkers of Various Body Fluids and Analysis Techniques

Body Fluid	Accession Number	Protein Marker	Technique Used	References
	O95274	Ly6/PLAUR domain-containing protein 3	Quadrupole time-of-flight mass spectrometry	Legg et al. (2017)
	P14780	Matrix metalloproteinase-9	Immunoassay	Short et al. (2021)
	P80188	Neutrophil gelatinase-associated lipocalin	MALDI-TOF/TOF	Tang et al. (2007)
	Q60WP8	Suprabasin	Western blotting	Celsi et al. (2022)
	P08670	Vimentin	Immunohistochemistry	Simons and Vintiner (2012)
	Q9HC84	Mucin-5B	LC-MS/MS	Fields et al. (2014)
	P35321	Cornifin-A	Liquid chromatography/electrospray ionisation time-of-flight mass spectrometry and peptide mass fingerprinting	Igoh et al. (2015)
Menstrual Fluid	P20810	Calpastatin	Immunoblotting	Singh et al. (2024)

unique to their respective protein biomarkers, effectively ionised, and generated a strong and distinctive MS/MS fragmentation pattern (Legg et al., 2017).

Another benefit of mass spectrometry-based proteomics is that sample preparation is compatible with DNA processing. Triple quadrupole-based MRM analysis of post-coital fluid that was 8 days old helped correlate the protein concentration of seminal fluid to the quantitative value of male DNA, thus proving the high utility of proteomics-based MS in grave crimes such as sexual assault (Niles, 2019).

17.2.3 Identification and analysis of proteins in forensic investigation

Current proteomics research is ongoing in various areas of forensics, ranging from fingerprinting to postmortem interval (PMI) detection. This section describes various studies in forensic proteomics. The objective of this section is to help readers understand and instigate further development of their interest in this growing avenue of forensic proteomics.

Proteomic analyses in forensic research have provided insights into the molecular transformations associated with cadaver decomposition, highlighting the influence of internal and external factors on this process (Mickleburgh et al., 2021). This is especially relevant for extended PMIs exceeding two years, during which remains are partially or fully skeletonised. Gent and coworkers compared various workflows for extracting, purifying, and analysing bone proteins using both tandem mass spectrometry and liquid chromatography. This study demonstrated enhanced peptide recovery and consistency when employing the data-independent acquisition mode, which captures all fragments within the chosen m/z range (Gent et al., 2024).

MALDI-TOF-MS is a technique for the clinical identification of bacteria and fungi that achieves over 90% accuracy for clinically relevant species. Forensic proteomics is an active research area that can identify and taxonomically characterise unknown samples of any complexity (Mortier et al., 2021). A case study on the serial passaging of Y. pestis revealed that differences in protein expression could distinguish wild isolates from laboratory-adapted pathogens

(Merkley et al., 2017). This study also addressed the determination of bacterial growth conditions based on protein expression data. This approach has direct applicability to bioterrorism-related pathogens as well.

In recent years, there has been a growing focus on detecting protein- or peptide-based performance-enhancing drugs in sports. As Thomas et al. outline in their article "Proteolysis and autolysis of proteases and the detection of degradation products in doping control", they sought:

> "...to identify degradation products directly from proteases of exogenous origin in urine samples using SDS-PAGE and capillary liquid chromatography—Orbitrap (tandem) mass spectrometry. Mass spectrometric identification was accomplished by a database search, considering the accurate monoisotopic precursor mass with data-dependent MS/MS analysis and no setting for specific cleavage site(s) or enzyme(s). The main characteristics of the method were Papain, Bacillolysin, Trypsin, and Subtilisin to obtain results for specificity, limit of detection, working range, precision, and recovery post-incubation."
>
> **Thomas et al. (2009).**

17.3 Lab to courtroom – Where proteins stand chance

Biological material recovered from a crime scene undergoes laboratory analysis for identification and individualisation. Furthermore, there is a need to identify the source of the biological material for source identification. Forensic proteomics can help achieve this objective. A representation of the forensic proteomics procedure from the crime scene to the courtroom is shown in **Figure 17.3**. Conventional methods for the identification of body fluids exploit the proteins present in these fluids. Haemoglobin present in the blood is targeted for field blood tests, such as the Kastle-Meyer test or Takayama test. These tests are based on the principle of peroxidase-like activity of haemoglobin. Similarly, amylase was targeted for saliva identification at crime scenes.

The major limitation of these tests is their discriminatory power for the identification of individuals. In recent years, protein science has explored its potential in the forensic domain for the identification of various body fluids and species identification. In recent years, protein science has been used to identify biological markers in forensic samples, which can aid in investigations related to drug abuse and environmental contamination.

Since the Daubert rule, the implementation of scientific principles in forensic science methodology and scrutiny has become a litmus test for cases involving forensic evidence in the United States. Daubert standard has been widely adopted, forensic science has undergone significant changes in terms of the reliability and validity of evidence presented in court.

To ensure the validity and reliability of the scientific evidence presented in legal proceedings, it has become essential for researchers to follow established scientific methodologies in their work. Therefore, researchers are expected to adhere to strict guidelines and protocols to ensure the accuracy and reliability of their findings, which can be used in legal practice settings. In addition to following strict guidelines and protocols, researchers should ensure that their methods are transparent and reproducible to further enhance the credibility of their findings.

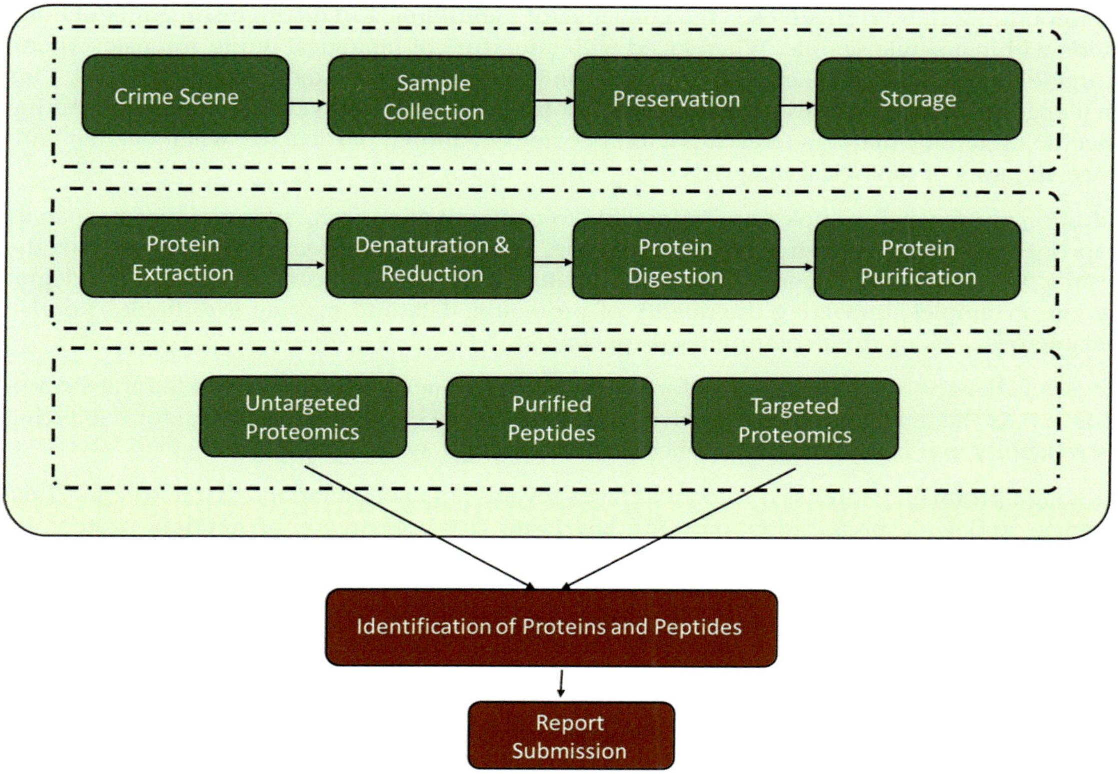

Figure 17.3 *Forensic proteomics – From crime scene to courtroom.*

17.3.1 Challenges and limitations of proteomics in forensic investigations

Forensic proteomics is an innovative field that applies proteomic techniques to criminal investigations and faces a range of significant challenges that can undermine its effectiveness and reliability. One of the foremost issues is the lack of standardised protocols across different laboratories. The absence of universally accepted methods often results in inconsistent results, making it difficult to compare and validate the findings across different settings. Additionally, the workflows in forensic proteomics are often described as "laborious and intricate", with each step from sample preparation to data analysis demanding meticulous attention and time, which can delay critical investigations.

Another significant hurdle is ensuring the sensitivity and specificity of targeted proteomic markers. The quest for markers that can "accurately detect minute quantities of proteins" while distinguishing between various bodily fluids and individuals is formidable and challenging. Achieving this delicate balance is crucial yet challenging because any deficiency in sensitivity or specificity can compromise the accuracy and reliability of forensic analyses.

The acceptance of proteomic evidence in court remains contentious. The legal system often views these techniques with scepticism, partly due to concerns over their reliability and reproducibility compared to more established methods such as DNA profiling. This scepticism results in a lower acceptance rate for proteomic evidence in legal proceedings, as courts tend to rely on methods with a more established track record.

Compounding these difficulties is the challenge of identifying and distinguishing mixed body fluids within a single sample. When faced with a mixture of biological fluids, forensic experts struggle to untangle and accurately identify components based on their protein profiles. This challenge is further exacerbated in cases involving multiple individuals, where attributing specific proteomic markers to each person becomes daunting, particularly when dealing with trace amounts of biological materials.

Additionally, forensic samples are vulnerable to contamination from various sources, such as improper handling or environmental exposure, which can compromise the integrity of the results. Environmental factors, such as temperature and humidity, can also lead to the degradation of samples, impacting the quality of proteomic data and making it difficult to derive reliable conclusions from compromised specimens.

These challenges highlight the need for ongoing advancements and refinement in forensic proteomics. As the field continues to evolve, addressing these issues will be crucial for enhancing its reliability and broadening its applicability in the legal system.

Although promising, forensic proteomics faces several notable limitations that hinder its effectiveness and widespread adoption. One key issue is the existence of multiple proteomics databases, none of which are specifically tailored for forensic applications. The absence of a dedicated forensic database complicates the task of matching proteomic profiles to individuals or crime scenes because the available resources are not optimised for forensic use.

Additionally, the field suffers from a lack of resources and infrastructure, which can impede progress and limit practical applications. The scarcity of funding and technological support has restricted the development and implementation of advanced proteomic techniques in forensic science.

Another significant challenge is the shortage of trained personnel and experts. The specialised nature of forensic proteomics requires a high level of expertise; however, there is a notable lack of professionals with the requisite skills and training. This gap in expertise affects the quality of analyses and overall reliability of the results.

Moreover, individual identification using proteomics is challenging. Unlike DNA profiling, which is well-established for distinguishing between individuals, proteomics has yet to achieve the same level of precision and reliability in personal identification. The complexity of protein profiles and their variability among individuals make it challenging to develop robust methods for unique identification.

These limitations highlight the need for continued research and development in forensic proteomics to address these challenges and enhance its legal applicability.

17.3.2 Current research status of forensic proteomics

Current proteomics research is ongoing in various areas of forensics, ranging from fingerprinting to PMI detection. In this section, we discuss various studies on forensic proteomics. The objective of this section is to help readers understand and instigate further development of their interest in this growing avenue of forensic proteomics.

In forensic studies, proteomic analyses have enabled the understanding of the molecular changes associated with cadaveric decomposition and how both internal and external factors can affect this process (Mickleburgh et al., 2021). This is particularly true for long-term PMIs (> 2 years) with partially or completely skeletonised remains. Gent et al. compared different workflows for extracting, purifying, and analysing bone proteins using liquid chromatography and tandem mass spectrometry, revealing improved peptide recovery and reproducibility when paired with the data-independent acquisition mode (all fragments were captured within the selected m/z ratio) (Gent et al., 2024).

MALDI-TOF-MS is a technique for the clinical identification of bacteria and fungi that achieves over 90% accuracy for clinically relevant species. Forensic proteomics is an active research area that can identify and taxonomically characterise unknown samples of any complexity (Mortier et al., 2021). A case study on the serial passaging of *Y. pestis* revealed that differences in protein expression could distinguish wild isolates from laboratory-adapted pathogens (Merkley et al., 2017). This study also addressed the determination of bacterial growth conditions based on protein expression data. This approach has direct applicability to bioterrorism-related pathogens as well.

Fingerprint studies occupy a special place wherever there is mention of forensics. Fingerprint secretions consist of eccrine sweat that contains proteins. A low quantity of proteins is masked by the electrolytes (Oonk et al., 2018). Researchers have been looking into establishing workflows for examining the protein signatures of groomed vs. ungroomed fingernails and those of males vs. females (with 80% accuracy) (Ferguson et al., 2012). *In situ* proteomics (Patel et al., 2015) is a "lab on a plate" technique that is in the developmental stage and can provide a molecular profile within minutes (Francese & Russo, 2019).

Forensic medicine-related mysteries and questions have been solved to a great extent using forensic proteomics. Some samples, such as bone (Procopio et al., 2018), muscle (Choi et al., 2019), brain, cerebrospinal fluid (Finehout et al., 2006), and decomposition fluid (Nolan et al., 2020), have been used to estimate the PMI. An approach was developed to probe the human digestome using MS-based proteomics to identify the nature of the last meal ingested at the time of death or disease (Pieri et al., 2018). Clinically important protein-based biomarkers can help diagnose diseases and determine whether death was caused by natural or unnatural reasons (Sacco & Aquila, 2023).

17.4 Future advancements and directions in forensic proteomics

The future scope of forensic proteomics includes the development of more sensitive and specific techniques for analysing protein samples, as well as the integration of proteomic data with other omics technologies to provide a more comprehensive understanding of forensic evidence. Advancements in bioinformatics tools will continue to play a crucial role in the interpretation of complex proteomic data and the identification of potential biomarkers for forensic applications. These advancements will ultimately enhance the accuracy and efficiency of forensic investigations, leading to more reliable results in criminal cases. Overall, the integration of proteomics with other technologies and the development of bioinformatics tools will revolutionise the field of forensic science in the coming years. By combining proteomic analysis with other forensic techniques, such as DNA analysis and trace evidence examination, investigators can create a more detailed picture of the events. This multidisciplinary approach strengthens the overall forensic process and increases the likelihood of accurately solving the cases. The integration of proteomics with artificial intelligence and machine learning algorithms will further enhance the analysis and interpretation of complex forensic data. This synergy will enable the development of predictive models for identifying potential suspects and reconstructing crime scenes based on proteomic profiles. Furthermore, the application of proteomics in forensic toxicology will expand, allowing for more precise detection and quantification of drugs and their metabolites in biological samples.

This advanced integration of proteomics into forensic science is likely to lead to the development of more sensitive and specific analytical methods for detecting and identifying biological substances at crime scenes. The increased accuracy and reliability of proteomic techniques may contribute to the establishment of new legal standards for admissible scientific evidence in the courtroom. As the field progresses, researchers may discover novel protein biomarkers that can provide insights into an individual's lifestyle, habits, or environmental exposure, further enhancing the potential of forensic proteomics.

17.5 Conclusion

Proteomics has established itself as one of the most effective and unprecedented tools that can be satisfactorily incorporated into the field of forensic science. It increases the accuracy and reliability of forensic analysis by deepening the understanding of the biological material found at crime sites to the molecular structure level. This chapter illustrates the wide range of proteomics applications, from forensic profiling and PMI estimation to biological source tracing. The combination of proteomics with other widely used forensic tools, such as DNA profiling, extends the capabilities and precision of investigations into crimes and their perpetrators.

Despite the prospects presented, the field of forensic proteomics has not been fully explored. Various issues, such as sample degradation, contamination, and protein data analysis, must be resolved to exploit the full range of the domain. With the consistent improvement of technology, especially in mass spectrometry and bioinformatics, these problems will probably be solved; hence, it is predicted that proteomics will become a more authoritative and complete domain of forensic science.

Furthermore, the development of deeper analytical methods with higher selectivity and sensitivity will be the direction of development in forensic proteomics. As research progresses, proteomics is anticipated to not only complement but may eventually surpass some traditional forensic methods in certain contexts. By continually pushing the boundaries of what is possible in criminal investigations, proteomics plays an increasingly crucial role in ensuring justice through science.

References

Aslam, B., Basit, M., Nisar, M. A., Khurshid, M., & Rasool, M. H. (2017). Proteomics: Technologies and their applications. *Journal of Chromatographic Science*, *55*(2), 182–196. https://doi.org/10.1093/chromsci/bmw167

Barranco, C. (2021). The human genome project. *Nature Research*.

Benkmann, H.-G., Hanssen, P., Ovenbeck, R., & Goedde, W. (2008). Distribution of alpha-l-antitrypsin and haptoglobin phenotypes in bladder cancer patients. *Human Heredity*, *37*(5), 290–293. https://doi.org/10.1159/000153720

Bonicelli, A., Mickleburgh, H. L., Chighine, A., Locci, E., Wescott, D. J., & Procopio, N. (2022). The 'ForensOMICS' approach for postmortem interval estimation from human bone by integrating metabolomics, lipidomics, and proteomics. *eLife*, *11*. https://doi.org/10.7554/eLife.83658

Bruce, L. J., Robinson, H. C., Guizouarn, H., Borgese, F., Harrison, P., King, M.-J., Goede, J. S., Coles, S. E., Gore, D. M., Lutz, H. U., Ficarella, R., Layton, D. M., Iolascon, A., Ellory, J. C., & Stewart, G. W. (2005). Monovalent cation leaks in human red cells caused by single amino-acid substitutions in the transport domain of the band 3 chloride-bicarbonate exchanger, AE1. *Nature Genetics*, *37*(11), 1258–1263.

Celsi, F., Monasta, L., Arrigoni, G., Battisti, I., Licastro, D., Aloisio, M., Di Lorenzo, G., Romano, F., Ricci, G., & Ura, B. (2022). Gel-based proteomic identification of suprabasin as a potential new candidate

biomarker in endometrial cancer. *International Journal of Molecular Sciences*, *23*(4), Article 4. https://doi.org/10.3390/ijms23042076
Choi, K.-M., Zissler, A., Kim, E., Ehrenfellner, B., Cho, E., Lee, S., Steinbacher, P., Yun, K. N., Shin, J. H., Kim, J. Y., Stoiber, W., Chung, H., Monticelli, F. C., Kim, J.-Y., & Pittner, S. (2019). Postmortem proteomics to discover biomarkers for forensic PMI estimation. *International Journal of Legal Medicine*, *133*(3), 899–908. https://doi.org/10.1007/s00414-019-02011-6
Demiralp, D. O., Peker, S., Turgut, B., & Akar, N. (2012). Comprehensive identification of erythrocyte membrane protein deficiency by 2D gel electrophoresis based proteomic analysis in hereditary elliptocytosis and spherocytosis. *Proteomics Clinical Applications*, *6*(7–8), 403–411. https://doi.org/10.1002/prca.201200010
Fanali, C., Inzitari, R., Cabras, T., Pisano, E., Castagnola, M., Celletti, R., Manni, A., & Messana, I. (2008). α-defensin levels in whole saliva of totally edentulous subjects. *International Journal of Immunopathology and Pharmacology*, *21*(4), 845–849. https://doi.org/10.1177/039463200802100409
Ferguson, L. S., Wulfert, F., Wolstenholme, R., Fonville, J. M., Clench, M. R., Carolan, V. A., & Francese, S. (2012). Direct detection of peptides and small proteins in fingermarks and determination of sex by MALDI mass spectrometry profiling. *Analyst*, *137*(20), 4686–4692. https://doi.org/10.1039/C2AN36074H
Fields, S., Song, B., Rasoul, B., Fong, J., Works, M. G., Shew, K., Yiu, Y., Mirsalis, J., & D'Andrea, A. (2014). New candidate biomarkers in the female genital tract to evaluate microbicide toxicity. *PLoS One*, *9*(10), e110980. https://doi.org/10.1371/journal.pone.0110980
Finehout, E. J., Franck, Z., Relkin, N., & Lee, K. H. (2006). Proteomic analysis of cerebrospinal fluid changes related to postmortem interval. *Clinical Chemistry*, *52*(10), 1906–1913. https://doi.org/10.1373/clinchem.2006.070508
Francese, S., & Russo, C. (2019). Fingermarks as a new proteomic specimen: State of the art and perspective of in situ proteomics. In E. D. Merkley (Ed.), *ACS Symposium Series* (Vol. 1339, pp. 91–105). American Chemical Society. https://doi.org/10.1021/bk-2019-1339.ch006
Gent, L., Chiappetta, M. E., Hesketh, S., Palmowski, P., Porter, A., Bonicelli, A., Schwalbe, E. C., & Procopio, N. (2024). Bone proteomics method optimization for forensic investigations. *Journal of Proteome Research*, *23*(5), 1844–1858. https://doi.org/10.1021/acs.jproteome.4c00151
Gobena, S., Admassu, B., Kinde, M. Z., & Gessese, A. T. (2024). Proteomics and its current application in biomedical area: Concise review. *The Scientific World Journal*, *2024*, 1–13. https://doi.org/10.1155/2024/4454744
Grocholska, P., Kowalska, M., & Bąchor, R. (2023). Qualitative and quantitative mass spectrometry in salivary metabolomics and proteomics. *Metabolites*, *13*(2), 155. https://doi.org/10.3390/metabo13020155
Hartenbach, F. A. R. R., Velasquez, É., Nogueira, F. C. S., Domont, G. B., Ferreira, E., & Colombo, A. P. V. (2020). Proteomic analysis of whole saliva in chronic periodontitis. *Journal of Proteomics*, *213*, 103602. https://doi.org/10.1016/j.jprot.2019.103602
Healy, D. A., Hayes, C. J., Leonard, P., McKenna, L., & O'Kennedy, R. (2007). Biosensor developments: Application to prostate-specific antigen detection. *Trends in Biotechnology*, *25*(3), 125–131. https://doi.org/10.1016/j.tibtech.2007.01.004
Hernández-Romero, D., Sánchez-Rodríguez, E., Osuna, E., Sibón, A., Martínez-Villanueva, M., Noguera-Velasco, J. A., & Pérez-Cárceles, M. D. (2020). Proteomics in deaths by drowning: Diagnostic efficacy of apolipoprotein A1 and α-1 antitrypsin, pilot study. *Diagnostics*, *10*(10), Article 10. https://doi.org/10.3390/diagnostics10100747
Hernández-Silva, G., Fabián López-Araiza, J. E., López-Torres, A. S., Larrea, F., Torres-Flores, V., & Chirinos, M. (2020). Proteomic characterization of human sperm plasma membrane-associated proteins and their role in capacitation. *Andrology*, *8*(1), 171–180. https://doi.org/10.1111/andr.12627
Hirtz, C., Chevalier, F., Centeno, D., Rofidal, V., Jean-Christophe, E., Rossignol, M., Sommerer, N., & Deville, D. (2005). MS characterization of multiple forms of alpha-amylase in human saliva. *Proteomics*, *5*, 4597–4607. https://doi.org/10.1002/pmic.200401316
Hitti, J., Lapidus, J. A., Lu, X., Reddy, A. P., Jacob, T., Dasari, S., Eschenbach, D. A., Gravett, M. G., & Nagalla, S. R. (2010). Noninvasive diagnosis of intraamniotic infection: Proteomic biomarkers in vaginal fluid. *American Journal of Obstetrics and Gynecology*, *203*(1), 32.e1–32.e8. https://doi.org/10.1016/j.ajog.2010.03.037
Hu, Y., Shah, P., Clark, D., Ao, M., & Zhang, H. (2017). *Reanalysis of global proteomic and phosphoproteomic data identified a large number of glycopeptide*. https://doi.org/10.1101/233247
Igoh, A., Doi, Y., & Sakurada, K. (2015). Identification and evaluation of potential forensic marker proteins in vaginal fluid by liquid chromatography/mass spectrometry. *Analytical and Bioanalytical Chemistry*, *407*(23), 7135–7144. https://doi.org/10.1007/s00216-015-8877-x

Institute of Medicine. (2012). Omics-based clinical discovery: Science, technology, and applications. In C. M. Micheel, S. J. Nass, & G. S. Omenn (Eds.) *Evolution of Translational Omics: Lessons Learned and the Path Forward*. National Academies Press (US). https://doi.org/10.17226/13297

Inzitari, R., Cabras, T., Rossetti, D. V., Fanali, C., Vitali, A., Pellegrini, M., Paludetti, G., Manni, A., Giardina, B., Messana, I., & Castagnola, M. (2006). Detection in human saliva of different statherin and P-B fragments and derivatives. *Proteomics*, *6*(23), 6370–6379. https://doi.org/10.1002/pmic.200600395

Karumuri, R., Shah, D., & Arnouk, H. (2022). Cornulin as a potential novel biomarker for cutaneous squamous cell carcinoma. *Cureus*, *14*(11), e31694. https://doi.org/10.7759/cureus.31694

Laemmli, U. K. (1970). Cleavage of structural proteins during the assembly of the head of bacteriophage T4. *Nature*, *227*(5259), 680–685. https://doi.org/10.1038/227680a0

Legg, K. M., Powell, R., Reisdorph, N., Reisdorph, R., & Danielson, P. B. (2017). Verification of protein biomarker specificity for the identification of biological stains by quadrupole time-of-flight mass spectrometry. *Electrophoresis*, *38*(6), 833–845. https://doi.org/10.1002/elps.201600352

Liori, B. (2018). *Salivary biomarkers in multiple sclerosis and autoimmune hepatitis explored by an integrate top-down and bottom-up platform*. https://iris.unica.it/handle/11584/255976

Lithanatudom, P., & Smith, D. R. (2016). Analysis of protein profiling studies of β-thalassemia/Hb E disease. *Proteomics Clinical Applications*, *10*(11), 1093–1102. https://doi.org/10.1002/prca.201600086

Lundwall, A., Bjartell, A., Olsson, A. Y., & Malm, J. (2002). Semenogelin I and II, the predominant human seminal plasma proteins, are also expressed in non-genital tissues. *Molecular Human Reproduction*, *8*(9), 805–810. https://doi.org/10.1093/molehr/8.9.805

Manconi, B., Liori, B., Cabras, T., Vincenzoni, F., Iavarone, F., Lorefice, L., Cocco, E., Castagnola, M., Messana, I., & Olianas, A. (2018). Top-down proteomic profiling of human saliva in multiple sclerosis patients. *Journal of Proteomics*, *187*, 212–222. https://doi.org/10.1016/j.jprot.2018.07.019

McKiernan, H. E., Brown, C. O., Arantes, L. C., Danielson, P. B., & Legg, K. M. (2019). NextGen serology: Leveraging mass spectrometry for protein-based human body fluid identification. In E. D. Merkley (Ed.), *ACS Symposium Series* (Vol. 1339, pp. 47–80). American Chemical Society. https://doi.org/10.1021/bk-2019-1339.ch004

Merkley, E. D., Sego, L. H., Lin, A., Leiser, O. P., Kaiser, B. L. D., Adkins, J. N., Keim, P. S., Wagner, D. M., & Kreuzer, H. W. (2017). Protein abundances can distinguish between naturally-occurring and laboratory strains of Yersinia pestis, the causative agent of plague. *PLoS One*, *12*(8), e0183478. https://doi.org/10.1371/journal.pone.0183478

Merkley, E. D., Wunschel, D. S., Wahl, K. L., & Jarman, K. H. (2019). Applications and challenges of forensic proteomics. *Forensic Science International*, *297*, 350–363. https://doi.org/10.1016/j.forsciint.2019.01.022

Mickleburgh, H. L., Schwalbe, E. C., Bonicelli, A., Mizukami, H., Sellitto, F., Starace, S., Wescott, D. J., Carter, D. O., & Procopio, N. (2021). Human bone proteomes Before and after decomposition: Investigating the effects of biological variation and taphonomic alteration on bone protein profiles and the implications for forensic proteomics. *Journal of Proteome Research*, *20*(5), 2533–2546. https://doi.org/10.1021/acs.jproteome.0c00992

Mirza, M. R., Sultan, R., Choudhary, M. I., Tahir, M., Larsen, M. R., Tariq, S., & Rahman, S. U. (2022). Label-free quantitation of the changes in salivary proteome associated with the chronic consumption of the betel nut (*Areca catechu*). *Molecular Omics*, *18*(2), 123–132. https://doi.org/10.1039/D1MO00391G

Mortier, T., Wieme, A. D., Vandamme, P., & Waegeman, W. (2021). Bacterial species identification using MALDI-TOF mass spectrometry and machine learning techniques: A large-scale benchmarking study. *Computational and Structural Biotechnology Journal*, *19*, 6157–6168. https://doi.org/10.1016/j.csbj.2021.11.004

Müller, O. F., & Fabricius, O. (1786). *Animalcula infusoria fluviatilia et marina que detexit, systematice descripsit et ad vivum delineari curavit Otho Fridericus Müller sistit opus hoc posthumum quod cum tabulis Aeneis L. in lucem tradit vidua ejus nobilissima cura Othonis Fabricii*. https://agris.fao.org/search/en/providers/122376/records/647472a779cbb2c2c1b2a429

NguyenHoang, S., Liu, Y., Xu, L., Chang, Y., Zhou, L., Liu, Z., Lin, Z., & Xu, J. (2016). High mucin-7 expression is an independent predictor of adverse clinical outcomes in patients with clear-cell renal cell carcinoma. *Tumour Biology: The Journal of the International Society for Oncodevelopmental Biology and Medicine*, *37*(11), 15193–15201. https://doi.org/10.1007/s13277-016-5375-5

Niles, S. F. (2019). *Improving semen identification and quantitation using protein mass spectrometry*. https://www.semanticscholar.org/paper/Improving-semen-identification-and-quantitation-Niles/452e71c906af877bf4f38524b99983956e0ad20d

Nolan, A.-N., Maker, G., Mead, R. J., Bringans, S., & Speers, S. J. (2020). Peptide analysis of mammalian decomposition fluid in relation to the post-mortem interval. *Forensic Science International*, *311*, 110269. https://doi.org/10.1016/j.forsciint.2020.110269

Oonk, S., Schuurmans, T., Pabst, M., De Smet, L. C. P. M., & De Puit, M. (2018). Proteomics as a new tool to study fingermark ageing in forensics. *Scientific Reports*, *8*(1), 16425. https://doi.org/10.1038/s41598-018-34791-z

Pandey, A., & Mann, M. (2000). Proteomics to study genes and genomes. *Nature*, *405*(6788), 837–846. https://doi.org/10.1038/35015709

Patel, E., Clench, M. R., West, A., Marshall, P. S., Marshall, N., & Francese, S. (2015). Alternative surfactants for improved efficiency of in situ tryptic proteolysis of fingermarks. *Journal of the American Society for Mass Spectrometry*, *26*(6), 862–872. https://doi.org/10.1007/s13361-015-1140-z

Patterson, S. D., & Aebersold, R. H. (2003). Proteomics: The first decade and beyond. *Nature Genetics*, *33*(S3), 311–323. https://doi.org/10.1038/ng1106

Patzelt, D. (2004). History of forensic serology and molecular genetics in the sphere of activity of the German society for forensic medicine. *Forensic Science International*, *144*(2–3), 185–191. https://doi.org/10.1016/j.forsciint.2004.04.053

Peng, J., & Gygi, S. P. (2001). Proteomics: The move to mixtures. *Journal of Mass Spectrometry*, *36*(10), 1083–1091. https://doi.org/10.1002/jms.229

Pieri, M., Lombardi, A., Basilicata, P., Mamone, G., & Picariello, G. (2018). Proteomics in forensic sciences: Identification of the nature of the last meal at autopsy. *Journal of Proteome Research*, *17*(7), 2412–2420. https://doi.org/10.1021/acs.jproteome.8b00159

Procopio, N., Williams, A., Chamberlain, A. T., & Buckley, M. (2018). Forensic proteomics for the evaluation of the post-mortem decay in bones. *Journal of Proteomics*, *177*, 21–30. https://doi.org/10.1016/j.jprot.2018.01.016

Sacco, M. A., & Aquila, I. (2023). Proteomics: A new research frontier in forensic pathology. *International Journal of Molecular Sciences*, *24*(13), 10735. https://doi.org/10.3390/ijms241310735

Sato, I., Kojima, K., Yamasaki, T., Yoshida, K., Yoshiike, M., Takano, S., Mukai, T., & Iwamoto, T. (2004). Rapid detection of semenogelin by one-step immunochromatographic assay for semen identification. *Journal of Immunological Methods*, *287*(1–2), 137–145. https://doi.org/10.1016/j.jim.2004.01.017

Schneider, T. D., Roschitzki, B., Grossmann, J., Kraemer, T., & Steuer, A. E. (2022). Determination of the time since deposition of blood traces utilizing a liquid Chromatography–Mass spectrometry-based proteomics approach. *Analytical Chemistry*, *94*(30), 10695–10704. https://doi.org/10.1021/acs.analchem.2c01009

Serrao, S. (2020). *Potential salivary biomarkers in mastocytosis: A proteomics approach*. https://iris.unica.it/handle/11584/284376

Short, C.-E. S., Quinlan, R. A., Wang, X., Preda, V. G., Smith, A., Marchesi, J. R., Lee, Y. S., MacIntyre, D. A., Bennett, P. R., & Taylor, G. P. (2021). Vaginal microbiota, genital inflammation and extracellular matrix remodelling collagenase: MMP-9 in pregnant women with HIV, a potential preterm birth mechanism warranting further exploration. *Frontiers in Cellular and Infection Microbiology*, *11*. https://doi.org/10.3389/fcimb.2021.750103

Shulman, S., Mamrod, L., Gonder, M. J., & Soanes, W. A. (1964). The detection of prostatic acid phosphatase by antibody reactions in gel diffusion1. *The Journal of Immunology*, *93*(3), 474–480. https://doi.org/10.4049/jimmunol.93.3.474

Simons, J. L., & Vintiner, S. K. (2012). Efficacy of several candidate protein biomarkers in the differentiation of vaginal from buccal epithelial cells. *Journal of Forensic Sciences*, *57*(6), 1585–1590. https://doi.org/10.1111/j.1556-4029.2012.02158.x

Singh, A., Tripathi, R., Gupta, R. K., Rashid, R., & Jha, R. K. (2024). Gonadotropin upregulates intraovarian calpains-1 and -2 during ovarian follicular recruitment in the SD rat model. *Reproductive Biology*, *24*(2), 100862. https://doi.org/10.1016/j.repbio.2024.100862

Tang, L.-J., De Seta, F., Odreman, F., Venge, P., Piva, C., Guaschino, S., & Garcia, R. C. (2007). Proteomic analysis of human cervical-vaginal fluids. *Journal of Proteome Research*, *6*(7), 2874–2883. https://doi.org/10.1021/pr0700899

Thomas, A., Kohler, M., Walpurgis, K., Schänzer, W., & Thevis, M. (2009). Proteolysis and autolysis of proteases and the detection of degradation products in doping control. *Drug Testing and Analysis*, *1*(2), 81–86. https://doi.org/10.1002/dta.20

Vickram, A. S., Anbarasu, K., Rao, K. A., Jayanthi, S., & Sridharan, T. B. (2018). Identification and in silico characterization of semenogelin II protein in semen - A marker for diagnosis of male infertility. *Current Proteomics*, *15*(4), 313–319. https://doi.org/10.2174/1570164615666180717145101

Walsh, M. D., & Jass, J. R. (2000). Histologically based methods for detection of mucin. In A. P. Corfield (Ed.), *Glycoprotein Methods and Protocols: The Mucins* (pp. 29–44). Humana Press. https://doi.org/10.1385/1-59259-048-9:029

Whaley, K. (1980). Biosynthesis of the complement components and the regulatory proteins of the alternative complement pathway by human peripheral blood monocytes. *Journal of Experimental Medicine*, *151*(3), 501–516. https://doi.org/10.1084/jem.151.3.501

Wilson, K., Hofmann, A., Walker, J. M., & Clokie, S. (2018). *Wilson and Walker's Principles and Techniques of Biochemistry and Molecular Biology*. Cambridge University Press.

Winter, N. A., Gibson, P. G., Fricker, M., Simpson, J. L., Wark, P. A., & McDonald, V. M. (2021). Hemopexin: A novel anti-inflammatory marker for distinguishing COPD from asthma. *Allergy, Asthma & Immunology Research*, *13*(3), 450–467. https://doi.org/10.4168/aair.2021.13.3.450

Yang, H., Zhou, B., Prinz, M., & Siegel, D. (2012). Proteomic analysis of menstrual blood. *Molecular & Cellular Proteomics*, *11*(10), 1024–1035. https://doi.org/10.1074/mcp.M112.018390

Yu, J.-T., & Tan, L. (2012). The role of clusterin in Alzheimer's disease: Pathways, pathogenesis, and therapy. *Molecular Neurobiology*, *45*(2), 314–326. https://doi.org/10.1007/s12035-012-8237-1

Zapico, S. C., & Adserias-Garriga, J. (2022). Postmortem interval estimation: New approaches by the analysis of human tissues and microbial communities' changes. *Forensic Sciences*, *2*(1), Article 1. https://doi.org/10.3390/forensicsci2010013

Zhu, H., Guo, D., Li, K., Yan, W., Tan, Y., Wang, X., Treiber, F. A., Chao, J., Snieder, H., & Dong, Y. (2008). Prostasin: A possible candidate Gene for human hypertension. *American Journal of Hypertension*, *21*(9), 1028–1033. https://doi.org/10.1038/ajh.2008.224

18

Unraveling Injustice
The Impact of DNA Exonerations

Avantika Johari and Purvi Pokhariyal

18.1 Introduction

One of the most devastating features of the criminal justice system is being convicted of an offence that the person did not commit. When an innocent person is wrongly convicted, it has a profound impact on them, their family, and friends, and even the actual victims of the crime. The ripple effects extend to everyone involved—attorneys, judges, witnesses, and investigators. It erodes trust of the public in the entire legal system, making people question its fairness and integrity.

For those who have been wrongfully convicted, the consequences are severe. They endure societal stigma, emotional trauma, and the agonizing loss of their freedom. Their personal relationships often suffer, as the burden of a wrongful consviction strains bonds with family and friends. Preventing these miscarriages of justice is crucial. To do so, we must understand the root causes. It is our collective responsibility to safeguard justice and uphold the public's trust in the legal system. In addition, the initial victims of crime do not receive proper closure since the real criminals are still continuing to pose a threat to society (Laporte, n.d.).

The involvement of forensic science in wrongful convictions is a complex and multifaceted issue. While forensic evidence serves as a potent instrument for uncovering the truth, its misapplication or misinterpretation can lead to significant errors. Forensic science encompasses a broad array of disciplines, including DNA analysis, fingerprinting, ballistics, and forensic pathology, each presenting unique challenges and potential for error (Laporte, n.d.).

In the quiet corridors of justice, where whispers of truth often go unheard, the Innocence Project stands as a beacon of hope and redemption. This non-profit organization, dedicated to the unravelling of the tangled web of wrongful convictions, has transformed the lives of countless individuals through the power of DNA technology. Imagine the heartache of an innocent soul, confined within the cold, unforgiving walls of a prison cell. Years pass, seasons change, and the world outside moves on, but for them, time stands still. It's a story that begins in despair but finds its hero in the relentless advocates of the Innocence Project. The Innocence Project's work is not just about setting the innocent free; it's about restoring faith in a system that sometimes falters and giving voice to those who have been silenced.

Since its establishment, the organization has diligently worked to leverage DNA evidence to exonerate the innocent and prevent future wrongful convictions. The work of this organization has not only led to the exoneration of numerous individuals but has also spurred significant reforms in forensic science practices and policies in the criminal justice system (*Innocence Project*, n.d.-b).

DOI: 10.4324/9781003474296-18

Wrongful convictions, elusive specters born of flawed eyewitness accounts, coerced confessions, negligent defense, and the sinister dance of official misconduct, haunt the sanctity of fairness. The scales of justice, meant to weigh truth with impartiality, sometimes tip under the weight of human error and deceit. However, forensic science plays a particularly significant role in many of these cases. Because jurors and judges often see forensic evidence as highly reliable, mistakes in its application can be especially damaging.

Most wrongful convictions involve complex investigations, a mix of different factors, and mistakes arising from outdated policies and practices. Jury decisions are also influenced by these complexities, and the absence of complete details or full trial transcripts can further cloud our understanding and introduce biases. Despite these challenges, it's essential to examine these cases and learn from them to improve forensic procedures and prevent future wrongful convictions.

Forensic scientists need to understand that their judgments and work, whether in trial testimony or forensic reports, have long-lasting effects on the freedom and the lives of people. They must strive to recognize and address their weaknesses. This awareness and continuous improvement are vital to ensure the accuracy and reliability of forensic evidence, ultimately upholding justice and protecting the innocent (Bonventre, 2021).

Erroneous convictions impact not only innocent individuals who are convicted but also their families, the original crime victims, and society overall. Forensic science plays an important role in both causing and rectifying these errors. By understanding how forensic science contributed to erroneous convictions and by using insights from past cases helps us to move towards a more precise criminal justice system. The Innocence Project has been pivotal in this effort by pushing reforms in forensic practices and taking efforts in advancing DNA technology to bring improvements.

18.2 DNA evidence

DNA evidence is important as it offers highly accurate and reliable tools for solving crimes and ensuring justice. Unlike conventional techniques, DNA technology has provided new avenues for investigation, especially in cases where traditional methods have fallen short. Before the advent of DNA technology, overturning wrongful convictions was incredibly challenging, even in the face of compelling evidence. The precision of DNA testing has become a critical tool for proving innocence, transforming the landscape of criminal justice (Vajpayee et al., 2020).

Organizations like the Innocence Project have harnessed DNA technology to successfully exonerate numerous individuals who were wrongfully convicted. These successes highlight the importance of DNA evidence to underscore the need for its continued application and improvement within the legal system. By leveraging the power of DNA, we can better protect the innocent and uphold the integrity of our justice system. These exonerations cases have not only freed the innocent people who are wrongfully convicted but also highlighted the urgent need to bring reforms in the criminal justice system (Wickenheiser, 2021).

Many individuals exonerated through DNA evidence have endured long periods of incarceration. Take Kirk Bloodsworth, for example. Convicted in 1985 for the harrowing crimes against a young girl, Kirk Bloodsworth endured nearly a decade behind bars before the tide of justice turned. In a pivotal moment echoing the power of scientific truth, DNA testing emerged as his savior, revealing his innocence beyond doubt. His story stands as a testament to the profound impact of technological advancement in unraveling the webs of wrongful conviction, granting freedom where despair once reigned. Kirk Bloodsworth's journey from darkness to redemption echoes across time, a poignant reminder of the enduring quest for truth and the resilience of the human spirit against all odds (Dash et al., 2020).

18.2.1 DNA evidence and its impact on the criminal justice system

The integration of DNA evidence in criminal cases has brought a lot of changes in the criminal system and altered it to a great extent by enhancing the investigation with more accuracy and playing a pivotal role in exonerating individuals who were wrongfully convicted. It has also helped in bringing advancements in forensic science and providing new approaches to identifying offenders (Bonventre, 2021).

In a landmark 2018 case in California, the echoes of injustice were finally silenced by the irrefutable voice of DNA evidence. After enduring a quarter-century of wrongful imprisonment for a crime he did not commit, a man found freedom as DNA testing unveiled the truth. This pivotal moment underscores the reliability and transformative power of DNA technology in the realm of justice, where it corrects grave errors and illuminates paths to truth. DNA databases, repositories of profiles from convicts and crime scenes alike, have become indispensable tools in law enforcement. By meticulously comparing DNA profiles, investigators unearth hidden connections and unmask perpetrators who have long evaded accountability. This method has not only led to countless convictions and the resolution of cold cases but also fortified the legal system's accuracy and efficacy. Take, for instance, the notorious Golden State Killer case, where decades-old crimes found resolution through DNA databases, ultimately exposing Joseph James DeAngelo as the elusive culprit. Such victories resonate beyond courtroom walls, restoring faith in justice and affirming its steadfast pursuit of truth and fairness (*DNA Evidence*, n.d.).

The incorporation of DNA proof into forensic science resulted in the development of various new methodologies and techniques, bettering the accuracy and reliability of forensic findings. These advancements in DNA technology have enabled the extraction of DNA from increasingly minute and degraded samples, thereby broadening the range of cases that can be resolved. This has significantly enhanced the criminal justice system's ability to solve crimes and exonerate the innocent.

However, despite these advantages, DNA evidence also carries the risk of misuse. Privacy concerns have emerged, as genetic profiles can be used to profile individuals, leading to potential discrimination. Additionally, the misinterpretation or mishandling of DNA evidence can result in wrongful convictions. It is essential to uphold ethical standards and ensure the responsible use of DNA evidence to mitigate these risks. Formulating rigorous rules and protocols for handling and analyzing DNA, while safeguarding privacy, is crucial to prevent these pitfalls.

The significant role of DNA evidence in exonerating wrongfully convicted individuals underscores its vital contribution to achieving justice. In the United States alone, DNA testing has facilitated the exoneration of over 375 individuals who were wrongfully convicted, demonstrating its profound impact on rectifying miscarriages of justice (*DNA Evidence*, n.d.).

18.2.2 The importance of post-conviction DNA testing

Post-conviction DNA testing stands as a beacon of hope for those ensnared by wrongful convictions, offering a second chance at truth and justice. In cases where DNA evidence was absent or overlooked during the initial trial, this forensic tool becomes pivotal. It unravels the knots of uncertainty, providing an opportunity to exonerate the innocent and unveil the identities of the real perpetrators who eluded justice.

With each strand of DNA analyzed, possibilities unfold: convictions overturned, lives reclaimed, and the course of justice redirected. This process not only rectifies past errors but also reaffirms the integrity of the legal system, ensuring that truth prevails despite initial setbacks. Post-conviction DNA testing stands not just as a scientific marvel but as a testament to resilience, offering redemption where doubt once clouded the path to righteousness. However, access to post-conviction DNA testing varies significantly among states, with some lacking the necessary legal framework to facilitate it (*DNA Evidence*, n.d.). This inconsistency can lead to prolonged struggles for wrongfully convicted individuals seeking to prove their innocence.

18.2.3 The limitations of DNA evidence

DNA evidence, while formidable in its impact, operates within defined boundaries dictated by the presence of biological material. Its potency shines brightest in cases where bodily fluids or biological traces exist, such as in sexual assaults or violent crimes. In these instances, DNA serves as a molecular fingerprint, linking individuals to events with a precision that transcends doubt. However, its limitations are stark where biological evidence is absent (Vajpayee et al., 2023). Crimes lacking such traces pose challenges for DNA-based investigations, compelling reliance on other forensic methodologies and evidence. This inherent constraint underscores the need for comprehensive investigative strategies that encompass diverse forensic disciplines, ensuring that justice is pursued with diligence and equity across all types of criminal cases. While DNA evidence stands as a cornerstone of modern forensic science, its power is most profound when complemented by a holistic approach that embraces the complexity of each case (*DNA Evidence*, n.d.). DNA cannot provide context on how the material was deposited at a crime scene, necessitating corroboration with other types of evidence to establish guilt or innocence comprehensively.

18.3 DNA evidence and eyewitness testimony

Eyewitness testimony has historically been a significant factor in wrongful convictions. DNA evidence has exposed the unreliability of eyewitness identifications, leading to a reevaluation of their weight in criminal trials. DNA testing has thus served as a corrective measure, exonerating those convicted on flawed eyewitness accounts and highlighting the need for more rigorous standards in eyewitness evidence.

18.3.1 The reliability of DNA evidence

DNA evidence stands as a pillar of reliability within the criminal justice system, distinguished by its unique ability to pinpoint individual genetic identities with unparalleled accuracy. The rarity of matching DNA profiles between unrelated individuals underscores their potency in identifying suspects and linking them to criminal acts. This dependability rests on the distinct genetic blueprint inherent to each person, deciphered through advanced testing techniques that leave little room for ambiguity.

Yet, amidst its robustness, DNA evidence is not immune to human fallibility. Errors during sample collection, storage, or analysis can introduce contamination or misinterpretation, potentially jeopardizing the evidential integrity. Rigorous protocols are therefore imperative, governing every step from collection to analysis in order to safeguard against such pitfalls. Laboratories adhere to stringent standards and meticulous evidence management strategies to preserve the pristine reliability of DNA findings, ensuring that justice is served with unwavering accuracy and fairness (Srivastava et al., 2022).

Despite these challenges, the standard reliability of DNA proof stays high, and its use has significantly bring the purpose of justice. The cautious software of DNA technology continues to be a cornerstone of efforts to right wrongful convictions and bring true perpetrators to justice.

18.3.2 DNA databases

The evolution and expansion of DNA databases have significantly enhanced the potency of DNA evidence in criminal investigations. These databases store profiles sourced from convicted offenders and individuals who voluntarily submit their DNA samples. Their pivotal role in resolving longstanding cases and identifying suspects in ongoing investigations cannot be overstated. A prime example is the Combined DNA Index System (CODIS), administered by the FBI, renowned as one of the largest and most effective DNA databases worldwide. CODIS serves as a centralized hub where DNA profiles from across jurisdictions are stored and compared, enabling law enforcement agencies to connect crimes, identify repeat offenders, and bring perpetrators to justice swiftly and accurately. Through the systematic integration of DNA databases like CODIS, the investigative landscape has been transformed, fostering collaboration and leveraging scientific advancements to uphold public safety and the integrity of criminal justice (Srivastava et al., 2022).

DNA databases have led to numerous breakthroughs in criminal investigations. For example, the Golden State Killer case, which had remained unsolved for decades, was finally cracked in 2018 using DNA evidence from a genealogical database. Such successes demonstrate the power of DNA databases to provide critical leads and solve crimes that would otherwise remain unresolved (Jack, 2024).

The utilization of DNA databases presents notable privacy concerns that warrant careful consideration. The possibility of DNA information being misused, whether through unauthorized access or data breaches, poses significant risks to individual privacy. The ethical quandaries surrounding DNA databases extend to the inclusion of profiles from individuals without criminal convictions, posing a delicate balance between leveraging these resources for crime-solving while safeguarding personal privacy. These dual imperatives demand rigorous regulation and vigilant oversight to ensure that the collection, storage, and utilization of DNA data uphold ethical standards and respect privacy rights.

Clear guidelines and stringent safeguards are crucial in navigating these complexities, ensuring transparency and accountability in the use of DNA technology. Efforts to establish robust frameworks for ethical conduct not only mitigate risks of misuse but also preserve public trust in the integrity of investigative practices. By striking this balance, we can harness the full potential of DNA databases in advancing justice while upholding the fundamental rights and dignity of every individual.

18.3.3 DNA evidence and wrongful convictions

DNA evidence has not only been pivotal in freeing numerous wrongfully convicted individuals but has also highlighted systemic flaws that contribute to these miscarriages of justice. Key elements include unreliable eyewitness testimony, false confessions, and flawed forensic evidence. DNA exonerations have exposed these issues, prompting urgent calls for systemic reforms to prevent future injustices. Eyewitness testimony, often persuasive in court, has been shown to be highly fallible.

False confessions are another troubling aspect. Psychological pressure, coercive interrogation tactics, and the mental state of the accused can all contribute to false admissions of guilt, even in serious crimes like murder. DNA exonerations have demonstrated instances where individuals falsely confessed, underscoring the essential need for better interrogation methods and safeguards against coercion.

18.4 The broader implications of wrongful convictions on the criminal justice system

1. ***Erosion of Public Trust:*** Erroneous convictions wield a profound and enduring influence on the criminal justice system, perhaps most significantly eroding public trust. These miscarriages of justice illuminate systemic vulnerabilities and demand critical reflection on how errors occur and how they can be prevented. The repercussions extend beyond the individual wrongfully convicted to impact societal perceptions of justice and the credibility of legal institutions. Rebuilding trust requires steadfast efforts to rectify past injustices, implement reforms to safeguard against future errors, and foster transparency and accountability at every level of the criminal justice system (Morris, 2023). This erosion of confidence is dangerous to the social contract, as residents expect and remember the justice gadget to guard their rights and hold social order.
2. ***Psychological and Social Impact on the Erroneously Convicted:*** For the individuals who are convicted wrongfully, the penalties are really disturbing and devastating. They go through psychological trauma and stress from being imprisoned for crimes for several years that they did not even commit. The stigma connected to a criminal conviction can comply with them even after they get freed, affecting their potential to discover employment, housing, and rebuild their lives.

18.4.1 The Innocence Project and DNA exonerations

Founded in 1992 by Barry Scheck and Peter Neufeld, the Innocence Project stands as a vanguard for justice, harnessing the power of DNA evidence to unravel wrongful convictions and champion reforms within the criminal justice system. Since its inception, this pioneering organization has been instrumental in exonerating over 375 individuals through rigorous DNA testing, each case a testament to the transformative impact of scientific truth.

Beyond securing freedom for the innocent, the Innocence Project's advocacy resonates in corridors of power, driving systemic reforms that address the root causes of wrongful convictions. Its efforts not only rectify past injustices but also shape a future where fairness and accuracy prevail. As a beacon of hope and resilience, the Innocence Project continues to redefine possibilities in the pursuit of justice for all

The Innocence Project's impact extends beyond individual cases, influencing broader reforms across the criminal justice landscape. By championing the use of DNA technology and raising awareness about wrongful convictions, the organization has spurred legislative efforts aimed at preventing future injustices. These reforms include initiatives like mandatory recording of interrogations, improved procedures for eyewitness identification, and the establishment of independent forensic science commissions to enhance the reliability of forensic evidence.

One notable outcome of the Innocence Project's tireless advocacy is the establishment of the National Commission on Forensic Science, a pivotal initiative aimed at enhancing the scientific rigor and reliability of forensic disciplines. This commission stands as a collaborative platform where experts, policymakers, and stakeholders converge to scrutinize and advance forensic practices. Through the commission's efforts, critical reforms are forged to bolster the accuracy and integrity of forensic evidence used in criminal investigations and trials. By setting rigorous standards, promoting research-driven methodologies, and fostering transparency, the commission plays a crucial role in fortifying the foundation upon which justice is pursued (Morris, 2023).

Despite the criminal justice system's fundamental aim to protect the innocent and hold the guilty accountable, the reality often reveals complexities and vulnerabilities to errors. The emergence of DNA testing in the late 20th century has illuminated the prevalence of wrongful convictions, with the Innocence Project at the forefront of efforts to rectify these injustices through rigorous advocacy and strategic use of forensic evidence (Furman, 2003). The Innocence Project's data (1989–2020) on wrongful convictions provides a detailed look at the factors contributing to these miscarriages of justice. Here are some key insights from their statistics:

Exoneration Details:

- **375** DNA exonerees.
- **37 states** have had exonerations.
- **14 years** was the average time served.
- **5,284 years** total years served.
- **26.6 years old** was the average age at wrongful conviction.
- **43 years old** was the average age at exoneration.
- **21 out of 375** people served time on death row.
- **44 out of 375** pled guilty to crimes they did not commit.

18.4.1.1 Eyewitness misidentification: A leading cause of wrongful conviction

One of the most significant contributors to wrongful convictions is eyewitness misidentification, implicated in 69% of DNA exoneration cases. The unreliability of eyewitness testimony has been well-documented, with various factors contributing to its fallibility (*DNA Exonerations in the United States (1989–2020)*, n.d.).

Breakdown of Misidentification Cases

- **Lineups and Photo Arrays**: 34% involved an in-person lineup, while 52% involved a photo array. Both methods are prone to suggestion and memory distortion.
- **Show-Up Procedures**: 16% of cases involved show-up procedures, where a single suspect is presented to a witness, often leading to biased identifications.
- **Composite Sketches and Hypnosis**: 27% involved composite sketches, and 2% involved hypnosis, both of which are scientifically questionable methods.
- **Multiple Procedures**: Notably, 77% of misidentification cases involved multiple identification procedures, compounding the risk of error.
- **Victim Misidentification**: 84% of misidentification cases involved a surviving victim, highlighting the emotional and psychological pressures on witnesses.
- **Cross-Racial Identification**: 42% involved cross-racial identifications, which are particularly prone to error due to difficulties in distinguishing the features of individuals from different racial backgrounds.

Eyewitness misidentification has a profound impact that cannot be overstated. It not only results in the wrongful conviction of innocent individuals but also enables the true perpetrators to evade justice, potentially leading to further crimes. This dual failure not only undermines the credibility of the justice system but also emphasizes the critical necessity for reform in eyewitness identification procedures (*DNA Exonerations in the United States (1989–2020)*, n.d.; Wixted, 2018).

18.4.1.2 The misapplication of forensic science

Forensic science, while a powerful tool for justice, has also been misapplied, contributing to 43% of wrongful convictions (*DNA Exonerations in the United States (1989–2020)*, n.d.). Various issues, such as erroneous methodologies, exaggerated claims, and human error, have afflicted forensic disciplines, from hair evaluation to chew mark comparisons.

- **Hair Analysis**: In several cases, microscopic hair assessment has been proven to be highly unreliable, leading to erroneous convictions primarily based on inaccurate matches (*DNA Exonerations in the United States (1989–2020)*, n.d.).
- **Bite Mark Analysis**: Similarly, bite mark evaluation has come underneath scrutiny for its lack of scientific validity, with various high-profile exonerations ensuing from the discrediting of this method (*Misapplication of Forensic Science*, n.d.).

The revelation of these flaws has brought about calls for rigorous requirements and oversight in forensic science. The institution of impartial forensic science commissions and the adoption of extra stringent certification methods are vital steps in the direction of making sure the reliability of forensic evidence.

18.4.1.3 False confessions: A disturbing reality

False confessions are some other major issue of erroneous convictions, concerned in 29% of DNA exoneration instances. The phenomenon of false confessions is complex, often resulting from intense police interrogation tactics, psychological pressure, and the vulnerability of certain individuals.

Demographics and Circumstances of False Confessions

- **Youth and Mental Capacity**: Within this grim landscape, stark statistics emerge: nearly half of those who falsely admit guilt are no older than 21 at arrest, with a third mere adolescents of 18 or younger. Amidst their ranks, a poignant minority, 9%, grapple with documented mental health or cognitive challenges (Gudjonsson, 2021).
- **Details in Confessions**: Alarmingly, at least 43 (52%) of the 83 DNA exonerees who falsely confessed included non-public facts in their confessions, indicating possible contamination or coercion during interrogations (Gudjonsson, 2021).

The consequences of false confessions are devastating, leading to the incarceration of innocent individuals and the continued threat posed by the real perpetrators. Preventing false confessions requires comprehensive reform of interrogation practices, including the mandatory recording of interrogations, the presence of legal counsel, and the use of less coercive techniques.

18.4.1.4 The role of informants

Informants, commonly referred to as snitches, have been implicated in approximately 17% of wrongful convictions. Their involvement in criminal cases introduces significant challenges, as informants frequently have incentives to offer false testimony in exchange for reduced sentences or other rewards. These troubles need comprehensive changes and reforms aimed at regulating and examining carefully the use of informants in criminal proceedings.

To reduce the threat of erroneous convictions stemming from unreliable informant testimony, stricter tips must be applied for their selection, handling, and testimony. Transparency in the informant technique is crucial, making sure that all agreements and exchanges are documented and disclosed to protection groups and the court. Moreover, informant testimony should be independently corroborated on every occasion feasible to affirm its accuracy and reliability.

These measures are necessary steps to uphold the integrity of the criminal justice system by safeguarding against wrongful convictions based on informant testimony and promoting fairness.

18.5 The human cost of wrongful convictions

The toll of wrongful convictions on individuals is incalculable. Exonerees endure the loss of years, often decades, of their lives, facing profound personal, social, and psychological consequences. Families are fractured, careers ruined, and the stigma of a wrongful conviction can linger long after their innocence is proven. These injustices highlight the urgent need for systemic reforms to rectify the flaws within the criminal justice system.

Demographic data on DNA exonerees reveals stark racial disparities. African Americans constitute 60% of exonerees, indicating pervasive issues of racial bias and discrimination within law enforcement, prosecution, and sentencing practices. Caucasians represent 31%, Latinx individuals 8%, and Asian Americans just 1%. These disparities underscore the imperative for comprehensive reforms aimed at addressing systemic inequities and ensuring fair treatment for all individuals within the criminal justice system (Gross et al., 2022).

18.5.1 The work of the Innocence Project

The Innocence Project stands as a stalwart advocate, reshaping the contours of justice through systemic reforms and profound revelations. By championing the preservation of biological evidence and refining eyewitness identification protocols, they fortify the bedrock of truth in criminal proceedings. Yet, their impact transcends procedural amendments; it pierces the heart of darkness that shrouds wrongful convictions. Through DNA exonerations, 165 innocent souls were liberated, unveiling a chilling sequel: the real perpetrators, unmasked. Astonishingly, these revealed criminals perpetrated 154 additional heinous acts, including sexual assaults and murders. This stark reality underscores not just the power of DNA in redemption but its pivotal role in ensuring that justice wields a sword keen enough to pierce through deceit and deliver closure to the aggrieved (Hampikian et al., 2011).

18.5.2 DNA exoneration global perspective

18.5.2.1 United States

In the United States, DNA evidence plays a crucial role in the criminal justice system through initiatives like the FBI's Combined DNA Index System (CODIS). CODIS serves as a national repository for DNA profiles from convicted offenders, crime scene evidence, and missing persons, facilitating comparisons to aid in investigations. The USA possesses robust legal frameworks and technological infrastructure that support extensive use of DNA in criminal cases.

However, the reliance on DNA evidence is not without controversy. Privacy concerns arise over the collection and storage of genetic information, while the potential for laboratory errors underscores the need for stringent quality controls and oversight (*CODIS and NDIS Fact Sheet*, n.d.).

The Innocence Project has successfully secured 252 exonerations, encompassing cases involving both DNA and non-DNA evidence. Out of these, 203 individuals were exonerated specifically through DNA evidence. On average, the wrongfully convicted were 27 years old at the time of their conviction and 45 years old when exonerated. In total, Innocence Project clients have spent 3,985 years imprisoned for crimes they did not commit. These figures underscore the profound impact of the Innocence Project in rectifying wrongful convictions and emphasize the pivotal role of DNA evidence in achieving justice for the innocent (*Innocence Project*, n.d.-a).

Eyewitness misidentification, false confessions, and insufficient criminal defense, these problems have been highlighted by this work. Beyond exoneration, the employer advocates for coverage reforms aimed at improving the equity and accuracy and proficiency of the legal system. These reforms consist of advocating for the obligatory recording of interrogations to forestall coercion, enhancing requirements in forensic science to decrease errors, and securing compensation for those who were erroneously convicted (*Innocence Project*, n.d.-c).

18.5.2.2 United Kingdom

The United Kingdom additionally closely depends on DNA evidence. The National DNA Database (NDNAD) is one of the biggest in the world, containing hundreds of thousands of profiles. The UK has implemented rigorous prison requirements for the collection, storage, and use of DNA evidence, ensuring its admissibility in court. The retention of DNA profiles, even from individuals not convicted of a crime, has raised moral and privacy issues. The UK has a lot of organizations committed to addressing erroneous convictions, such as the Innocence Network UK (INUK) and Inside Justice(Wallace, 2006). These organizations work to become aware of potential miscarriages of justice, to evaluate cases and grant legal and investigative assistance to acquit or free up the innocent. The Criminal Cases Review Commission (CCRC) performs essential function in referring instances to the Court of Appeal ("About the CCRC," n.d.).

18.5.2.3 Canada

Canada makes use of DNA proof extensively, supported by means of the National DNA Data Bank (NDDB). Canadian regulation emphasizes the importance of acquiring DNA samples legally, frequently requiring judicial authorization. The country's criminal machine additionally places widespread emphasis on the potential for DNA proof to exonerate the wrongfully convicted, with countless fantastic cases of exoneration or freeing up of wrongfully convicted attributed to advances in DNA testing. Canada's Innocence Canada, previously known as the Association in Defence of the Wrongly Convicted (AIDWYC), has successfully acquitted several individuals. The corporation collaborates with legal professionals, investigators, and forensic experts to evaluate instances and recommend for judicial reforms or changes. Canada also has the Criminal Conviction Review Group (CCRG), which evaluation functions from individuals claiming erroneous convictions (Milot et al., 2013).

18.5.2.4 Australia

Australia's use of DNA evidence is regulated by way of each federal and kingdom law, with the National Criminal Investigation DNA Database (NCIDD) facilitating the alternation of DNA information throughout jurisdictions. The criminal framework ensures strict protocols for the collection and use of DNA evidence, reflecting a stability between fixing crimes and safeguarding individual rights. The Innocence Project in Australia operates in a similar fashion to its counterparts in the USA and UK, focusing on exonerating the erroneously convicted and merchandising legal reforms. Australian Innocence Projects frequently collaborate with universities, legal professionals, and forensic specialists to discover and look at possible instances of erroneously conviction (*DNA Database Systems*, n.d.; *Media Statement: National Criminal Investigation DNA Database | Australian Criminal Intelligence Commission*, n.d.).

18.5.2.5 South Africa

South Africa's journey with DNA evidence illuminates a transformative path in criminal justice. Enacted in 2013, the Criminal Law (Forensic Procedures) Amendment Act mandated a national DNA database, marking a pivotal leap in forensic capabilities. Amidst challenges like pervasive crime and resource constraints, DNA has emerged as a powerful ally, pivotal in resolving cases and securing justice. South Africa's Innocence Project South Africa (IPSA) focuses on exonerating the wrongfully convicted through DNA testing and other forensic methods. The country's unique social and economic challenges impact its criminal justice system, making the work of IPSA particularly crucial. The organization also advocates for policy changes to prevent future miscarriages of justice (*Innocence Project Starts in South Africa — DNA Project – South Africa*, n.d.).

The work of The Innocence Project and the revelations from DNA exonerations have catalyzed a broader movement for criminal justice reform. However, significant challenges remain to prevent future wrongful convictions and to ensure the fair and accurate operation of the justice system.

As DNA evidence continues to play an increasing number of distinguished functions in criminal investigations, it is critical to think about its future implications. While DNA has been pivotal in exonerating the erroneously acquitted or convicted, it additionally raises necessary worries involving privacy, accuracy, and fairness (Pandey & Tiwari, 2017).

The growth of DNA databases raises significant privacy concerns, including risks of unauthorized access and misuse of genetic information beyond criminal investigations (Ahmed, 2019). To mitigate these risks, DNA evidence should be limited to essential cases for securing convictions. Strong legal frameworks are needed to safeguard genetic privacy, clearly define access rights, and establish strict controls to prevent misuse. These measures ensure the ethical use of genetic data while balancing its role in justice with the need for privacy protection.

While DNA evidence is usually regarded as noticeably reliable, there have been situations of mishandling or misinterpretation resulting in erroneous convictions. Upholding the accuracy of DNA evidence needs stringent requirements in its collection, processing, and analysis. This necessitates complete coaching for forensic experts, rigorous protocols for evidence handling, and normal audits of forensic laboratories. Moreover, structures that have to be in an area must be in place for ongoing retesting and reevaluation of evidence as new applied sciences and methodologies emerge, making sure the adoption of the most particular and contemporary practices in forensic analysis.

While DNA evidence is profoundly attainable for justice, its extended use necessitates careful consideration of privacy protections, accuracy assurances, and the ongoing enhancement of forensic practices. Addressing these challenges is integral for constructing a criminal justice system that operates equitably and efficaciously in the years to come (Dedrickson, 2017).

The use of DNA evidence or proof in criminal justice raises large questions about equity and the potential for discrimination in collecting, analyzing, and interpreting. While DNA can both incriminate and acquit individuals, making sure equity requires strong mechanisms to forestall bias and ensure a detailed and thorough evaluation of all evidence.

To promote equity, unbiased oversight bodies should be installed to display DNA analysis approaches and make certain adherence to standardized protocols. It may also additionally be beneficial to mandate the involvement of more than one professional in reviewing DNA evidence earlier than any conclusions are drawn, thereby minimizing the danger of blunders or biases influencing outcomes. However, this improvement additionally raises profound moral worries involving the manipulation of genetic records. To tackle these moral concerns, rigorous policies must be enacted to govern the ethical use of gene modifying and different rising applied sciences in forensic contexts. These regulations should prioritize accountable use, inclusive of safeguards to forestall misuse or misconduct and guard the integrity of genetic evidence. This method would assist in making certain that development in science and technology makes positive contributions to justice while upholding moral requirements and preserving public faith in the legal and criminal justice system.

18.5.3 Common causes of wrongful convictions and their prevention

Erroneous convictions not only impact the lives of those who are wrongly imprisoned but additionally undermine public self-assurance in the criminal justice system and the legal system as a whole. Understanding the root motives of erroneous convictions is necessary for developing effective prevention strategies.

18.5.3.1 Eyewitness misidentification

Eyewitness misidentification stands as a principal contributor to erroneous convictions, despite its perceived reliability, due to numerous factors:

- **Stress and Anxiety:** High-stress situations, such as witnessing a crime, can distort memory and perception.
- **Weapon Focus Effect:** The presence of a weapon can divert attention from accurately identifying the perpetrator.

- **Suggestive Identification Procedures:** Improperly conducted police lineups and photo arrays can lead to erroneous identifications (*Eyewitness Misidentification*, n.d.; Marr et al., 2021).

Prevention Strategies:

- **Double-Blind Lineups:** Conduct lineups where the administering officer does not know the suspect's identity to avoid influencing the witness.
- **Clear Instructions:** Inform witnesses that the perpetrator may not be present in the lineup and that it is acceptable not to make an identification (McGough, n.d.).

18.5.3.2 False confessions

False confessions make contributions extensively to erroneous convictions, acting in over 25% of DNA exoneration cases and instances (Gudjonsson, 2021), due to a range of reasons:

- **Coercion:** Intensive interrogation techniques can induce individuals to confess falsely.
- **Mental Health Issues:** Those with mental illnesses are particularly susceptible to providing false confessions (*False Confessions*, n.d.).
- **Misunderstanding or Attention-seeking:** Some individuals may confess falsely due to misunderstanding the situation or seeking attention (Leo, 2009).

Prevention Strategies:

- **Proper Interrogation Techniques:** Train law enforcement officers in non-coercive methods of interrogation.
- **Recording Interrogations:** Recording all interrogations can establish a clear record and prevent coercion.
- **Informed Rights:** Ensure suspects are knowledgeable certainly of their rights, inclusive of the right to stay silent and to legal and criminal representation (Drizin & Leo, 2004).

18.5.3.3 Junk science

The use of unvalidated or improperly utilized forensic techniques, termed junk science, is another substantial cause of erroneous convictions:

- **Hair Analysis:** Traditionally used to match suspects to crime scenes, hair analysis has been discredited for its lack of reliability.
- **Bite Mark Analysis:** Similarly, bite mark analysis lacks scientific validity and has been shown to lead to erroneous conclusions (*FBI Hair Analysis Errors Led to Convictions New Study Finds - CBS News*, 2015; *FBI Testimony on Microscopic Hair Analysis Contained Errors in at Least 90 Percent of Cases in Ongoing Review — FBI*, n.d.; Laporte, n.d.; *Misapplication of Forensic Science*, n.d.).

Prevention Strategies:

- **Education and Training:** Provide judges and attorneys with ongoing education on current scientific research and methodologies to enable critical evaluation of forensic evidence.

- **Scientific Validation:** Admit only scientifically validated methods as evidence in court.

18.5.3.4 Prosecutorial misconduct

Prosecutorial misconduct, such as withholding exculpatory proof or coercing witnesses, poses a giant task of erroneous convictions:

- **Detection Challenges:** Misconduct can be hard to notice, however, has profound impacts on case results (*Official Misconduct*, n.d.).

Prevention Strategies:

- **Enhanced Oversight and Accountability:** Implement robust oversight mechanisms and hold prosecutors accountable for misconduct (*Official Misconduct*, n.d.; Spolin, 2022).
- **Ethics Training:** Provide ongoing ethics training for prosecutors, emphasizing the pursuit of justice over conviction rates.

18.5.3.5 Inadequate defense

Inadequate defense, frequently stemming from elements like overworked public defenders or insufficient resources (Poggio, 2023), appreciably contributes to erroneous convictions:

- **Public Defender Caseloads:** Overburdened public defenders may lack the time and resources to mount an effective defense (Pace et al., 2017).
- **Resource Constraints:** Defendants without financial means may lack access to essential resources such as expert witnesses or investigators (Andersson, 2022).
- **Quality of Legal Representation:** Some defendants may receive representation from inexperienced or unqualified attorneys (*Inadequate Defense*, n.d.).

Prevention Strategies:

- **Increased Funding:** Allocate additional resources to public defender offices to reduce caseloads and enhance defense quality ("*Making Justice Equal*," 2016).
- **Training and Support:** Provide continuous training and support for public defenders to improve their effectiveness (*Adequate Defense Counsel Can Prevent Wrongful Convictions*, n.d.).
- **Equitable Access to Resources:** Ensure all defendants, regardless of financial status, have access to necessary resources for a robust defense (Andersson, 2022).

18.6 Conclusion and recommendation

Wrongful convictions or acquittals expose profound systemic screw ups inside the criminal justice and legal system, necessitating complete reforms and changes aimed at stopping future miscarriages of justice. Key areas requiring interest encompass eyewitness identification procedures, forensic science practices, legal representation, prosecutorial integrity, adoption of technology, and the broader social influence of erroneous convictions.

18.6.1 Improved eyewitness identification procedures

Eyewitness misidentification is a main purpose of erroneous convictions, influenced by means of factors such as suggestive identification strategies and the fallibility of human memory (Marr et al., 2021). Reforms should center of attention on:

- **Double-Blind Lineups:** Implementing tactics where neither the administrator nor the witness is aware of the suspect's identification to stop inadvertent influence (Wells & Bradfield, 1998; Wells & Quigley-McBride, 2016).
- **Clear Instructions:** Ensuring witnesses apprehend that the perpetrator may also no longer be present in the lineup, lowering strain to make an identification (Clark, 2005).

18.6.2 Enhanced forensic science practices

The reliability of forensic proof has been puzzled in erroneous convictions due to flawed or overstated forensic testimony. Reforms need to prioritize:

- **Standardization and Validation:** Establishing rigorous requirements and validation procedures for forensic strategies to make sure accuracy.
- **Independent Oversight:** Implementing unbiased oversight of forensic laboratories to hold great and reliability of analyses (Council, 2009; *Ten Years Later*, n.d.).

18.6.3 Ensuring adequate legal representation

Inadequate criminal defense(Poggio, 2023), especially for indigent defendants, contributes substantially to wrongful convictions. Reforms have to include:

- **Manageable Caseloads:** Ensuring public defenders have manageable caseloads to provide high quality representation (Zalman, 2018).
- **Access to Resources:** Providing assets such as specialist witnesses and investigators necessary for a sturdy defense (Furst, 2019).
- **Accountability:** Implementing mechanisms to preserve protection attorneys responsible for providing able representation (*Defense Function*, n.d.; *Strickland v. Washington*, n.d.).

18.6.4 Prosecutorial accountability and integrity

Prosecutorial behavior profoundly affects case results and outcomes. Reforms must promote:

- **Mandatory Disclosure:** Requiring prosecutors to divulge exculpatory evidence to prevent wrongful convictions (*Justice Manual | 9-5.000 - Issues Related To Discovery, Trials, And Other Proceedings | United States Department of Justice*, 2015).
- **Transparency:** Adopting open-file discovery insurance policies and organizing the independent oversight to tackle prosecutorial misconduct (*NACDL - Discovery Reform*, n.d.).

18.6.5 Adoption of technology and data analytics

Modern science and technology and fact analytics can pick out patterns resulting in erroneous convictions and put in force preventive measures. Reforms need to include:

- **Advanced Data Analysis:** Utilizing statistics to realize anomalies in conviction costs and judicial practices (Jackson, 2023).
- **Recording Interrogations:** Implementing science to report interrogations to prevent coercion and uphold suspects' rights (*Recording Interrogations Can Prevent Wrongful Convictions*, n.d.).

18.6.6 Long-term consequences and broader social impact

Wrongful convictions have far-reaching penalties past people and families, affecting neighborhood protection and perpetuating social inequalities. Reforms need to address:

- **Resource Allocation:** Ensuring criminal justice assets are efficiently and effectively alloted to enhance public safety (*To Prevent Wrongful Convictions, Invest in Public Defense Solutions – The Hill*, n.d.).
- **Equity Issues:** Addressing disparities in erroneous convictions affecting marginalized groups through intersectional reforms (*How Racial Bias Contributes to Wrongful Conviction*, n.d.).

Addressing erroneous convictions requires legislative reforms, policy changes, and cultural shifts. Education, training, and public advocacy are essential to foster systemic reforms, ensuring fairness, preventing future wrongful convictions, and restoring public trust.

Bibliography

About the CCRC. (n.d.). Criminal Cases Review Commission. Retrieved January 16, 2025, from https://ccrc.gov.uk/what-we-do/

Adequate Defense Counsel Can Prevent Wrongful Convictions. (n.d.). Innocence Project. Retrieved January 17, 2025, from https://innocenceproject.org/adequate-defense-counsel-can-prevent-wrongful-convictions/

Ahmed, A. (2019, January 14). Ethical concerns of DNA databases used for crime control—Bill of health. Retrieved February 17, 2025 from https://blog.petrieflom.law.harvard.edu/2019/01/14/ethical-concerns-of-dna-databases-used-for-crime-control/

Andersson, E. (2022, March 8). If you care about freedom, you should be asking why we don't fund our public defender systems. American Civil Liberties Union. Retrieved February 17, 2025 from https://www.aclu.org/news/criminal-law-reform/if-you-care-about-freedom-you-should-be-asking-why-we-dont-fund-our-public-defender-systems

Bonventre, C. L. (2021). Wrongful convictions and forensic science. *WIREs Forensic Science*, 3(4), e1406. https://doi.org/10.1002/wfs2.1406

Clark, S. E. (2005). A re-examination of the effects of biased lineup instructions in eyewitness identification. *Law and Human Behavior*, 29(4), 395–424. https://doi.org/10.1007/s10979-005-5690-7

CODIS and NDIS Fact Sheet. (n.d.). Federal Bureau of Investigation. Retrieved January 16, 2025, from https://www.fbi.gov/how-we-can-help-you/dna-fingerprint-act-of-2005-expungement-policy/codis-and-ndis-fact-sheet

Council, N. R. (2009). Strengthening forensic science in the United States: A path forward. CrimRxiv. https://doi.org/10.21428/cb6ab371.b2d683d2

Dash, H. R., Vajpayee, K., & Agarwal, R. (2020). A glimpse of famous cases in history solved by DNA typing. In H. R. Dash, P. Shrivastava, J. A. Lorente (Eds.) *Handbook of DNA Profiling*, 1–17. Springer

Dedrickson, K. (2017). Universal DNA databases: A way to improve privacy? *Journal of Law and the Biosciences*, 4(3), 637–647. https://doi.org/10.1093/jlb/lsx041

Defense Function. (n.d.). American Bar Association. Retrieved January 18, 2025, from https://www.americanbar.org/groups/criminal_justice/resources/standards/defense-function/

DNA Database Systems. (n.d.). ALRC. Retrieved January 16, 2025, from https://www.alrc.gov.au/publication/essentially-yours-the-protection-of-human-genetic-information-in-australia-alrc-report-96/43-dna-database-systems/dna-database-systems/

DNA Evidence: A Beacon of Exoneration. (n.d.). FasterCapital. Retrieved January 15, 2025, from https://fastercapital.com/content/DNA-Evidence–A-Beacon-of-Exoneration.html

DNA Exonerations in the United States (1989–2020). (n.d.). Innocence Project. Retrieved January 15, 2025, from https://innocenceproject.org/dna-exonerations-in-the-united-states/

Drizin, S. A., & Leo, R. A. (2004). The problem of false confessions in the post-DNA world. *North Carolina Law Review, 82*(3), 891–1007.

Educate. (n.d.). Northern California Innocence Project. Retrieved January 18, 2025, from https://ncip.org/educate/

Eyewitness Misidentification. (n.d.). Innocence Project. Retrieved January 16, 2025, from https://innocenceproject.org/eyewitness-misidentification/

False Confessions. (n.d.). Innocence Project. Retrieved January 16, 2025, from https://innocenceproject.org/false-confessions/

FBI Hair Analysis Errors Led to Convictions New Study Finds—CBS News. (2015, April 20). Retrieved February 17, 2025 from https://www.cbsnews.com/news/fbi-hair-analysis-errors-led-to-convictions-new-report-finds/

FBI Testimony on Microscopic Hair Analysis Contained Errors in at least 90 Percent of Cases in Ongoing Review—FBI. (n.d.). Retrieved January 17, 2025, from https://www.fbi.gov/news/press-releases/fbi-testimony-on-microscopic-hair-analysis-contained-errors-in-at-least-90-percent-of-cases-in-ongoing-review

Furman, H. P. (2003). Wrongful convictions and the accuracy of the criminal justice system. *The Colorado Lawyer, 32*(9), 11–22.

Furst, B. (2019). A fair fight achieving indigent defense resource parity. Brennan Center for Justice at New York University School of Law.

Gross, S. R., Possley, M., Otterbourg, K., Stephens, K., Paredes, J., & O'Brien, B. (2022). Race and wrongful convictions in the United States 2022. SSRN Electronic Journal. https://doi.org/10.2139/ssrn.4245863

Gudjonsson, G. H. (2021). The science-based pathways to understanding false confessions and wrongful convictions. *Frontiers in Psychology*, 12, 633936. https://doi.org/10.3389/fpsyg.2021.633936

Hampikian, G., West, E., & Akselrod, O. (2011). The genetics of innocence: Analysis of 194 U.S. DNA exonerations. *Annual Review of Genomics and Human Genetics*, 12(1), 97–120. https://doi.org/10.1146/annurev-genom-082509-141715

How Racial Bias Contributes to Wrongful Conviction. (n.d.). Innocence Project. Retrieved January 18, 2025, from https://innocenceproject.org/how-racial-bias-contributes-to-wrongful-conviction/

Inadequate Defense. (n.d.). Innocence Project. Retrieved January 17, 2025, from https://innocenceproject.org/inadequate-defense/

Innocence Project Starts in South Africa—DNA Project – South Africa. (n.d.). Retrieved January 16, 2025, from https://dnaproject.co.za/innocence-project-starts-in-south-africa/

Innocence Project: By the Numbers. (n.d.-a). Innocence Project. Retrieved January 16, 2025, from https://innocenceproject.org/exonerations-data/

Innocence Project: Restoring Freedom. (n.d.-b). Innocence Project. Retrieved January 15, 2025, from https://innocenceproject.org/restoring-freedom/

Innocence Project: The Issues. (n.d.-c). Innocence Project. Retrieved January 16, 2025, from https://innocenceproject.org/the-issues/

Jack, M. (2024, January 28). Golden state killer's arrest: The DNA breakthrough that cracked the case wide open! Medium. Retrieved February 17, 2025 from https://medium.com/@mjack/golden-state-killers-arrest-the-dna-breakthrough-that-cracked-the-case-wide-open-40628ed44d9b

Jackson, M. (2023, April 17). McCourt school students use data analytics to help overturn wrongful convictions. McCourt School of Public Policy. Retrieved February 17, 2025 from https://mccourt.georgetown.edu/news/data-analytics-overturn-wrongful-convictions/

Justice Manual | 9-5.000—Issues Related to Discovery, Trials, and other Proceedings | United States Department of Justice. (2015, February 19). Retrieved February 17, 2025 from https://www.justice.gov/jm/jm-9-5000-issues-related-trials-and-other-court-proceedings

Laporte, G. (n.d.). Wrongful convictions and DNA exonerations: Understanding the role of forensic science. *National Institute of Justice Journal, 279*, 1–16

Leo, R. A. (2009). False confessions: Causes, consequences, and implications. *The Journal of the American Academy of Psychiatry and the Law*, 37(3), 332–343.

Making Justice Equal. (2016, December 8). Center for American Progress. Retrieved February 17, 2025 from https://www.americanprogress.org/article/making-justice-equal/

Marr, C., Otgaar, H., Sauerland, M., Quaedflieg, C. W. E. M., & Hope, L. (2021). The effects of stress on eyewitness memory: A survey of memory experts and laypeople. *Memory & Cognition*, 49(3), 401–421. https://doi.org/10.3758/s13421-020-01115-4
McGough, M. (n.d.). Archived | To err is human: Using science to reduce mistaken eyewitness identifications through police lineups | National Institute of Justice. Retrieved January 16, 2025, from https://nij.ojp.gov/topics/articles/err-human-using-science-reduce-mistaken-eyewitness-identifications-through-police
Media Statement: National Criminal Investigation DNA Database | Australian Criminal Intelligence Commission. (n.d.). Retrieved January 16, 2025, from https://www.acic.gov.au/media-centre/media-releases-and-statements/media-statement-national-criminal-investigation-dna-database?utm_source=chatgpt.com
Methods of Reform | Innocence Project of Florida. (n.d.). IPF. Retrieved January 18, 2025, from https://www.floridainnocence.org/methods-of-reform
Milot, E., Lecomte, M. M. J., Germain, H., & Crispino, F. (2013). The National DNA Data Bank of Canada: A Quebecer perspective. *Frontiers in Genetics*, 4. https://doi.org/10.3389/fgene.2013.00249
Misapplication of Forensic Science. (n.d.). Innocence Project. Retrieved January 15, 2025, from https://innocenceproject.org/misapplication-of-forensic-science/
Morris, K. L. (2023, November 28). How wrongful convictions destroy lives and shatter trust in the judicial system. Medium.
NACDL - Discovery Reform. (n.d.). NACDL - National Association of Criminal Defense Lawyers. Retrieved January 18, 2025, from https://www.nacdl.org/Landing/Discovery
Official Misconduct. (n.d.). Death Penalty Information Center. Retrieved January 17, 2025, from https://deathpenaltyinfo.org/policy-issues/policy/prosecutorial-accountability
Pace, N. M., Brink, M. N., Lee, C. G., & Han, L. (2017). *A Critical Review of the Forensic DNA Analysis Process in the U.S.* National Academy Press. https://doi.org/10.17226/24841
Pandey, H., & Tiwari, A. (2017). Evidential value of DNA: A judicial approach. *Bharati Law Review*, *6*(1), 12–35.
Poggio (2023) https://innocenceproject.org/inadequate-defense/
Preventing Wrongful Convictions—Why Reform Matters. (n.d.). National Conference of State Legislatures. Retrieved January 16, 2025, from https://www.ncsl.org/research/civil-and-criminal-justice/wrongful-convictions.aspx
Quasi-Military Solutions for Forensic Analysis. (2017, May 24). Retrieved February 17, 2025 from https://journal.nationalgeographic.com/quasi-military-solutions-for-forensic-analysis
Recording Interrogations Can Prevent Wrongful Convictions (n.d.) https://innocenceproject.org/news/recording-interrogations-can-prevent-wrongful-convictions/; https://thehill.com/changing-america/opinion/586305-to-prevent-wrongful-convictions-invest-in-public-defense-solutions/
Robertson, J. E., & Hervey, T. K. (2011). Law and the ethics of forensic science. International Review of Law, 3(3), 55–72.
Role of Forensic Science in Criminal Justice Reform: A Conversation. (2023, January 14). National Institute of Justice.
Spolin (2022) https://spolinlaw.com/blog/2022/11/21/what-is-prosecutorial-misconduct/
Srivastava, A., Harshey, A., Das, T., & Nigam, K. (2022). Impact of DNA evidence in criminal justice system: Indian legislative perspectives. *Egyptian Journal of Forensic Sciences*, 12(1). https://doi.org/10.1186/s41935-022-00309-y
Strickland v. Washington (n.d.) https://supreme.justia.com/cases/federal/us/466/668/
Ten Years Later (n.d) https://innocenceproject.org/news/lasting-impact-of-2009-nas-report/
The Innocence Project: The Impact of Wrongful Convictions. (2019, October 25). Retrieved January 16, 2025, from https://innocenceproject.org/the-impact-of-wrongful-convictions/
The Innocence Project and Wrongful Convictions. (2022, October 9). The Law Works. Retrieved February 17, 2025 from https://www.thelawworks.com
The Innocence Project: Facts. (2023, December 2). Retrieved February 17, 2025 from https://www.innocenceproject.org/facts-about-the-innocence-project/
The Role of Forensic Science in Wrongful Convictions. (n.d.). American Academy of Forensic Sciences. Retrieved January 17, 2025, from https://www.aafs.org
The United States Criminal Justice System. (2024, January 18). Bureau of Justice Statistics. Retrieved February 17, 2025 from https://www.bjs.gov
Vajpayee, K., Dash, H. R., Parekh, P. B., & Shukla, R. K. (2023). PCR inhibitors and facilitators-their role in forensic DNA analysis. *Forensic Science International*, 349, 111773.
Vajpayee, K., Sagar, D., & Dash, H. R. (2020). Forensic DNA typing: Inception, methodology, and technical advancements. In P. Shrivastava, H. Ranjan Dash, J. A. Lorente, & J. Imam (Eds.) *Forensic DNA Typing: Principles, Applications and Advancements*, 3–26. Springer

Wallace H. (2006). The UK National DNA Database. Balancing crime detection, human rights and privacy. *EMBO reports*, *7*(S1), S26–S30. https://doi.org/10.1038/sj.embor.7400727

Wells, G. L., & Bradfield, A. L. (1998). "Good, you identified the suspect": Feedback to eyewitnesses distorts their reports of the witnessing experience. *Journal of Applied Psychology*, *83*(3), 360–376. https://doi.org/10.1037/0021-9010.83.3.360

Wells, G. L., & Quigley-McBride, A. (2016). Applying eyewitness identification research to the legal system: A glance at where we have been and where we could go. *Journal of Applied Research in Memory and Cognition*, *5*(3), 290–294. https://doi.org/10.1016/j.jarmac.2016.07.007

Wickenheiser R. A. (2021). The value of forensic DNA leads in preventing crime and eliminating the innocent. *Forensic Science International: Synergy*, *3*, 100201. https://doi.org/10.1016/j.fsisyn.2021.100201

Wixted, J. T., Mickes, L., & Fisher, R. P. (2018). In the DNA exoneration cases, eyewitness memory was not the problem: A reply to Berkowitz and Frenda (2018) and Wade, Nash, and Lindsay (2018). *Perspectives on Psychological Science*, *13*(3), 343–345. https://doi.org/10.1177/1745691618769859

Zalman, M. (2018). Wrongful convictions: Comparative perspectives. In A. J. Treviño (Ed.), *The Cambridge Handbook of Social Problems* (pp. 449–472). Cambridge University Press.

Index

Note: Locators in *italics* represent figures and **bold** indicate tables in the text.

AAFS, *see* American Academy of Forensic Sciences
ABO blood group system, 43–44, 61, 303
Acer macrophyllum, 212
Active DNA demethylation, 256
Adenosine triphosphate, 83
Admixture models, 233
aDNA, *see* Ancient DNA
Advanced sequencing technologies, 166
Affinity chromatography, 300
AFLP, *see* Amplified fragment length polymorphism
AFSN, *see* Asian Forensic Sciences Network
Age estimation, 261–262
Age-related CpG markers (AR-CpG), 262
AHRR gene, 266
AI, *see* Artificial intelligence
AIDWYC, *see* Association in Defence of the Wrongly Convicted
AIMs, *see* Ancestry-informative markers
AISNPs, *see* Ancestry-informative SNPs
Ajnala massacre (mass murder case 165 years ago), 199–201
ALAD gene, 264
Alkaline lysis method, 49
Allele, 62
Allelic dropout, 113, 118, 131
Amelogenin gene, 123
American Academy of Forensic Sciences (AAFS), 153
Amplified fragment length polymorphism (AFLP), 68–69, *68*
AmpliType PM, 125
Anal samples, 24
Ancestry-informative markers (AIMs), 144, 196
Ancestry-informative SNPs (AISNPs), 139, 230
Ancient DNA (aDNA), 81
Anthrax attacks (USA) (2001), 185
Anthropology, forensic, biotechnology applications in, 12–13
ANZFSS, *see* Australian and New Zealand Forensic Science Society
Apolipoprotein A1 (ApoA1), 302
Applications of forensic biotechnology
  in body fluid identification, 11–12
  in environmental investigations, 13–14
  in forensic anthropology, 12–13
  in microbial forensics, 10–11
  in toxicology, 8–9
Applied biosystems, 84
  Ion Torrent, 84
  SOLiD, 84
Applied Biosystems RapidHIT ID System, 247–249
Aptamer-based assays, 7–8
Ara macao, 212
Archaeology, forensic, 195
  Ajnala massacre (mass murder case 165 years ago), 199–201
  biogeographical ancestry in forensic archaeology of India, 196–198, *197*
  forensic archaeology of a 400-year ago murder, 198–199
AR-CpG, *see* Age-related CpG markers
Argonaute protein complex, 265
ArmedXpert, 115
  features, 116
  limitation, 116
Arrestee index, 142
Artificial intelligence (AI), 187, 228, 233, 236, 246
  and 3D printing, 7
Asian Forensic Sciences Network (AFSN), 153
Assays, aptamer-based, 7–8
Association in Defence of the Wrongly Convicted (AIDWYC), 325
aSTRs, *see* Autosomal STRs
ATR-FTIR spectroscopy, 210
Australia, DNA exoneration in, 325
Australian and New Zealand Forensic Science Society (ANZFSS), 153
Autosomal STRs (aSTRs), 72, 138

Bacillus anthracis, 175, 177, 178, 185
Barcode of Life Data Systems (BOLD), 163, 167, 215
Barcodes, 10, 164
Barcoding gap, 161
Baton Rouge Serial Killer (2003), 231
Bayesian framework, 108
Bayesian inference, 118
BGA analysis, *see* Biogeographical ancestry analysis
Biocolonialism, 149

Biogeographical ancestry (BGA) analysis, **143**, 144, 230
in forensic archaeology of India, 196–198
Bioinformatics, 4, 5, 180
Biological fluids, 23
Biological samples
packaging, 26–27
preservation and storage, 29–30, **30**
transportation, 27–29
Biological samples, collection of, 21
biological fluids, 23
cadaveric remains, 24
bodies in good condition, 25
carbonized bodies, 25
putrefied bodies, 25
crime scene evidence, 21
non-transportable objects, biological stains on, 22
transportable objects, biological stains on, 22
reference samples, 25
blood samples, 25
hair samples, 26
saliva samples, 26
transfused subjects, 26
victim's body, samples from, 23
biological stains, 23
clothes, 24
evidence from nails, 23
hairs and fibers, 24
saliva from bite marks, 23
sexual assault evidence, 24
Biological stains, 23
on non-transportable objects
non-porous surfaces, 22
porous surfaces, 22
on transportable objects, 22
Bioremediation, 13
Biosensors, 9, 14
and immunochromatographic methods, 53–54
Biotechnology and forensic science, blending of, 1
Bioterrorism and biocrime investigations, 176
Bisulfite restriction analysis, combined, 270
Bite mark analysis, 322, 327
Bite marks, saliva from, 23
Blood, 35, 36, 49
tests for
blood typing, 43–44
confirmatory tests, 41–43, **41**
presumptive tests, **40–41**
source identification, **40**
Blood samples, 25, 38
Bodies
carbonized, 25
in good condition, 25
putrefied, 25
Body fluid, *34*
advanced identification methods for, 52
biosensors and immunochromatographic methods, 53–54
epigenetic markers, 52–53
RNA-based methods, 52
spectroscopic techniques, 53
collection and preservation of, 37
blood samples, 38
saliva samples, 38
semen samples, 39
urine samples, 39
DNA testing in identification of, 47
DNA isolation and extraction, 48
types of DNA extraction methods, 48–52
forensic protein markers of, 303–305
fundamental identification methods for, 39, *39*
blood, tests for, 40–44
saliva, tests for, 44
semen, tests for, 45
urine, tests for, 45–46
identification, 260–261
forensic biotechnology in, 11–12
using biotechnology, 12
latest developments in identification of, 54–57, *55*
types and importance of, 36–37
Body surface, 24
BOLD, *see* Barcode of Life Data Systems
Bones and teeth, 25

Cadaveric remains, 24
bodies in good condition, 25
carbonized bodies, 25
putrefied bodies, 25
Canada
DNA exoneration in, 324–325
Innocence Canada, 325
Capillary array DNA sequencing, 287
Capillary electrophoresis (CE), 87, 123, 139, 245, 249, 288
Carbonized bodies, 25
CBOL, *see* Consortium for the Barcode of Life
CCD camera, *see* Charged-coupled device camera
CCRC, *see* Criminal Cases Review Commission
CCRG, *see* Criminal Conviction Review Group
CE, *see* Capillary electrophoresis
Cell lysis, 51
Cervix, 24
Chain of custody, 129
electronic recording of, 32
and legal considerations, 181–182, 182, 183
Challenges and future directions, forensic biotechnology, 14–15
Chaotropic salts, 65–66
Charged-coupled device (CCD) camera, 123
Chelex extraction method, 49, 64, *64*
Chemical preservation, 30
Chemometrics, 54
Chisholm Cold Case (2013), 231
Chromatography method, 42
Circadian clock genes, 265
circRNA, *see* Circular RNA
Circular RNA (circRNA), 52

Class characteristics, 104
Clostridium botulinum, 167
Clothes, 24
Cloud computing, 94
Clustered Regularly Interspaced Short Palindromic Repeats (CRISPR) technology, 6–7, 77
CMOS sensor array chip, *see* Complementary metal-oxide semiconductor sensor array chip
COBRA, *see* Combined Bisulfite Restriction Analysis
CODIS, *see* Combined DNA Index System
COI gene, *see* Cytochrome c oxidase subunit I gene
Combined Bisulfite Restriction Analysis (COBRA), 261, 270
Combined DNA Index System (CODIS), 2, 47, 89, 142, 247, 319, 324
Combined probability of inclusion (CPI), 109
Complementary metal-oxide semiconductor (CMOS) sensor array chip, 84
Complete Genomics sequencing, 83
Complexity paradox, 119
Confirmatory tests
  for blood, 41–43, **41**
  for saliva, **44**
  for semen, **45**
  for urine, 46, **46**
Connectomics, 298
Consortium for the Barcode of Life (CBOL), 162–163
Contamination issues, 30–32
Conventional body fluid identification techniques, 11
CpG dinucleotides, 256, 271
CpG islands (CGIs), 271, 272
CPI, *see* Combined probability of inclusion
Crime investigations, role of DNA methylation and miRNA in, 259
  age estimation, 261–262
  authentication of forensic DNA samples, 262–263
  body fluid identification, 260–261
  cause of death analysis, 264–265
  discrimination of monozygotic twins, 263
  illegal drug abuse identification, 266
  lifestyle identification, 266–267
  maternal or paternal allele source identification, 264
  post-mortem interval (PMI) estimation, 265–266
Crime scene evidence, 21
  biological stains on non-transportable objects
    non-porous surfaces, 22
    porous surfaces, 22
  biological stains on transportable objects, 22
Crime scene investigation, 168
  and evidence collection, 206
    poached wild animals, 206–207, **207**
    seized wildlife products, 207–208
Crime scene isolation and evidence collection, 20
  scene documentation before evidence collection, 20–21
Criminal Cases Review Commission (CCRC), 324
Criminal Conviction Review Group (CCRG), 325
Criminal investigations, potential applications of RAPID DNA typing in, 253–254
Criminal justice system, wrongful convictions on, 320
  Innocence Project and DNA exonerations, 320
    eyewitness misidentification, 321–322
    false confessions, 322
    forensic science, misapplication of, 322
    informants, role of, 323
CRISPR technology, *see* Clustered Regularly Interspaced Short Palindromic Repeats technology
Cross-contamination, 22, 24, 26, 31
Cutting method, 37
Cybergenetics, 111
Cybergenetics TrueAllele Casework system, 112
Cytochrome *b* (Cyt *b*) gene, 211
Cytochrome c oxidase subunit I (COI) gene, 162, 163, 164, 211
Cytosine, 268

Data access and protection, 147
Data-independent acquisition mode, 309
Data privacy, 147
Daubert Standard, 128, 306
Daubert v. Merrell Dow Pharmaceuticals, 128
DCFDA dye, *see* 2′7′-Dichlorofluorescein di-acetate dye
ddNTPs, *see* Di-deoxynucleotide triphosphates
Death analysis, cause of, 264–265
Denaturation, 2
Deoxynucleotide triphosphates (dNTPs), 82
DGCR8, 258, *258*
2′7′-Dichlorofluorescein di-acetate (DCFDA) dye, 57
Di-deoxynucleotide triphosphates (ddNTPs), 82
Differentially methylated regions (DMRs), 54, 257
Direct matching and familial DNA searching, 147–149
Direct-to-consumer (DTC) genetic testing, 142
Disaster victim identification (DVI), 73
DMRs, *see* Differentially methylated regions
DNA (deoxyribonucleic acid), 287
DNA, ethics, and legal experts, 152–153
DNA adsorption to silica, 65
DNA amplification, 2
DNA analysis, forensic, 20
  biological samples
    packaging, 26–27
    preservation and storage, 29–30, **30**
    transportation, 27–29
  biological samples, collection of, 21
    biological fluids, 23
    cadaveric remains, 24–25
    crime scene evidence, 21–22
    reference samples, 25–26
    victim's body, samples from, 23–24
  chain of custody, 32
  contamination issues, 30–32
  crime scene isolation and evidence collection, 20

scene documentation before evidence collection, 20–21
DNA barcoding, 10, 25, 161, 163, *165*
challenges and limitations, 169
concept of, 164
definition and scope, 161
ethical and legal considerations, 168–169
forensic investigations, applications in, 167
crime scene investigation, 168
wildlife forensics, 168
future directions, 169–170
historical background, 161–163, *162*
real-world applications, 170
techniques, 164
advanced sequencing technologies, 166
reference databases, 167
standardized DNA barcode regions, 164–166
DNA contamination, 125–126
DNA databases, 2, 15, 212
DNA degradation, 124–125
DNA elimination database, 31
DNA evidence, 316
and eyewitness testimony, 318
DNA databases, 319
reliability of DNA evidence, 318–319
wrongful convictions, 319–320
and its impact on criminal justice system, 317
limitations of, 318
post-conviction DNA testing, importance of, 317–318
DNA exoneration global perspective
Australia, 325
Canada, 324–325
South Africa, 325–326
United Kingdom, 324
United States, 324
DNA extraction, 4, 63
Chelex extraction method, 64, *64*
DNA isolation with organic (phenol-chloroform) solvents, 63, *63*
FTA paper, 64–65, *65*
solid-phase extraction method, 65–66, **65**, *66*
DNA fingerprinting, 123, 136
amplified fragment length polymorphism (AFLP), 68–69, *68*
discovery of, 137–138
restriction Fragment Length Polymorphism (RFLP), 69, *70*
Short Tandem Repeat (STR), 70–72
Variable Number Tandem Repeat (VNTR), 69–70
DNA hybridization, 161
DNA immunoprecipitation (DIP), *see* Methylated DNA immunoprecipitation
DNA legislation, 149–150
DNA markers, 138
short tandem repeat (STR), 138–139
single-nucleotide polymorphism (SNP), 139
DNA methylation clock, 257
DNA methyltransferases, 256, 263
DNAMIX, 117–118
DNAmixtures, 118
DNA polymerase, 64
DNA precipitation, 51
DNA profiling, 13, 15, 61, 227
advantages of, 74–75
application of emerging technologies
CRISPR-based technologies, 77
microfluidic-based DNA analysis, 76–77
next-generation sequencing (NGS), 75
single-molecule sequencing, 76
disadvantages of, 75
forensic DNA sampling and collection, 62
mitochondrial DNA (mt-DNA) analysis in, 73
principle of, 62
single-nucleotide polymorphisms (SNPs), 74, **75**
DNA quantitation, 66
end-point PCR, 67
fluorescent interchelating dye assay, 67
real-time PCR, 67–68
slot blot assay, 66–67
yield" gel method, 67
DNA reference database, 215–216, *215*
DNA samples, mixed, 127
DNA sequence variation, 139
DNA sequencing, 94, 178
DNA technology, forensic, 1, 137
evolution in the use of, 142, **143**
kinship-based database searches, 144–146, **145**
predictive applications of DNA testing, 144
genetic databases, types of, 140
forensic databases, 142
population reference databases, 140–142, *141*
public databases, 142
DNA testing; *see also* DNA fingerprinting
in body fluid identification, 47
DNA isolation and extraction, 48
types of DNA extraction methods, 48–52
predictive applications of, 144
types of, 139–140
DNA typing, *see* DNA profiling
DNA·VIEW software, 116
limitations, 117
dNTPs, *see* Deoxynucleotide triphosphates
Double-blind lineups, 327, 329
Double Swabbing method, 37
DQA1 systems, 125
Drosha enzyme, 258
Dry samples, 27
Dry storage, 29
DTC genetic testing, *see* Direct-to-consumer genetic testing
DVI, *see* Disaster victim identification

Edman's degradation method, 299
eDNA, *see* Environmental DNA
Electropherogram (EPG), 3, 113, 123
Electrophoresis, 42, 83

Electrospray ionization (ESI), 210, 288–289, 291
Electrospray ionization mass spectrometry (ESI-MS), 291
Electrospray ionization-time-of-flight mass spectrometry (ESI-TOF-MS), 291
Elephant Preservation Act (1879), 217
ELISA, *see* Enzyme-linked immunosorbent assay
Emerging technologies in forensic biotechnology, 6
  aptamer-based assays, 7–8
  artificial intelligence and 3D printing, 7
  CRISPR (Clustered Regularly Interspaced Short Palindromic Repeats) technology, 6–7
em-PCR, *see* Emulsion PCR
Emulsion PCR (em-PCR), 83
End-point PCR, 67
ENFSI, *see* European Network of Forensic Science Institutes
Enhanced forensic science practices, 329
Environmental and agricultural monitoring, 177
Environmental DNA (eDNA), 5, 13–14, 166, 171, 180, 205, 220
Environmental impacts investigations, 181
Environmental investigations, forensic biotechnology in, 13–14
Enzyme-based assays, 303
Enzyme-linked immunosorbent assay (ELISA), 42
EPG, *see* Electropherogram
EpiDirect® methylation assay, 270
Epigenetic factors, role of, in forensic science, *see* MicroRNA analysis, DNA methylation and
Epigenetic markers, 52–53
Epigenetics, integration of, 236
Epigenome-wide association study (EWAS), 267
Epigenomics, 298
EpiPrimer®, 270
Equitable access to forensic DNA technology, 152
ESI, *see* Electrospray ionization
ESI-MS, *see* Electrospray ionization mass spectrometry
ESI-TOF-MS, *see* Electrospray ionization-time-of-flight mass spectrometry
ESS, *see* European Standard Set; Expert System Software
Ethanol, 30
Ethical and normative considerations of forensic DNA technology, 146
  data access and protection, 147
  data privacy, 147
  forensic database operations
    direct matching and familial DNA searching, 147–149
    DNA legislation, 149–150
  genetic data, potential misuse of, 150
    secondary use of non-forensic genetic data, 150–151
    technological determinism and misuse in legal proceedings, 151
  informed consent, 146–147
  personal autonomy, 146
  recommendations, 152
    DNA, ethics, and legal experts, 152–153
    equitable access to forensic DNA technology, 152
    international forensic ethics guidelines, 153–154
    stakeholder and public education, 154
EuroForMix, 113, 118
European Network of Forensic Science Institutes (ENFSI), 21
European Standard Set (ESS), 148
EVCs, *see* Externally visible characteristics
Evidence collection, 31
  scene documentation before, 20–21
EWAS, *see* Epigenome-wide association study
Exclusion probability, 117
Expert System Software (ESS), 247
Exportin-5/Ran-GTP complex, 258
Externally visible characteristics (EVCs), 144, 226
Eyewitness identification procedures, improved, 329
Eyewitness misidentification, 321–322, 326–327
Eyewitness testimony, DNA evidence and, 318
  DNA databases, 319
  reliability of DNA evidence, 318–319
  wrongful convictions, 319–320

F2RL3 gene, 266
Facial reconstruction, 13
Faith Hedgepeth Murder Case (2012), 232
False confessions, 322, 327
Familial reference index, 142
Familial searching (FS), 129–130, **143**
FBI, *see* Federal Bureau of Investigation
FDP, *see* Forensic DNA phenotyping
Federal Bureau of Investigation (FBI), 247
FGPI, *see* Forensic Genetics Policy Initiative
Fibers, hairs and, 24, 27
FIGG testing, *see* Forensic investigative genetic genealogical testing
Finger puncture, 25
First-generation sequencing, 82
Fluorescein-derived 2′7′-dichlorofluorescein di-acetate, 57
Fluorescent interchelating dye assay, 67
Foodomics, 298
ForenSeq FGx System, 144
Forensic anthropology, biotechnology applications in, 12–13
Forensic applications of NGS, 88
  forensic epigenetics, 93
  microbiome analysis, 93
  microRNA assessment, 92
  Short Tandem Repeat (STR) and single-nucleotide polymorphism (SNP) analysis, 88–92
Forensic archaeology of a 400-year ago murder, 198–199
Forensic biotechnology, 1–2
Forensic database operations
  direct matching and familial DNA searching, 147–149
  DNA legislation, 149–150

Forensic databases, 142
Forensic DNA analysis, 104–105
Forensic DNA phenotyping (FDP), **143**, 226–228
  challenges and limitations
    accuracy and reliability, 233–234
    ethical and legal concerns, 234–235
    population-specific issues, 235
  future directions in, 235
    epigenetics, integration of, 236
    ethical and legal frameworks, 237
    expanding trait predictions, 236
    interdisciplinary collaboration, 237
    machine learning and AI applications, 236
    population-specific models and diverse databases, 236–237
    sequencing technologies, advancements in, 235–236
  successes and applications of, 231
    challenges and future directions, 233
    global adoption and ethical considerations, 232–233
    law enforcement, impact on, 232
  techniques in
    HIrisPlex-S system, 229
    machine learning algorithms in FDP, 230–231
    next-generation sequencing (NGS), 230
    single nucleotide polymorphisms (SNPs), 228–229
Forensic DNA profiling technique, 3, 61, 123
  challenges for, 124
    DNA contamination, 125–126
    DNA degradation, 124–125
    inhibitors, presence of, 126–127
    mixed DNA samples, 127
  emerging challenges
    advancing technology, impacts of, 134
    synthetic biology and fabricated DNA samples, 133–134
  forensic DNA sampling and collection, 62
  legal and ethical challenges, 128
    DNA evidence, admissibility of, 128–129
    familial searching and privacy concerns, 129–130
    reliability and validity, standards of, 129
  limitations of
    low-quantity DNA samples, 130–131
    partial DNA profiles, 131
    short tandem repeat (STR) analysis, 132–133
  principle of, 62
Forensic DNA samples, authentication of, 262–263
Forensic entomology, wildlife forensics, 214–215
Forensic epigenetics, 93
Forensic Genetics Policy Initiative (FGPI), 152, 153
Forensic genomics, 144
  application of next-generation sequencing on, 87–88
Forensic investigative genetic genealogical (FIGG) testing, 140, *141*, **143**, 144, 145–146, **145**
Forensic metabolomics, 5
Forensic palynology, 167
Forensic proteomics; *see also* Proteomics and metabolomics, 5–6
Forensic science, misapplication of, 322
FPIC, *see* Free, Prior, and Informed Consent
Francisella tularensis, 167, 177
Fraunhofer lines, 42
Free, Prior, and Informed Consent (FPIC), 147, 149
Freeze—thaw cycles, 29
Freezing, 30
French Homicides (2012), 231
Frequentist approach, 108
Frye test, 128
FS, *see* Familial searching
FTA paper, 64–65, *65*
FTIR spectroscopy, 55
Future directions, forensic biotechnology, 14–15

GA, *see* Genome Analyzer
GE Healthcare MegaBACETM system, 83
GenBank, USA, 167
Genetic admixture, 196
Genetic data, potential misuse of, 150
  non-forensic genetic data, secondary use of, 150–151
  technological determinism and misuse in legal proceedings, 151
Genetic databases, types of, 140
  forensic databases, 142
  population reference databases, 140–142, *141*
  public databases, 142
Genetic discrimination, 237
Genetic surveillance state, 233
Genome Analyzer (GA), 81
Genomic imprinting, 264
Genomics, 298
Genomics and next-generation sequencing, 86–87
Geographical information systems (GIS), 201
Geographic origin determination, 205
Geolocation identification, 180
GIS, *see* Geographical information systems
GlobalFiler Express chemistry, 251
Global microbial databases, expansion of, 187–188
Glutamate rich 1 (*ERICH1*) gene, 267
Glutamine-fructose-6-phosphate transaminase 2 (*GFPT2*) gene, 267
Glycomics, 298
Golden State Killer case (2018), 231, 317, 319
Gold nanoparticles (AuNPs), 56
Guanidinium salts, 65

Hair, 36
  and fibers, 24, 27
  samples, 26
Hair analysis, 322, 327
Handling and storage, 31
Haplogroup nomenclature, 92
Hardy–Weinberg equilibrium, 117–118
HBA, *see* Hemoglobin alpha
Hebert, Paul, 162–163, *163*
Hematin/Hemin Test, *see* Teichmann Crystal test

Hemochromogen or Pyridine Hemochromogen Test, *see* Takayama Test
Hemoglobin, 51
Hemoglobin alpha (HBA), 11
HERC2 gene, 228
Heteroplasmy, 92
Heterozygous alleles, 62
Hierarchical Bayesian probability model, 112
High molecular weight DNA, 63
High-performance liquid chromatography (HPLC), 300
High-throughput sequencing (HTS), 166, 168
HIrisPlex-S system, 226, 229
HIrisPlex system, 227
Homogeneous mass extension system (Sequenom), 293
Homozygous alleles, 62
HpaII tiny fragment Enrichment by Ligation-mediated PCR (HELP) methylation assay, 270–271
HPLC, *see* High-performance liquid chromatography
HTS, *see* High-throughput sequencing
Human Genome Project, 227
Human-wildlife conflict (HWC), 214
Humidity, 29
HWC, *see* Human-wildlife conflict
5-Hydroxymethylcytosine (5hmC), 256, 273
Hyper Variable Regions (HVRs), *see* Variable Number Tandem Repeat

I-BGA, *see* Individual's biogeographical ancestry
iBOL, *see* International Barcode of Life
Identification of Oregon's Missing Person (1959/2023), 231
Identity SNPs (iSNPs), 139
IEF, *see* Isoelectric focusing
Illegal drug abuse identification, 266
Illumina GAIIx platform, 91
Illumina HiSeq and Illumina MiSeq, 83–84
Illumina Infinium Methylation Assay, 272
Immunoaffinity chromatography, 300
Immunochromatographic method, 53–54
Immunoelectrophoresis, 42
Immunological tests, 303
Inadequate defense, 328
InDels, *see* Insertion/deletion markers
India's commitment to wildlife conservation, 206
Indigenous peoples (IPs), 140, 147, 149
Individual's biogeographical ancestry (I-BGA), 196
Informants, role of, 323
Informed consent, 146–147
Inhibitors, presence of, 126–127
Innocence Canada, 325
Innocence Network UK (INUK), 324
Innocence Project
  and DNA exonerations, 320
    eyewitness misidentification, 321–322
    false confessions, 322
    forensic science, misapplication of, 322
    informants, role of, 323
  work of, 323
Innocence Project South Africa (IPSA), 325
INSDC, *see* International Nucleotide Sequence Database Collaboration
Insertion/deletion markers (InDels), 4, 90
In situ proteomics, 309
Intelligent Bio-Systems Inc. system, 83
Internal transcribed spacer (ITS), 165–166
International Barcode of Life (iBOL), 162–163, 215
International forensic ethics guidelines, 153–154
International Nucleotide Sequence Database Collaboration (INSDC), 167
International Society for Forensic Genetics (ISFG), 153
INUK, *see* Innocence Network UK
Investigatory officer (IO), 216
IO, *see* Investigatory officer
Ion exchange chromatography, 300–301
Ion suppression, 289
Ion Torrent, 84
IPs, *see* Indigenous peoples
IPSA, *see* Innocence Project South Africa
Iroquois Homeobox A (*IRX2*) gene, 266
ISFG, *see* International Society for Forensic Genetics
iSNPs, *see* Identity SNPs
Isoalleles, 138, 139
Isoelectric focusing (IEF), 299
ISO/IEC 17020 standards, 21
ISO/IEC 17025 standards, 31
Isotope analysis, 200
Isotope and chemical analysis, 180
  environmental impacts investigations, 181
  human identification, 181
  laboratory strains vs. wild isolates, 181
  source attribution, 181
  transmission route identification, 181
Isotopes analysis, wildlife forensics, 213
ITS, *see* Internal transcribed spacer

Jeffreys, Alec, 1, 137
Junk science, 327–328

Kastle–Meyer test, 306
Kinship analysis, 253
Kinship-based database searches, 144–146, **145**
Kinship Simulation, 117
Kirk Bloodsworth case, 316–317

Lab-on-chip systems, 15, 56
Laboratory analysis, wildlife forensics, 208
  forensic entomology, 214–215
  isotopes analysis, 213
  microscopic analysis, 208–209
  molecular analysis, 211–213, **212**
  spectroscopic and chromatographic analysis, 210
  toxicological analysis, 213–214
Laboratory environment, 31

Laboratory Information Management Systems (LIMS), 32
Laboratory strains vs. wild isolates, 181
Lab Retriever software, 114
limitation, 115
program parameter, 115
Law enforcement in wildlife forensics, 216
Elephant Preservation Act (1879), 217
Wild Birds Protection Act (1887), 217
Wildlife (Protection) Act (WPA) (1972), 217–218
LC-MS, *see* Liquid chromatography-mass spectrometry
LD, *see* Linkage disequilibrium
Legal representation, adequate, 329
Library preparation, 83
Lifestyle identification, 266–267
Ligation-mediated PCR (LM-PCR), 271
Likelihood ratio (LR), 107–108
LikeLTD, 113
limitation, 114
LIMS, *see* Laboratory Information Management Systems
Lineage markers, 90–92
Linkage analysis, 70
Linkage disequilibrium (LD), 196
Lipidomics, 298
Liquid biological samples, 27
Liquid chromatography-mass spectrometry (LC-MS), 289
LM-PCR, *see* Ligation-mediated PCR
Long non-coding RNAs, 256
Long-term consequences and broader social impact, 330
Low-quantity DNA samples, 130–131
Low-template DNA analysis, 246, 253
LR, *see* Likelihood ratio
LRmix, 110
features of, 110–111
limitations, 111
procedure, 111
Luciferin, 83
Lyophilization, 30

Machine learning (ML), 179, 187, 226, 228, 230–231, 236
Magnetic bead-based extraction, 49, 246
MALDI, *see* Matrix-assisted laser desorption/ionization
MALDI-TOF-MS, *see* Matrix-assisted laser desorption ionization-time-of-flight mass spectrometry
Marking evidence, 20
Markov Chain Monte Carlo (MCMC), 109
Massively parallel sequencing (MPS) technology, 81, 90, 138, 139–140, 228; *see also* Next-generation sequencing
Mass spectrometry in forensic DNA analysis, 287
applications, 290–293
electrospray ionization (ESI), 288–289
matrix-assisted laser desorption/ionization (MALDI), 289–290
Maternal or paternal allele source identification, 264
matK barcode, *see* Maturase K barcode
MATLAB, 112
Matrix-assisted laser desorption/ionization (MALDI), 210, 288, 289–290, 301
Matrix-assisted laser desorption ionization-time-of-flight mass spectrometry (MALDI-TOF-MS), 288, *290*, 292–293, 301, 305, 309
Maturase K (matK) barcode, 165
MBD-seq, *see* Methyl-binding domain sequencing
MC1R gene, 228
MCMC, *see* Markov Chain Monte Carlo
Medicinal Materials DNA Barcode Database (MMDBD), 167
MeDIP, *see* Methylated DNA immunoprecipitation
Menstrual blood, steps for DNA extraction from, 51
Menstrual blood DNA extraction, special considerations for, 51–52
Messenger RNAs (mRNAs), 11, 52
Metabarcoding, 166, 168
Metabolomics, 298
Metagenomics, 179
geolocation identification, 180
individual identification, 179
post-mortem interval (PMI) estimation, 179
tissue or fluid identification, 180
Metal chelate chromatography, 300
Methylated DNA immunoprecipitation (MeDIP), 273
Methylation-sensitive high resolution melting (MS-HRM), 270
Methylation-sensitive restriction enzymes, 273
Methylation-specific PCR (MSP), 261, 269–270
Methyl-binding domain sequencing (MBD-seq), 272–273
5-Methylcytosine (5mC), 256–257, 263, 267, 268, 273, 274, 277
MethyLight MSP, 269
MICADAS AMS technology, *see* Mini Carbon Dating system AMS technology
Microarray-based DNA methylation and miRNA analysis, 271–272
Microbial data collection, informed consent and data ownership in, 185
Microbial diversity and mutation rates, 183
Microbial ecology, 87
Microbial forensics, 5, 175
bioinformatics, 180
case studies
anthrax attacks (USA) (2001), 185
Bacillus anthracis, use of, in mailed letters, 185
forensic genomics and research lab linkage, 185–186
SARS-CoV-2 origin investigation, 186

chain of custody and legal considerations, 181–182, *182*, *183*
challenges in, 183
limited databases of microbial genomes, 184
microbial data collection, informed consent and data ownership in, 185
microbial diversity and mutation rates, 183
microbial genomic data, beneficial uses of, 184–185
microbial genomic data, malicious uses of, 185
natural vs. deliberate outbreaks, 183
DNA sequencing, 178
forensic biotechnology in, 10–11
future directions, 187–188
isotope and chemical analysis, 180
environmental impacts investigations, 181
human identification, 181
laboratory strains vs. wild isolates, 181
source attribution, 181
transmission route identification, 181
key microbial agents, 177
metagenomics, 179
geolocation identification, 180
individual identification, 179
post-mortem interval (PMI) estimation, 179
tissue or fluid identification, 180
polymerase chain reaction (PCR), 178–179
scope and applications, 176
bioterrorism and biocrime investigations, 176
environmental and agricultural monitoring, 177
outbreak tracing, 176–177
Microbial genomic data
beneficial uses of, 184–185
limited databases, 184
malicious uses of, 185
Microbial science, application of next-generation sequencing in, 87
Microbial signatures, 10
Microbial surveillance, 184
Microbiome analysis, 93
Microcrystal assay, 42
Microfluidic-based DNA analysis, 76–77
Microfluidic devices, 246
MicroRNA (miRNA), 36, 92, 258
MicroRNA (miRNA) analysis, DNA methylation and, 259
age estimation, 261–262
body fluid identification, 260–261
challenges and limitations associated with, 275
crime scene samples, DNA degradation, and contamination, 275
epigenetic complexity and individual variability, 275
legal challenges in court, 276
technical limitations, standardization, validation, and interpretations, 276
death analysis, cause of, 264–265
forensic DNA samples, authentication of, 262–263
future perspective, 276–278
illegal drug abuse identification, 266
lifestyle identification, 266–267
maternal or paternal allele source identification, 264
methods and techniques of, 267
combined bisulfite restriction analysis, 270
HELP methylation assay, 270–271
Illumina Infinium Methylation Assay, 272
methylated DNA immunoprecipitation (MeDIP), 273
methylation specific PCR, 269–270
methyl-binding domain sequencing (MBD-seq), 272–273
microarray-based analysis, 271–272
nanopore sequencing, 274
sequencing-based analysis, 268–269
single-base extension (SBE) methylation analysis, 274
single molecule, real-time (SMRT) sequencing, 273–274
monozygotic twins, discrimination of, 263
post-mortem interval (PMI) estimation, 265–266
Microsatellite genotyping, 212
Microsatellites, *see* Short Tandem Repeat
Microscopic analysis, wildlife forensics, 208–209
Mini Carbon Dating system (MICADAS) AMS technology, 213
MinION™, 85
Minisatellite marker, *see* Variable Number Tandem Repeat
MiniSTR analysis, 37
miRNA, *see* MicroRNA
Misapplication of forensic science, 322
Miscarriages of justice, 315, 317, 319, 320, 321, 324
Mitochondrial DNA (mtDNA), 2, 4, 37, 73, 91, 133, 144, 199–200, 211–212, 230
Mitochondrial SNPs (mtDNA SNPs), 139
Mixed profile, understanding of, 108–109
ML, *see* Machine learning
MMDBD, *see* Medicinal Materials DNA Barcode Database
Molecular analysis, wildlife forensics, 211–213, **212**
Molecular epidemiology, 186
Molecular systematics, 161
Monozygotic twins, discrimination of, 263
MPS technology, *see* Massively parallel sequencing technology
mRNAs, *see* Messenger RNAs
MS-HRM, *see* Methylation-sensitive high resolution melting
M/S Ivory Trader and Manufacturer Association and Other vs. Union Bank of India and Other, 218
MSP, *see* Methylation-specific PCR
mtDNA, *see* Mitochondrial DNA
Mullis, Kary, 2
Multiplexed PCR, 89
Multiplexing, 12

Multiplex PCR, 3
-based short tandem repeats, 123
Multiplex STR typing, 124
Muscle tissue, 25
Mycobacterium tuberculosis, 167

N6-methyladenine (m6A), 256, 258
Nails, evidence from, 23
Nano-electrospray ion trap MS, 303
Nanopore based systems, 85
Nanopore sequencing, 274
National Centre for Biotechnology Information (NCBI), 167
National Commission on Forensic Science, 321
National Crime Record Bureau (NCRB), 205
National Criminal Investigation DNA Database (NCIDD), 325
National DNA Data Bank (NDDB), 324
National DNA Database (NDNAD), 142, 324
National DNA Index System (NDIS), 247
National Institute of Justice, 181
National Institute of Standards and Technology (NIST), 21, 115
Natural vs. deliberate outbreaks, 183
NCBI, *see* National Centre for Biotechnology Information
NCIDD, *see* National Criminal Investigation DNA Database
NCRB, *see* National Crime Record Bureau
NDDB, *see* National DNA Data Bank
NDIS, *see* National DNA Index System
nDNA, *see* Nuclear DNA
NDNAD, *see* National DNA Database
Necrobiome, 93
Negative controls, 125–126
Neobalanocarpus heimii, 213
Nested PCR, 3
Next-generation sequencing (NGS), 2, 4–5, 75, 80, 92, 139, 166, 176, 205, 226, 230; *see also* Massively parallel sequencing technology
advantages and limitations, 94–95
applications of, 85
on forensic genomics, 87–88
genomics and next-generation sequencing, 86–87
in microbial science, 87
first-generation sequencing, 82
forensic applications of, 88
forensic epigenetics, 93
microbiome analysis, 93
microRNA assessment, 92
STR and SNP analysis, 88–92
future prospects, 95–96
legitimacy, acceptance, and dependability, issues with, 93–94
second-generation sequencing, 82
applied biosystems, 84
Illumina HiSeq and Illumina MiSeq, 83–84
Roche 454, 83
third-generation sequencing, 84–85
Next-next-next generation (NNN), *see* Third-generation sequencing
NGS, *see* Next-generation sequencing
NIST, *see* National Institute of Standards and Technology
Non-coding RNAs, 256, 261
Non-forensic genetic data, secondary use of, 150–151
Non-porous surfaces, biological stains on, 22
Non-transportable objects, biological stains on
non-porous surfaces, 22
porous surfaces, 22
Notes and sketches, 21
Nuclear DNA (nDNA), 211
Nutrigenomics, 298

Omics technologies, 298
ONT, *see* Oxford Nanotechnology
Open reading frames (ORFs), 178
Optimized protocols for low-template DNA, 246
Oral cavity, 24
ORFs, *see* Open reading frames
Ouchterlony double diffusion, 303
Oxford Nanopore MinION, 187
Oxford Nanopore Technologies, 85
Oxford Nanotechnology (ONT), 274

P16 gene, 264
Pacific Biosciences (PacBioR), 85
Packaging and transportation, 31
Pantholops hodgsonii, 212
Parabon NanoLabs, 231
Partial DNA profiles, 131
Patatin-like phospholipase domain-containing protein 3A (PFN3A), 261
PCBs, *see* Polychlorinated biphenyls
PCDDs, *see* Polychlorinated di-benzodioxins
PCDFs, *see* Polychlorinated dibenzofurans
PCR, *see* Polymerase chain reaction
PCR-STR, *see* Polymerase chain reaction-short-tandem repeats
Peptide mass fingerprint (PMF), 210
Personal autonomy, 146
Personal protective equipment (PPE), 30, 31
PFN3A, *see* Patatin-like phospholipase domain-containing protein 3A
Pharmacogenomics, 298
Phenol-chloroform extraction, 48
Phenotype-informative SNPs (piSNPs), 139, 144, 228, 230
pH levels, 29
Photography, 20
Picking method, 38
piSNPs, *see* Phenotype-informative SNPs
Pitchfork, Colin, 137–138
PMF, *see* Peptide mass fingerprint
PMI, *see* Post-mortem interval
Poached wild animals, 206–207, **207**

Polonator, 83
Polychlorinated biphenyls (PCBs), 214
Polychlorinated di-benzodioxins (PCDDs), 214
Polychlorinated dibenzofurans (PCDFs), 214
Polymerase chain reaction (PCR), 1, 94, 123, 138, 164, 178–179, 205
- contamination, 125–126
- end-point, 67
- and its variants, 2–3
- real-time, 67–68

Polymerase chain reaction-short-tandem repeats (PCR-STR), 227
Population match probability (Pm), 70
Population reference databases, 140–142, *141*
Porous surfaces, biological stains on, 22
Portable sequencing devices, 187
Post-conviction DNA testing, 317–318
Post-mortem interval (PMI), 1
- detection, 305
- estimation, 179, 265–266

PPE, *see* Personal protective equipment
Precision oncology, 87
Presumptive tests
- for blood, **40–41**
- for saliva, **44**
- for semen, **45**
- for urine, **45**

Primary packaging, 27
Principal component analysis, 196
PRM1 (protamine 1), 11
Probabilistic algorithms, 128
Probability, 106–107
Prosecutorial accountability and integrity, 329
Prosecutorial misconduct, 328
Prostate-specific antigen (PSA), 11
Proteinase K, 48–51, 63
Proteins
- at scene of crime, 303
- -based biomarkers, 303
- forensic protein markers of body fluids, 303–305
- identification and analysis in forensic investigation, 305–306

Proteoforms, 299
Proteomics, 12, 296–299, *298*
- challenges and limitations of, 307–308
- current research status of, 308–309
- future advancements and directions in, 309–310
- proteins
  - at scene of crime, 303
  - forensic protein markers of body fluids, 303–305, **304–305**
  - identification and analysis in forensic investigation, 305–306
- tools and techniques used in
  - affinity chromatography, 300
  - ion exchange chromatography, 300–301
  - MALDI-TOF-MS, 301
  - SDS-PAGE, 299
  - 2D-PAGE, 299–300

PSA, *see* Prostate-specific antigen
Psychogenomics, 298
Pubic area, 24
Public databases, 142
Public defender caseloads, 328
Public health surveillance, 184–185
Public prosecutor, 216
Putrefied bodies, 25

QAS, *see* Quality Assurance Standards
qMSP, *see* Quantitative methylation-specific PCR
QPCR (real-time PCR), 2
Quality Assurance Standards (QAS), 247
Quantitative methylation-specific PCR (qMSP), 268
Quantum dots, 56

Radioisotopes, 67
Raman spectroscopy, 53, 55
Random forest algorithm, 54
Random match probability (RMP), 91, 107
Rapid DNA Analysis System (RDAS), 244
RAPID DNA typing, 244
- analysis software, 251–252
- Applied Biosystems RapidHIT ID System, 247–249
- challenges and limitations of, 254
- consumables, 251, *251*
- fundamentals, 245
  - data processing techniques, 245
  - instrumentation, 244
  - sample preparation, 244–245
- future applications in forensic DNA analysis, 252–253
- instrument, 249–250, *250*
- potential applications in criminal investigations, 253–254
- RapidHIT ID System, 247–249, *248*, *249*
- significant improvements in DNA extraction tools for, 245–246

RapidHIT ID System, 247–249, *248*, *249*
RapidLINK v2.0, 252
Rapid lysis techniques, 246
rbcL, *see* Ribulose-1,5-bisphosphate carboxylase/oxygenase
RDAS, *see* Rapid DNA Analysis System
rDNA genes, *see* Ribosomal DNA genes
Real-time PCR techniques, 67–68, 127
Real-time quantitative PCR (RQ) techniques, 11
Recording of interrogations, 320, 322, 324
Reduced representation bisulfite sequencing (RRBS), 268
Reference databases, 167
Reference samples, 25
- blood samples, 25
- hair samples, 26
- saliva samples, 26
- transfused subjects, 26

Relative fluorescence units (RFUs), 123
REs, *see* Restriction endonucleases

Restriction endonucleases (REs), 69
Restriction fragment length polymorphism (RFLP), 69, *70*, 123, 137, 227
Reverse transcription endpoint PCR, 52
Reverse transcription PCR (RT-PCR), 11, 179
RFLP, *see* Restriction fragment length polymorphism
RFUs, *see* Relative fluorescence units
Rh blood group system, 44
Rhinoceros DNA Index System (RhODIS), 215
RhODIS, *see* Rhinoceros DNA Index System
Ribosomal DNA (rDNA) genes, 165
Ribulose-1,5-bisphosphate carboxylase/oxygenase (rbcL), 165
RMP, *see* Random match probability
RNA-based assays, 42
RNA-based body fluid identification, 11
RNA-based methods, 52
RNA degradation, 52
RNA-induced silencing complex (RISC), 258
RNA molecules, 11
Robotic liquid handling systems, 245
Roche 454, 83
RRBS, *see* Reduced representation bisulfite sequencing
RT-PCR, *see* Reverse transcription PCR

S. and Marper v. United Kingdom, 142, 147
SA, *see* Sensitivity analysis
Saliva, 35, 36, 49–50
  from bite marks, 23
  samples, 26, 38
  tests for, 44
    confirmatory tests, **44**
    presumptive tests, **44**
SANE, *see* Sexual assault nurse examiner
Sanger, Fred, 161
Sanger sequencing, 178
Sarbah, Christiana, 137
SARS-CoV-2 origin investigation, 186
  tracing the virus's origins, 186
SBE methylation analysis, *see* Single-base extension methylation analysis
SBS, *see* Sequencing by synthesis
Scanning electron microscopy (SEM), 209
Scene documentation before evidence collection, 20–21
Scrapping method, 37
SDS, *see* Sodium dodecyl sulfate
SDS-PAGE, *see* Sodium dodecyl sulphate-polyacrylamide gel electrophoresis
Secondary packaging, 27
Second-generation sequencing, 80, 82
  applied biosystems, 84
    Ion Torrent, 84
    SOLiD, 84
  Illumina HiSeq and Illumina MiSeq, 83–84
  Roche 454, 83
Seized wildlife products, 207–208
SEM, *see* Scanning electron microscopy
Semen, 35, 36, 50
  samples, 39
  tests for, 45
    confirmatory tests, **45**
    presumptive tests, **45**
Semenogelin, 11
Semen specific antigen, 57
Sensitivity analysis (SA), 111
Sequencing-based DNA methylation and miRNA analysis, 268–269
Sequencing by Oligo Ligation Detection (SOLiD), 81, 84
Sequencing by synthesis (SBS), 83
Sequencing technologies, advancements in, 235–236
Sexual assault evidence, 24
Sexual assault nurse examiner (SANE), 24
Short tandem repeat (STR), 70–72, 87, 88, 123, 196, 245
  limitations of, 132
    mitochondrial DNA (mtDNA) analysis, 133
    Y-STR analysis, 132–133
  mass analysis, 291
  STR DNA markers, 138–139
  STR profiling, 2, 3
Silica-based methods, 49
Simple sequence repeat (SSR), *see* Short tandem repeat
Single-base extension (SBE) methylation analysis, 274
Single molecule, real-time (SMRT) sequencing, 85, 273–274
Single-molecule sequencing (SMS), 76, 82
Single nucleotide mutation (SNM), 293
Single nucleotide polymorphisms (SNPs), 4, 74, **75**, **76**, 81, 88, 196, 226, 228–229, 287, 289
  SNP DNA markers, 139
Single-source DNA samples, 247
Single-stranded DNA recovery, 64
Slipped-strand mispairing, 72
Slot blot assay, 66–67
SMRT sequencing, *see* Single molecule, real-time sequencing
SMS, *see* Single-molecule sequencing
SNaPshot™ minisequencing, 228–229
SNM, *see* Single nucleotide mutation
SNPs, *see* Single nucleotide polymorphisms
Sodium dodecyl sulfate (SDS), 63
Sodium dodecyl sulphate-polyacrylamide gel electrophoresis (SDS-PAGE), 299, 300
Soft ionization techniques, 288
Solexa's Genome Analyzer, 81
SOLiD, *see* Sequencing by Oligo Ligation Detection
Solid-phase extraction method, 65–66, **65**, *66*
South Africa, DNA exoneration in, 325–326
Spectroscopic and chromatographic analysis, wildlife forensics, 210
Spectroscopic examination, 42

Spectroscopic techniques, 53
Stable isotope analysis, 205
Stakeholder and public education, 154
Standardized DNA barcode regions, 164–166
Staphylococcus aureus, 265
State of Madhya Pradesh v. Rajesh Garg & Others, 219
STATH (statherin), 11
Statistical software in forensic DNA typing, 104
  ArmedXpert, 115
    features, 116
    limitation, 116
  case studies
    federal criminal trail, 120
    mysterious death of a young boy, 120
  DNAMIX, 117–118
  DNAmixtures, 118
  DNA·VIEW, 116
    limitations, 117
  EuroForMix, 118
  forensic DNA analysis, 104–105
  Lab Retriever software, 114
    limitation, 115
    program parameter, 115
  legal and ethical consideration, 119
  LikeLTD, 113
    limitation, 114
  LRmix, 110
    features of LRmix, 110–111
    limitations, 111
    procedure, 111
  mixed profile, understanding of, 108–109
  role of, 105
  STRmix, 109
    limitations, 110
    STRmix™ working, 110
  TrueAllele, 111
    features, 112
    limitation, 113
    TrueAllele system, 112–113
  validation of, 118–119
Statistics, fundamental of, 106
  Bayesian framework, 108
  likelihood ratios, 107–108
  probability, 106–107
Stochastic effects, 113
STR, *see* Short tandem repeat
Strain typing, 179
Strangulation, forensic evidence of, 198
STR loci, 123, 132
STRmix, 109
  limitations, 110
  working, 110
Substrate controls, 125
Sweat, 35, 36, 50–51
Synthetic biology, 133–134, 185

Takayama test, 42, 303, 306
Tape-lifting method, 37–38
TCMs, *see* Traditional Chinese medicines
tDMRs, *see* Tissue-specific differentially methylated regions
TE buffer, *see* Tris-EDTA buffer
Technology and data analytics adoption, 330
Teeth, 25
Teichmann Crystal test, 42
Temperature, 29
Ten-eleven translocation (TET) enzymes, 256, *257*
Tertiary packaging, 27
TET enzymes, *see* Ten-eleven translocation enzymes
Thanato-microbiome analysis, 93
Thermostable DNA polymerases, 138
TH gene, *see* Tyrosine hydroxylase gene
Third-generation sequencing, 84
  nanopore based systems, 85
  Pacific Biosciences (PacBioR), 85
3D bioprinting, 7
3D printing, 7
Tissue-specific differentially methylated regions (tDMRs), 261
Touch DNA, 32
Touchdown PCR, 2–3
Toxicogenomics, 298
Toxicological analysis, wildlife forensics, 213–214
Toxicology, forensic biotechnology in, 8–9
Traditional Chinese medicines (TCMs), 216
Transcriptomics, 298
Transfused subjects, 26
Transmission route identification, 181
Transportable objects, biological stains on, 22
Tris-EDTA (TE) buffer, 30
TrueAllele, 111
  features, 112
  limitation, 113
  system, 112–113
Tumour heterogeneity, 86
Two-dimensional polyacrylamide gel electrophoresis (2D-PAGE), 299–300
Tyrosine hydroxylase (*TH*) gene, 266

Ultra-performance liquid chromatography-mass spectroscopy (UPLC/MS), 210
Ultraviolet (UV) light damage to DNA, 29
UNDRIP, *see* United Nations Declaration on the Rights of Indigenous Peoples
United Kingdom, DNA exoneration in, 324
United Nations Declaration on the Rights of Indigenous Peoples (UNDRIP), 147
United States, DNA exoneration in, 324
United States Army Medical Research Institute of Infectious Diseases (USAMRIID), 186
UPLC/MS, *see* Ultra-performance liquid chromatography-mass spectroscopy
Urine, 35, 36, 50
  samples, 39
  tests for, 45

confirmatory tests, 46, **46**
presumptive tests, **45**
Ursus thibetanus, 212
USAMRIID, *see* United States Army Medical Research Institute of Infectious Diseases
US Combined DNA Index System, 142
U.S. Federal Bureau of Investigation, 247
UV light damage to DNA, *see* Ultraviolet light damage to DNA

Vaginal cavity, 24
Vaginal fluid, 35, 51
Variable number tandem repeat (VNTR), 69–70, 123, 293
Variola major, 167
Vein puncture, 25
Vibert's Reagent, 42
Victim's body, samples from, 23
biological stains, 23
clothes, 24
evidence from nails, 23
hairs and fibers, 24
saliva from bite marks, 23
sexual assault evidence, 24
Visual User Interface (VUIer), 112
VNTR, *see* Variable number tandem repeat
VUIer, *see* Visual User Interface
Vulvar region, 24

Wagenaar Test, 42
Western blot, 43
Wet samples, 26
WGBS, *see* Whole genome bisulfite sequencing
WGS, *see* Whole genome sequences
Whole genome bisulfite sequencing (WGBS), 268
Whole genome sequences (WGS), 3, 96, 136
Whole mitochondrial genome sequencing, 91
Wild animals, poached, 206–207
Wild Birds Protection Act (1887), 217
Wildlife (Protection) Act (WPA) (1972), 216, 217–218
Wildlife forensics, 168, 204
crime scene investigation and evidence collection, 206
poached wild animals, 206–207, **207**
seized wildlife products, 207–208
DNA reference database, 215–216, *215*
future prospects and challenges, 220
history of, 205
India's commitment to wildlife conservation, 206
laboratory analysis, 208, *208*
forensic entomology, 214–215
isotopes analysis, 213
microscopic analysis, 208–209
molecular analysis, 211–213, **212**
spectroscopic and chromatographic analysis, 210
toxicological analysis, 213–214
landmark judgments and case studies, 218
M/S Ivory Trader and Manufacturer Association and Other vs. Union Bank of India and Other, 218
State of Madhya Pradesh v. Rajesh Garg & Others, 219
Wildlife Protection Society of India vs. Ministry of Environment and Forest & Others, 219–220
Wildlife vs. Ashok Kumar & Ors., 218–219
law enforcement, 216
Elephant Preservation Act (1879), 217
Wild Birds Protection Act (1887), 217
Wildlife (Protection) Act (WPA) (1972), 217–218
trial of cases, 216
Wildlife Offence Report (WLOR), 206
Wildlife products, seized, 207–208
Wildlife Protection Society of India (WPSI) vs. Ministry of Environment and Forest & Others, 219–220
Wildlife vs. Ashok Kumar & Others, 218–219
WLOR, *see* Wildlife Offence Report
Woese, Carl, 162
Wood's lamp, 24
World Trade Center attacks, 117
WPA, *see* Wildlife Act
Wrongful convictions, 323
common causes and prevention of, 326
eyewitness misidentification, 326–327
false confessions, 327
inadequate defense, 328
junk science, 327–328
prosecutorial misconduct, 328
DNA evidence and, 319–320
DNA exoneration global perspective
Australia, 325
Canada, 324–325
South Africa, 325–326
United Kingdom, 324
United States, 324
Innocence Project, work of, 323
recommendation, 328
adequate legal representation, 329
enhanced forensic science practices, 329
improved eyewitness identification procedures, 329
long-term consequences and broader social impact, 330
prosecutorial accountability and integrity, 329
technology and data analytics adoption, 330
Wrongful convictions on criminal justice system, 320
Innocence Project and DNA exonerations, 320
eyewitness misidentification, 321–322
false confessions, 322
forensic science, misapplication of, 322
informants, role of, 323

X chromosome, 12
X-chromosome linked STRs, 72
X chromosome Short Tandem Repeats (XSTRs), 139
XSTRs, *see* X chromosome Short Tandem Repeats

Y-chromosomal single-nucleotide polymorphisms (YSNPs), 139
Y chromosome, 4, 12, 212
Y-Chromosome Haplotype Reference Database (YHRD), 140–142
Y-chromosome linked STRs, 72
Y chromosome Short Tandem Repeats (YSTRs), 56, 132–133, 138–139
Y chromosome typing, 2
Yersinia pestis, 177
Y-haplogroups, 133
YHRD, *see* Y-Chromosome Haplotype Reference Database
Yield gel, 124
"Yield" gel method, 67
YSNPs, *see* Y-chromosomal single-nucleotide polymorphisms
YSTRs, *see* Y chromosome Short Tandem Repeats

Zero-mode waveguides (ZMWs), 85, 273

For Product Safety Concerns and Information please contact our
EU representative GPSR@taylorandfrancis.com Taylor & Francis
Verlag GmbH, Kaufingerstraße 24, 80331 München, Germany